KV-456-430

MOLECULAR

MICROBIOLOGY

MOLECULAR MICROBIOLOGY

J. B. G. Kwapinski
S. G. Bradley
E. R. Brown
P. R. Burton
H. Fraenkel-Conrat
N. Kordová
W. M. O'Leary
F. J. Reithel
M. R. J. Salton
R. Storck
E. E. Woodside

A WILEY BIOMEDICAL-HEALTH PUBLICATION

JOHN WILEY & SONS, New York • London • Sydney • Toronto

Library of Congress Cataloging in Publication Data

Main entry under title:
Molecular microbiology.

"A Wiley Biomedical-Health publication."
Includes bibliographies.
1. Microbiology. 2. Molecular biology.
I. Kwapinski, Jerzy B. G., 1920– [DNLM: 1. Microbiology. QW4 M718 1974]

QR60.M64 1974 576'.11'92 73–15950
ISBN 0-471-51109-9

Printed in the United States of America

10 9 8 7 6 5 4 3 2 1

AUTHORS

S. G. BRADLEY, Professor and Chairman, Department ot Microbiology,
The Commonwealth of Virginia Medical College,
Richmond, Virginia

E. R. BROWN, Professor and Chairman, Department of Microbiology,
University of Health Sciences, The Chicago Medical School,
Chicago, Illinois

P. R. BURTON, Professor, Department of Physiology and Cell Biology,
University of Kansas,
Lawrence, Kansas

H. FRAENKEL-CONRAT, Professor, Department of Molecular Biology,
University of California at Berkeley,
Berkeley, California

N. KORDOVÁ, Associate Professor, Department of Medical Microbiology,
University of Manitoba,
Winnipeg, Manitoba, Canada

J. B. G. KWAPINSKI, Professor, Department of Medical Microbiology,
University of Manitoba,
Winnipeg, Manitoba, Canada

W. M. O'LEARY, Professor, Department of Microbiology,
Cornell University,
New York, New York

F. J. REITHEL, Professor, Department of Chemistry,
University of Oregon,
Eugene, Oregon

M. R. J. SALTON, Professor and Chairman, Department of Microbiology,
New York School of Medicine,
New York, New York

R. STORCK, Professor, Department of Biology,
Rice University,
Houston, Texas

E. E. WOODSIDE, Associate Professor, Department of Microbiology,
University of Louisville School of Medicine,
Louisville, Kentucky

ASSOCIATE AUTHORS

R. A. ALBACH, Professor, Department of Microbiology,
University of Health Sciences, The Chicago Medical School,
Chicago, Illinois

L. W. ENQUIST, Department of Microbiology,
The Commonwealth of Virginia Medical College,
Richmond, Virginia

M. J. GERBER, Assistant Professor, Department of Microbiology,
University of Health Sciences, The Chicago Medical School,
Chicago, Illinois

CONTENTS

MOLECULAR

MICROBIOLOGY

CHAPTER ONE

The Concept and Criteria of Molecular Microbiology

J. B. G. KWAPINSKI

The concept of molecular microbiology is based on the consideration of microorganisms as highly organized, dynamic entities of live matter, the particles and forces of which constantly change and interact with each other and with the environment.

The biological processes occurring in the various forms of microorganisms are governed by the general laws of physical and chemical entities. Thus the synthesis of macromolecules in the various forms of microorganisms and their replication and reproduction depend on the same laws of chemistry that control the behavior of macromolecules outside the established structural entities. Since identical laws apply to the synthesis, function, and composition of macromolecules inside and outside the cells, the composition and function of macromolecules occurring within substructural and structural microbial entities can be studied in the isolated state *in vitro*.

The molecules occurring in microorganisms are probably identical with the molecules existing in other anatomical and nonassembled forms of biocolloids. The constitutional and structural formulas of the molecules ought to be investigated under determinable conditions so that the influence of the environment on the molecular structure and function, as well as the changes occurring within the molecule, may be examined. Subsequently, it may be possible to deduce what the structure and function of the molecules would be like in their biological assembly or environment.

In a more conventional approach to studies on molecular microbiology, the primary, secondary, tertiary, and quaternary structures of selected parts of larger organized systems may be investigated with special attention focused on the chemical bonding, chemical kinetics, and other physical and chemical intra- and intermolecular forces. Although the research thus would be focused on one selected part of a larger, organized microbial system at a time, its ultimate objective would be to elucidate the relationship of selected structural and functional parts to the physicochemical organization of a biological entity or phenomenon. Alternatively, the chemical organization and systematization of macromolecules can often be explained by reference to another, similarly organized chemical system, since the structures of small molecules and macromolecules, derived from any source, are fundamentally alike.

An intimate insight into, and understanding of, the specific structures and functions of biomolecules may be acquired by investigating the internuclear and interelectronic relationships displayed at the submolecular level. The biochemical and immunochemical specificity of molecules and biomolecules, their interactions, and their life processes may be explained in terms of quantum biology. By the application of the quantum principle, these phenomena may be considered as specific internuclear and interelectronic reactions that utilize and transform appropriate quanta of static and kinetic energy. Consequently, the study of biomolecules would rely ultimately on investigation of the electron affinities and energy coefficients of the highest filled molecular orbital.

In contrast to the ever-changing kinetic states of energy of biomolecules, the ground state is probably just a longer episode maintained between the electron-promoted, excited state and configurations, the sum of which occupies a much larger time and space in intramolecular reactions than in kinetic states.

The appearance and structure of the dynamic biocolloids, composed of different macromolecules, rest upon the intricate two- and three-dimensional patterns of their molecular constituents. The biopolymers and macromolecular components of different biochemical systems and processes expand and contract during the biochemical interactions, and in a natural environment their structure and form continuously change because of the internal rearrangement of their chain constitutents and the overall molecular shape.

Chemical changes and mechanical functions in the morphologically definable entities are interdependent. Therefore a change in dimension is followed by a modification in the reactivity of the functional system and the subsequent alteration of the extent of a chemical reaction. The reactivity of macromolecules may be modified by the application of a mechanical force that alters the distances between the interacting groups involved in a chemical transformation. An even greater cooperative change is induced by a reagent interacting with several groups present on a polymeric chain.

Various bioentities, including microorganisms, grow and reproduce by the flow of chemical energy, which usually passes through enzymatic pathways and chemical constituents identical for all live matter. In various live forms, however, certain stereochemical differences and individualization of deoxyribonucleic acid molecules cause the same chemical constituents to link together in different sequences, positions, quantities, and arrangements, thus giving rise to individual and diverse forms of life.

How the biocolloids of microorganisms evolved during the evolution of organic matter is largely a subject of speculation, although substantial conclusions may be drawn from experiments on the activation and assembly of molecules under electric discharge (1) and the polymerization of organic molecules in the presence of heat energy and catalysts (2, 4, 5). From these experiments, it may be surmised that, when methane, ammonia, water, carbon dioxide, and hydrogen, the earliest molecules present on earth, were subjected to ionizing energy or ultraviolet light, the bonds of carbon–carbon, hydrogen–hydrogen, carbon–hydrogen, nitrogen–hydrogen, and hydrogen–oxygen were broken to form free atoms. By reorganization, these atoms yielded more complex molecules, such as formic acid, acetic acid, succinic acid, and glycine. If, subsequently, one of the carbon-bond hydrogen atoms of glycine was exchanged for any of a group of other atoms, about 20 different amino acids would be formed. These organic molecules could be converted, with the participation of heat energy and heavy metals acting as catalysts, to building blocks for polymers and biomolecules, both in the experiments and during the chemical evolution (1–3, 5). Thus it was observed that different, moderately complex organic compounds, such as adenine and ribose, were formed when a mixture of hydrogen, cyanide, and ammonia was heated at 70°C for 25 days. Various bases and nucleotides were synthesized by the action of UV rays on hydrogen cyanide, whereas ribose and deoxyribose were formed by the irradiation of formaldehyde with UV or gamma rays (6, 7). Nucleic acids were synthesized by the interactions between nucleotides and polyphosphate esters at 50 to 60°C.

The heating of concentrated mixtures of amino acids in molten glutamic acid in the presence of polyphosphoric acid, or the heating of amino acids in aqueous ammonia, caused in them intermolecular reactions that yielded polypeptides of considerable size. As described by Fox (1), the heteropolyamino acids, or thermal protenoids, were obtained by direct heating of amino acids at such a temperature that water boiled off first and allowed the dry residue to undergo polymerization under the then thermodynamically appropriate condition of a low content of water in the material. It is known that highly polymerized peptides and proteins may acquire a characteristic helical structure because of a specific arrangement of carbon, hydrogen, nitrogen, and oxygen atoms in the amino acid chains. Filamentous proteins are able to aggregate in a characteristic, ordered array, thus generating a highly organized order in a subcellular material. Thus the proteinoids, possessing limited heterogeneity and capable of interacting selectively with certain enzyme substrates, were assembled into organized microsystems that showed many of the characteristics of contemporary cells (3, 4, 8). When the proteinoids were heated in aqueous solution and the clear, hot solution was allowed to cool, numerous microspheres were produced, each containing about 10^{10} molecules. The proteinoid microsystems thus formed communicate with each other through junctions that they form, and are capable of proliferation.

Postulating on the experimental evidence discussed above, one may assume that under natural conditions the molecules synthesized predominantly from methane and containing at least six carbon atoms, as well as other molecules possessing additional SH and NH_2 groups, condensed on the surface of the air–water interface and were cross-linked to form an insoluble product by the effect of electric discharges and heat energy. Consequently, surface films of predominantly hydrocarbon nature might have arisen and acquired multiple SH, NH_2, and O groups on the aqueous site, at a later stage of biomolecular evolution.

The emergence and diversification of organic compounds linked together by electric forces, originating and gradually perfected during the evolution of microsystems, produced molecules of peptides, saccharides, polysaccharides, lipids, and nucleic acids. Interaction between proteins and nucleic acids led to the development of a pattern for protein synthesis that thus was directed by nucleic acid.

The synthesized macromolecules, consisting of different compounds, were gradually linked together and arranged in a specific pattern by the highly specialized informational macromolecules. The evolving microsystems were equipped with the ability to perform complex and interconnected sequences of biochemical reactions, leading to the synthesis of subunits, to be assembled into specific configurations of primitive forms of life, which provided for the most efficient movement of compounds along the pathways. In this way, various forms of biological molecules were exploited and ramified in the ascent of primitive biocolloids to the particulate and unicellular levels of organization as viroids, virions, ribosomes, mitochondria, protoplasts, and protocells. The assembly or rearrangement of biological structures consisting of many interacting parts has been a cooperative process dependent on bond energy and activation energy.

Not surprisingly, biological differences existing between the fundamentally similar macromolecules occurring in different organisms seem to depend on a very fine chemical bonding or a small and unusual molecule, attached to a larger unit, which is detectable only by highly sophisticated physicochemical and immunochemical procedures. Predictably, very specific biological and live processes occurring in microorganisms will eventually be explained in terms of intimate interactions of hadrons, leptons, gamma quanta, and other families of particles still to be discovered. However, until that stage of scientific progress has been reached, our contemporary experiments in molecular microbiology will have to rely on the utilization of a vital and a rather large part of a functional, living system as a template needed to replicate another or a missing part of the same or a closely related living system.

REFERENCES

1. Fox, S. W., *Ann. N.Y. Acad. Sci.* **194**, 71 (1972).
2. Fox, S. W., K. Harada, G. Krampitz, and G. Mueller, *Chem. Eng. News* **48**, 80 (1970).
3. Harada, K., *Nature,* **214**, 479 (1967).
4. Hsu, L. L., S. Brooke, and S. W. Fox, *Curr. Mod. Biol.* **4**, 12 (1971).
5. Miller, S. L., and H. C. Urey, *Science* **130**, 245 (1959).
6. Orö, J., and J. Han, *Science* **153**, 1393 (1966).
7. Ponnamperuma, C., F. Woeller, J. Flores, M. Romiez, and W. Allen, *Amer. Chem. Soc. Advan. Chem.* **80**, 280 (1969).
8. Rohlfing, D. D., and S. W. Fox, *Advan. Catal.* **20**, 373 (1969).

CHAPTER

TWO

Molecular Arrangement and Assembly

J. B. G. KWAPINSKI

1. CONFIGURATION AND CONFORMATION OF MOLECULES

In terms of quantum biology, the reactivity of a molecule is described as a group electron problem, assuming a fixed core (52).

The electronic structure of a molecule is expressed by the momentum and position of the electrons in the molecule. The momentum refers to the energy, and the position applies to the distribution of energy or electrons in relation to the appropriate nucleus. The probability distribution of positions in a molecule is given by the square of a function of the free-space coordinates of the nuclei and electrons, and of the spin coordinate relevant to each particle.

In order to describe molecules in terms of quantum mechanics, it has been necessary until now to use approximations that do not completely elucidate the intimate structure of molecules. For example, according to Pauli's uncertainty principle, the momentum and the position of a particle or molecule cannot be determined simultaneously. Customarily, therefore, only the momentum of electrons is determined accurately, and, insteadd of position measurement, the probability distribution, which is expressed by the square of the wave function, is calculated. Accordingly, the electronic description of a molecule possessing a given momentum or energy is based on the probability assumption that an electron in a given element of volume is described by the square of its wave function.

The wave function is the solution of the Schrödinger equation, which is the basic equation of quantum mechanics. For a simple case of a single particle of mass m moving in a region in which the force acting on it can be represented by a potential energy function V, the Schrödinger equation is as follows:

$$-\frac{h^2}{8\pi^2 m}\left(\frac{\gamma^2\psi}{\gamma x^2}+\frac{\gamma^2\psi}{\gamma y^2}+\frac{\gamma^2\psi}{\gamma z^2}\right)+V\psi=\frac{\lambda\gamma\psi}{2\pi i\gamma r}$$

If it were possible to solve the Schrödinger equation for an assembly of any number of particles, any molecule could be described accurately, and almost complete information about any molecule would be available. The structure of a molecule would primarily be determined by the geometric arrangement, expressed by the space coordinates of each of its atoms, so that the energy of the molecule as a whole is at a minimum. The vibrational force constants would be obtained from the variation of the energy with the variation of the nuclear coordinates (32). However, the Schrödinger equation can be solved only for systems consisting of one to three particles.

Although the exact form of the molecular potential energy is unknown, the total molecular potential energy in the repeating chemical units is represented by harmonic potentials, torsional barriers, the Lennard–Jones potential function, and a compensatory constraint (42). The potential energy may be approximated by the following equation:

$$\mathrm{E}=\underset{\text{All bonds}}{\sum\tfrac{1}{2}K_o(b-b_0)^2}+\underset{\text{All bond angles}}{\sum\tfrac{1}{2}K_r(r-r_0)^2}+\underset{\text{All dihedral angles}}{\sum\tfrac{1}{2}K_o\{1+\cos(n^0-\delta)\}}$$

$$+\underset{\text{All nonbonded pairs}}{\sum \mathrm{E}_{ij}\left(\frac{r_{ij}^0}{r_{ij}}\right)^{12}-2\left(\frac{r_{ij}^0}{r_{ij}}\right)^{6}}+\underset{\text{All atomic coordinates}}{\sum\tfrac{1}{2}w(x_i-x_i^0)^2}$$

where K_e is the bond force constant; b, bond length; b_0, equilibrium bond length; K_γ, bond angle bending force constant; γ, bond angle; γ_0, equilibrium bond length; K_o, torsional barrier; θ, dihedral angel; n, periodicity of rotational function; δ, phase; E_{ij}, depth of nonbonded minimum; r_{ij}, distance between atoms i and j; $r^0{}_{ij}$, distance of nonbonded minimum; w, constraining force for all atoms; x_i, atomic Cartesian coordinate; and $x^0{}_i$, experimental coordinate.

A detailed and precise description of the molecule is attained by determination of its stereochemical formula, which expresses the spatial arrangement of the atoms in a molecule, as determined by accurate measurement of the lengths of the bonds linking these atoms and the angles between the bonds. The positions of the atoms and electrons relative to each other are revealed precisely by the analysis of X-ray, electron, or neutron diffraction patterns. This analysis is used to investigate mutual atomic positions and interatomic distances, as well as the arrangement and combination between atoms, conferred by homopolar, single, or double bonds. From such data, electrondensity maps (Fig. 1) are drawn; these show the deviations of atoms from pure sphericity arising from the usual thermal motions. Stereochemical analysis also provides indications as to the location of the electrons in single or double bonds.

Fig. 1. Electron density map of a molecule (4).

Macromolecules may possess alternative positions of the atoms within the same bond patterns; these are determined by the conformational properties of a molecule, dependent on the interactions of distant parts of the molecule over and above those due to conjugated double bonds. Thus configurational patterns estimated by stereochemical investigations can be supplemented by examination of the conformational properties of the macromolecules. The molecular conformation is determined by the temperature-sensitive van der Waals forces and by mutual interaction of parts of a molecule other than those immediately related by primary valence bonds (4). Similar forces are involved in antigen–antibody reactions (12).

Macromolecules possess a wide range of possible conformations because of their flexibility. The flexibility of a molecule depends on the rotational and torsional degrees of freedom of its bond. The conformation of the isolated molecule is determined by the sum of the effects of the interactions of the more distant parts present in the molecule. The distribution of different conformations is greatly influenced by environmental factors, for example, the presence of similar molecules or solvent molecules. These factors are especially effective in long, flexible macromolecules. The completly stretched-out and the coiled forms of the same molecule represent extreme types of conformation.

Biomolecules originating from various sources or from the same cell may possess similar qualitative and quantitative patterns, yet they differ in atomic detail. Biomolecular structures can be studied in atomic detail by the ap-

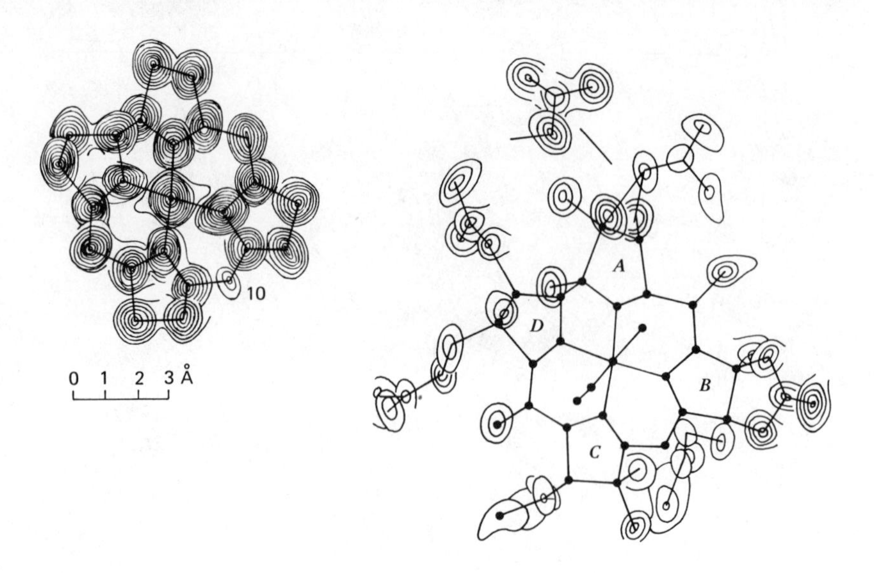

Fig. 2. Electron image of a molecule (4).

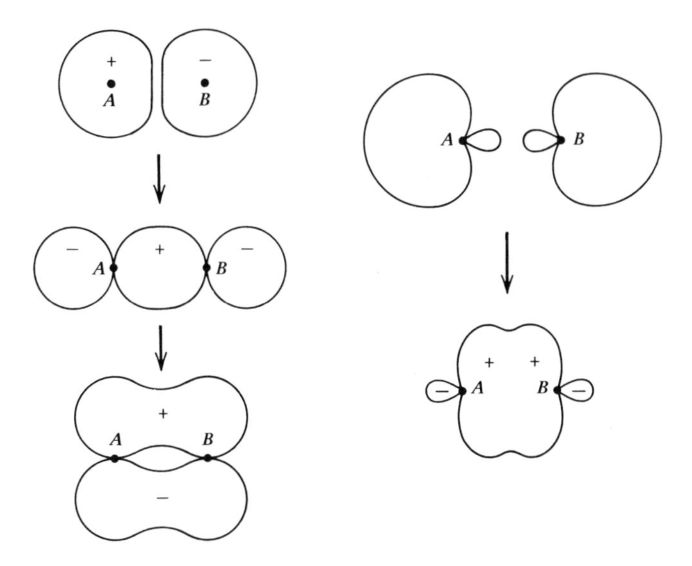

Fig. 3. Emergence of molecules by hybridization of electron orbitals of atoms *A* and *B*; a deformation of molecular symmetry (32).

plication of atomic resolution, which describes the three-dimensional structures. In this method, the positions of heavy atoms in the protein molecule are revealed and measured on an automatic three-circle diffractometer and compared with the positions of heavy atoms in standard crystalline molecules (62). The image of a biomolecule can be obtained by photographing the electron densities representing the arrangement of atoms in a biomolecule (Fig. 2).

Before a molecule is formed by a reaction between atoms, certain orbitals of each atom are combined to form molecular orbitals. Depending on certain symmetry properties, only some orbitals of two atoms mix or hybridize, and individual atomic orbitals are replaced by hybrid orbitals, as shown in Fig. 3. The hybridization of atoms or molecules is associated with a change in energy; for example, the promotion of a fraction of an electron from one atomic orbital to another requires considerable energy. In some cases, the energy can be compensated for by increased binding energy or reduced repulsion of electrons.

2. MOLECULAR ASSEMBLY

The molecular assembly is an orderly association of atoms and interatomic energy bonds, forming a functional molecular structure that bears a conformational stability relative to the environmental forces and the time factor.

Groups of atoms joined by covalent bonds and displaying relatively constant amounts of kinetic energy are assembled in the form of molecules. The nature and dimensions of the atoms occurring in the molecules are specified by limited distances, directions, and rotations. These properties, together with the physical nature and the kinetic energy of the atoms, determine the conformation of a molecule.

Through the association of molecules by binding energy, macromolecules are formed. As their complexity and degree of polymerization increase, supramolecular subunits evolve, and these, upon activation, are assembled into configurational and conformational entities.

Instrumental in the assembly, polymerization, and interactions of macromolecules are the physicochemical processes accompanied by an overall decrease in free energy. The rate of intermolecular reactions is determined by the energy of activation needed for transitions. Only certain types of macromolecules are selected for a specific function in the dynamic self-reproducing systems of biological structures. Interactions between different types of the selected macromolecules, leading to a characteristic assembly and arrangement of molecules, are coordinated by control mechanisms that are intrinsic properties of the assembled structures. Macromolecular and biomolecular organization is controlled by the thermodynamic principle of minimum free energy and the biological principle of natural selection.

Molecular interactions are controlled by two mechanisms: the equilibrium and the kinetics of the interactions. Any reaction participating in the assembly of macromolecules possesses a characteristic equilibrium constant that is determined by the reacting molecules. The reactions are enhanced by enzymes which lower the activation energy required for the process. In the interactions in higher and complex biological macromolecules and polymers, especially those leading to the formation of morphologically definable structures such as the virus particle, changes in the state of macromolecules may control the equilibrium constant. All biological regulation can be expressed as the result of a coordinated control of equilibria and rates, which depend on the physical properties of the interacting components (10).

In biological entities, the molecular constituents of the supramolecular and conformational entities do not seem to exist randomly but rather are hierarchically ordered in a dynamic complex population or a system of mutually interdependent, heterogeneous components which incessantly interact, change their form and location, divide and merge, break down, and are replaced. The freedom of molecules is restrained by a hierarchical, intramolecular framework of order, so as to maintain, for a time, the recognizable structural and morphological characteristics of a particulate or conformational entity thus produced. The arrangement of heterogeneous parts within a conformational entity is ever-changing, but the constituents assume dynamically proper mutual space and functional activities.

Although individual species of molecules possess inherent physicochemical properties, they express themselves dynamically only through cooperative interdependence and collective interaction. The dependence between molecules and molecular systems is expressed in cooperation, interference, or unilateral parasitism.

In this dynamic view of molecular assembly, a virion or a bacterium is a systematically organized community of diverse molecular populations or supramolecular entities, endowed for a time with a specific, hierarchical framework of order provided by the interactions of the constituent, interdependent molecules.

In live entities, as in nonliving matter, the polymer molecules are arranged in coils, in straight chains, or in helical form. The helical type of molecular arrangement may involve a low-angled helix, a ring form, or an open-tube helix (Fig. 4). Long molecules can be arranged by coiling, folding, or twining. Each

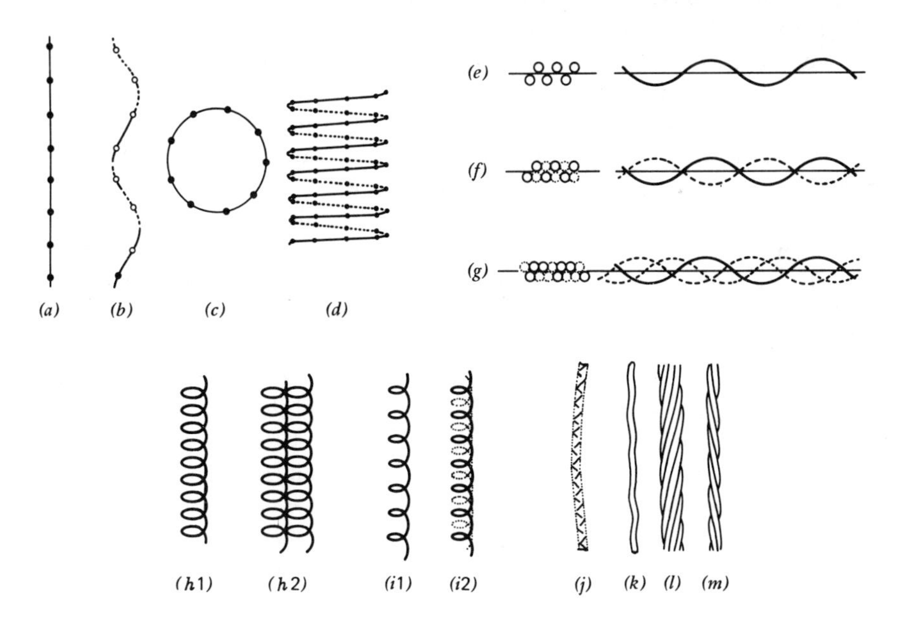

Fig. 4. Types of molecular arrangement (4). (*a*) Straight-chain polymer. (*b*) High-angled helix. (*c*) Low-angled helix. (*d*) Open-tube helix. (*e*) Single, high-angled helix. (*f*) Double, high-angled helix. (*g*) Treble, high-angled helix. (*h*1,2) Low-angled helices. (*i*1,2) Juxtaposed arrangement. (*j*) Coiled-coil, showing internal helix. (*k*) Twined coiled-coil. (*l*) Sevenfold twist of twined coiled-coils. (*m*) *Threefold twist of twined coil-coils.*
coiled-coils. (*m*) Threefold twist of twined coiled-coils.

of these arrangements may occur as a unit or in a sequence; for example, a molecule may first be coiled twice and then folded, depending on the dimensions of the groups forming a polymer molecule. In the process of twining, some linear molecules are arranged as constituents of a row, forming a set of helices, which are joined together to produce a thicker unit.

Separate coils and folded structures may be complexed further to form aggregate molecules. The molecular aggregates are larger units consisting of several identical or very similar molecules, held together by secondary valences and existing for a time as independent units (4). Aggregate molecules occur in pigments and in larger proteins, such as globulins.

The polymer aggregates, arranged in the form of extended parallel molecules or in bundles, produce a fibrous structure. Large structures of molecules that are coiled and folded and either aggregated or nonaggregated produce globular assemblies. A globular arrangement can be changed into a fibrous arrangement, since these two types of aggregates and arrangements are physicochemically closely interrelated. Molecules of polymers reacting with a liquid medium or environment produce either a maximum or a minimum surface and therefore occur in coiled-up or extended form.

The energy required for the biosynthesis of polymers derives from the "activation" building units, which possess the energy needed for anabolic reactions. The activation results from a chemical binding of building components to a carrier. To provide activated groups that can be incorporated into molecules of polymers, numerous catabolic reactions are utilized.

The activation energy required to initiate assembly in protein association reactions is relatively high, and the size of the aggregates is determined by the kinetics of the nucleation. In this type of cooperative association of molecular units, the number of bonds increases much more rapidly initially than at a later stage. In such assemblies, a maximum ratio of three bonds per unit in the lattice surface is found, since any plane lattice may have a maximum of six nearest neighbors. The fraction of the units at the perimeter of a compact planc array of n units that have formed less than the maximum number of bonds is proportional to $n^{-1/2}$. The fraction of units at the surface of a crystallite in a three-dimensional lattice consisting of n units is proportional to $n^{-1/3}$ (51). In a lattice, the negative change in free energy per unit increases with the number of units if the bond energy is constant. In closed structures, such as rings or shells, more bonds are formed when the last unit is added than for any of the previous ones.

At the various stages of organization, the process of assembly is controlled by changes in the equilibrium constants for the functional interactions in response to changes in the environment. The energies of interaction between various molecules vary from 1 to 60 kcal/mole. Short-range interactions between molecules in aqueous systems are regulated by covalent, hydrogen, "hydrophobic," ionic, disulfide, or London–van der Waals linkages, or by a combination of these bonds. Long-range macromolecular interactions depend on long-range forces, which are of two types: the London–van der Waals forces, and charge pattern forces. The hydrogen bond is a permanent dipole interaction energy bond, since it depends on the small size of the proton which permits the close approach of dipoles; and it varies inversely with the third power of the distance of separation of molecular groups.

Charge patterns of macromolecules carrying many groups capable of being positively or negatively charged or uncharged are established on the interacting molecules as a direct result of mutual electrical polarizations. An electrostatic attractive energy, which depends on the established charge patterns on the interacting molecules, decreases in proportion to the distance. Interactions through charge fluctuation at short distances seem to be essential to the participation of protein moieties in the mechanism of hydrolytic enzyme activity (37). Charge fluctuation and charge patterns, resulting from interactions between antigenic molecules and the corresponding active sites on immunoglobulins, may represent fundatmental physicochemical phenomena and mechanisms of immunochemical reactions.

The surface of a macromolecule is defined chiefly by groups of small molecules interacting through van der Waals forces. These groups must be complementary to each other so that the stereochemical configuration on the surface of one macromolecule binds a complementary configuration present on the surface of another, interacting molecule. The smaller the surface area of interaction, the more subtle is the combination between the interacting surface sections. Stable molecular interactions are produced if the total interaction energy is about $10^{15} \times kT$, where kT stands for kinetic energy. Since the energies involved in chemical or immunochemical reactions are highly sensitive to distance, alterations in the stereomeric configuration and molecular structure cause imperfect steric fit and nonspecific reactions or no reaction at all. Thus, by minute imperfections in the complementation, large alterations are evoked in the interaction energy and product.

The fundamental types of energy and of forces involved in molecular interactions, which depend on the long- and short-range patterns of interacting groups, are apparently responsible for the specificity of size, shape, and chemical and biological differentiation of interacting molecules. These forces and interactions are instrumental not only in the mutual transcendence of structure and function but also in the development of aggretates, arrangements, biostructures, and entities of live matter.

Specific molecular interactions participate not only in the formation of new complexes, which may vary in size, but also in the production of distinct morphological constituents in microorganisms (e.g., cell wall, cytoplasmic membrane, flagellum, viral coat). The polar and/or the nonpolar portion of the molecules forming certain functional and morphological parts or layers of microorganisms, for example, bacterial membranes or flagella, must be arranged in a specific manner in order to provide an energy barrier for permeability. The nonpolar portion of lipid molecules, for instance, seems to represent a high-energy barrier to the penetration of water. In contrast, the development of a microbial structure or component (e.g., cytoplasmic inclusions) which does not carry any critical physical function requires very little specificity in the assembly of macromolecules.

The synthesis of specific small molecules (e.g., monosaccharides, amino acids, nucleotides) and of macromolecules of relatively simple homopolymers, such as polysaccharides or peptides, seems to be kinetically controlled by coupled enzymatic reactions. The energy required for these reactions is usually supplied by phosphorylated compounds, but the specificity depends on the equilibrium association of substrate and enzyme.

The synthesis of specific macromolecules is directed by a template, which arranges the component residues by equilibrium association through noncovalent bonds. Amino acids, for example, are enzymically linked to the transfer RNA.

The template system functions as a catalyst which enhances the synthesis of a polypeptide and the attainment of a particular sequence. The folding of polymers and the assembly of a macromolecular structure, such as a virion, are controlled by the specific bonding properties of the constituent parts, which are capable of self-assembly. More complex biological structures require a kinetic control mechanism. This mechanism may act as a template in selecting specific forms of the energetically possible designs, or it may function like an enzyme, accelerating the rate of production of an individual type of a noncovalent bond (10).

The assembly of macromolecules and biomolecules into any ordered structure is promoted by the formation of specific, stabilizing bonds and by the elimination of thermodynamically unfavorable interactions, such as the ordering of a solvent. Consequently, a considerable amount of energy is used for the diversion of the bonded groups to a solvent.

The stability of any biomolecular, condensed assembly consisting of many repetitive units or a number of different units depends greatly on the relative free energies of the units. Therefore the equilibrium transition between different possible states occurs abruptly at critical environmental conditions.

Interactions occurring between the constituents of a highly polymerized structure, which depend on the transfer of kinetic energy between electrons and protons, contribute directly to the order of arrangement of the structure. Two general types of order are represented by crystals and liquids. The order of arrangement in the crystal consists of regularly repeating patterns, whereas the arrangement of constituents in a noncrystal is more flexible.

Because of the participation of coordinated, very weak forces in the assembly of biological macromolecules, their stable conformations are sensitive to the composition of the molecules and to the interactions with the external environment. The presence of weak bonds is instrumental in the selection of dynamic molecules for specific functions that would not be possible in the presence of strong bonds. The stability of a biostructure depends not only on the forces existing between the assembled molecules but also on entropy, which may result from interaction with a solvent.

The three-dimensional arrangement of macromolecules, forming morphologically substructural and structural microbial entities, is derived from a natural, thermodynamically acceptable organization of atoms and molecules at the two-dimensional level of the particles and from molecular interactions occurring within the proximity of dimensions of interacting molecules. Subcellular colloidal aggregates, which emerge from these processes, interplay with each other, mutually modifying their activities and producing an entity of live matter. During this process, specific molecules are selected for the interactions. Examples of such specific interactions are antigen-antibody and enzyme–substrate reactions; synthesis of proteins, polysaccharides, nucleic acids, and phospholipids; and construction of virions, bacteria, and protozoa. The interactions between macromolecules are essentially similar to the short-range interactions occurring between small molecules and submolecular groups of atoms possessing different interaction properties and at suitable distances and locations. Only the more highly polymerized, condensed, and organized complexes of macromolecules are specifically selected and arranged to form structural morphological entities that can be detected by electron microscopy or X-ray diffraction.

Biological assemblies are composed of configurationally complementary molecules possessing specific bonding properties. Consequently, any structure consisting of identical parts has some regularity, as the possible designs are restricted by the nature of the assembly process. Unlike nonbiological structures, biological structures are based on the designs selected for functions in the dynamic, self-reproducing, biological system. Along the definite patterns, the organized biological structures are assembled by specific associations of various components and constituents, synthesized separately by a subassembly process. Each stage and period of synthesis and assembly is directed by a different type of control mechanism.

In the synthesis of large organized biological systems, the specificity of noncovalent bond interactions occurring between macromolecules is greatly utilized. These interactions determine the specific associations of enzymes with their substrates. After the component parts have been made, identical structure units self-assemble into a biological structure without a template or other specific external genetic control. Large organized structures, such as a virus particle or bacterial wall, consist of a large number of subunits, but only a small variety of subunits is available for the assembly. All levels of biological organization are controlled by two fundamental physicochemical principles, the thermodynamic principle of minimum energy and the biological principle of natural selection. Thus the self-assembly structure owes its stability to correct bonding, which leads to a state of the lowest realizable free energy for the system. If the difference in free energy between different bonding states is only slight, incorrect bonding may occur during the assembly, yet the most stable bonding pattern will eventually be established through dissociation and reassociation. The size of biological self-assembly structures that can be built efficiently is restricted by self-limited designs, selected for specific functions. The self-limitation may arise from coordinated interactions between two or more different types of components, or it may be an essential property of a stable bonding pattern of a single type of subunit.

Molecular interactions occur at dimensions of 3 to 5 Å or less. The minimum self-replicating unit is a body that is capable of replication in the absence of other living cells. The smallest free-living forms have linear dimensions that are approximately 1000 times larger than the linear dimensions of an atom. The smallest free-living cell, the *Mycoplasma* organism, consists of an amount of specific protoplasm, surrounded by a unit membrane. It possesses less than 10^9 atoms in the nonaqueous portion, and a genome consisting of a double-stranded DNA molecule 2.28 nm in size, and it contains approximately 400 ribosomes. The protein mass is of the order of 5×10^{-15} g and is the equivalent of about 60,000 protein molecules, each possessing a moiecular weight of 50,000. The unit membrane has a thickness of approximately 75 Å (6).

The biosynthesis of molecules in microorganisms is mediated and controlled by approximately 700 different enzymes; an additional 200 to 300 enzymes are involved in the biosynthesis of polymer macromolecules. During the synthesis, micromolecules seem to be in constant movement, shuffled by enzymes. The biosynthesis of bacterial constituents requires a suitable supply of low-molecular-weight compounds such as saccharides, amino acids, and phosphates, which serve as precursors of raw materials for biosynthesis. Aside from this, the biosynthesis and the nature of biomolecules of specific heteropolymers arising from the interactions between molecules of different microbial homopolymers or between microbial homopolymers and other molecular systems are very little known.

The probability that a biomolecule and a ligand will join and form part of a heteropolymer complex under certain environmental conditions (temperature, pressure) is determined by the "probability of binding" (68), according to the following equation:

$$P = \frac{\text{actual concentration of biomolecule–ligand complexes}}{\text{maximum concentration of biomolecule–ligand complexes}}$$

where P stands for probability of binding.

Binding reactions may be classified as simple, multiple, or cooperative. A heterogeneous protein population binding a single mole of ligand per mole of protein can be the origin of multiple binding, and the binding of several moles of ligand to the same protein may involve the simple type of reaction. Facilitation in the binding of successive molecules of ligand once the first molecule has been bound is usually attributed to cooperative binding.

Cooperative effects in binding may arise in equilibria among tautomeric forms of protein when 2 or more moles of ligand are bound in stationary states set up among protein tautomers, and the mean activity of the system depends on the probability of binding. Cooperative effects may be observed, for example, in 10^{-6} M protein solutions at high ionic strength, where the intermolecular distances are of the order of 500 Å.

3. ASSEMBLY OF MOLECULAR SUBUNITS

A molecular assembly at any level of molecular organization is derived from the interaction energy between atoms. The strongest interaction forces are covalent and coionic bonds, which result from the sharing of electrons. A very high activation energy is needed for forming or breaking covalent bonds. The noncovalent molecular interactions are electrostatic and emerge from forces interacting between some combination of charge, dipole, or induced dipole. The energy is inversely proportional to the effective dielectric constant between the interacting pairs. The energy depends on the product of the magnitude of the interacting groups and the orientation of the dipoles or the polarizability of vectors in relation to each other and the direction joining them.

The specific bonding energy, which is instrumental in the structure and conformation of macromolecules, uniquely determines the design and stability of molecular subunits. Under appropriate environmental conditions quaternary structures of macromolecules can assemble spontaneously into more complex and morphologically distinct entities, without special genetic information. In this process, macromolecules of different homo- and heteropolymers, formed in

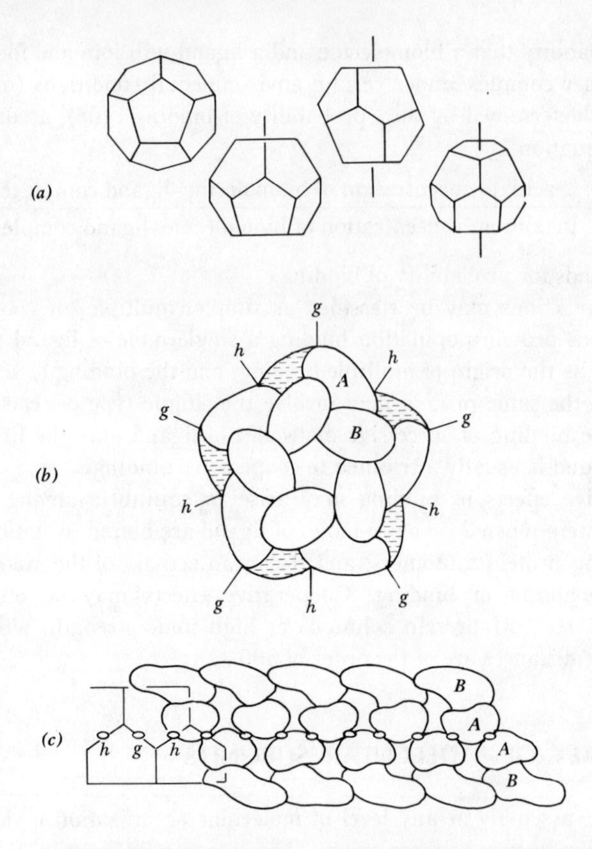

Fig. 5. A protein assembly consisting of 20 subunits (29). (*a*) Regular pentagonal dodecahedron. (*b*) Distribution of 20 chiral subunits in a pentagonal dodecahedron. (*c*) A cylindrical projection from the fivefold axis (the combined *A* and *B* subunits constitute a protomer of the oligomeric protein macromolecule). *gh*-two types of isologous domains of bonding.

the specific sites of synthesis, are transferred to the site of construction of subunits for final assembly.

Many biostructures, and especially microbial structures, are built by self-assembly of molecular subunits (protomers) or asymmetric units, rather than symmetric units. Molecular subunits are polymer entities, consisting of several linked, identical molecules of a homo- or heteropolymer, which are larger than a single polypeptide chain but smaller than a protein molecule.

The regular association of identical subunits observed in the higher level of organization, which may have been derived by selection, may have its roots in the economical use of genetic information, the efficiency of construction, and the capacity of substructures to respond to alterations in the environment by massive, cooperative changes.

A symmetrical structure is built from equivalently related parts, each composed of smaller, identical units (protomers), which are connected regularly by bonding in a closed space. A small assembly may enlarge by binding a set of identical units, the same contact pattern is then repeated, and a larger symmetrical structure emerges. For example, subunits possessing dihedral point-group symmetry occur in the structure of arginine decarboxylase of *Escherichia coli* (6). This assembly is composed of a ring of 10 protomers (Fig. 5). Each protomer has two distinct isologous binding sets, and the axes associated with these linkages define the plane of the ring (29). A similar, symmetric structure is found in acetoacetate decarboxylase of *Clostridium acetobutylicum*. This assembly has 12 constitutionally identical subunits (65).

A symmetrical assembly remains relatively stable as long as it does not acquire sufficient activation energy to prevent spontaneous transition to a modified and usually more stable structure. Because of the great functional activity of biological structures, their assembly is subject to rather small, nonrandom variation in a regular bonding pattern, so that bonds between identical or equal units are systematically deformed in slightly different ways. The resulting modified structure is often more stable than a symmetrical structure based on strictly regular bonding, since it possesses many very stable bonds and little free energy. This quasi-equivalent bonding is found in the icosahedral shell design of viruses. Biological structures more often possess quasi-symmetry relations rather than ideal symmetry.

Since the molecular and structural properties of viruses, rickettsiae, bacteria, and fungi are presented in detail in Chapters 7, 9, 10, and 11, respectively, only certain fundamental aspects of the molecular assembly leading to the formation of these biostructures will be discussed in the following paragraphs.

The apparent regularity of the shape of small viruses is a consequence of the geometry of the packing of identical subunits, which tend to arrange themselves like piles of cannon balls. The subunits in most small viruses are arranged at equidistant points over the surface, with the resulting geometric configuration of an icosahedron, which has twelve corners with five faces. This geometrical configuration in space seems to contribute to the chemical and physical stability required for the virions to exist in an environment for a time. The physicochemical stability of a virus is essential for processes of minimum energy and selection.

The physicochemical properties of the subunits composing a morphological structure, such as a virus shell, are not necessarily absolutely identical. The chemical characteristics of an individual subunit or a group of subunits may differ from others, depending on the position of electrons relative to the adjacent subunits and the quantum of energy carried by the subunits at a time. Differences arise from the icosahedral pattern, since twelve units or groups of subunits are shared among five faces, a variable number are shared between two faces, and the remaining units lie within one face. It is also possible that final arrangement is never achieved completely, or at least that the final arrangement is a modification of a protein that is produced normally.

There are only a few possible types of efficient design for the framework of a virus, and particularly its coat. The design of the protein coat is determined by the specific bonding properties of the identical structural units comprising the coat. An asymmetric shell or virus coat can possess one of the three types of cubic point-group symmetry: the tetrahedral, octahedral, or icosahedral, which requires 12, 24, or 60 subunits, respectively.

The virus coat in small spherical or cylindrical virions is built by self-assembly of identical protein subunits arranged in the form of icosahedral symmetry. The icosahedral symmetry of a particle forming a triangular net represents the optimum design for a state of minimum energy of a closed shell composed of regularly bonded, identical molecular subunits (11). The number of structural units in a substructure is always of the form $60T$, where T is the triangulation number.

The subunits on an icosahedral surface are arranged in the shape of pentagons (pentamers) and hexagons (hexamers). The number of pentamers is always 12, whereas from 12 to 240 hexamers per morphological unit may be found in various viruses. The icosahedral structure is stable despite a constant, systematic deformation of the bonds connecting equal units.

The helical pattern arises from the assembly of identical subunits, held together by their specific binding properties. Helical designs occur in various rod-shaped structures, for example, rod-shaped viruses, tails of certain bacteriophages, flagella, and pili.

The virus subunits seem to contain all the information required to produce, by a system of self-assembly, the structure of virus particles with its characteristic shape and size. Genetic information is needed, however, to produce the final, stable structure. This additional information is given by supplementary morphopoietic factors which interact specifically with the subunit or with a polymerized product arising from combinations of the subunits. Through this interaction, the supplementary factors determine the final shape of virions. A morphopoietic factor may or may not be integrated into the final product, which may possess helical form as determined by the factor. If the existence of morphopoietic cores is accepted, one envisions that these three-dimensional bodies guide the assembly of subunits on their surfaces and counteract the intrinsic tendency toward curvature. Variations in shape and size may be regarded as variations in the order of morphopoiesis.

Subunits of the phage T4 head can be assembled into one of the following four structures: a short icosahedral structure, a prolate icosahedron, a two-tailed biprolate icosahedron, and a tubular polyhead, which can have open or closed ends. The different types of molecular arrangements and morphopoiesis of the phage may arise from a slight conformational or physical alteration of the subunits forming the phage particles. The subunits are presumably associated with an allosteric effector, which modifies their conformation and prevents spontaneous association into polyheads or polysheaths. The conformation for sheath subunits would enable the subunits to settle on the central tube, functioning as a morphopoietic core. The conformation for head subunits, which have a molecular weight of 55,000, would enable them to settle on a morphopoietic core in order to produce normal capsids. Association with the allosteric effector is rather unstable because of the disintegration of the effector by a specific enzyme.

The stability of the quaternary structure of the T-even phage and the attachment of structural parts to one another in the complete virion depend mainly on noncovalent interactions. During chemical degradation, the contracted sheaths of the phage are progressively degraded from one or both ends; this process resembles the dissolution of a three-dimensional crystal in a solvent, which starts from the surface.

The merids, which are regular assemblies, each composed of a single type of protomer, can be dissolved by strong, protein-denaturating agents, causing the destruction of the tertiary and secondary structures of the subunits. The intramerid interactions are hydrophobic, resembling those involved in the tertiary and secondary structures.

Sheath contractions represent one of the ways in which the structural constituents of a phage particle can be dissociated. In a medium containing anionic detergents, urea at neutral pH, or glycine–HCl, the phage sheath moves, together with the base plate, away from the distal part of the tube. Normally the sheaths are attached to the proximal part of the tube by relatively strong interactions. The sheath is also linked to the base plate, which is in turn attached to the tube.

The interactions holding identical protomers and chemical subunits within the surface lattices of structural parts of T-even bacteriophages are as follows, starting with the interaction of least strength and arranged in order of increasing strength: those between base plates, tail tubes, capsids, tail fibers, and contracted sheaths.

Dissociated subunits can probably be reassembled into any structural part of the phage particle. The tail tube, however, can be reassembled into a merid most easily, since only a single type of fully shaped specif subunit and a weak interaction are involved in this reassociation.

The physicochemical structures of virions of adenoviruses and papilloma viruses are easily dissociated into capsomeres. The capsomeres are further dissociated into chemical subunits by rather drastic agents.

Subunits forming the protein coat of polyoma virus have a molecular weight of 50,200. The protein molecule forming a part of the subunits of the tobacco mosaic virus (TMV) is a cyclic trimer with a molecular weight of 17,530 and consists of a single chain of 157 amino acids, folded to produce a compact unit, measuring 6 × 2 × 2 nm. Protein molecules are linked to each other by hydrophobic bonds, and are bound to ribonucleic acid molecules by salt linkages, which increase the cohesion between the subunits. The protein of TMV, dissociated from the nucleoprotein complex, may polymerize, giving a helical structure of a double disc containing 32 subunits. The double discs can further polymerize in stacks to form long rods (10). Nonpolar residues of the elliptical protein subunit attract each other in a stereospecific manner. Polymerization of the protein from the molecular subunits is controlled by carboxyl–carboxylate binding. The subunit ellipsoids do not form salt linkages, hydrogen bonds, or covalent bonds with neighboring ellipsoid subunits, but certain zones of the units can interact with water. The virions of TMV reassemble *in vitro* from nucleic acid and protein subunits in a physicochemical process resembling crystallization.

Virions of influenza virus A contain at least three different types of nucleoprotein subunits, recognized as the 50S, 60S, and 70S components, and probably two more subunits (17).

The protein shell of tumor viruses is assembled in two polymorphic forms, the wide and the narrow tube. In the wide tube, hexamers are arranged on a hexagonal-plane lattice, rolled up into a cylindrical tube (38). Pentamers, bonded across the twofold axes of a surface lattice, are arranged in a pentagonal tessellation in the narrow-tube assembly. The pentagonal tessellation has twofold axes of symmetry, which are perpendicular to the plane.

Selected specific molecular subunits are assembled into different bacterial and protozoal organelles and morphological structures, such as flagella, pili, cell membranes, spore coats, mitotic apparatus, ribosomes, and granules.

Bacterial flagella consist of about 70% protein, 20% lipid, and 10% polysaccharide. The protein part is composed of protein subunits. The protein subunits termed flagellin have a diameter of about 5 nm and a molecular weight ranging from 20,000 to 40,000. The flagellin macromolecules may be assembled to form a flagellum in two different fashions: (1) the subunits are wound around a common center with about 8 subunits per turn, thus producing a single-or double-helix arrangement, or (2) the subunits are assembled in the form of a vertical rod (51). The flagellin subunits are sometimes linked to carbohydrate molecules.

Polymerization of the flagellin in some bacteria seems to be controlled by tiny, 0.2-nm fragments of the flagellum or smaller nuclei which serve as nucleating agents for condensation. The nucleus grows into a polymer by addition of monomers. No nuclei at all are needed to initiate polymerization in other species of bacteria.

The self-assembly of flagellin subunits to produce a flagellum is influenced by certain environmental factors. For example, the flagella are formed at a temperature below 45°C but not above this critical temperature.

Polymerization of flagellin, a protein occurring in bacterial flagella, and of an actinlike protein occurring in plasmodia is determined by constituents, regulators, structure selectors, and energy sources involving covalent or noncovalent bonding. Polymerization of protein molecules may lead to one or more stable polymer structures. If a special energy source is needed to produce a change in the polymer structure, nucleotides can serve as the source. This process does not normally require any special energy source, however, and is easily reversible. Different types of structure arise from diverse conformations of monomer proteins and different ways of interaction between the proteins. The protein molecules possess various regulators of the rate and the extent of polymerization and adequate thermodynamic conditions for interaction with a neighboring subunit.

At temperatures exceeding 28°C flagellin is converted reversibly to a conformational isomer. The dissociated or disperse components of flagella reassemble into stable structures by a process of quasi crystallization (2). The reassembly of flagella occurs only in the presence of a piece of organized flagellum that acts as a model or crystal focus.

Pilin, a protein subunit with a molecular weight of about 17,000, contains a preponderance of nonpolar side chains. The pilin subunits are arranged in the form of a single helix containing 3 1/8 subunits per turn. Pili grow by the assembly of previously synthesized subunits utilizing pools of pilins.

The plasmodium *F-actin* appears in a salt-free solvent as a monomer that is transformed into a fibrous polymer by the addition of monovalent salts. The polymerization and transformation of the polymer structure are reversible. The general configuration of the *Mycoplasma laidlawii* membrane is a monolayer or bilayer of lipid molecules, and this configuration is maintained below and above the transition, that is, at higher temperatures. The thermal change in the membrane is a phase transition of the fatty acyl chains from a closely packed hexagonal array to a more fluid state, which occurs over a 6 to 10°C range. The hydrocarbon chains are hexagonally packed at a center-to-center spacing of 4.28 Å, similarly to the hexagonal phase of the long-chain paraffins (18).

The cell wall of *Candida albicans* is assembled from globular aggregates, consisting of smaller spherical bodies, which contain a glucomannan–protein complex (48). The polysaccharide constituent contains mannose and glucose in a ratio of approximately 1:1, whereas the protein part of the complex is composed of 17 different amino acids, with glutamic acid and aspartic acid as the predominant ones. The protein and polysaccharide parts are joined through ester linkages or other covalent linkages, which also hold together other components of the wall structure.

Two "unit membranes," each probably consisting of different molecular subunits, constitute the cell wall in bacteria: the inner, semirigid peptidoglycan mucopeptide laver, and the outer, interdigitated lipophilic complex. The outer, lipophilic material is covalently linked to the "bag-shaped" peptidoglycan, sacculus, which may represent 12% of the total cell wall (44).

The outer layer of the cell membrane structure consists of a heteropolymer complex of a lipopolysaccharide, phospholipid, and proteins. Although the linkages between constituents of the heteropolymers or its parts are not known, presumably the subunits are arranged in layers. The lipopolysaccharide is a long-chain, phosphorus-containing heteropolymer, composed of a lipid covalently linked to a core polysaccharide to which antigenically different oligosaccharide units are attached; it functions as endotoxin.

The peptidoglycan complex is composed of two different *N*-acetyl hexosamines, their 3-*O*-D-lactic acid ether, termed *N*-acetylmuramic acid, and a few different amino acids. The amino acids forming a peptide are alanine, glutamic acid, and a dibasic amino acid, which most frequently is lysine or diaminopimelic acid. The peptide chains are cross-linked to form a net, which surrounds the bacterial cell.

In Gram-positive bacteria, the two "unit membrane" structures are an externally located, nonpeptidoglycan complex and the inner peptidoglycan. The nonpeptidoglycan component is either a protein, a polymer of polyol phosphate, teichoic acid, a polymer of *N*-acetylgalactosamine and glucuronic acid, termed teichuronic acid, or a polymer of L-rhamnose and *N*-acetyl-D-glucosamine. The nonpeptidoglycan possesses chemical determinants for antigenic specificity receptors for interaction with phages. Some nonpeptidoglycan components function as endotoxins or other factors of virulence. The peptidoglycan is built up of polysaccharide chains, cross-linked through peptides. Its glycan portion is composed of alternating β-1,4-linked units of *N*-acetylglucosamine and *N*-acetylmuramic acid, arranged in linear chains (58).

In the *N*-acetylmuramic acid residues, carboxyl groups are bound to terminal L-alanine residues of the peptide moiety by amide linkages. The peptide portion is composed of peptide subunits, consisting of L-alanyl, γ-D-isoglutaminyl, and L-lysyl-D-alanyl, cross-linked by peptide bridges that extend from the carboxyl group of the terminal L-alanine of one peptide subunit to the ϵ-amino group of the lysine residue of another peptide subunit.

The glycan portion of the *Staphylococcus aureus* cell wall peptidoglycan is composed of alternating β-1, 4-linked *N*-acetylglucosamine and *N*-acetylmuramic acid, which occur in the pyranose ring form. The polysaccharide in the peptidoglycan probably contains constituents composed of 4 to 100 hexosamine residues. The glycan portion of the *M. roseus* peptidoglycan seems to be similar to, if not identical with, the glycan occurring in the cell walls of *Staph. aureus*. Both complexes resemble substituted chitin, that is, a polymer of β-1,4-linked *N*-acetylglucosamine residues in which every other sugar is substituted by a 3-*O*-D-lactyl group.

The peptide part of the cell wall glycopeptide seems to be composed of units of tripeptide, consisting mainly of L-alanine-γ-D-glutamic acid (α-$CONH_2$)-L-lysine.

The glycan and peptide portion linked together are essential for the insolubility and mechanical strength of the bacterial cell wall. The third dimension of the peptidoglycan structure has not been elucidated.

The peptide subunit of peptidoglycans occurring in bacterial cell walls may contain a set of lysine, and one or more of the LL-, DD-, or meso-isomers of diaminopimelic acid. Diaminopimelic acid is found in virtually all Gram-positive and some Gram-negative bacteria. The peptidoglycans of a few bacterial cell walls contain L-2,4-diaminobutyric acid, L- and possibly D-ornithine, hydroxylysine, or 2,6-diamino-3-hydroxypimelic acid instead of diaminopimelic acid and lysine.

The length and physicochemical nature of the peptide bridges that form the peptide subunits vary, depending on the bacterial species: a pentaglycine bridge, for example, in *Staph. aureus*, tri-L-alanine or tri-L-alanine-L-threonine in *M. roseus* strains, and an L-alanine bridge in *Arthrobacter*. No cross-linking peptide bridges exist in the diaminopimelic type of peptidoglycan, and D-alanine is linked directly to diaminopimelic acid in the peptidoglycans of *Corynebacterium diphtheriae* and *E. coli*.

Peptide-linked oligomers of the size of octomers are present in peptidoglycans of *Staph. aureus*, whereas the peptide-linked oligomers found in the cell walls of *E. coli* and *C. diphtheriae* are dimers. Some membrane peptidoglycans occur as enormous molecules, which form a tight network encompassing the bacterium.

Membrane complexes are excreted during the growth of *E. coli* and *Salmonella typhimurium*, and the excretion increases after cessation of protein synthesis (58). Membrane complexes consisting of mucopolysaccharides and a phospholipid are excreted during the growth of mycobacteria, nocardiae, dermatophili, and streptomycetes, as demonstrated by immunochemical methods (40, 41).

A protein containing a relatively large ratio of cystine and a small amount of lipid and glycopeptide constitute the structure of spore coats, whereas the endospore cortex forms the basis of a glycopeptide. The spore protoplast contains protein, DNA, RNA, and dipicolinic acid.

Proteins and transfer RNA represent the main macromolecular material of the soluble cytoplasmic content of bacteria. The transfer RNA consists of single polynucleotide chains containing 70 to 80 nucleotide residues. Each RNA molecule appears as a cloverleaf-type structure with three major loops in which unpaired nucleotide residues occur in a certain sequence. The residues that are not part of a loop are paired. The end of the molecule that accepts the amino acids always has the following sequence: —C—C—A.

Specific ribonucleoprotein particles, the ribosomes, function as centers for protein synthesis. In this process they appear as clusters of particles termed polysomes, which have a molecular weight of 3.5×10^6. Ribosomes, which in the bacterium may be as numerous as 10,000, occur in the form of round or slightly elongated bodies possessing an average diameter of 100 or 200 Å. Ribosomes in *E. coli* form two different types of particles, having sedimentation constants of 30S and 50S and molecular weights ranging from 7500 to 50,000, respectively. The 30S ribosomal particle consists of 1 molecule of 16S RNA and 10 molecules of protein. The 50S ribosome particle is composed of 1 molecule of 5S RNA and 20 molecules of protein. The two constituent particles of the *E. coli* ribosome were found to consist of 55 protein species (64).

Ribosomal RNA is a single-stranded molecule containing double-helical regions, held together by hydrogen-bonded base pairs (22). Although the mechanisms of the synthesis of ribosomal RNA and protein have not been fully elucidated, it seems that the synthesis of a bibosome occurs in three successive stages: (1) the synthesis of a common precursor, the eosome; (2) the assembly of a small eosome into RNA of a slightly bigger size (16S for the 30S neosome, 23S for the 43S neosome): proteins are present at this stage, as they derive from a pool of preexisting molecules; and (3) the conversion of neosomes to ribosomes, which is done by the addition of newly synthesized ribosomal protein to the unmatured neosomal particles (57). The ribosomal protein seems to exist in a bacterium in the form of a pool, from which it is incorporated into eosomes and neosomes without apparent participation of an enzyme. The biogenesis of bacterial ribosomal units from ribosomal subunits of 50S and 30S is illustrated in Fig. 6.

The 50S ribosomal subunit in *E. coli* is produced by two sequential intermediates of 30S–32S and 40S–43S, termed the (30S)-nascent and (40S)-nascent ribosomal particles. The intermediate particles of 50S ribosome formation have been isolated from exponentially growing cells of *E. coli* (51).

The reconstitution of 30S ribosomes from RNA and proteins *in vitro* requires a relatively high activation energy of 37.8 kcal/mole, which is much higher than is needed for most common enzyme-catalyzed chemical reactions.

Poly-β-hydroxybutarate granules, occurring in the cytoplasm of *Bacillus megaterium*, consist of a stereo regular polymer with the following formula:

$$\left[\begin{array}{l} CH-CH_2-C-O \\ | \qquad\qquad\quad \| \\ CH_3 \qquad\quad\;\; O \end{array}\right]_n$$

The extended polymeric chains are arranged in the form of aggregated fibrils, possessing a diameter of 100 to 150 Å. This unique morphological structure develops because of the simultaneous synthesis and crystallization taking place during a stage of granule formation (19). The synthesis and crystallization seem to occur on the surface of the polymerizing enzymes. The polymerizing system, in the form of protein subunits, aggregates into a unicellular form.

In aqueous acetate fibrils may reorganize into fundamentally altered morphopoietic units, which occur in the form of 50-Å-thick ribbons. These ribbon-like structures can be transformed into single crystal lamellae having folded chains.

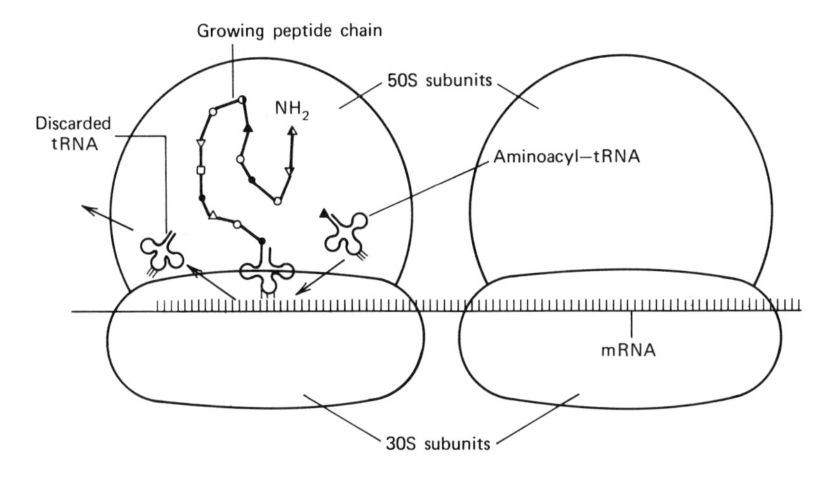

Fig. 6. Biogenesis of ribosomal units (7).

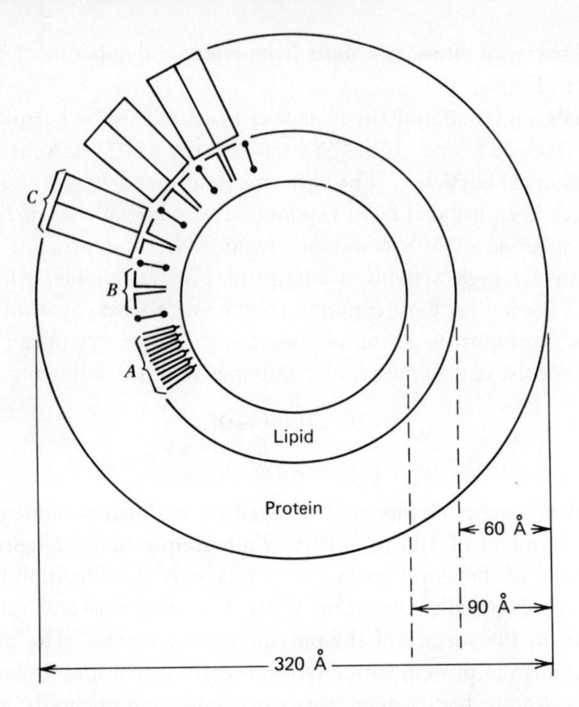

Fig. 7. Ultrastructure of chromatophore (9). *A*-phospholipid molecule monolayer, *B* pigment molecules, *C*-protein molecule layer.

Volutin granules are made of a polymetaphosphate, which may serve as a store for phosphate and energy.

Chromatophores, the photochemical organelles of bacteria, are composed mainly of protein and lipid. The chromatophores in *Rhodospirillum spheroides* contain two complexes of bacterial chlorophyll, one absorbing maximally at 850 nm and the other, smaller complex absorbing maximally at 870 nm.

The chromatophore of *Chromatium* appears as a hollow sphere measuring approximately 320 Å in diameter and possessing a cortex area, which is about 90 Å thick. The cortex region holds about 1.3×10^{-17} g of protein, or approximately 67,000 amino acid residues. The physicochemical structure of the chromatophore is assembled from structural subunits, composed of 1 carotenoid molecule, 2 bacteriochlorophyll molecules, 10 phospholipid molecules, and a protein molecule containing 220 amino acids. The protein is folded and is related directly to the 2 chlorophyll molecules and indirectly to the 1 carotenoid and 10 phospholipid molecules. The subunits of the chromatophores are linked and maintained in a relatively stable structure by disulfide bridges (Fig. 7).

The complete chromatophore consists of 300 carotenoid molecules, 600 chlorophyll molecules, 3000 cephalin molecules, and the protein equivalent to 67,000 amino acid residues.

4. BIOMOLECULAR INTERACTIONS

Interactions between microbial biomolecules and the structural components of more complex live entities may produce the following effects: (1) molecular and structural lesions, and (2) induction of specific immunoglobulins.

Molecular lesions result from direct interactions between two or more types of specific biomolecules, subunits, or units, bringing about the alteration or destruction of one or more types of biomolecules or systems. Specific sites and components of the microbial macromolecules and larger structures that cause noticeable physicochemical, functional, and structural changes in a higher live unit, such as man, may be termed the *factors of virulence*.

Determinant sites on microbial and other macromolecules, which interact with certain unknown sites on molecular structures in animal cells, have been elucidated to a considerable extent. These interactions, occurring under as yet unknown environmental conditions, lead to the emergence of specific immunoglobulins, which are equipped with reactive sites fitting the determinant groups present on antigenic molecules.

A. Molecular Factors of Virulence

Chemical mutation and selection, which probably occurred at an early stage of chemical evolution, were the causes of the first, primitive, parasitic molecules. Unilateral parasitism has been perpetuated among different molecules, molecular systems, and configurational and conformational entities, representing a quite natural form of mutual, dynamic interdependence and collective interaction. Although different forms and levels of virulence have gradually developed, with perhaps the most common type of parasitism occurring between microorganisms and higher hosts, the basic mechanism of the host–parasite relationship is a specific reaction at the molecular level, which brings about detrimental changes in both the host and the parasite.

No harmful effect, however, is caused by the interaction between a virulent agent and a mutant or variant host that has lost one or more of its enzymes or sites essential for the parasite–host association and for reproduction of the virulent agent. This agent is often harbored in the mutant or variant host

Table 1. Periodic Table of Pathogens[a]

Parasitism			Reproduction Site			Replication Mode			Physicochemical Quality			Pathogen Category	
Molecular	Intracellular	Extracellular	Nucleus	Cytoplasm	Outer Cell	Chemotransformation	Binary Fission	Life Cycle	Filterability at 250 mμ	Cell Wall	Growth in Artificial Medium	Symbolic	Taxonomic
A	B	C	D	E	F	G	H	I	J	K	L	ADGJ	Viroids
A	B	C	D	E	F	G	H	I	J	K	L	BDGJ	"Nuclear viruses"
A	B	C	D	E	F	G	H	I	J	K	L	BEGJ	"Cytoplasmic viruses"
A	B	C	D	E	F	G	H	I	J	K	L	BEGHJK	Chlamydiaceae
A	B	C	D	E	F	G	H	I	J	K	L	BE(G)HJK	Leprosy agent
A	B	C	D	E	F	G	H	I	J	K	L	BEHJK	Coxiellae
A	B	C	D	E	F	G	H	I	J	K	L	BEHK	Rickettsiae
A	B	C	D	E	F	G	H	I	J	K	L	BEHKL	*Bartonella*
A	B	C	D	E	F	G	H	I	J	K	L	BEHKL	*Trypanosoma*
A	B	C	D	E	F	G	H	I	J	K	L	BEIK	*Plasmodium*
A	B	C	D	E	F	G	H	I	J	K	L	CFHK	Spirochaetales
A	B	C	D	E	F	G	H	I	J	K	L	CFHL	*Leptospira*
A	B	C	D	E	F	G	H	I	J	K	L	CFHKL	Bacteria
A	B	C	D	E	F	G	H	I	J	K	L	CFIKL	Microfungi
A	B	C	D	E	F	G	H	I	J	K	L	CFIL	Mycoplasmatales

[a] Kwapinski, unpublished data, 1969.

without the latter being subjected to its detrimental action, unless, through serial loss mutations, frank parasitic interactions are unleashed.

Considering the emergence of virulence in the light of the evolution of biomolecules and their interactions with the environment, and relating these fundamental biological properties of microorganisms to their modes and sites of reproduction and to their essential physicochemical characteristics, Kwapinski proposed in 1969 a "Periodic Chart of Pathogens" (Table 1). By application of the criteria used in this chart, the corresponding pathogen categories of microorganisms created in the biomolecular evolution may be determined or deduced. On the basis of this chart, the existence of subviruses has been predicted. It has also been suggested that one of the subviruses is the agent of human leukemia, whose unique nucleic acid macromolecules seemingly established relationships with the receptor molecules of a compatible predecessor of man (probably in the era of ruling reptiles), the property having been passed to some descendants during each subsequent stage of evolution. Existence of the subviruses predicted by the "Periodic Chart" in 1969 was substantiated in 1971 by the detection of viroids by T. O. Diener (14). These specific particles of nucleic acid, possessing a molecular weight of 50,000, were discovered in potato spindle tuber disease.

It appears that virulence depends essentially on interactions between the conformationally unique subunits of (pathogenic) microorganisms and the complementary subunits in a sensitive host. As a result of these interactions, important biomolecules and biostructures of the host are destroyed. However, the molecular nature of virulence has not been adequately investigated.

It seems that the presence or absence of suitable energy sources and compounds in different cells and tissues is a decisive factor for the affinity between the two live systems, the unicellular microorganisms and the multicellular macroorganisms. Generally, virulent strains of viruses possess greater ability than avirulent strains to grow in susceptible cells, but the molecular and biochemical natures of these discrepancies are unknown. Frequently virulent strains, in contrast to nonvirulent strains, of a species have an exclusive or a greater power to produce biologically highly important metabolites or enzymes. Thus exclusive production of powerful toxins by certain clostridia and corynebacteria is synonymous with their virulence, but the molecular interactions preceding the penetration of mucous membranes by microbial secretions or constituents have not been fully elucidated in most cases. It is known, however, that bacterial hyaluronidase interacts with tissue mucopolysaccharides, causing depolymerization of these heteropolymers and thus aiding the penetration.

Coagulase, as well as DNase and phosphatase, is synthesized by virulent staphylococci. Highly virulent strains of *P. tularensis* exhibit greater oxidizing capacity than strains of lower virulence. These quantitative differences in virulence seem to be dependent on alterations of the enzyme participating in the tricarboxylic acid cycle (70). The virulence of *P. pestis* is also related to catalase activity, the rate of synthesis of capsular material, and the VW-antigens (8).

In contrast with these correlations between virulence and enhanced biochemical potency, the degrees of virulence in mycobacteria and brucellae seem to be inversely proportional to their enzyme activities (5, 39).

The role of membranes in virulence is uncertain. Thus protoplasts, as well as bacteria possessing fully developed cell walls, are resistant to phagocytosis (33). In contrast, streptococci appearing in the L form, devoid of cell wall, are avirulent although still capable of hemolytic activity (21).

Macromolecular composition and the action of chemical components of microorganisms are more important for virulence than the physical resistance of anatomical constituents. For example, a macromolecular glycopeptide complex, occurring in the streptococcal cell walls, seems to function as a virulence factor that causes skin lesions.

Purified proteins of *Mycobacterium tuberculosis* produce delayed hypersensitivity, whereas its polysaccharides function as haptens for lysozymes or antibodies, thus diverting them from their defense of the host. Proteins and phospholipids induce the formation of tubercles or similar lesions. A carbohydrate–lipid complex of tubercle bacilli produces pulmonary lesions and tuberculin sensitivity, and the destructive action of their lipids seems to depend partly on the presence of trehalose-6,6-dimycolate. The virulence of *C. albicans* also appears to be related to a specific lipopolysaccharide (42).

Virulence is often correlated with bacterial aggressins, and particularly with toxins. *Aggressins*, which immobilize phagocytic cells and inhibit the bactericidal action of β-lysins, complement, lysozyme, and basic polypeptides in host tissues, are mainly specific peptide polymers. Antibactericidins, synthesized by certain virulent bacteria, have been found to be either toxic polyglutamic acid polymers or cell wall complexes composed of protein, carbohydrate, lipid, and formyl residues.

Endotoxins usually consist of a single type of toxic heteropolymer, but occasional endotoxin preparations may contain several chemically and biologically active molecular complexes.

Macromolecules of endotoxic heteropolymers are composed of normally harmless constituents, that is, carbohydrates, amino acids, short- and long-chain carboxylic acids, amines, and phosphorus. However, the assembly of these compounds into a macromolecular structure, controlled by a biologically specific DNA or RNA, and the resulting, probably unique sequence, distribution, and stereochemical arrangement of certain functional groups, a change in the distance between these groups, or the masking of these groups through complex formation lead to detoxification (49). The constituent micromolecules of endotoxin complexes are arranged in three major zones in the macromolecular structure, occurring as either polysaccharide- and lipid-rich moieties or amino-acid-rich moieties. The backbone of this structure is probably a polysaccharide consisting of different carbohydrates, such as hexoses, heptoses, octonic acid derivatives, deoxy sugars, amino sugars, and their derivatives. This backbone carries molecules of amino acids, linked probably through amino sugars and carboxylic acid, which are ester-bound to hydroxy groups or amide-bound to amino groups of carbohydrates. Phosphorus, in the form of phosphoric acid residues, occurs mainly in the lipid-rich zones, but some amount of phosphorus is present also in the polysaccharide moiety.

The following amino acids—aspartic acid, glutamic acid, valine, leucine, alanine, serine, arginine, lysine, and cysteine—are predominant types in the peptide moiety. The lipid moiety contains carboxylic acid in the form of usual, even-numbered, saturated and unsaturated fatty acids, but odd-numbered acids are occasionally detectable. Hydroxy acids are probably the most characteristic constituents of all endotoxins.

The chemical structure of the lipid moiety in endotoxin macromolecules is unique. Particularly specific, however, are the fatty acid–carbohydrate linkages, which have not been found in any other natural product.

The stereochemical arrangement in the endotoxin biomolecules has not been elucidated, but it appears that specific functional groups are distributed unequally on the molecule in the form of active sites and are responsible for different biological and immunochemical reactions of endotoxins. Thus determinant sites of the polysaccharide moiety determine the antigenic specificity of endotoxin biomolecules. The peptide, or the amino acids present in the peptide, are also immunodeterminants and enhance the immunogenicity of the macromolecules. All or most of the active toxic sites are presumably distributed within the lipid moiety, which is also instrumental in pyrogenicity. The cleavage of ester-bound carboxylic acid, oxidation, deacylation, and acylation alter the number of carboxylic acid molecules surrounding some regions of the endotoxin macromolecule, causing the loss of toxicity.

The complete biocolloidal structure of endotoxin is built of associated subunits, which possess a certain amount of specific activity even in the dissociated state. The sites on the subunits that are involved in subunit aggregation are the same sites through which complexes with the targets of endotoxin action are formed. The *O*-acyl-bound carboxylic acids especially seem to participate in endotoxin reactions. Probably the carboxylic acids form complexes between endotoxin molecules and target molecules. However, the actual interactions of these moieties, as well as the possible roles of other components of the biomolecules (e.g., phosphoric acid and carbohydrates), have not been investigated.

A certain degree of polymerization (and, not fully understood, micellar organization if there are inert subunits in the endotoxin macromolecule) contributes directly to the biological activity of endotoxin. For example, the endotoxin of *Serratia marcescens* dissociates into free subunits under treatment with sodium lauryl sulfate; but the degraded endotoxin macromolecule reaggregates spontaneously after the removal of the SLS by dialysis, emerging as a pyrogenic compound possessing a molecular weight of 500,000 to 1,000,000 (56).

Endotoxin macromolecules have a marked tendency to form aggregates and complexes with other macromolecules. For example, endotoxins can be complexed with casein under alkaline conditions, or with gelatin; and these new complexes do not have the characteristic biological properties of the original toxin (72).

The virulence of *Bacillus anthracis* appears to depend on the production *in vivo* of a lethal toxin, a massive influx of bacteria, and the synthesis of a capsular glutamic acid polymer. The toxin macrocomplex is derived from a specific arrangement of at least three nontoxic components: two are proteins and one is a chelating agent containing protein, carbohydrate, phosphorus, and a group absorbing at 260 nm; the exact biochemical action of the toxin is unknown.

The virulence of *Clostridium perfringens* is correlated with the production of alpha toxin, which has the biochemical activity of a specific phosphatases or lecithinase.

The toxins of *Cl. botulinum* occur in the form of globulinlike protein molecules. The toxin is inactivated by tryptophan and substances capable of releasing serotonin into the circulation. Tryptophan may be involved in a spatial arrangement, which is essential for the toxicity. The oxidation of tryptophan results in the inactivation of the toxin, apparently because of cleavage of the peptide bond adjacent to the tryptophan.

A toxin produced by lysogenic strains of *C. diphtheriae* is a crystallizable, simple protein, which has a molecular weight of 72,000, but possesses no unusual property to account for its toxicity. The toxin exerts a primary and reversible effect at the susceptible cell surface by inhibiting the cytochrome-linked phosphorylation concerned with inorganic phosphate transfer across the cell membrane (36).

In the group A streptococci, the following metabolites function as factors of virulence: the O- and S-hemolysins, leucotoxin, erythrogenic toxin, fibrinolysin, proteinase, hyaluronidase, deosyribonuclease, diphosphopyridine nucleotidase, hyaluronic acid, and a type-specific protein. The nicotinamide–ribose linkage of DPN is broken (9) by the action of diphosphopyridine nucleotidase.

The following toxic metabolites are synthesized by *Staph. aureus*: enterotoxin, exotoxin, leucocidin, hemolysin, coagulase, fibrinolysin, hyaluronidase, and deoxyribonuclease.

The mechanisms of *viral virulence* consist essentially in the formation of large quantities of viral macromolecules, subunits, or toxinlike substances (e.g., neuraminidase), which induce cytopathic change in cells without concomitant viral multiplication.

A host cell is sometimes destroyed by an unknown injurious mechanism of association with the viral coat, devoid of nucleic acid. Although the exact mechanism of the virus–cell interactions leading to cell destruction is unknown, it is significant that the injury sustained by host cells is often derived from the synthesis and considerable accumulation of viral DNA and protein. In other cases of virus–cell interactions, the cell injury results from two different mechanisms: the accumulation of virus-specific materials, leading to the formation of inclusion bodies, and the action of a toxinlike protein.

Also largely unknown are the molecular factors of *fungal virulence*, although fungi in their yeastlike phase are able to produce aggressins or virulence factors in the form of biologically specific polysaccharides that interfere with phagocytic ingestion. The large size and rapidity of reproduction of fungal cells in host tissues may also be regarded as virulence factors.

The toxic mechanisms of virulent fungi are equally obscure. However, it is known that an alpha toxin is produced by certain species of *A. fumigatus* outside the cell, and that specific peptidases are probably responsible for the toxic action of dermatophytes. Certain hitherto ill-defined chemical components produced by fungi are injurious to tissues by means of the hypersensitivity elicited by these bodies.

Selective affinities existing between fungi and some animal tissues apparently depend on certain chemical constituents that have not been fully investigated. An example of such relations is the association between *A. fumigatus* and the plancenta of domestic animals that leads to abortion, and that may depend on nutritional factors occurring in the placenta.

The virulence of *protozoa* seems to depend on such factors as the life cycle, morphological variability, and antigenic plasticity. Paraminobenzoic acid influences malaria infections (31). Virulent protozoa may be expected to produce aggressins interfering with extracellular lysis, ingestion by phagocytes, or intracellular killing. Thus far, protozoal aggressins have not been identified, with the exception of a leucocidal factor produced by entamoebae (34).

B. Molecular Factors of Immunochemical Specificity

The immunochemical properties of a molecule consist of two main factors:

1. The immunogenicity, or its ability to elicit the formation of specific immunoglobulins with sites complementary to specific regions of the molecule's surface.
2. The combining capability, directed strictly to stereochemical sites on the immunoglobulin macromolecule complementary to the conformation-inducing molecule.

Molecules smaller than a copolymer consisting of two amino acid residues or of 20 saccharides and having a molecular weight below 1000 are usually unable to induce the production of complementary sites on antibody molecules. However, evidence currently being accumulated points to the immunogenicity of compounds having molecular weights below 500, for example, L-tyrosine-azobenzene-*p*-arsonate (1).

As defined by quantum immunochemistry (41), an immune response is initiated when the electrons of atoms in an immunocompetent cell have absorbed the energy from an antigen molecule and are then excited to a higher energy quantum. The energy absorbed is quantitized, and the quantum is then passed to the progeny cells. In response, the electrons of the nascent immunoglobulins are excited to a specific energy quantum compatible with the dynamic energy state of the original antigen. Essentially, the above phenomena rely on the absorption and emission of specific frequencies of electromagnetic radiation. Only the antigen and antibody molecules formed by the absorption and transmission of energy quanta and differing in respect to the outer-orbital energies of their electrons may interact, yielding products of an immunoreaction.

The significance of the conformation and spatial arrangement of atoms forming a molecule for immunogenicity is illustrated by the following data from immunochemical studies on synthetic polyamino acids. Homopolymers, composed of molecules of aromatic amino acid or of cystine coupled to the ϵ-amino group of lysine residues of gelatin, elicit the biosynthesis of immunoglobulins which combine with the molecules of polytyrosine, polytryptophan, polyphenylalanine, and polycystine (61). In contrast, linear homopolymers of different amino acids are nonimmunogenic. The ring structure of molecules seems to be important for immunogenicity; thus the hydroaromatic cyclohexane of polycyclohexylalanine, combined with a gelatin molecule, also induced the formation of specific reactive sites on immunoglobulins. In contrast, a copolymer of glutamic acid and lysine exerted a weak immunogenic effect, and linear homopolymers of amino acids were nonimmunogenic. Tyrosine molecules enhance the immunogenic activity of any homopolymer, heteropolymer, or copolymer, but the mechanism of this action has not been elucidated. Thus poly-L-tyrosine, coupled with molecules of poly-L-glutamic acid and attached to the ε-amino groups of a linear poly-L-lysine chain, elicits potent antibody response. The accessibility of the tyrosine residues is a prerequisite for the antigenic action of the polymers, since the saturation of free amino groups of the poly-L-tyrosine site chains results in the loss of antigenicity. The antigenicity of the copolymers is also enhanced by an increased heterogeneity or randomness of the polyamino acids. It may be assumed that a certain degree of rigidity, resulting from the electrostatic attraction between the negatively charged glutamyl and the positively charged lysyl side chains, is one of the factors in immunogenicity. The determinant position of the glutamyl–lysine copolymers is the negatively charged side chain of the glutamyl residues, since the immunogenicity does not decrease on acetylation, methylation, or deamination of the ε-amino groups of lysyl residues.

Copolymers prepared from the D-amino acids possess a different immunochemical specificity. Copolymers consisting of L-aminoacyl residues and D-aminoacyl residues elicit the formation of two types of antibodies, which are reactive with the corresponding L- or D-aminoacyl residues (63). The immunological specificity of antibodies produced in response to the LD-polymers is directed against the N-terminal amino acid residues of the peptides.

The size of the antigenic determinants in polyalanyl polyamino acids is approximately 23 × 11 × 6.5 Å.

The antigenicity of the protein and peptide molecules depends mainly on the presence of certain amino acid molecules, located on the surface of the molecule of protein or peptide. But the immunogenic sites present inside a macromolecule may also be made accessible to the antibody-producing cells after exposure of the large molecule to proteolytic enzymes, which break up the macromolecule and bring hidden determinant sites to the surface.

Simple sugars and oligosaccharides are not immunogenic but are converted into immunogens if coupled to a carrier containing at least one ring-structured molecule. Polysaccharide polymers elicit antibody production very seldom, and then only in an extremely restricted environment. An example of an immunogenic polysaccharide is the pneumococcal polymer consisting of molecules of galacturonic acid, galactose, fucose, and glucosamine, which induces antibodies in man. The determinant groups of these polysaccharides and of synthetic polyglucosans are in the form of one or two monosaccharide units, whereas the determinant groups in dextran consist of slightly more than a single glucosyl residue (e.g., an isomaltose unit) (36).

Fatty acids, triglycerides, and other pure lipids do not seem to have immunogenic properties, although molecules of these compounds appear to nonspecifically increase the antigenic action of other substances.

Terminally located monosaccharides play an important part in the specificity of polysaccharide and lipopolysaccharide molecules.

The immunogenicity of a soluble substance is enhanced by the polarity of its determinant groups, their stability, their susceptibility to enzymatic degradation, and the size of their carrier molecules. The antibody formation can be increased by simultaneous administration of antigens and of a substance that stimulates multiplication of lymphoid cells.

The antigenic character of a molecule is determined by the arrangement, spatial configuration, and sequence of amino acids or monosaccharides on the surface of a macromolecule. The specificity of natural proteins depends on antigenic determinanats, represented by groups of four or five amino acid residues, particularly those containing distinctive structures, (e.g., the benzene ring), arranged in a specific manner on the surface of the macromolecule. The average size of an antigenic determinant is comparable to less than the surface area of a peptide with a molecular weight of 700. An immunochemically active site on a protein macromolecule is a site in which certain amino residues come within bond distance of approximately 0.2 nm of the nascent immunoglobulin molecule.

Although the three-dimensional structure of an antigen molecule is determined by its primary sequence and the environment, its precise immunochemical specificity is described by small constellations of groups called antigenic determinants that appear on the surface of the folded molecule (12).

A protein molecule contains a number of different and complex determinants, each eliciting the formation of antibody molecules possessing a complementary reactive site. Physicochemically identical proteins either may consist of different molecules or of identical molecules occurring in unequal proportions and sequence or may differ by a quantum of kinetic energy. The relation of the small molecules comprising a macromolecular structure to its surface is reflected in the microheterogeneity, immunogenicity, and immunochemical specificity of the macromolecule. Thus a single molecule of a chemically pure substance may possess several antigenic determinants of diverse specificity. The multideterminancy depends on specific groups of atoms present on the molecule's surface. Thus protein molecules coupled to diverse acid radicals possess different immunological specificities, imparted by a small sulfonic, arsonic, or carboxyl group attached to a surface site on the macromolecule. Tertiary structures, which support the combining sites and determinant sites of an antigen molecule, are not directly involved in the sites themselves, but they contribute to the microheterogeneity.

The mode of the stereo arrangement of hydrogen and hydroxyl around an asymmetric carbon atom in a side chain on a macromolecule also alters the specificity of protein molecules. The manner of spatial arrangement of acid radicals on a molecule's surface and the length of the aliphatic chains coupled to a macromolecule are distinctly reflected in their immunochemical specificities.

The hapten groups are surface constituents capable of taking on a wide spectrum of structure, depending on the environment. In relation to the conformational properties of a carrier, the hapten group either protrudes outward from its surface or lies embedded under the surface, bound by the van der Waals contact forces.

Determinant groups on macromolecules of nucleic acids are formed in part by nucleotide bases. Purine and pyrimidine bases are simple hapten molecules, which can induce antibody production when coupled physicochemically to a carrier.

The immunochemical specificity of polyglycerol phosphates depends predominantly on the presence of free hydroxyl groups in the molecule.

The combining site of immunoglobulin molecules is visualized in the form of a shallow cavity, possessing dimensions of about 700 Å. The architecture, conformation, dimensions, and electric charge of the antibody-combining sites may vary, depending on the antigen determinant that induced formation of the reactive sites and on environmental conditions. Differences in the electric charge contribute to the immunochemical specificity, since charge interactions play an important role in the combination between molecules of antigen and antibody. Thus the reactive group of anti-positive-ion antibody molecules possesses a negative charge, whereas the combining region of an antibody molecule directed against a negatively charged determinant (e.g., a benzoic group) possesses a positive charge. The antibody molecule which carries combining sites with a positive charge has been found to be capable of combining not only with the relatively large, negatively charged benzoic group but also with the negatively charged groups smaller than carboxylate.

Interactions between determinant sites of the hapten or antigen molecule and the reactive sites on the antibody molecule depend greatly on the nature of the spatial orientation and the binding force which enter into the immunochemical specificity. The factors involved in the interactions, as well as the interactions themselves, depend on weak forces which lead to the formation of antigen–antibody bonds of 4 to 12 kcal/mole. The following forces participate in the immunochemical reactions: (1) van der Waals forces of very short range, which are essentially attractive and vary inversely as the seventh power of interatomic distances; (2) polar forces, including hydrogen bonds, which are present between nonionic groups in which the positive and negative centers appear as dipoles, and vary inversely as the sixth power of the distance between groups; and (3) Coulombic forces between charged groups, which vary inversely as the square of the distance between groups. Several weak bonds are needed for relatively firm binding between the antigen determinant site and the antibody reactive site. The multibonded, short-range forces are instrumental in immunochemical complementarity and specificity. Between two and ten combining sites or complementary regions may be found on an antibody molecule, depending on the class of immunoglobulins.

Before an antibody and an antigen molecule can combine, the complementary reacting groups must approach each other very closely through mutual adaptation in space. In these circumstances, the short-range London forces occurring between nonpolar groups increase to a level exceeding the electrostatic interaction, thus leading to a mutual combination of antigen and antibody molecules. If the antigen and antibody molecules are differently charged, the charge enhances the affinity between the two types of molecules. From a similarity in the stereochemical configuration and conformation of antigen and antibody molecules, at the complementary combining site, the immunochemical cross reaction originates.

For more information on immunological actions of molecules consult Kwapinski's (42) Molecular Immunology.

REFERENCES

1. Alkan, S. S., E. B. Williams, D. E. Nitecki, and J. W. Goodman, *J. Exp. Med.* **135,** 1228.
2. Anderson, N. S., J. W. Campbell, M. M. Harding, D. A. Rees, and J. W. B. Samuel, *J. Mol. Biol.* **45,** 85 (1969).
3. Beer, H., A. I. Braude, and C. C. Brinton, Jr., *Ann. N.Y. Acad. Sci.* **133,** 450 (1966).
4. Bernal, J. D., The Structure of Molecules, in *Comprehensive Biochemistry*, Vol. 1, M. Florkin and E. H. Stotz (eds), Elsevier, Amsterdam, 1962.
5. Bloch, H., *Ann. N.Y. Acad. Sci.* **88,** 1075 (1960).
6. Boeker, E. A., and E. E. Snell, *J. Biol. Chem.* **243,** 1678 (1968).
7. Britten, R. J., *Ann. N.Y. Acad. Sci.* **108,** 283 (1963).
8. Burrows, T. W., *Nature* **179,** 1246 (1957).
9. Carlson, A. S., A. Kellner, A. W. Bernheimer, and E. B. Freeman, *J. Exp. Med.* **106,** 15 (1957).
10. Caspar, D. L. D., Design and Assembly of Organized Biological Structures, in *Molecular Architecture in Cell Physiology*, T. Hayashi and A. G. Szent-Györgyi (eds.), Prentice-Hall, Englewood Cliffs, N.J., 1966.

11. Caspar, D. L. D., Design Principles in Organized Biological Structure, in *Principles of Biomolecular Organization*, G. E. W. Wolstenholme and M. O'Connor (eds.), Churchill, London, 1966.

12. Dasinger, B. L., Oxidation of Glutamic Acid by *B. abortus* Strains of Low and High Virulence, Doctoral Thesis, University of Wisconsin, Madison, Wis., 1960.

13. Dawson, R. M. C., The Nature of the Interaction Between Protein and Lipid During the Formation of Lipoprotein Membranes, in *Biological Membranes*, D. Chapman (ed.), Academic Press, London, 1968.

14. Diener, T. O. Potats spindle tuber virus with properties of free nucleic acid. Virology, **43,** 75 (1971).

15. De Santis, P., E. Giglio, A. M. Liquori, and A. Ripamonti, *Nature* **206,** 456 (1956).

16. Dietrich, C. P., M. Matsuhasi, and J. L. Strominger, *Biochem. Biophys. Res. Commun.* **21,** 619 (1965).

17. Duesburg, P. H., *J. Mol. Biol.* **42,** 485 (1969).

18. Engelman, D. N., *J. Mol. Biol.* **47,** 115 (1970).

19. Ellar, D., D. G. Lundgren, K. Okamura, and R. H. Marchessault, *J. Mol. Biol.* **35,** 489 (1968).

20. Ellwood, D. C., J. Keppie, and H. Smith, *Brit. Exp. Pathol* **48,** 28 (1967).

21. Freimer, E. H., R. M. Krause, and M. Macarthy, *J. Exp. Med.* **110,** 853 (1959).

22. Fresco, J., B. Alberts, and P. Doty, *Nature* **188,** 98 (1960).

23. Friedman, R. M., and I. Pastan, *J. Mol. Biol.* **40,** 107 (1969).

24. Gross, F., The Cell Machinery, in *Molecular Biophysics*, B. Pullman and M. Weissbluth (eds.), Academic Press, London, 1968.

25. Grossberg, A. L., and D. Pressman, The Chemical Nature of Antibody Sites, in *Protides of Biological Fluids*, Vol. II, H. Peters (ed), Elsevier, Amsterdam, 1963.

26. Gurd, F. R. N., Association of Lipids with Proteins, in *Lipid Chemistry*, D. J. Hanachan (ed.), Wiley, New York, 1960.

27. Guthrie, C., and M. Nomura, *Nature* **219,** 232 (1968).

28. Haggis, G. H., *Introduction to Molecular Biology*, Longmans, Green, London, 1967.

29. Hanson, K. R., *J. Mol. Biol* **38,** 133 (1969).

30. Heidelberger, N., and F. E. Kendal, *J. Exp. Med.* **53,** 65, 1931.

31. Howie, J. W., and A. J. O'Hea, *Mechanisms of Microbial Pathogenicity*, University Press, Cambridge, England, 1955.

32. Jaffé, H. H., An Electronic Theory of Organic Molecules, in *Comprehensive Biochemistry*, Vol. 1, M. Florkin and E. H. Stotz (eds.), Elsevier, Amsterdam, 1962.

33. Janssen, W. A., E. D. Beesley, and M. J. Surgally, *Bacteriol. Proc.*, p. 114, 1960.

34. Jarumilints, R., and F. Kradolfer, *Ann. Trop. Med*, p. 58, 1964.

35. Johnson, M. W., and G. J. Hills, *Virology* **21,** 517 (1963).

36. Kabat, E. A., *J. Immunol.* **77,** 377 (1956).

37. Kirkwood, J. G., *J. Cell. Comp. Physiol.* **49** (Suppl. 1), 59 (1969).

38. Kiselev, N. A., and A. Klug, *J. Mol. Biol.* **40,** 155 (1969).

39. Kwapinski, J. B. G., *Pam. Zj. Pgr.*, p. 40, 1947.

40. Kwapinski, J. B. G., *Zentrbl. Bakteriol. Parasitol.* I. *Orig.* **200,** 80 (1966).

41. Kwapinski, J. B. G., *Res. Immunochem. Immunobiol.* **2,** 1 (1972).

42. Kwapinski, J. B. G., *Molecular Immunology.* University Park Press, Baltimore, 1973.

43. Levitt, M., and S. Lifson, *J. Mol. Biol.* **46,** 269 (1969).

44. Lüderitz, O., A. M. Staub, and O. Westphal, *Bacteriol. Rev.* **30,** 192 (1966).

45. Lüderitz, O., K. Jann, and R. Wheat, Somatic Encapsular Antigens of Gram-Negative Bacteria, in *Comprehensive Biochemistry*, M. Florkin and E. H. Stotz (eds.), Elsevier, Amsterdam, 1968, pp. 105–228.

46. Milner, K. C., R. L. Anacker, K. Fukushi, W. T. Haskins, M. Landy, D. Malgren, and E. Ribi, *Bacteriol. Rev.* **27,** 352 (1963).

47. Morowitz, H. J., The Minimum Size of Cells, in *Principles of Biomolecular Organization*, G.E.W. Wolstenholme and M. O'Connor (eds.), Churchill, London, 1966.

48. Nickerson, W. J., G. Falcone, and G. Kessler, Polysaccharide–Protein Complexes of Yeast Cell Walls, in *Macromolecular Complexes*, M. V. Edds (ed.), Ronald Press, New York, 1961, pp. 205–250.

49. Nowotny, A., *Bacteriol. Rev.* **33,** 72 (1969).

50. Oosawa, F., M. Kasai, S. Hatano, and S. Asakura, Polymerization of Actin and Flagellin, in *Principles of Biomolecular Organization*, G. E. W. Wolstenholme and M. O'connor (eds.), Churchill, London, 1966.

51. Oosawa, S., E. Otaka, T. Itoh, and T. Fukui, *J. Mol. Biol.* **40,** 321 (1969).

52. Pullman, B., Aspects of the Electronic Structure of the Nucleic Acids and Their Constituents, in *Molecular Biophysics*, B. Pullman and W. Weissbluth (eds.), Academic Press, New York, 1965.

53. Ramachandran, G. N., C. Ramakrishan, and V. Sasisekharan, in *Aspects of Protein Structures*, G. N. Ramachandran (ed.), Academic Press, New York, 1963, p. 121.

54. Ramachandran, G. N., and V. Sasisekharan, *Advan. Protein Chem.* **23,** 283 (1968).

55. Rhoades, M., and C. A. Thomas, Jr., *J. Mol. Biol.* **37,** 41 (1968).

56. Ribi, E., L. Anacker, R. Brown, W. T. Haskins, B. Malgren, K. C. Milner, and J. A. Rudbach, *J. Bacteriol.* **92,** 1493, 1509 (1966).

57. Roberts, R. B., R. J. Britten, and B. T. McCarty, in *Molecular Genetics,* Part I, J. H. Taylor (ed.), Academic Press, New York, 1963, p. 291.

58. Rogers, H. J., *Symp. Soc. Gen. Microbiol.* **15,** 186 (1965).

59. Rothfield, L., M. Takeshita, M. Pearlman, and R. Horne, *Fed. Proc.* **25,** 1495 (1966).

60. Rothfield, L., and M. Pearlman-Krothnecz, *J. Mol. Biol.* **214,** 477 (1969).

61. Schechter, I., and M. Sela, *Biochim. Biophys. Acta* **104,** 301 (1965).

62. Scouloudi, H., *Proc. Royal Soc.* **A258,** 181 (1960).

63. Sela, M., *Advan. Immunol.* **5,** 30 (1966).

64. Shands, J. W., Jr., J. A. Graham, and K. Nath, *J. Mol. Biol.* **25,** 15 (1967).

65. Tagaki, W., J. P. Guthrie, and F. H. Westheimer, *Biochemistry* **7,** 905 (1968).

66. Wald, G., *Sci. Amer.*, p. 47, 1954.

67. Waugh, D. F., Molecular Interactions and Structure Formation in Biological Systems, in *Macromolecular Complexes*, M. V. Edds (ed.), Ronald Press, New York, 1961, pp. 3–60.

68. Weber, G., The Binding of Small Molecules to Proteins, in *Molecular Biophysics*, B. Pullman and W. Weissbluth (eds.), Academic Press, New York, 1965.

69. Weiner, J. M., T. Higuchi, L. Rithfield, M. Saltamarsh-Andrew, M. J. Osborn, and B. L. Horecker, *Proc. Natl. Acad. Sci. U.S.* **54,** 228 (1965).

70. Weinstein, I., M. L. Guss, and R. A. Altenbern, *J. Bacteriol.* **83,** 1010 (1962).

71. Wilson, J. B., and B. L. Dasinger, *Ann. N.Y. Acad. Sci.* **88,** 1156 (1960).

72. Woodside, E. E., and C. W. Fishel, *Fed. Proc.* **26,** 583 (1967).

CHAPTER

THREE

Microbial Nucleic Acids

S. G. BRADLEY AND L. W. ENQUIST

1. PROPERTIES AND INTERACTIONS OF NUCLEIC ACIDS

A. Structure of Nucleic Acids

Nucleic acids are long polymers consisting of alternating phosphate groups and five-carbon sugar residues (pentoses). Characteristic heterocyclic substances, the purines and the pyrimidines, are attached to each pentose (Fig. 1). In deoxyribonucleic acid (DNA) the pentose is 2-deoxyribose, the purines are adenine (A) and guanine (G), and the pyrimidines are cytosine (C) and thymine (T). Usually DNA is found as a duplex molecule consisting of two interlocking corkscrews or spiraled springs. The purines and pyrimidines protrude inward, and the pentose–phosphate chains constitute the outer boundary of the helical duplex polymer. The two DNA strands of the duplex molecule coil so that the purine-base adenine of one strand is adjacent to the pyrimidine thymine of the other strand; similarly the purine-base guanine of one strand is

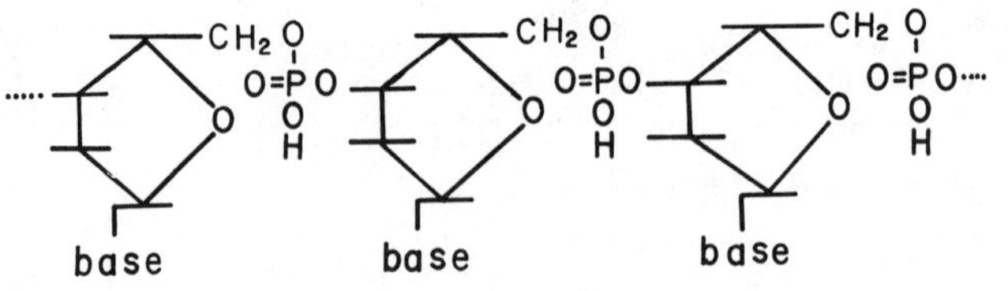

Fig. 1. The pentose–phosphate backbone of the nucleic acids.

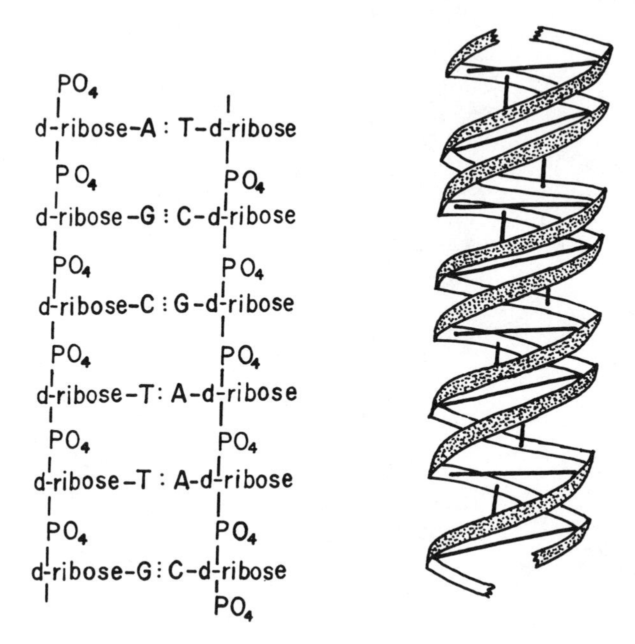

Fig. 2. DNA helical duplex molecule stabilized by complementary pairing between the bases.

paired with the pyrimidine-base cytosine of the other strand (Fig. 2). Such a structure is said to have complementary base pairing; this is possible because hydrogen bonds can form specifically between adenine and thymine and between guanine and cytosine (Fig. 3). The amount of adenine in DNA, therefore, is equal to that of thymine, and, similarly, the amounts of guanine and cytosine are equal. There are 10 base pairs for every complete turn of the helix.

The primary structure of DNA is generally the same for every DNA molecule; however, secondary and tertiary configurations may vary considerably. For example, some viral genomes are single-stranded (e.g., ϕX174), and others have specifically located interruptions in the double-helical backbone (e.g., coliphage T5). A commonly observed structure is the circular DNA molecule. Such molecules apparently play an important role in the life cycles of many diverse prokaryotic and eukaryotic life forms. Several bacteriophages (*Escherichia coli* phages ϕX174, P1, P2, and λ; *Salmonella* phage P22; *Bacillus* phage ϕ29), fertility and resistance transfer factors, bacteria (*E. coli* and *B. subtilis*), animal viruses (SV40 and polyoma), mitochondria, and

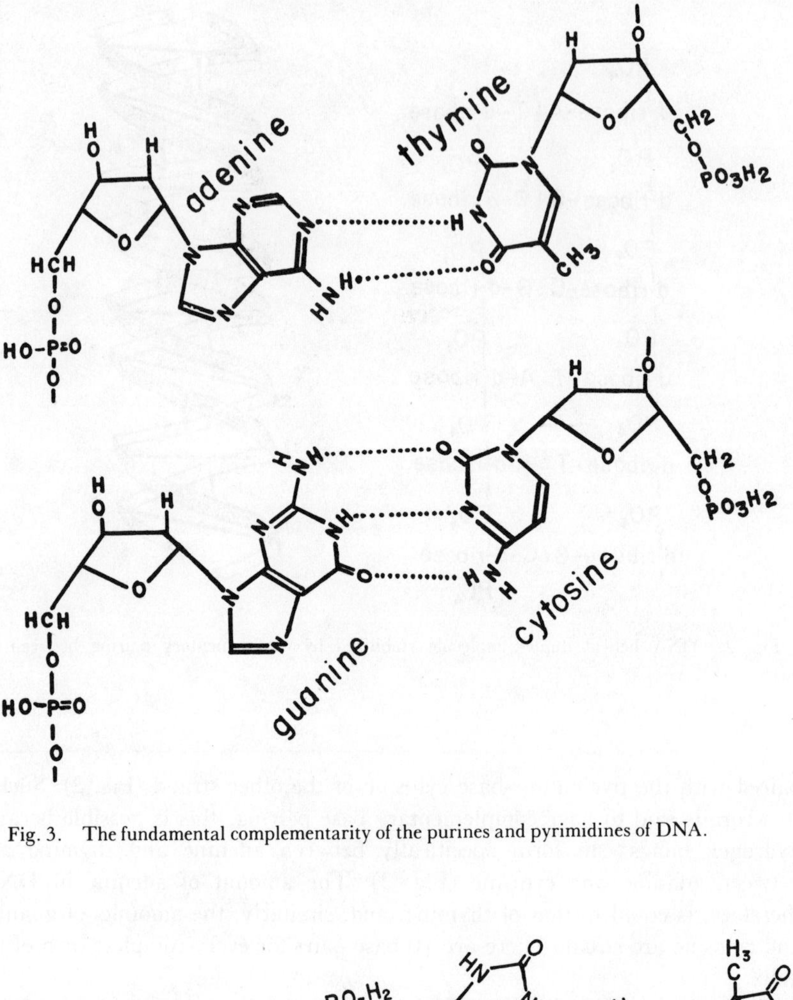

Fig. 3. The fundamental complementarity of the purines and pyrimidines of DNA.

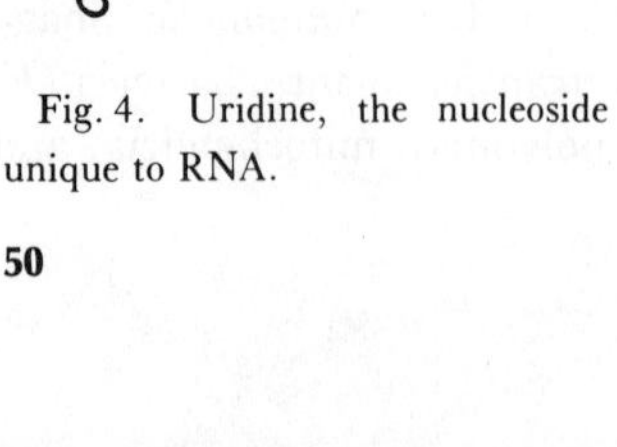

Fig. 4. Uridine, the nucleoside unique to RNA.

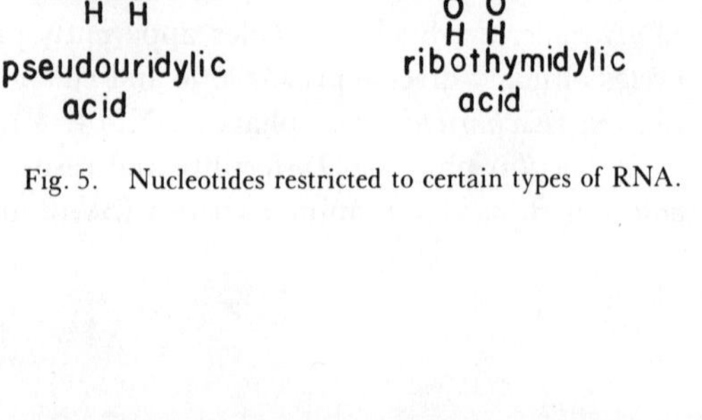

Fig. 5. Nucleotides restricted to certain types of RNA.

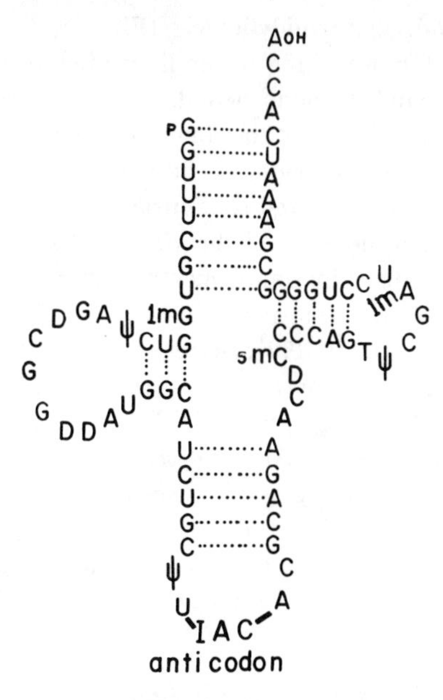

Fig. 6. Structure of yeast valine-tRNA, showing conformation due to complementary base pairing. I = inosine; 1mA = 1 methyladenine; 1mG = 1 methylguanine; 5mC = 5 methylcytosine; ψ = pseudouracil.

chloroplasts have circular DNA molecules at one time or another during their respective life cycles.

In ribonucleic acid (RNA) the pentose is ribose, the purines are adenine and guanine, and the pyrimidines are cytosine and uracil (Fig. 4). The pyrimidine uracil of RNA replaces the thymine (5-methyluracil) of DNA. Certain types of RNA also contain a significant proportion of unusual nucleotides such as

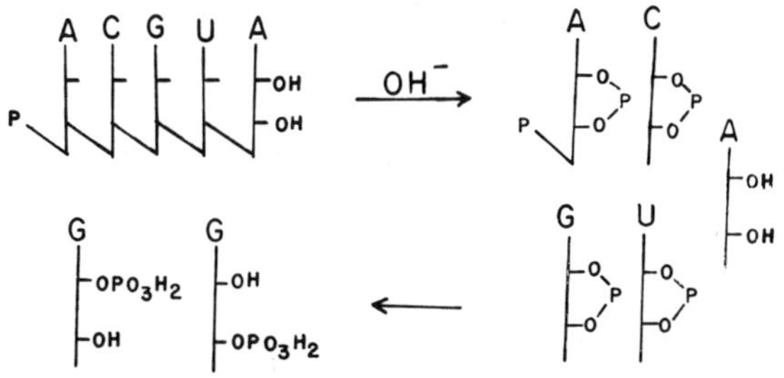

Fig. 7. Hydrolysis of RNA by alkali. All four cyclic esters can be hydrolyzed to give mixtures of the 2′- and 3′-phosphate derivatives.

pseudouridylic acid and ribothymidylic acid (Fig. 5). As with DNA, no evidence has been found for branching in the linear polymer. Most often, RNA molecules are single-stranded polynucleotide chains; therefore the amount of adenine does not necessarily equal the amount of uracil, and the amount of guanine equals the amount of cytosine only by chance. In solution RNA molecules assume a partially ordered configuration, consisting of short regions of double helix formed by looping of the single RNA chain. The secondary structure of RNA is stabilized by complementary interactions between adenine and uracil and guanine and cytosine (Fig. 6). The 2′-OH group of ribose renders RNA sensitive to alkali. The 5′-ester linkage of the polyribonucleotide is hydrolyzed, allowing the 3′-phosphate to form a cyclic ester with the 2′-OH (Fig. 7). Whereas RNA is alkali labile, DNA is alkali resistant because it lacks the 2′-OH on the pentose. Although DNA can be denatured by alkali without any other appreciable change, it is sensitive to acid because the glycosidic bonds between deoxyribose and the purines are quite unstable. The purines can be released quantitatively by acid and measured spectrophotometrically.

B. Biosynthesis of Deoxyribonucleic Acids

The basis for exact replication of DNA is the base complementarity between adenine and thymine and between guanine and cytosine. It must be emphasized that the two polynucleotide chains are antiparallel, that is, the 3′-5′ phosphodiester linkages of one strand run in the direction opposite to the 3′-5′ linkages in the complementary strand. Synthesis of new DNA begins with the separation of the two strands of a preexisting double-stranded DNA molecule. Strand separation begins at one point and proceeds sequentially down the length of the molecule. The precursors for DNA synthesis are not the nucleotides themselves but are believed to be their triphosphate derivatives. Polymerization consists of condensation of the nucleotide triphosphate, properly aligned along the DNA template, with concurrent release of pyrophosphate. The enzyme catalyzing this condensation is DNA polymerase.

Each bacterial genome, so far as is known, consists of a circular DNA double helix. The genome has a specific site for initiation of DNA synthesis. Inhibition of protein synthesis usually allows DNA replication to continue until the genome is completely copied, but a new cycle of DNA synthesis cannot be initiated. A protein, therefore, is a structural or functional component of the initiator region of the genophore. Recent evidence also indicates that RNA synthesis is required for initiation of DNA replication. Moreover, RNA has been implicated as the primer for initiation of DNA replication. The synthesis of DNA in *E. coli* is virtually continuous during the exponential phase of growth; accordingly, the majority of the DNA from rapidly growing cultures is in some intermediate stage of duplication. Remarkable micrographs of autoradiographs of the replicating genophore of *E. coli* have been obtained; these resemble the line drawings in Fig. 8. Both the initiation point and the replicating or growing point can be recognized in the original micrographs. It is now believed that DNA replication in *E. coli* is bidirectional; that is, two replicative forks travel in opposite directions from a common initiation site. Bidirectional replication has been documented for several bacteriophages, including λ and T7. In addition, evidence is accumulating that the chromosomes of higher organisms contain many discrete initiation sites from which bidirectional replicative forks originate.

Because bacterial DNA replicates sequentially from one end of the molecule to the other, the enzymic system catalyzing this synthesis must travel the length of the DNA molecule or the DNA molecule must be pulled through fixed replicating or growing points. In either event, there is a maximum rate of DNA synthesis for each growing point. Under appropriate nutritive conditions certain bacteria have generation times less than the time required to duplicate the DNA with a single growing point. In order to decrease the generation time, the number of growing points must be increased (Fig. 9).

Enzyme systems that will synthesize DNA *in vitro* have been isolated from a variety of sources, including calf thymus and bacteria. The DNA polymerase I

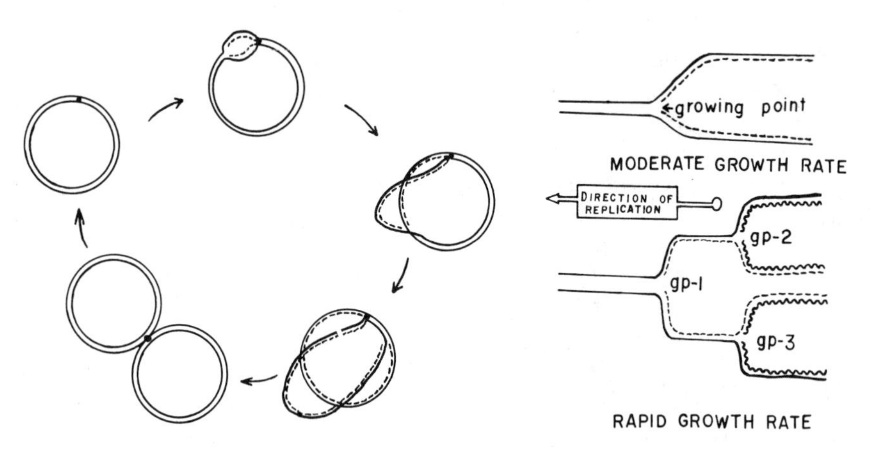

Fig. 8. Diagrammatic representation of the replicating bacterial genophore.

Fig. 9. Diagrammatic representation of the replicating genophore in a bacterium growing at a moderate rate and one growing very rapidly.

(DNA nucleotidyltransferase) from *E. coli* incorporates the nucleotide residues from deoxyribonucleoside–5′-triphosphates into a new DNA strand complementary to the DNA template provided (47). The reaction is reversible, and DNA polymerase I may act as a nuclease. The DNA polymerase from *E. coli* infected with coliphage T2 and that from calf thymus require single-stranded DNA as a template. *Escherichia coli* polymerase I and the similar *Bacillus subtilis* DNA polymerase may use either single-stranded or double-stranded DNA as a template, depending on the extent of the enzyme purification. The DNA polymerase I of *E. coli* condenses nucleotides to the 3′-hydroxyl end of the growing chain. The properties of *E. coli* DNA polymerase I differ in two important respects from the properties expected of the biologically active DNA polymerase: (*a*) there is no evidence that the *E. coli* DNA polymerase can catalyze DNA replication by adding nucleotides to the 5′-hydroxyl end of a DNA chain, and (*b*) the rate of condensation *in vitro* is too slow to account for the rate of DNA synthesis in the intact cell. Rapidly growing bacteria contain a genome 1 mm long composed of 4 to 5×10^6 nucleotide pairs. In order to produce a new genome every 40 min, the bacterium must add 2×10^3 nucleotides per second to each growing nucleotide strand. The rate of condensation catalyzed by the DNA polymerase may be much greater when the enzyme and DNA are functioning in the structured and integrated environment of the intact cell. Even the lack of dual polarity of the enzyme can be overcome.

New DNA may be synthesized at the growing point only by addition to the 3′-hydroxyl end of the new chain (Fig. 10). To achieve this result, a growing point may be formed by uncoiling a short region of the parental double helix, thereby creating a fork consisting of two single-stranded arms and the unaffected portion of the double-stranded parental genophore. New synthesis would start by aligning a complementary nucleotide triphosphate with the terminal (free 3′-hydroxyl) nucleotide of the single-stranded arm of the growing fork. Thereafter a new DNA strand, complementary to one of the parental strands, would be synthesized by addition to the 3′-hydroxyl end of the new strand. The elongating DNA strand would replicate toward the point of strand separation, would continue around the branching point, and would continue to extend along the other parental strand until it reached the terminal (free 5′-phosphate) nucleotide. The result is a Y-shaped region in which each arm consists of one parental DNA strand and one new DNA strand. Now, the newly synthesized DNA strand must break at the growing point so that the previously unaffected portion of the double-stranded parental DNA can separate, creating a new growing fork. The newly synthesized DNA strand with a free 3′-hydroxyl would be elongated by addition of complementary nucleotide triphosphates. Again, the elongating DNA strand would replicate

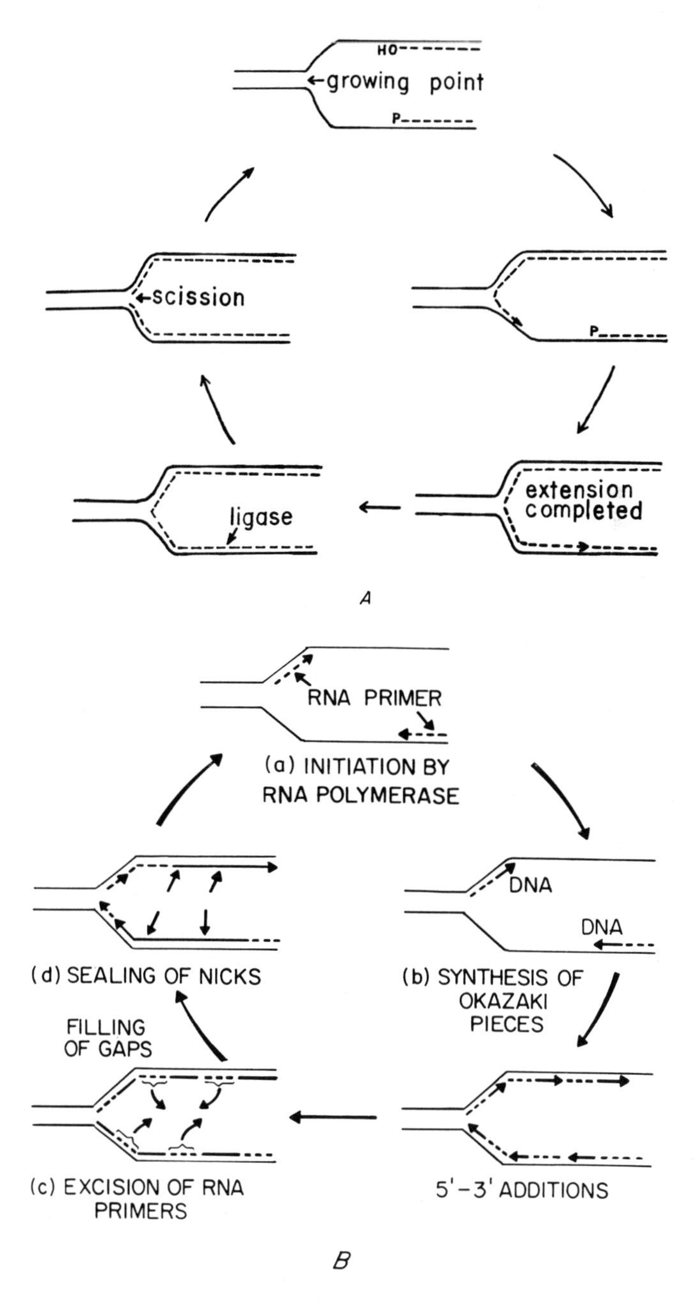

Fig. 10. Model for extending DNA by adding only to the 3′-hydroxyl end; (*A*) the general model and (*B*) the Okazaki model.

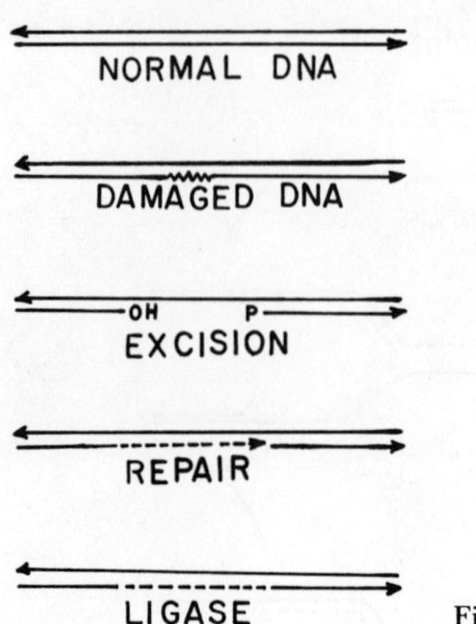

Fig. 11. DNA repair.

toward the point of strand separation, would continue around the branching point, and would continue to extend along the other parental strand until it reached the free 5′-phosphate of the DNA strand formed by the previous replicative fork. The two new DNA strands are joined by an enzyme called polynucleotide ligase; about the same time, the newly synthesized DNA strand would break at the growing point, allowing for formation of another replicative fork.

It is equally probable that the known DNA polymerases are used to repair damaged DNA rather than to replicate new genomes. The DNA may be damaged by ultraviolet radiation, ionizing radiation, alkylating chemicals, or free radicals. The damaged region in one strand is excised enzymically and the gap closed by new synthesis, using the undamaged strand as the template (Fig. 11). There are similarities between DNA repair, integration of a new genetic sequence into the genome as a result of recombination, and the normal DNA replicative process (18). Further studies are needed to determine whether these DNA syntheses are catalyzed by the same or different polymerases.

DeLucia and Cairns (22) gave strong impetus to uncovering other DNA polymerases from *E. coli* when they isolated an *amber* mutant (*pol*Al) which lacked DNA polymerase I activity yet synthesized DNA normally. Since this startling discovery, two other DNA polymerases, polymerase II and polymerase III, have been found. Polymerase II is distinct from polymerases I and III but does not seem to be required for normal DNA replication. Polymerase III is clearly distinct from polymerases I and II and appears to be essential for replication of *E. coli* DNA. Estimates of the relative amounts of polymerase I, polymerase II, and polymerase III have been made and are roughly 400:40:4, molecules per cell, respectively. Although polymerase I apparently is not required for *E. coli* replication, it is absolutely essential for replication of the *E. coli* 15 plasmid and the colicinogenic factor *Col*EI. These DNA molecules cannot use polymerase I or III. Obviously replication complexes are not identical for every genome. Further analyses of the subtleties of DNA replication can be found in reviews by Klein and Bonhoeffer (45) and Gross (32).

Despite the apparent complexities of DNA replication, it is possible to construct a model that fits most of the experimental data. The model is the discontinuous synthesis hypothesis first presented by Okazaki et al. (82). Four distinct reactions are postulated: (*a*) the initiation of polynucleotide chains, (*b*) polymerization of short pieces of DNA on one or both strands of gaps, and (*d*) sealing of the nicks restoring genophore integrity. The initiation reactions presumably involve the *dna*A, C, and G gene products. The short fragments (ca. 3000 nucleotides) formed during reaction *b* have been called Okazaki pieces and have been described in both prokaryotic and eukaryotic DNA replication. The synthesis of these fragments is initiated on an RNA primer formed by reaction *a*. The DNA-RNA joint is apparently unique and has the sequence p(rPy)p(rA)p(rU or rC)p(dC)p (83). The RNA primer is probably removed during the gap-filling reaction *c*, possibly by the gap-filling polymerase itself. All three known *E. coli* DNA polymerases could conceivably carry out reactions *b* and *c*, but the absolute requirement in *E. coli* of only polymerase III and not polymerases I and II indicates that polymerase III can do both. Polymerase I seems to be uniquely qualified to take part in reaction *c*. In polymerase I deficient strains, Okazaki pieces are joined very slowly. Moreover, of the three known polymerases, only polymerases, only polymerase I possesses a nuclease activity capable of removing the RNA primer from the 5′-end of the Okazaki fragment, a special requirement of reaction *c*. The final reaction *d* has been suggested from experiments using both *in vitro* and *in vivo* replication systems. If one uses nicotinamide mononucleotide, a specific inhibitor of *E. coli* DNA ligase, Okazaki fragments are not joined and accumulate (81). Ligase defective mutants have been constructed and have been shown to be deficient in the postulated joining reaction.

C. Biosynthesis of Ribonucleic Acids

The basis for exact replication of RNA, like that of DNA, is complementary nucleotide pairing. For RNA synthesis, however, the template may be either an RNA or a DNA molecule. In many viruses, the genophore is RNA. Al-

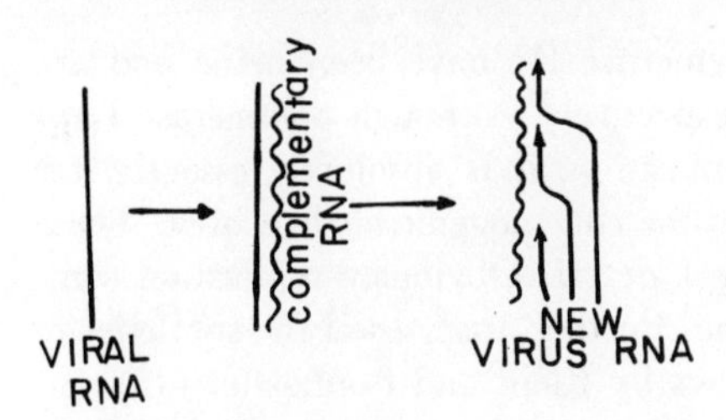

Fig. 12. Replication of viral RNA.

though RNA usually exists as a single-stranded molecule, the RNA in the free reovirus particle is double stranded and segmented. In general, viral RNA replication involves the synthesis of a complementary strand, which then serves as the template for production of new viral RNA (Fig. 12). In RNA synthesis, A pairs with U and G with C (75).

Not only can RNA direct the synthesis of new RNA, but also it can produce complementary DNA. Although the prevalence of this reaction is not known, it occurs in cells infected with Rous sarcoma virus, and the RNA-dependent DNA polymerase has been found in nearly a dozen other virus-infected cells. The discovery that RNA can be transcribed to make DNA helps to account for latency or masking of RNA viruses and provides a sound theoretical basis for the hereditable progression of virus-induced oncogenesis (1).

The most abundant cellular RNA is that in the ribosomes. It constitutes about 80% of the bacterial RNA. Bacterial ribosomes are complex particles, consisting of a 30S subunit and a 50S subunit, each of which in turn contains many different proteins and RNA molecules. The composite 70S bacterial ribosome is approximately 60% RNA and 40% protein and is 15 nm in diameter. The ribosomes of eukaryotes have a sedimentation coefficient of 80S and are composed of 40S and 60S subunits. The eukaryotic ribosomes are nearly 50% protein, in contrast to less than 40% protein in prokaryotic ribosomes. The RNAs of both bacterial and eukaryotic ribosomes are transcribed from DNA. In bacteria, about 0.3% of the DNA is devoted to coding for the 16S and 23S RNA subunits. The eukaryotic nucleolus is an aggregate of ribosomal precursors. The loci controlling ribosomal synthesis can be located cytologically under selected conditions and with appropriate material.

The next most abundant RNA is soluble or transfer RNA (tRNA). Transfer RNA makes up about 10% of the bacterial RNA, has an average molecular weight of 25,000 daltons, has a sedimentation coefficient of 4S to 5S, and is coded for by about 0.02% of the bacterial DNA. There are about 60 different species of tRNA. Although tRNA molecules are single stranded, there are complementary sequences that may dictate a "clover-leaf" conformation for the molecule (Fig. 13). Transfer RNA contains many unique purines and pyrimidines. The primary nucleotide sequence, however, is determined by the usual dA = rU, dT = rA, G = C complementarity, and the nucleotide bases are modified after the precursor tRNA molecule has been condensed.

The third class of RNA is messenger RNA (mRNA), which directly determines the amino acid sequence in proteins. Messenger RNA is highly heterogeneous; its molecular weight ranges between 10^5 and 1.5×10^6 daltons, and its sedimentation coefficient between 8S and 30S. Messenger RNA makes up only 1 to 2% of the bacterial RNA, but the greater part of the bacterial DNA is committed to mRNA synthesis. At any given moment, only 15 to 40% of the DNA may be involved in transcribing mRNA from the DNA. Only one strand of the DNA double helix is involved in the synthesis of a particular mRNA, but some regions of each strand are transcribed. Although the same DNA-dependent RNA polymerase is apparently used to synthesize ribosomal RNA, tRNA, and mRNA, its specificity is strictly controlled by small protein factors, including the σ factor and cyclic AMP binding protein (52).

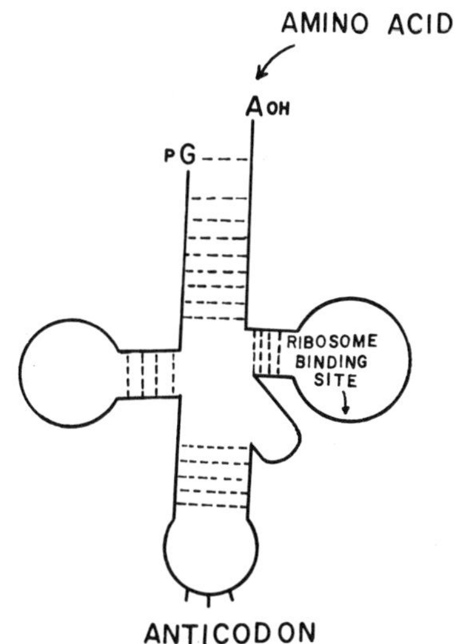

Fig. 13. Generalized conformation of transfer RNA.

2. GENETIC CONTROL OF METABOLISM

A. Protein Synthesis

The phenotype of the cell is determined and manifested both directly and indirectly by its structural and enzymic proteins. Consideration of the genetic control of protein synthesis, therefore, is tantamount to study of the genetic control of the cell's phenotype. Protein synthesis must be considered in light of the total cellular environment: energy must be generated, precursors synthesized, and the intermediates transported to the appropriate biosynthetic sites. In the following description of protein synthesis, an adequate supply of appropriate essential precursors and cofactors will be assumed.

The first step in protein synthesis is amino acid activation (Fig. 14). Aminoacyl AMP is not found free in the cell because it remains bound to the activating enzyme. Specific activating enzymes have been demonstrated for all 20 essential amino acids. Next, the activated amino acid is transferred to a specific tRNA acceptor. Each tRNA is specific for a single amino acid. The enzyme complex responsible for this two-stage reaction is called an aminoacyl–tRNA synthetase. The activated amino acid is transferred to the terminal adenylate of the appropriate tRNA (Fig. 15). The genes that specify the structures of the various aminoacyl–tRNA synthetases are scattered throughout the genophore of *E. coli*. Although mutations drastically altering the structures of these synthetases would be lethal, many temperature-sensitive synthetases have been recognized as a consequence of conditionally lethal mutations. These synthetases are functional at one temperature (e.g., 33°C) but are inactive at another incubation temperature (e.g., 42°C). A specific aminoacyl–tRNA or charged tRNA is selectively bound to the 50S subunit ribosome according to the specific trinucleotide codon bound to the 30S subunit. One loop of all tRNA molecules contains the sequence guanylate–ribothymidylate-pseudouridylate-cytidylate-guanylate (Fig. 16). This loop is presumed to be the ribosome-binding site.

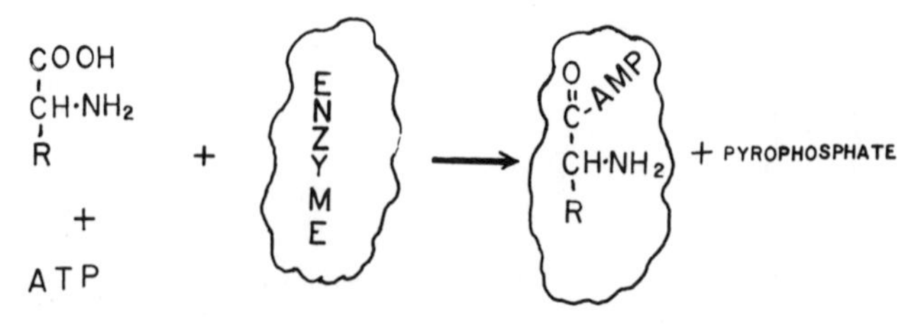

Fig. 14. Amino acid activation.

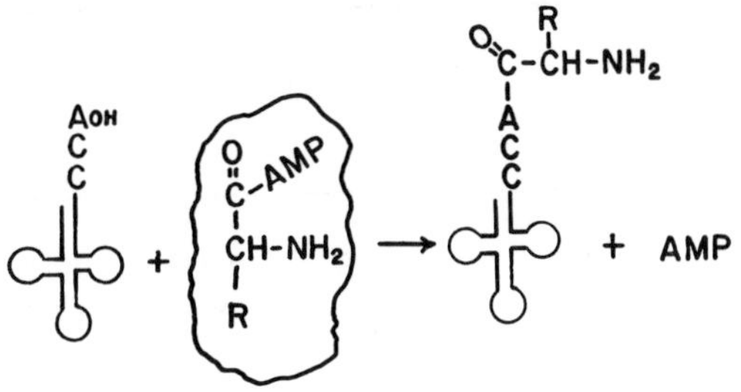

Fig. 15. Synthesis of aminoacyl–tRNA from aminoacyl–adenylate.

Messenger RNA specifies the sequence of amino acids in a protein. The ribosomes are relatively nonspecific surfaces upon which the charged tRNA molecules bind, transfer their amino acids to the nascent polypeptide, and are released as uncharged tRNA. The information specifying the incorporation of a given amino acid is contained in the sequence of three nucleotides along the mRNA polymer. There are 64 possible triplet sequences for the four nucleotides A, C, G, and U. Because there are only 20 amino acids to be specified, it is obvious that some amino acids have multiple codons. The codon dictionary for all 64 triplets has been deciphered and, so far as is known, is

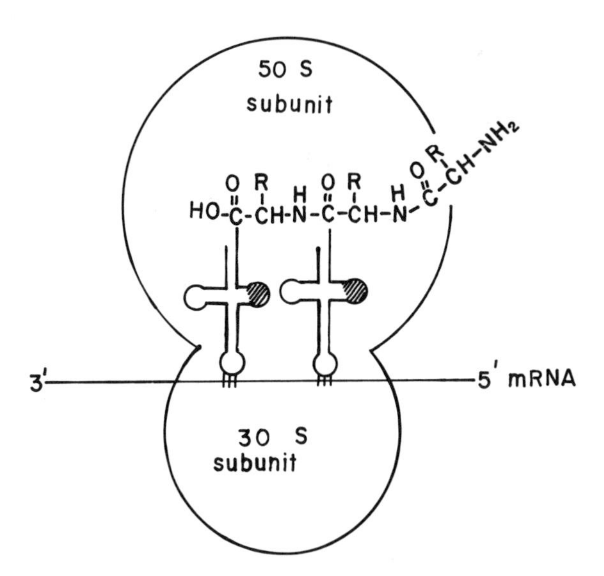

Fig. 16. Relationship of aminoacyl-tRNA to the ribosome. The presumed ribosome-binding site (shaded) of the aminoacyl–tRNA contains the sequence G–T–ψ–C–G.

First Base	Second Base U	C	A	G	Third Base
U	PHE	SER	TYR	CYS	U
	PHE	SER	TYR	CYS	C
	LEU	SER	X	X	A
	LEU	SER	X	TRP	G
C	LEU	PRO	HIS	ARG	U
	LEU	PRO	HIS	ARG	C
	LEU	PRO	GLN	ARG	A
	LEU	PRO	GLN	ARG	G
A	ILE	THR	ASN	SER	U
	ILE	THR	ASN	SER	C
	ILE	THR	LYS	ARG	A
	MET,F-MET	THR	LYS	ARG	G
G	VAL	ALA	ASP	GLY	U
	VAL	ALA	ASP	GLY	C
	VAL	ALA	GLU	GLY	A
	VAL	ALA	GLU	GLY	G

Fig. 17. The genetic code; X denotes the terminating codons.

essentially universal for all living beings (Fig. 17). The mRNA is translated, starting at the free 5′-phosphoester and proceeding to the free 3′-phosphoester. Translation proceeds in the same direction as transcription. The polypeptide is initiated with the amino group free and terminates with a free carboxyl group (Fig. 18). In bacteria, but not eukaryotic organisms, polypeptide synthesis begins with the codon AUG, which leads to the incorporation of *N*-formylmethionine (fMet).

The initiation complex of *E. coli* consists of a 30S ribosomal subunit, mRNA, and fMet–tRNA. Subsequently, a 50S ribosomal subunit is attached, forming the 70S ribosome. The next step in protein synthesis is the formation

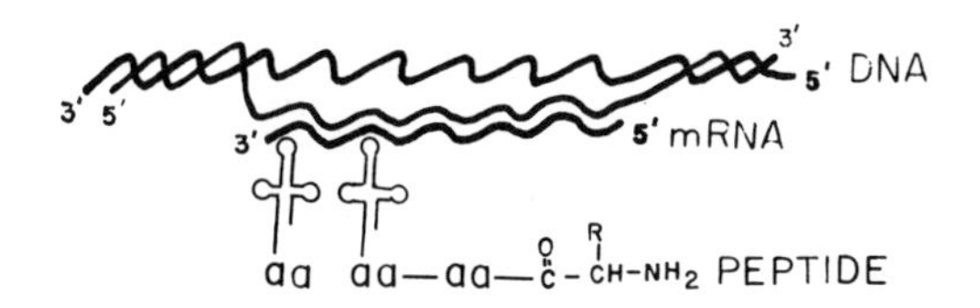

Fig. 18. The direction of transcription and translation.

of a peptide bond between the carboxyl group of fMet–tRNA and the α-amino group of another aminoacyl–tRNA bound to the ribosome. The discharged tRNA is released from the ribosome, and the peptidyl–tRNA is translocated from the aminoacyl (A) site to the peptidyl (P) site. Translocation requires guanosine triphosphate. Simultaneously with translocation, the ribosome moves along the mRNA by the length of one codon. The codon aligned with the A site will bind the appropriate aminoacyl–tRNA. The α-amino group of the aminoacyl–tRNA is linked by a peptide bond to the carboxyl group of the peptidyl–tRNA. The elongated peptidyl–tRNA is translocated to the P site, and the stage is set for the addition of another aminoacyl residue. Each addition involves the following steps: aminoacyl–tRNA binding, peptide bond formation, and translocation (51). These processes are repeated, and polypeptide chain elongation continues until the ribosome encounters a chain-terminating codon.

In principle, binding of fMet–tRNA is sufficient to terminate a polypeptide chain because *N*-formylmethionine cannot form a peptide bond with the pre-

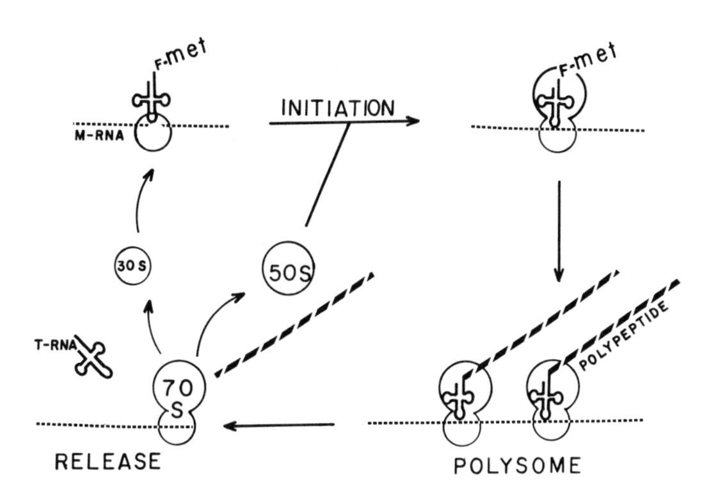

Fig. 19. Ribosome recycling.

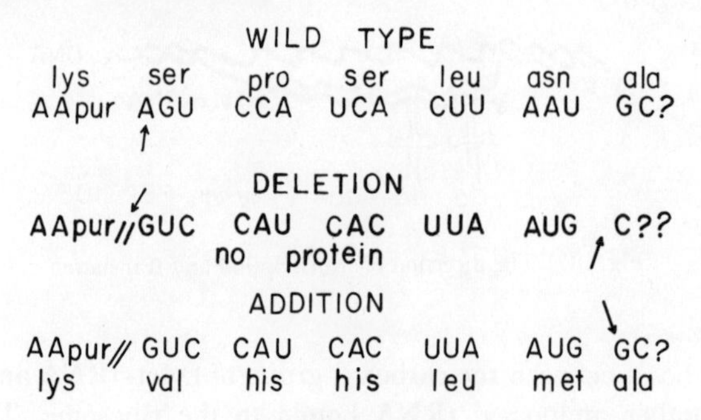

Fig. 20. Action of a suppressor mutation in the lysozyme locus of coliphage T4. The revertant arose by adding a base near the site of the previous deletion.

ceding amino acid specified by the mRNA (27). The chain-terminating signals, however, seem to be the codons UAA, UAG, and UGA. After release of the nascent polypeptide, the mRNA–ribosome complex falls apart, giving rise to a 30S and a 50S ribosomal subunit (Fig. 19). Ribosomal dissociation is necessary to provide the 30S ribosomal subunit needed for a new initiation event (66). The *N*-formylmethionine residue which served as the initiating N-terminal amino acid is either deformylated or cleaved. *N*-Formylmethionine may be removed from a nascent protein in two steps: (*a*) deformylation, followed by (*b*) hydrolytic cleavage of the methionine residue; alternatively, it may be cleaved as *N*-formylmethionine per se.

The code is read from the initiating point to its terminating codon with three nonoverlapping nucleotides specifying each amino acid. The deletion of a single nucleotide because of a mutation usually results in the formation of a chain-terminating codon nearby, leading to a premature interruption of polypeptide elongation. These incomplete proteins are inactive unless the mutation occurred very near the end of its structural gene and unless the amino acid residues deleted are not involved in the active site or conformation of the protein. A mutation adding a nucleotide near the site of the previous deletion may permit a functional protein to be made (Fig. 20). Such mutations constitute one class of suppressor mutations.

A particular aminoacyl–tRNA may bind to the 50S subunit of the ribosome only if the appropriate codon in the mRNA is bound to the 30S subunit of the ribosome. The loop of the tRNA molecule containing the anticodon has been tentatively identified. A modified adenylate, frequently isopentenyladenosine, is found adjacent to the anticodon. The anticodon frequently does not contain the standard complementary bases; instead, inosinate, ribothymidylate, and pseudouridylate have been found in the anticodon. These unique nucleotides frequently but not invariably occupy the free 3′-phosphate end of the triplet.

Although four codons are known for alanine (pGpCpA, pGpCpC, pGpCpG, and pGpCpU), only one alanyl–tRNA has been detected and characterized. Its anticodon reads CpGpIp (note that the antiparallel reading of the anticodon might be stated as pIpGpC). This indicates that inosinate can pair with A, G, C, or U, at least when it occupies the third position of the codon. The ability of one anticodon to pair with several codons specifying the same amino acid potentially enables a cell to survive with less than one tRNA species per codon. At present, only one tRNA is known for the two codons for phenylalanine, but multiple tRNA species are known for leucine (six codons, five known tRNA species), arginine, and serine. Conversely, there are two distinctly different tRNA molecules for the codon pApUpG, both having the anticodon pCpApU (or UpApCp to shown antiparallel and complementary orientation). One tRNA directs the incorporation of methionine into protein; the other serves to initiate protein synthesis with *N*-formylmethionine.

Because there are multiple tRNA species for a given amino acid, or because the third position in the triplet may pair with a nonexacting nucleotide in the anticodon, a mutation in a structural gene does not necessarily lead to an altered protein. Such mutations are called neutral mutations. If the mutation is to a codon that specifies the same amino acid, but the tRNA corresponding to the mutant codon is present at a lower concentration, growth of the mutant may be retarded. Of the 549 possible single nucleotide substitutions, 134 of these are to synonymous codons. Mutations in structural genes leading to an amino acid substitution are called missense mutations, and mutations forming one of the three chain-terminating codons are called nonsense. A missense mutation may be lethal, may impair growth, or may have undetectable effects if it occurs in a locus concerned with a nonessential function or if the amino acid substitution has little effect on the conformation of the resulting protein. Missense mutations are commonly "leaky", that is, they reduce the growth rate under prescribed conditions rather than cause all-or-none responses. Nonsense mutations are usually lethal unless they occur in nonessential loci. Missense and nonsense mutations may be corrected by mutations affecting the anticodon of a tRNA. In *E. coli*, a mutant carrying a nonsense mutation in a structural gene, converting one codon from pUpApC to pUpApG, was isolated. Subsequently, a suppressor mutation was discovered in which the anticodon of the tyrosinyl–tRNA was changed from pGpUpA to pCpUpA. Suppressor mutations may affect the corresponding tRNA locus or may occur in the mutant locus itself (see Fig. 20).

B. Regulation of Protein Synthesis

Not all of the genotypic potential of an organism is expressed at any one time. Genotypic expression is controlled and regulated by several different processes.

Enzymic activity may be inhibited by the end product of the pathway. In feedback inhibition, the end product acts on the first enzyme unique to the biosynthesis of the end product. The catalytic site of an enzyme susceptible to end-product inhibition is different from the region that binds the end product. Such enzymes are called allosteric enzymes. Feedback inhibition is generally reversible, that is, the bound end product can dissociate from the enzyme, thereby restoring enzymic activity. Feedback inhibition allows a cell to consume precursors and energy from a metabolic pool at a rate essential for efficient biosynthesis, but no faster. Allosteric inhibition must be distinguished from isosteric or competitive inhibition, in which a substrate and an inhibitor compete for the same catalytic site on the enzyme.

Metabolic activity is also controlled at the transcriptional level. Genes do not continuously make mRNA but are switched off or on by repressors. Classically, the structural genes controlling a biosynthetic pathway are repressed when the product is supplied exogenously. Repression must be clearly differentiated from end-product inhibition; repression affects enzyme synthesis, whereas allosteric inhibition affects enzyme action. The genes subject to repression have a complex organization. The region to which DNA-dependent RNA polymerase attaches is the *promoter* region. The region that determines whether or not the RNA polymerase will make mRNA is the *operator*. Contiguous to these two controlling regions are the corresponding structural genes which code for from a few to many enzymes. The overall genetic unit, which consists of the promoter, operator, and structural genes, has been designated the *operon* (Fig. 21). In repression, a regulatory gene, which is usually some distance away from the operon, produces an inactive or unstable repressor. The aporepressor (or inactive repressor) must react with the anabolic product or a derivative of it (the corepressor) to make an active, stable repressor that can bind to the operator. When active repressor is bound to the operator, the RNA polymerase is prevented from transcribing the DNA. When the appropriate anabolic product is not provided exogenously, the aporepressor cannot bind to the operator and the RNA polymerase is able to direct the synthesis of new mRNA. Several repressors have now been isolated and partially characterized. The *Lambda* repressor and the *lac* repressor of *E. coli* are weakly acidic proteins, both of which contain four subunits of about 30,000 daltons each. Another phage repressor which is a weakly basic protein has been isolated.

An mRNA that codes for many peptides and, therefore, several enzymes is termed a *polycistronic* messenger. After a polycistronic mRNA is released from the DNA, a ribosome attaches to the initiating end of the molecule and begins to synthesize protein. As the ribosome moves down the mRNA, other ribosomes attach and begin to translate the messenger. A single mRNA may

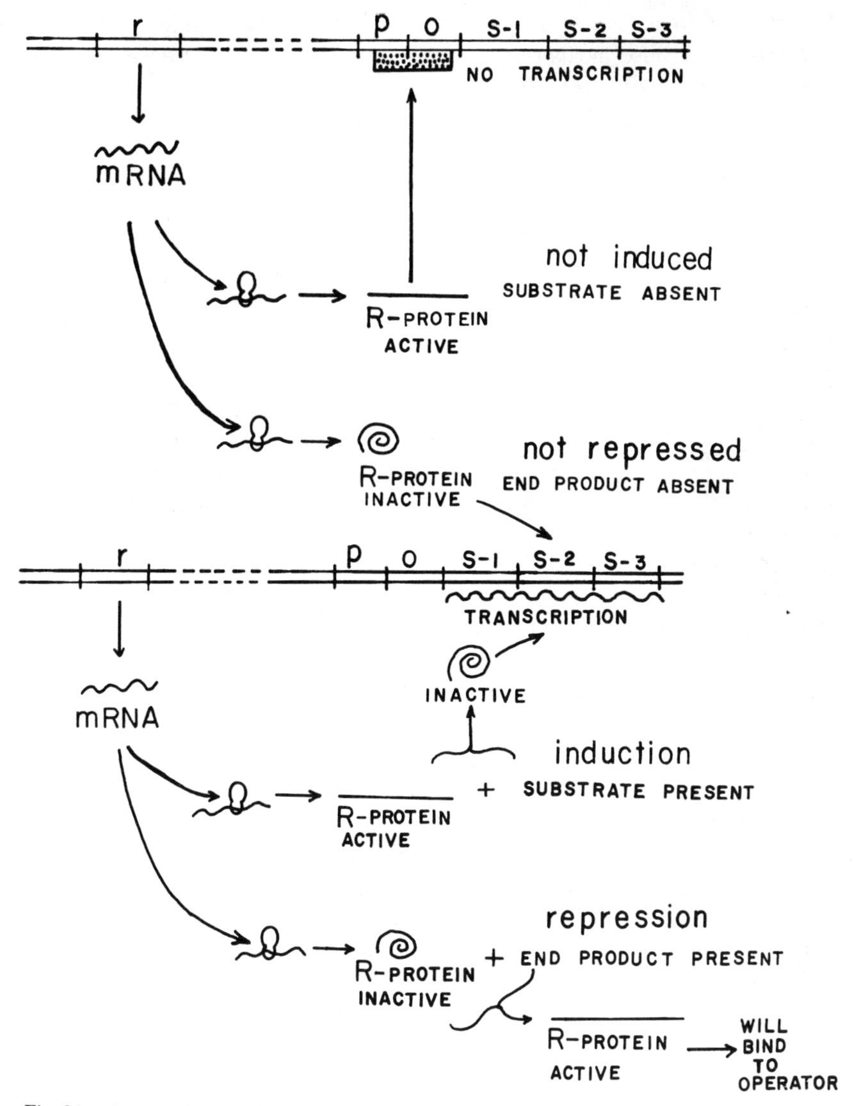

Fig. 21. A general model for induction and repression of enzyme synthesis.

be attached to as many as a dozen ribosomes. Such aggregates are called *polysomes*. Because a ribosome may detach from the mRNA, more protein molecules are made from the portion of the mRNA corresponding to the structural genes near the operator than for the structural genes distal from the operator. Nonsense mutations frequently cause the ribosomes to detach from the mRNA, thereby preventing synthesis of the enzymes coded for by the distal portion of the polycistronic mRNA, as well as of the particular enzyme affected directly by the mutation. Such mutations are said to be *polar*.

Repression may regulate catabolic as well as anabolic processes. Many bacteria can utilize particular nutrients (e.g., amino acids) as either precursors or energy sources. If an alternative energy source is supplied (e.g., glucose), the genes controlling the biosynthesis of enzymes that degrade the precursor substrate are repressed. This effect has been termed glucose repression or catabolite repression. Catabolite repression, like end-product repression, affects enzyme synthesis and not enzyme activity. Catabolite repression is specific and presumably analogous to end-product repression in biosynthetic pathways.

Exogenously supplied metabolites may increase the amount of enzyme produced; such enzymes are said to be induced. Induction and repression are merely different manifestations of a similar control mechanism. In induction, the structural genes controlling a catabolic pathway are repressed unless the appropriate substrate is added. This is so because the regulatory gene normally produces an active stable repressor that binds to the operator region, thereby preventing the RNA polymerase from transcribing the DNA. Added inducer, or a derivative of it, must react with the active stable repressor to convert it to an inactive form that cannot bind to the operator region. After the active repressor is removed or destroyed, the RNA polymerase is able to direct the synthesis of new mRNA. It should be noted that an inducer may be, but is not necessarily, a substrate for the induced enzyme. Moreover a substrate for the induced enzyme may be, but is not necessarily, an inducer.

Constitutive enzymes are those whose concentrations do not fluctuate appreciably under a variety of growth conditions. An induced enzyme system may become constitutive by mutation in either the operator region or the regulatory gene. A mutation that alters the operator in such a way that it can no longer bind the active repressor will result in constitutive enzyme formation. Alternatively, a mutation in the regulatory gene that prevents repressor synthesis or makes the repressor unable to react with the corresponding inducer (or its derivative) will lead to constitutive enzyme formation. Similar mutations in the operator region or regulatory gene controlling an anabolic pathway will result in derepression of the pathway. Such derepressed mutants, which overproduce an end product, are used extensively in the fermentation industry to produce amino acids and vitamins.

Regulation in mammalian cells differs in many respects from the control mechanisms operative in bacteria. In contrast to bacterial systems, mammalian mRNA is long-lived, polycistronic messengers have not been found, there is no evidence for the existence of operons, and mammalian DNA contains many repeated nucleotide sequences. Nevertheless, there are common denominators, one of which is cyclic adenosine monophosphate (AMP). Cyclic AMP exerts a multiplicity of effects at several metabolic levels. How these multiple effects are interrelated is not yet clear. The formation of cyclic AMP from ATP is catalyzed by the enzyme adenyl cyclase. The reaction can be reversed, that is, cyclic AMP can adenylate pyrophosphate to yield ATP. Adenyl cyclase appears to be ubiquitous. Cyclic AMP is hydrolyzed to 5′-AMP by cAMP-P-diesterase; this enzyme is also widespread. Accordingly, cellular cyclic AMP levels can be controlled by (*a*) altering the activity of adenyl cyclase, (*b*) altering the activity of phosphodiesterase, or (*c*) releasing cyclic AMP into the medium. The activity of adenyl cyclase is affected by a wide variety of pharmacologically active substances, for example, insulin, prostaglandin E_1, histamine, and serotonin.

Exogenous cyclic AMP overcomes catabolite repression in *E. coli*, indicating that cyclic AMP can act at the level of transcription. Cyclic AMP and rifampicin seem to act at the same metabolic site, which is the initiation of mRNA synthesis. That cyclic AMP affects the rate of mRNA initiation rather than nucleotide polymerization is consistent with the observation that some mutants of *E. coli* with deletions in the promoter region of the *lac* operon do not respond to added cyclic AMP (by release of catabolite repression). As discussed previously, the promoter region is thought to be the site at which RNA polymerase binds to an operon in order to initiate transcription. Cyclic AMP does not act directly on DNA or RNA polymerase, but functions in concert with a protein designated as cyclic AMP receptor protein (68). Evidence for the stimulation of transcription by cyclic AMP in eukaryotes is indirect and inconclusive.

Cyclic AMP has also been purported to affect the rate of translation of mRNA. This view is corroborated by the observation that the ribosomal G factor, which is a protein that participates in the translocation of the nascent polypeptide chain, binds cyclic AMP in the presence of guanosine triphosphate. Cyclic AMP does modulate the activity of a number of enzymes and is an allosteric cofactor for phosphorylase b kinase, which catalyzes the phosphorylation of seryl residues in inactive phosphorylase b kinase, converting the enzyme to its active form. In addition, cyclic AMP regulates glycogen metabolism by stimulating the conversion of active glycogen synthetase to its inactive phosphorylated form. Concurrently, glycogen phosphorylation, leading to glycogenolysis, is activated by two phosphorylation reactions, at least one of which is mediated by cyclic AMP. Perhaps cyclic AMP modulates enzyme activity and the initiation of mRNA synthesis by means of these phosphorylation reactions (38).

Cyclic AMP is the chemotactic substance of the cellular slime mold *Dictyostelium discoideum*. Because morphogenesis in *Dictyostelium* is correlated with a number of alterations in enzyme synthesis and activity, and because cyclic AMP can regulate enzyme synthesis and activity, a causal relationship may be involved. It must be noted, though, that mutants of *E. coli* having defective adenyl cyclase or defective cyclic AMP receptor protein are viable; therefore cyclic AMP regulation does not appear to be crucial for survival. It is significant that the intracellular concentration of cyclic AMP in *Salmonella* cells infected with phage P22 controls the decision to lysogenize or lyse the bacterium. In *E. coli* cells infected with phage λ, however, the decision to lysogenize is not appreciably controlled by the concentration of cyclic AMP.

3. GENE TRANSFER

A. Transformation

Genetic transformation involves the transfer of genetic information as free DNA from a disrupted cell to an intact one. Although transformation may be obtained with DNA spontaneously released by a lysing donor cell, extracted and purified DNA is more often used. Obviously, DNA extracts from a bacterial mass contain more than one genetic determinant; in fact, essentially every genetic locus can be transformed in a suitable recipient if there is an adequate selective system to detect the recombinant type. Genetic transformation is definitively established for a number of bacterial genera: *Diplococcus, Hemophilus, Bacillus, Neisseria, Rhizobium*, and *Streptococcus*. Among these genera not all species can be transformed, and within a transformable species not all strains can be transformed.

The general protocol for detecting transformation can be summarized as follows. The donor cells are lysed by procedures that minimize the degradation of the DNA polymer. Protein is removed from the crude DNA preparation by enzymic digestion or by treating the extract with a denaturing reagent. The purified DNA is sterilized during the final ethanol precipitation. In general, filtration is not only ineffective but also deleterious. The stock DNA is stable in 0.15 *M* or 2 *M* NaCl and in 0.01 *M* EDTA at 5°C for years. The DNA is added to a suitable recipient at a concentration of about 1 μg/ml; after 10 to 30 min, deoxyribonuclease is added. Claims for transformation based on experiments using several magnitudes more of DNA and treatment periods of several days should be viewed with skepticism. The recipient bacteria, after exposure to the donor DNA, must be given adequate time for integration and expression of the new genetic determinants.

The principal factor limiting the detection of transformation is the genetic and physiologic competence of the recipient to incorporate DNA. Physiologic competence develops only under particular cultural conditions, which must be empirically determined for each system. Experience gained with one species does not necessarily apply to another. Physiologic competence affects DNA uptake primarily. Once the donor DNA has entered the recipient cell, it may be restricted. Restricted DNA is degraded by the nucleases of the recipient cell. Donor DNA that is not recognized as foreign must be able to integrate into the genophore of the recipient. Some bacterial strains lack the enzymes necessary to insert the new genetic determinants. Such organisms, called Rec^-, are also exquisitely sensitive to ultraviolet radiation, leading to the suggestion that the enzymes for DNA repair and for recombination are the same.

B. Transduction

Genetic transduction is the process by which a limited amount of genetic information is transferred from a disrupted bacterium to a recipient cell by a bacteriophage vector. When a bacterial cell is infected with a bacterial virus, the phage may initiate uncontrolled growth, resulting in cell death and release of more infective virus particles. Alternatively, the viral genome may become a normal cell component, replicating in concert with the bacterial genophore. The viral genophore is inserted into the bacterial genophore in some but not all lysogenized bacteria. In general, the host and the virus share a small region of genetic homology, that is, they have a common nucleotide sequence which facilitates physical incorporation of the phage genome into the bacterial genome. Another factor controlling the insertion of temperate phage DNA into a particular locus on the bacterial genophore is a phage-coded enzyme with specificity for the integration site. Frequently, viral DNA is restricted by the host, that is, the viral DNA is recognized as foreign and is degraded by the host's nucleases. During unrestrained viral replication, whether in the lytic cycle of a virulent phage or after release of temperate phage DNA from host control, the maturing phage particles may incorporate into themselves host genetic determinants. Such phage particles are usually defective in that some viral genes are omitted to accommodate the host genetic determinants. A bacterium infected with a phage particle carrying donor bacterial genes may incorporate the donor genetic determinants into its genophore. Genetic transduction has been definitively established in *Salmonella, Escherichia, Staphylococcus, Pseudomonas*, and *Bacillus*.

Temperate phages such as λ, which inserts into a specific site on the bacterial genophore, are able to transfer only genes adjacent to the phage-insertion region. Lambda phage in *E. coli* can transduce genes controlling galactose fermentation or biotin dependence, and only very rarely other genetic determinants. Mainly temperate phages that have been induced to unrestrained growth in the lysogenic bacterium are effective in limited transduction. Temperate phages such as phage P1 of *E. coli* do not have specific attachment sites in the bacterial genophore. These phages are able to transduce essentially all bacterial genetic determinants, albeit only a few at a time. Both lytically propagated and induced temperate phages are effective in *generalized* transduction. Occasionally the bacterial genes introduced by transduction are unable to incorporate into the genophore of the recipient but are able to function. This phenomenon is called *abortive* transduction. Such recipients undergo semiclonal inheritance, in which the daughter cell lacks the donor gene but the parental cell manifests the donor characteristic. When the characteristic examined is a nutritional requirement, abortive transduction gives rise to microcolonies on a chemically defined medium because only the original recipient cell can grow and divide.

C. Conversion

Bacteriophages per se may alter the phenotype of a bacterium, in contrast to transduction, wherein the bacterial virus is merely a vector for the transfer of bacterial genes. The altered phenotype may be manifested in the lysogenic bacterium when the bacteriophage DNA is reduced to a normal cell component or may be manifested only when the bacteriophage reenters the lytic growth cycle. Most bacteria acquire two new phenotypes upon lysogenization: (*a*) the bacterium has the ability to release phage particles sporadically, and (*b*) the cell is immune to lytic infection by exogenous homologous phage particles. Phage genes affect many other properties of the host; for example, they play a prominent role in the determination of the antigenic structure of the somatic antigens or O-antigens of *Salmonella*. In *Corynebacterium*, toxin production is the result of induction of the appropriate temperate phage.

Episomes and Plasmids. *Episomes* are genetic determinants that may exist integrated into the bacterial genophore or may replicate autonomously apart from it. *Plasmids* are genetic determinants that replicate autonomously apart from the bacterial genophore and for which there is no known integrated stage. The characteristics controlled by episomes and/or plasmids include (*a*) the fertility factor of *E. coli*, (*b*) infectious drug resistance in enteric bacteria, (*c*) penicillin resistance in some *Staphylococcus* strains, (*d*) the capability to produce temperate bacterial viruses, and (*e*) the production of bacteriocins, which are species-specific protein antibiotics (or defective phage components). Episomes in the autonomous state and plasmids may be transferred from one cell to another upon contact. Episomes in the integrated state cannot be transmitted by contact. A cell may be freed of episomes and plasmids by cultivating it in the presence of proflavin, quinacrine, or ethidium bromide. Episomes and plasmids, like other genetic determinants, can undergo mutation and recombine with others of their kind to produce new gene combinations. Episomes may retain a minute amount of the bacterial genophore when they disengage from it and enter the autonomous state. The fertility factor in *E. coli*, for example, may carry with it the gene controlling lactose fermentation.

D. Syncytic Recombination

The transfer of many genetic determinants from a living donor to a recipient has been called conjugation or syncytic recombination. Syncytic recombination has been definitively demonstrated in *Escherichia, Salmonella, Shigella, Pseudomonas, Nocardia*, and *Streptomyces*. To detect syncytic recombination, mutant strains having different nutritional requirements are grown together, and recombinants with new combinations of nutritional requirements (independence or dependence) are selected. The first step in syncytic recombination is cell fusion. In the enteric bacteria, the pilus seems to be the conjugal organelle. The DNA is transferred from a donor cell to the recipient cell, and frequently the complete donor genophore is not transferred. A transitory heterogenomic or merozygotic organism is established. Because there are no nuclear membranes to restrict interactions between different genomes, classical heterokaryosis and diploidy do not exist in the prokaryotes. Ultimately, crossing over occurs between the donor and the recipient genomes, leading to new, stable nonparental types. Homogenomy is established in enteric bacteria and pseudomonads as a consequence of binary fission, in nocardiae by fragmentation, and in streptomycetes by sporulation.

4. GENETIC MAPPING

A. Complementation Mapping

Metabolically blocked mutants are useful for the analysis of biosynthetic pathways. After accumulating a large number of auxotrophic mutants whose growth requirements can be satisfied by a common metabolite, it is necessary to determine which mutants are identical. To achieve this, the mutants are mixed in all possible pairwise combinations. The mixtures are then assayed,

Complementation responses of ten mutants

	A	B	C	D	E	F	G	H	I	J
A	-	-	-	+	+	+	+	-	+	+
B	-	-	-	+	-	+	+	-	+	+
C	-	-	-	-	-	-	+	-	-	+
D	+	+	-	-	-	-	+	+	-	+
E	+	-	-	-	-	+	+	+	-	+
F	+	+	-	-	+	-	-	+	-	-
G	+	+	+	+	+	-	-	+	+	-
H	-	-	-	+	+	+	+	-	+	+
I	+	+	-	-	-	-	+	+	-	+
J	+	+	+	+	+	-	-	+	+	-

Fig. 22. Typical results of complementation tests used for constructing a complementation map.

enzymically or by growth response, to determine whether a pair complementarily produces an enzyme that neither mutant could produce alone. Because growth is controlled by many factors and is not always accurately and precisely measurable, failure to complement is not proof of identity. Ability to complement, however, is proof of nonidentity.

As an example, let us derive a complementation map based upon 10 mutants designated as *A, B, C, D, E, F, G, H, I,* and *J*. As shown in Fig. 22, mutants *D, E, F, G, I,* and *J* complement mutant *A*; therefore, *A* is nonallelic with these genes. Mutants *B, C,* and *H* did not interact with *A* to produce the wild type; hence they may be identical. By comparing the interactions of *A, B, C,* and *H* with the other strains, it is seen that *A* and *H* are alike but *B* and *C* behave differently with respect to mutant *E*. Now, considering *B* more thoroughly, it should be noted that *B* complements *D, F, G, I*, and *J* but not *A, C, E*, and *H*. The former group of mutants is definitely not allelic with *B*, but the latter group may be. Because each of the other mutants differs from *B* in at least one response, *B* should be considered a unique mutant. By continuing this analysis, the mutants are divided into seven unique classes.

Inasmuch as *H* is identical to *A, G* to *J*, and *D* to *I*, mutants *H, I*, and *J* can be omitted from further consideration. Because mutants *B* and *C* do not interact successfully with *A*, the genetic deficiencies of *B* and *C* overlap *A*. Mutant *C* reacts positively only with *G*; therefore it overlaps the entire region except for site *G*. Determinants for *D, E, F*, and *G* do not overlap the *A* region (Fig. 23, step 1). Strain *B* fails to correct the metabolic lesion in strain *E*; thus region *E* overlaps region *B* but not regions *A* and *F* (Fig. 23, step 2). It should be noted that combinations of *D* and *E* or *D* and *F* do not interact successfully, but combinations of *B* and *D* or *B* and *F* do. Accordingly, the map can be refined to indicate that *D* overlaps *E* and *F* but not *A, B*, or *G*. Finally, mutants *F* and *G* are not complementary, but *G* complements all of the other mutants; for these reasons, one must conclude that *F* and *G* overlap (Fig. 23, step 3).

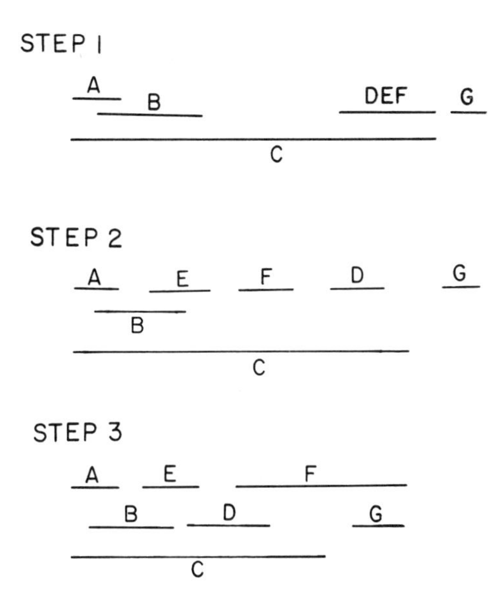

Fig. 23. Steps for constructing a complementation map based on the data in Fig. 21.

$Cys^- Pro^- Ade^+ Str\text{-}s$ X $Cys^+ Pro^+ Ade^- Str\text{-}r$

Selected $Pro^+ Ade^+$

recombinant phenotypes	%
$Cys^+ Str\text{-}r$	10
$Cys^+ Str\text{-}s$	79
$Cys^- Str\text{-}r$	0
$Cys^- Str\text{-}s$	11

Selected $Cys^+ Ade^+$

recombinant phenotypes	%
$Pro^+ Str\text{-}r$	42
$Pro^+ Str\text{-}s$	9
$Pro^- Str\text{-}r$	0
$Pro^- Str\text{-}s$	49

Fig. 24. Frequency of unselected markers among recombinant progeny.

B. Linkage to a Selected Marker

The frequency with which various combinations of markers are recovered among the progeny is classically taken as evidence for linkage and crossing over. The frequency of recombination between linked markers is used to define the unit of distance in genetic mapping. Obviously, genes linked closely to selected markers will appear frequently among the recombinant progeny. As an example, let us consider data from a cross involving an adenine-dependent, streptomycin-resistant streptomycete ($Cys^+ Pro^+ Ade^-$ Str–r) and a cystine- and proline-dependent, streptomycin-sensitive streptomycete (Cys^- $Pro^- Ade^+$ Str–s). When the selected markers are Pro^+ and Ade^+, 90% of the recombinants are Str–s and 89% are Cys^+ (Fig. 24). This indicates that the locus for streptomycin sensitivity is closely linked to the locus of one of the selected markers; in this instance, the selected marker in the Str–s parent is adenine independence. Because the percentage of recombination between the loci for adenine and streptomycin is 10%, the genetic map is initiated by placing *ade* and *str* 10 units apart. Similarly, 89% of the recombinants are Cys^+, indicating that the *cys* locus is linked to the *pro* locus. Accordingly, the linkage map can be extended by placing the *cys* locus 11 units from the *pro* locus. At this stage of the analysis, the two linkage groups have not been shown to be connected.

The parental combinations of the *cys* locus and *str* locus are Cys^-Str–s and Cys^+Str–r; these phenotypes make up 11% and 10%, respectively, of the total recombinant population. The nonparental types, Cys^+Str–s and Cys^-Str–r, make up 79% and less than 1%, respectively, of the recombinant colonies. Because the parental and nonparental phenotypes do not occur with equal frequency, it is clear that the loci for *cys* and *str* are linked. This conclusion can be demonstrated more dramatically by analyzing the results when the selected markers are Cys^+ and Ade^+. There are nearly equal proportions of Pro^+ and Pro^- phenotypes. This indicates that the *pro* locus is either unlinked to *cys* and *ade* or that it is approximately midway between the two selected markers. Because the previous cross established that *cys* and *pro* are linked, the only arrangement compatible with both results is *cys pro ade str*.

C. Linkage between Unselected Markers

An analysis of recombination frequencies between pairs of unselected markers constitutes a reliable basis for formulating linkage relationships. As an example of this approach to genetic mapping, let us consider a cross between a cystine- and proline-dependent streptomycete and a nicotinamide- and adenine-dependent streptomycete (Fig. 25). When Cys^+ and Ade^+ are the selected phenotypes, 84% of the recombinants have the parental combinations of the unselected markers (Nic^-Pro^+ and Nic^+Pro^-). These data establish that *nic* and *pro* are linked. There is an excess of the Nic^- phenotype, indicating that *nic* is linked to *cys*. There are two possible arrangements of the three linked loci: *cys nic pro* and *nic cys pro*. The order *nic pro cys* is not possible, however, because it predicts that *pro* is linked more closely to *cys* than to *nic*; this prediction is contrary to the data. Since the *ade* locus is weakly linked to the other three loci, it can be placed to the far right or the far left of the cluster of three linked loci. These observations lead to four possible arrangements of the four loci. By analyzing the consequences of crossovers in each of the four models, it is possible to rule out certain arragements. For example, if the *nic* and *cys* loci are adjacent (Fig. 25, B-1), the mandatory crossover between *cys* and *ade* might be expected to yield abundant Nic^-Pro^+ and Nic^+Pro^+ recombinants. The prototrophic Nic^+Pro^+ recombinant, however, is not found. Unless it is assumed that *nic* and *cys* are very close, the arrangement *ade nic cys pro* must be ruled out. Because recombination does occur between *nic* and *cys* 33% of the time, the two loci cannot be very close.

Similarly, if the *pro* and *cys* loci are adjacent (Fig. 25, B-2), the mandatory crossover between *cys* and *ade* might be expected to yield abundant Nic^-Pro^- and Nic^-Pro^+ recombinants. However, the second most frequent class is Nic^+Pro^-, contrary to the predictions of this model. When the selected markers are distal to one another (Fig. 25, B-3), there are three possible crossovers. To obtain the prototrophic recombinant, a triple crossover would be required. This model is consistent with the data. If the selected markers are adjacent to one another (Fig. 25, B-4), the most prevalent phenotype should be Nic^-Pro^+, which is consistent with the data. On the basis of previous data (Fig. 24), the sequênce *cys nic pro ade* is preferred. It should be noted that models B-3 and B-4 in Fig. 25 are the same if the genophore is converted into a circular model by joining the ends.

$Cys^-Pro^-Nic^+Ade^+$ X $Cys^+Pro^+Nic^-Ade^-$

Part A. selected Cys^+Ade^+

Nic^+Pro^+	0%
Nic^+Pro^-	33%
Nic^-Pro^+	51%
Nic^-Pro^-	16%

Part B. possible gene sequences

1. ade nic cys pro
2. nic cys pro ade
3. cys nic pro ade
4. ade cys nic pro

Fig. 25. Frequency of unselected markers among recombinant progeny and the possible gene sequences.

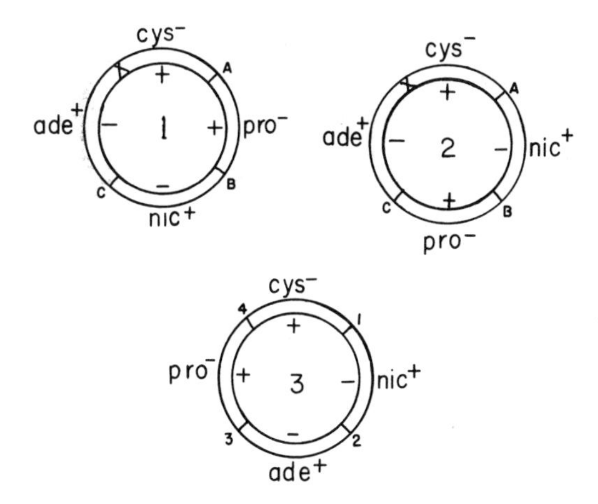

Fig. 26. Development of a circular genetic map. In models 1 and 2, the mandatory crossover is indicated by X; another crossover must occur in region *A*, *B*, or *C*. In model 3, crossover must occur in region 1 or 2 and in region 3 or 4.

D. Mapping with Circular Models

All bacterial genophores analyzed in detail consist of a single, circular linkage group. Accordingly, double circular models can be drawn for each of the possible arrangements of four loci (Fig. 26). Using the data from Fig. 25, we can discard model 1 because prototrophic recombinants would be expected to occur rather frequently, but in fact they were not found. Similarly, model 3 allows for the genesis of all four combinations of the unselected markers, including the prototrophic class. Only model 2 is consistent with the data. When the selected loci are not adjacent (Fig. 26, model 3), accepted arrangements must show the following relationship:

$$\frac{\text{No. of crossovers in regions 1 and 3}}{\text{No. of crossovers in regions 1 and 4}} = \frac{\text{No. of crossovers in regions 2 and 3}}{\text{No. of crossovers in regions 2 and 4}}$$

5. SYSTEMATIC IMPLICATIONS OF DNA ANALYSES

The genetic potential (genotype) of an organism is encoded in the linear order of the four nucleotide bases in its deoxyribonucleic acid (DNA). These sequences are translated into colinear sequences of amino acids in structural or enzymic proteins, which directly or indirectly constitute the phenotype of the cell. Accordingly, evolutionary divergence from a common ancestor proceeds as the progeny accumulate base substitutions in their DNA. Recent evidence strongly indicates that remnants of an organism's evolutionary history are retained, inscribed in the genetic determinants themselves. Because of our increased understanding of the molecular architecture of DNA, approaches to microbial classification other than classical determinative systematics can be developed. In fact, the evolutionary approach to bacterial classification, long hindered by the lack of a recognized fossil record, now seems feasible at several molecular levels. The documentation of the "fossil record" inscribed in the molecules of cells has been reviewed by Mandel (53). An evolutionary classification, therefore, can be formulated for a group of microorganisms by analyzing their DNA. Accordingly, this section is concerned with the isolation of DNA, the base composition of DNA, the renaturation of DNA, evolutionary divergence, and the development of quantitative statements about relatedness.

A. Isolation of DNA

A widely used method for the isolation of DNA from microorganisms is that described by Marmur (55). In this method, lysozyme or a detergent disrupts the cells, releasing the cellular components. Protein is dissociated from the nucleic acids by treating the lysed suspension with 1 *M* sodium perchlorate. Most proteins are denatured by gently shaking the viscous extract with chloroform and isoamyl alcohol (Sevag's mixture); ribonucleic acid (RNA) contamination is effectively removed by treatment with ribonuclease. Isopropanol, in the presence of acetate anions, selectively precipitates crude DNA fibers, leaving RNA fragments in solution. Chelating agents, such as ethylenediamine tetraacetate and citrate, are added to the buffers used during the isolation procedure because they effectively minimize endogenous deoxyribonuclease activity. Moreover, sodium dodecyl sulfate acts as a protein denaturant during the lysis step. Despite the utility of Marmur's method, it has some limitations. Under optimum conditions, less than half of the total cellular DNA is recovered. Furthermore, the frequent use of precipitation steps and the Sevag's deproteinization procedure shears the DNA to a size of approximately 10^6 daltons. Adsorption of DNA to the denatured protein and selective loss of small molecules of DNA have been reported (71).

On the other hand, there are few descriptions in the literature of methods improving on Marmur's technique for actual isolation of DNA. Various modifications of the phenol method of Kirby (42, 43) replace the Sevag's deproteinization step. Aqueous phenol itself arrests incipient nuclease activity, and, significantly, phenol extraction requires very little shaking. However, phenol accumulates harmful peroxides which must be eliminated by frequent distillation or treatment with 8-hydroxyquinoline. Moreover, the high ultraviolet absorbance of phenol interferes with routine analysis of DNA; therefore care must be taken for the complete removal of phenol from DNA preparations. Interestingly, there is evidence that phenol deproteinization causes a selective loss of minor DNA components (74).

An alternative to the Marmur method offering much in speed, efficiency, and resolution is CsCl density gradient centrifugation of lysates in a fixed-angle rotor (25). Because protein tends to float and RNA tends to be sedimented in dense CsCl, relatively pure DNA can be isolated with little pretreatment. Moreover, the concentrating effect of the CsCl gradient allows quantitative recovery of DNA from small samples of microbial cells.

The use of hydroxyapatite as described by Britten, Pavich, and Smith (16) is another valuable alternative to the Marmur method. Hydroxyapatite is a modified calcium phosphate gel that is effective in fractionating double-stranded and single-stranded DNA (4, 13, 64). Lysates in 0.24 *M* phosphate buffer are applied to a column of hydroxyapatite. At this salt concentration double-stranded DNA binds to the column, while protein, RNA, and other cellular material pass through it. Because double-stranded DNA is not bound to hydroxyapatite in 0.4 *M* phosphate buffer, washing the column with this salt concentration elutes the partially purified nucleic acid.

B. Base Composition of DNA

The Watson and Crick model of double-stranded DNA predicts that the adenine (A) content of a DNA molecule equals its thymine (T) content; likewise, the guanine (G) and cytosine (C) contents are also equal. However, the ratio of $(A + T)/(G + C)$ or the mole per cent of guanine plus cytosine (%GC) may vary from one species of organism to another. Marmur, Rownd, and Schildkraut (57) reviewed the arguments for considering comparative analysis of DNA base compositions as a cardinal criterion in microbial taxonomy. In brief, they concluded that a necessary but not sufficient condition for substantial genetic relatedness between pairs of organisms was overall similarity in DNA base composition. In practice, however, the principal value of the %GC content is its use as an exclusionary determinant in the formulation of taxonomic groups. Only differences in %GC are significant. Similarity in %GC does not necessarily indicate genomic similarity; for example, the GC ratios of DNA from *Saccharomyces cerevisiae, Bacillus subtilis*, and man are all about the same (37). Moreover, there are theoretical reasons to believe that highly similar organisms must have very similar GC ratios (19). In the final analysis, DNA base composition is useful as a first approximation of relatedness, but it cannot serve as a quantitative measure of evolutionary divergence.

The techniques routinely used to determine base composition can be categorized as methods involving the following: hydrolysis and chromatography as reviewed by Bendich (3), melting temperature (T_m) determinations from the sharp increase in absorbance at 260 nm upon heating a DNA in aqueous solutions (56), and determinations of the bouyant density of DNA in isopycnic cesium chloride centrifugation (72). Other methods less commonly used include the A_{260}/A_{280} ratio in 0.1 *M* acetic acid (26), depurination in dilute acid, followed by dialysis and spectrophotometric analysis (35), and quantitative gas–liquid chromatography (29).

It should be realized that, of the routine methods used, only hydrolysis and chromatography (and possibly the depurination method) give direct estimates of base composition. The calculation of nucleotide composition from measurements of buoyant density in CsCl or of T_m is predicated on an empirically established relationship with chemical data. The values derived by the different techniques may differ because of the uncertainty of the formulas used in the interconversion of thermal transition and buoyant density data to chemical composition (28). The GC content of DNA can be useful for diagnostic and taxonomic purposes only when the determinations are truly comparative. It must be emphasized that the techniques, chemical supplies, and equipment used in different laboratories and even in a given laboratory can differ enough to produce minor discrepancies in observed %GC results. The reader must refrain, therefore, from placing undue weight on differences in GC contents determined in various laboratories by diverse methods (54).

A number of variables can influence experimental T_m values. Cations such as Mg^{2+}, Ba^{2+}, Mn^{2+}, Co^{2+}, Ni^{2+}, and Zn^{2+} increase the T_m, while other cations (e.g., Cu^{2+}, Cd^{2+}, and Pb^{2+}) decrease it (44). Molecules such as urea, guanidinium HCl, salicylate, formamide, dimethyl formamide, dimethyl sulfoxide, tetramethylurea, formaldehyde, aminonaphthol, hydroxylamine, and a variety of alcohols lower the T_m (44). Marmur and Doty (56) found small differences in the T_m of bacteriophage DNA due to the presence of glucosylation or hydroxymethylcytosine.

Buoyant density measurements are not as sensitive to experimental variation as are T_m measurements. However, RbCl contamination of the CsCl can cause low buoyant density values. Schildkraut, Marmur, and Doty (72) suggest that the RbCl concentration should be less than 0.01%. Buoyant density analysis is generally dependent on the density of a known reference DNA. Most investigators use *E. coli* DNA as the reference; but, as discussed by Schildkraut et al. (72), there are at least two published values for the density of *E. coli* DNA. Reported data must include the reference DNA and the assumed density if valid comparisons are to be made (54).

Table 1. DNA Base Compositions of Representative Bacterial Genera[a]

Genus	Guanine + Cytosine (mole %)	Genus	Guanine + Cytosine (mole %)
Gram-Positive		Gram-Negative (Continued)	
Actinomyces	58–68	*Bordetella*	68–69
Arthrobacter	62–64	*Brucella*	55–58
Bacillus	32–66	*Caulobacter*	62–67
Clostridium	30–45	*Chromobacterium*	61–72
Corynebacterium	45–60	*Citrobacter*	50–53
Lactobacillus	33–59	*Derxia*	64–72
Leuconostoc	39–42	*Enterobacter*	52–54
Listeria	37–39	*Erwinia*	50–57
Micrococcus	62–74	*Escherichia*	50–52
Mycobacterium	60–70	*Flavobacterium*	32–42
Nocardia	62–72	*Hemophilus*	38–42
Propionibacterium	66–71	*Klebsiella*	52–59
Staphylococcus	30–40	*Moraxella*	39–45
Streptococcus	33–42	*Myxococcus*	67–70
Streptomyces	67–75	*Neisseria*	40–52
		Pasteurella	35–48
Gram-Negative		*Photobacterium*	43–47
		Pseudomonas	36–70
Acetobacter	54–56	*Rhizobium*	59–66
Aerobacter	50–59	*Salmonella*	50–54
Aeromonas	51–63	*Serratia*	54–63
Agrobacterium	58–66	*Shigella*	49–54
Alcaligenes	67–70	*Thiobacillus*	58–70
Azotobacter	54–66	*Vibrio*	46–49
Bacteroides	40–45	*Xanthomonas*	62–68

[a] Data taken from ref. 37.

Szybalski (77) reviewed the techniques for buoyant density determination. The rare nucleotides, that is, those other than adenylate, guanylate, cytidylate, and thymidylate, and the substituted nucleotides have diverse properties in cesium gradients. The density of DNA containing glucosylated hydroxymethylcytosine, for example, decreases in CsCl gradients, whereas the density increases in Cs_2SO_4 gradients. This is due to the buoyant density of glucose, which corresponds to 1.6 g/cm^3 in CsCl and 1.5 g/cm^3 in Cs_2SO_4. In general, methylation decreases the buoyant density of DNA in both CsCl and

Cs_2SO_4 gradients. This decrease is approximately 1 mg/cm^3 in CsCl and 0.4 mg/cm^3 in Cs_2SO_4 per 1% of methylated base.

Several chemotherapeutic dyes and antibiotics, including actinomycin, anthracyclines, chromomycins (olivomycin, mithramycin), and ethidium bromide, form with native or denatured DNA complexes the buoyant densities of which are lower than the value for free DNA, both in CsCl and Cs_2SO_4 gradients (40).

Methods for determining nucleotide composition and the resulting values for a number of bacterial DNA preparations (Table 1) have been reviewed by Jones and Sneath (37), DeLey (20), Mandel (53), and Hill (33). The nucleotide compositions of a number of fungal DNA preparations have been compiled by Storck and Alexopoulos (76). These tabulations are valuable when used properly, but indiscriminant use of comparative data gleaned from the literature can lead to erroneous conclusions (28).

C. Reassociation of DNA

If we assume that the evolutionary development of an organism is reflected in the sequence of bases in its DNA, comparison of the DNA sequences of organisms should give a complete phyletic evaluation of their present relatedness. Although this proposal is extremely provocative, complete, direct sequence analysis of a DNA molecule is not currently possible. Fortunately, the complementary nature of the DNA double helix itself can be used to circumvent these technical difficulties. It is well known that the two strands of DNA can be separated and specifically reassociated. The utility of this reaction for comparing DNA sequences can be realized by mixing DNAs from two different species and allowing them to form double-stranded interspecific DNA duplexes. By determining the extent of hybrid reassociation, the relative number of sequences held in common can be calculated. By measuring the thermal stabilities of these interspecific DNA duplexes, their exactness of fit can be estimated. Finally, by examining the kinetics of the reaction, the concentrations of reactants (i.e., the presence or absence of sequence repetition) can be determined.

The extent, fidelity, and rate of DNA reassociation depend strongly on experimental conditions. The unwitting selection of unsuitable conditions results in misleading or useless data. On the other hand, by carefully choosing the proper experimental conditions, it is possible to determine the relative amounts of "closely" or "distantly" related nucleotide sequences. Thus, with rigorously defined criteria, the discerning investigator can not only assess quantitative relatedness (extent of reaction), but also estimate the degree of evolutionary divergence (sequence mismatching).

D. Methods of Reassociation

A difficult problem in nucleic acid reassociation is the selection of a system that will give the desired information. There are two general methods to choose from: reassociation carried out with denatured DNA from one source immobilized in an agar matrix or on a membrane filter surface, and reassociation performed with both test nucleic acids free in solution. Either procedure will yield specific and reproducible data when properly applied. Nevertheless, both have limitations, and certain applications are better suited to one or another of these techniques. Factors influencing the choice of methods can be found in the literature, by experience, and from an understanding of some of the basic phenomena involved (39).

Free-solution reassociation. This system initially posed two major difficulties. First, each species of nucleic acid present was free to react with itself, and, second, the product of the heterologous reaction was difficult to quantitate and almost impossible to isolate (57). However, recent progress in understanding the kinetics of free-solution reassociation has resulted in a revival of interest in the system. It is now apparent that in free solution the concentration of each DNA species and the time of incubation can be readily adjusted, and thereby the type and extent of duplex formed can be controlled. Furthermore, the use of hydroxyapatite to fractionate single-stranded and double-stranded DNA has probably been a major factor in the reemergence of free-solution systems. Duplex nucleic acid molecules are bound by hydroxyapatite in 0.12 *M* phosphate buffer, whereas single-stranded nucleic acids are not (4, 64). In 0.4 *M* phosphate buffer, the double-stranded molecules are eluted (9, 15). Thermal stability profiles of reassociated DNA duplexes are generated by washing the hydroxyapatite with an elution series at increasing temperatures. The advantages of hydroxyapatite have been summarized by Brenner et al. (11):

1. It is not necessary to immobilize the unlabeled DNA, and one need not be concerned with reassociated (hybrid) DNA leaching out of the agar or from the filter in thermal elution studies.

2. The binding of labeled bacterial DNA fragments to unlabeled DNA from the same source is routinely 20 to 40% in agar, 10 to 70% on filters, and from 75 to 95% in free solutions.

3. Unlabeled DNA is not immobilized; thus its absorbancy can be assayed, providing a valuable internal control.

4. Kinetics in free solution are typically uncomplicated second-order, whereas kinetics in agar and filters are more complex.

Brenner et al. (11) do point out one disadvantage; competition experiments cannot be done with hydroxyapatite. Moreover, citrate and potassium ions greatly diminish the ability of hydroxyapatite to bind DNA. Formamide and dimethyl sulfoxide cannot be used with hydroxyapatite unless their concentration is less than 1% because they apparently destroy the cross linking of the hydroxyapatite (62).

A new method described by DeLey, Cattoir, and Reynaerts (21), based on the rate of DNA reassociation in solution, allows quantitative determination of the relative similarities of DNA from selected bacteria. The progress of DNA annealing is monitored optically by measuring the decrease in absorbance at 260 nm. Optical reassociation of DNA constitutes a powerful experimental tool, both for estimating genome sizes and for determining the degree of shared nucleotide sequences in DNA samples from bacteria whose DNA can be labeled only with great difficulty (7). In addition, the total amount of time required to assess the relatedness between two DNA samples is less than that needed for isotopic binding methods and may be as short as 60 min per assay. Optical reassociation methods should have broad utility for many microorganisms.

Immobilized DNA. Systems that allowed single-stranded DNA to be immobilized and yet remain available for binding complementary polynucleotides overcame many of the disadvantages of the early free-solution systems. Possibly the most obvious advantage was that single strands of DNA in or on an insoluble matrix could not self-reassociate to form duplexed regions (5).

The first advance in techniques utilizing immobilized DNA came with the introduction by Bautz and Hall (2) of nitrocellulose columns to which glucosylated DNA was attached. Bolton and McCarthy (5) quickly realized that mechanical immobilization per se would suffice. They developed agar columns containing unlabeled, high-molecular-weight, denatured DNA trapped in the gel. Labeled, sheared DNA was denatured, added to the DNA agar, and incubated long enough to ensure maximum reaction between labeled and unlabeled DNA. After the column had been washed thoroughly to remove all unreacted labeled DNA, only the interspecific reassociated label remained. This was easily removed by increasing the temperature of the rinsing solution and by decreasing its ionic strength. Alternatively the thermal stability of the duplexes could be determined by a series of elutions at increasing temperatures. The agar–gel technique has, however, some definite limitations. The system requires relatively large amounts of unlabeled and labeled DNA. Moreover, the reassociated (unlabeled with labeled) DNA has a marked tendency to leach out of the agar. To complicate matters, temperatures in thermal elution studies with agar–gel cannot exceed 85°C because the Oxoid No. 2 agar commonly used melts at this temperature. Hoyer and Roberts (34) have reviewed the extensive work done with the agar–gel system.

The nitrocellulose filter method is similar in principle to the DNA–agar method. Denhardt (23) modified the technique of Gillespie and Spiegelman (30) whereby single-stranded, high-molecular-weight unlabeled DNA is nearly irreversibly bound to a nitrocellulose membrane filter. This is accomplished by slowly filtering dilute denatured DNA solutions dissolved in a salt solution composed of 0.9 *M* NaCl and 0.09 *M* sodium citrate through the filters. Thorough drying fixes the DNA to the filters. The exact reason why denatured DNA, but not double-stranded DNA or any kind of RNA, is bound to nitrocellulose filters is not known. By incubating the DNA-filters in a solution containing free denatured, sheared, labeled DNA, Denhardt was able to detect DNA by DNA interactions. Denhardt's method relies on a preincubation of DNA-filters in a solution (0.02% each of Ficoll, polyvinylpyrrolidone, and bovine albumin) to prevent nonspecific binding of the denatured labeled DNA. Warnaar and Cohen (79) simultaneously described a similar method which requires no albumin preincubation. Their procedure makes use of the fact that single-stranded DNA is eluted from nitrocellulose with buffers of low ionic strength and high pH (10^{-3} *M* trishydroxymethylaminomethane, pH 9.4), whereas the hybridized DNA is not. Moreover, the background levels of nonspecifically bound DNA have been reported to be lower by 1 order of magnitude than with the method of Denhardt (79).

The principles for quantification of the duplexes formed between the mobile DNA and nitrocellulose-immobilized DNA and for determination of the thermal stability of these duplexes are similar to those applying to the agar method. The methods using filter-immobilized DNA have a number of inherent disadvantages that limit their application. The kinetics of reaction of labeled DNA with fixed DNA, for example, are somewhat different from those of free solution (61). Moreover, the total reassociation of free DNA with DNA fixed to filters rarely exceeds 50% (11). It is possible that the observed binding is not representative of the entire genome. Nevertheless, the rates of reaction determined with this system do reflect the complexity of the DNA. Leaching of reassociated DNA from filters can also occur. This apparently was not a problem in the original work of Gillespie and Spiegelman (30), Denhardt (23), and Warnaar and Cohen (79). However, in some instances, leaching from the DNA-filters seriously limits the assays (67).

To overcome this undesirable elution of fixed DNA at high temperature, McConaughy, Laird, and McCarthy (62), using a method developed by Bonner, Kung, and Bekhor (6), added formamide to the incubation mixture. They found that 1% formamide reduced the optical T_m of *B. subtilis* DNA by 0.72°C. With this method, high specificity and rates of reaction were achieved, utilizing incubation temperatures of 37°C or less. The thermal elutions were complete at 40 to 50°C with no loss of fixed DNA. McConaughy and his coworkers compared the rates of reaction in free solution with the reaction rates on filters both with and without formamide and concluded that the values obtained were identical. They also showed that this system could be adapted for use with hydroxyapatite provided that the reassociating solution containing formamide was diluted with 0.12 *M* phosphate buffer so that the formamide concentration was below 1% before application to the hydroxyapatite. Legault-De'mare et al. (49) showed that the temperature of renaturation of DNA on membrane filters could be lowered if 30% (v/v) dimethyl sulfoxide (DMSO) was incorporated into the incubation solution. Rogul et al. (70) used DMSO as a solvent for the Denhardt method. The high background obtained with the original Denhardt procedure was considerably reduced, and specific binding could be obtained at lower incubation temperatures. The modified system employing DMSO, however, gave an unacceptably large experimental error.

E. Specificity of DNA Reassociation

The parameters affecting DNA reassociation have been adequately summarized by Kennell (39), McCarthy and Church (60), Brenner (8), Mandel (53), and Brenner et al. (10). Their importance cannot be overemphasized, and we therefore repeat them here.

1. The G-C base pairs exhibit greater thermal stability than the A-T base pairs; thus, if a given DNA duplex contains more G-C pairs, its thermal stability will be higher. Moreover, the sites for initiation of reassociation appear to involve sequences rich in G and C. McCarthy and Church (60) point out that initial reaction products are rich in G-C pairs.
2. The size of the DNA fragments affects DNA reassociation in free solution, larger fragments reassociating faster than smaller ones (13). Moreover, below a chain length of about 15 nucleotides (in bacteria) there is no specific duplex formation (63). Ideally, we want to compare specific DNA sequences at the exclusion of all others; however, most current methods for producing DNA sequences generate a random population of fragments of different sizes. In fact, Brenner, Fournier, and Doctor (12) note that the DNA fragments used in

their experiments sedimented as a broad band in Cs_2SO_4 and alkaline sucrose density gradients. Production of fragments is usually accomplished by mechanical shearing. The most common method involves passing DNA through a needle valve. The size of the fragments produced is governed by the pressure drop used. Brenner et al. (12) report production of fragments with an average molecular weight of 1.25×10^5 daltons by the use of a 50,000-lb/in.^{2} pressure drop. Unless modified, the ordinary French press will withstand up to about 20,000 lb/in.^{2} and will generate fragments in the range of 3 to 5×10^5 daltons. Because the rate of reassociation is inversely proportional to the viscosity of the solution (80), the reaction is affected by chain length. This effect can be controlled, however, by using uniformly sheared DNA fragments. Large fragments may produce other undesirable effects. If a particular fragment contains an internal sequence capable of forming a duplex with the other DNA species under the conditions employed, this reaction and others may be influenced by the effect of free terminal stretches of single-stranded DNA (78).

3. The most common procedure for producing a single-stranded DNA is heating aqueous solutions to 4 to 5°C above the T_m, followed by quick-cooling and increasing the salt concentration. For organisms of GC content less than 50%, this method is probably acceptable. However, for DNA samples of high GC content, high molecular weight, and high concentration, complete strand separation may not occur at 100°C even in dilute buffers. Mandel (53) states that, if separation is not complete and cross links occur, intrastrand and interstrand reassociations will decrease the number of available sites for interspecific duplex formation. Brenner et al. (10) describe a simple method for removing cross-linked and partially reassociated DNA from the fragment preparation. A hydroxyapatite column is equilibrated at a suitable temperature (ca. 30°C below the T_m) with 0.14 *M* phosphate buffer, and DNA fragments in 0.14 *M* phosphate buffer are passed through the column. Single-stranded DNA passes through the column, while cross-linked DNA is bound. Heating DNA during denaturation and during reassociation may also produce undesirable effects. Greer and Zamenhof (31) showed that DNA can be depurinated and ultimately degraded by heating at high temperatures in dilute buffers. Shapiro and Klein (73) observed that cytidine and cytosine are deaminated at 95°C in a variety of aqueous buffers. Because most DNA studies are performed in a saline–citrate buffer, Shapiro's and Klein's observation that the rate of deamination increased with increasing molarity of citric acid–citrate buffers is disturbing. A logical alternative to heat denaturation is denaturation by NaOH. Another approach could be the use of denaturants such as formamide and dimethyl sulfoxide. In any event, the choice of methods for producing and fragmenting single-stranded DNA can drastically influence DNA reassociation.

4. The rate of reassociation is highly dependent on salt concentration; moreover, the thermal stability of reassociated DNA increases as the ionic strength increases. Brenner (8) points out that one can easily shift the midpoint temperature of strand separation by 20°C or more by changing the salt concentration.

5. The optimal temperature for reassociation is about 30°C below the T_m of a given DNA (57). As inferred by Martin and Hoyer (58) and postulated by McCarthy (59), the studies of Johnson and Ordal (36) and of Brenner and Cowie (9) confirmed the existence of a class of nucleotide sequences which can reassociate at permissive incubation temperatures but cannot reassociate under more exacting incubation conditions. McCarthy (59) suggests that the duplexes formed at less exacting incubation temperatures are distantly related and can be used as a measure of evolutionary divergence. This suggestion, with its subsequent corroboration, raised the status of DNA reassociation from a laboratory curiosity to a powerful tool for discerning molecular relationships.

6. To obtain meaningful reassociation data, one must carefully choose the concentrations of labeled and unlabeled DNA. For studies in which one DNA species is immobilized in agar or on a filter, a 100:1 ratio of unlabeled to labeled DNA is sufficient to provide an excess of available sites for the labeled DNA to reassociate. Most reported experiments use incubation times varying between 15 and 20 hr. In free-solution reactions, the DNA concentrations and incubation times are especially critical. Since labeled by labeled DNA cannot be distinguished from labeled by unlabeled DNA, a large excess of unlabeled DNA is used (usually from 4000- to 8000-fold excess). This usually ensures that little reassociation occurs between two labeled DNA strands.

F. The Cot Concept

Britten and Kohne (14, 15) showed that specific hybrid formation is a function of the initial concentration of each DNA species and the time of incubation. They introduced the acronym *Cot*, derived from the product of initial concentration (c_0) and time (t). The units are generally moles of nucleotides per liter and seconds. Cot controls the reassociation of DNA when the temperature, salt concentration, and fragment size are defined. If we assume that 1 μg of DNA has an absorbance at 260 nm of 0.024, the Cot units are readily calculated, using the initial A_{260} and incubation time (13):

$$\text{Cot} = 1/2(A_{260})\ (\text{incubation time in hr}) = (\text{moles of nucleotide/liter})\ (\text{seconds})$$

It is generally assumed that renaturation of DNA follows second-order reaction kinetics because the process involves the collision of two complementary strands. A graphic representation of the relationships between the extent of reassociation and Cot allows an investigator to decide whether the reaction rates deviate significantly from second-order kinetics. For this purpose, the per cent reassociation is shown on the ordinate in arithmetic units and the Cot on the abscissa in logarithmic units. The curve generated by an uncomplicated second-order reaction (Fig. 27) is reasonably symmetrical, is sigmoid-shaped, and makes a relatively straight transition from the completely denatured state to the completely reassociated state over a 100-fold range in Cot values (15).

One of the most convenient ways to measure per cent reassociation is as absorbance at 260 nm. Because dissociated DNA absorbs more ultraviolet light than reassociated DNA does, simply following the decrease in absorbance with time will provide the necessary information on degree of reassociation. Another useful method for determining per cent reassociation is by the use of hydroxyapatite. Under proper ionic conditions duplex DNA can be separated from single-stranded DNA when passed over hydroxyapatite. By measuring the amount of double-stranded DNA binding to the gel, one can easily calculate the per cent reassociation. The value of hydroxyapatite lies in the fact that DNA sequences can be fractionated on a preparative scale. Using this method, Kohne (46) isolated ribosomal cistrons from *E. coli* and *Proteus mirabilis*. Likewise, Brenner et al. (12) isolated and characterized tRNA cistrons from *E. coli*.

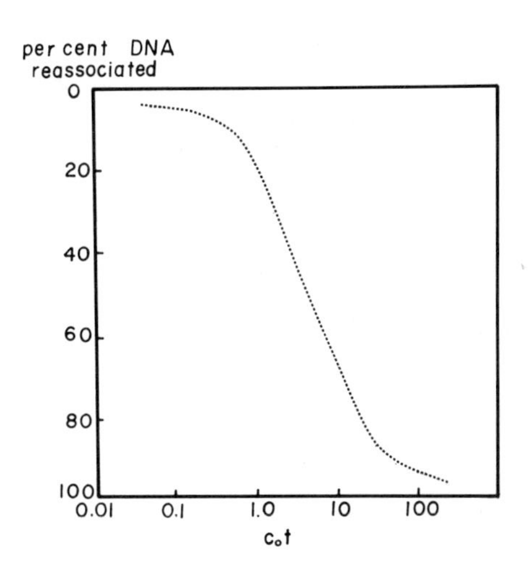

Fig. 27. Idealized time course for DNA reassociation.

Additional information about genome structure can be inferred from the time course of DNA reassociation. A useful point is the Cot value at which half of the initially denatured DNA has reassociated. This point has been designated as $Cot_{1/2}$ or Cot/2. Some of the implications of Cot/2 can best be symbolized mathematically.

The rate of disappearance of denatured DNA should be

$$-\frac{dc}{dt} = kc^2$$

Where c = concentration of denatured DNA
t = time of renaturation

By integrating and evaluating over t = 0 ($c = c_0$), we obtain—

$$\frac{c}{c_0} = \frac{1}{1 + k(c_0t)}$$

When the DNA is half-renatured

$$\frac{c}{c_0} = \frac{1}{2} = \frac{1}{1 + k(c_0t)}$$

Solving—cot = 1/k at half-renaturation, or Cot = 1.

It is important to realize that the DNA of each organism may be characterized by the value of Cot/2. Since *k* is inversely proportional to the complexity of the DNA, Cot/2 is directly proportional to the genome size. Cairn's measurement of the size of the *E. coli* genome (4.5×10^6 nucleotide pairs) is frequently used as a reference value. Thus, if an organism's DNA has a Cot/2 twice that of *E. coli* DNA, the organism has a genome size of 9×10^6 nucleotide pairs. Britten and Kohne (15) point out that the linear relationship between Cot/2 and genome size is true only in the absence of repeated sequences. These workers also emphasize that the Cot/2 value measured by optical methods is different from that measured in hydroxyapatite. In fact, the latter method gives a Cot/2 of about 50% of the Cot/2 determined optically. This is to be expected because the fraction of fragments reassociated is measured by hydroxyapatite, whereas the fraction of total strand length reassociated is determined optically. Nevertheles. by using standards of known genome size to calibrate each system, excellent agreement between the results obtained by the two methods is obtained.

G. Quantitative Relatedness and Evolutionary Divergence

In order to make taxonomic inferences based on nucleic acid reassociation data, the number of DNA sequences held in common between a test DNA and a reference DNA must be measured. The number of common sequences becomes a quantitative index of relatedness. At first, it seemed that "one number" would define the absolute relatedness of one organism to another. However, with the recognition that DNA samples from related organisms often contain identical sequences, a spectrum of partially matched sequences, and totally dissimilar sequences, the inadequacy of the one-number concept became obvious (8). Although somewhat of a dilemma at first, analysis of this spectrum of sequence matching provides a basis for deducing relationships among organisms.

The precision of base pairing, as well as the extent of reassociation between single-stranded DNA preparations from different species, can be measured experimentally. In one method used extensively, reassociation is allowed to proceed at two different incubation temperatures. At the higher temperature, only well-matched sequences should form duplexes. At the lower temperature, partially matched and exactly matched sequences should form duplexes. Many observations corroborate these general principles and premises, but selection of the proper experimental conditions is difficult.

As a general rule, the extent of reassociation reaches a maximum about 30°C below the T_m of the native DNA. As the incubation temperature is increased, the amount of duplex formed between partially matched nucleotide sequences decreases. At a reaction temperature 15°C below the T_m of the native reference DNA, duplex formation between all but well-matched sequences is precluded. These assumptions are consistent with results of measurements of the thermal stabilities of DNA duplexes formed at various incubation temperatures. In general, the thermal stability of these duplexes is increased, although the absolute amount of binding is decreased (9, 36). In our laboratory, we call incubation temperatures that allow distantly related sequences to react *nonexacting*. This is in contrast to *exacting* incubation temperatures, which allow only closely related sequences to react. We assume that the reduced thermal stability of the "distantly related" sequences reflects the proportion of unpaired bases within the interspecific duplex. It is also possible that preferential binding of AT-rich sequences at a lower reassociation temperature may contribute to the reduced thermal stability of duplexes formed under nonexacting conditions. Brenner (8), Brenner and Cowie (9), and Kingsbury et al. (41) however, have presented evidence that most instability observed in enterobacterial DNA duplexes is due to the presence of unpaired bases. This assumption leads to some interesting inferences. The thermal stability of reassociated duplex is characterized by its $T_{m,e}$ (elution temperature at which 50% of the DNA duplexes have been disassociated). The difference between the $T_{m,e}$ value of an interspecific duplex and that of the homologous reference reaction is designated as the $\Delta T_{m,e}$ value. There appears to be a direct correlation between $\Delta T_{m,e}$ and the percentage of unpaired bases in an interspecific duplex.

Laird, McConaughy, and McCarthy (48) reviewed studies with artificial polymers and presented data based on natural DNA polymers suggesting that a 1.5°C decrease in thermal stability results from 1% of unpaired bases within a DNA duplex. This correlation constitutes a bridge between quantitative reassociation and nucleotide divergence. Brenner (8) attempted to categorize enterobacterial relationships from this standpoint. His approach was to multiply the ΔT_{me} by 1.5 to obtain the fraction of unpaired bases within duplexes. This value is called the per cent divergence. It should be noted that this is divergence with respect to the reference DNA (*E. coli* K12).

When surveying the DNA from a large group of organisms, it is time consuming to determine the thermal stability of each reaction at a number of incubation temperatures. In our laboratory we find it convenient to use initially a graphical approach to indicate relationships between the test organisms and the reference organisms. The data are obtained by using two incubation temperatures, exacting and nonexacting. Usually the temperatures are chosen by setting the nonexacting temperature at 25 to 30°C less than the T_m of the reference DNA, and the exacting temperature 10 to 15°C less than the T_m of the reference DNA. The per cent reassociation relative to the homologous reaction is determined for each incubation temperature. By dividing the amount of relative binding at exacting conditions by that at nonexacting conditions, a useful ratio is obtained. This ratio has been called the divergence index (DI) by Enquist and Bradley (24) and the thermal binding index (TBI) by Brenner (8)and Brenner et al. (10). Divergence index values are useful in gauging the presence or absence of closely related genetic material. A number close to 1.00 indicates that all of the sequences that bind the reference DNA are almost identical to it, whereas a value approaching 0.00 indicates that the test DNA shares almost no regions of similarity with the reference DNA. Although not rigorously proved, there appears to be a direct correlation between $\Delta T_{m,e}$ and DI; the lower the DI, the greater the $\Delta T_{m,e}$ (8).

Divergence index values can be interpreted at the molecular level in terms of the distribution of nucleotide divergence (Fig. 28). We refer to the nucleotide divergence occurring more or less randomly throughout the genome as *dispersed divergence*. This is in contrast to *localized divergence* or *localized conservation*, where changes occur in specific regions only (24).

The duplexes formed during nonexacting conditions are most easily thought of as being composed of both incompletely matched and closely matched sequences. Under the more exacting incubation conditions, the duplexes formed are those of closely matched sequences.

Symbolically, let a = per cent of incompletely matched duplexes, b = per cent of closely matched duplexes, and c = per cent of unreacting DNA sequences. Then the nonexacting incubation gives $a + b$, while the exacting incubation gives b.

$$\text{The divergence index (DI)} = \frac{b}{a + b}$$

The values of $a + b$ and b are subject to the following limitations:

$$100\% > a + b \geq 0, \qquad b \ngtr a + b$$

The total number of sequences available to react is defined as

$$a + b + c = 100\%$$

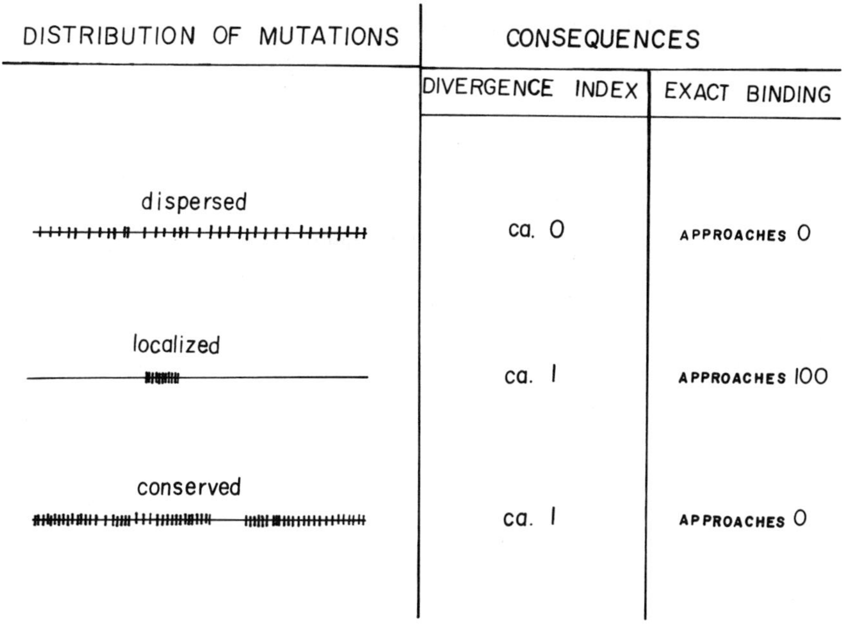

DISTRIBUTION OF MUTATIONS	CONSEQUENCES	
	DIVERGENCE INDEX	EXACT BINDING
dispersed	ca. 0	APPROACHES 0
localized	ca. 1	APPROACHES 100
conserved	ca. 1	APPROACHES 0

Fig. 28. Distribution of mutations and the effects of different patterns of nucleotide divergence on reassociation assays.

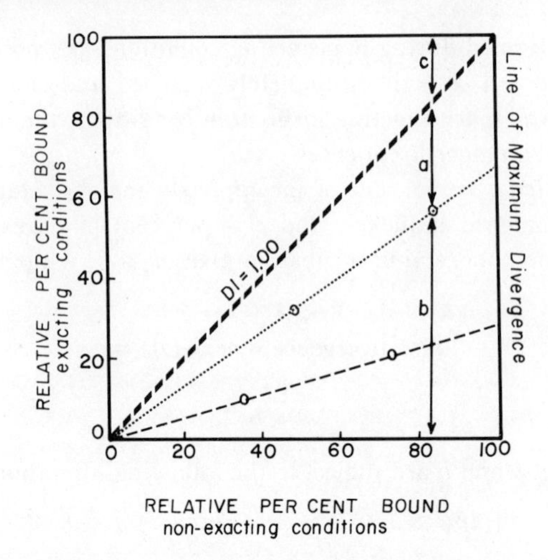

Fig. 29. Determination of divergence patterns: a = per cent of incompletely matched duplexes; b = per cent of well-matched duplexes; c = per cent of unreacting DNA sequences. The divergence index (DI) is defined as $b/(a + b)$. All points on a line drawn through the origin have the same DI. The line for DI = 1.00 divides the graph into an upper half containing values of c and a lower half containing values of a and b. Points near the line maximum divergence indicate that the respective DNA preparations have undergone extensive dispersed divergence.

When binding at exacting conditions *(b)* is plotted against binding at nonexacting conditions $(a + b)$, we obtain a graphical presentation of the data relative to the reference DNA (Fig. 29). The diagonal (45°) line separates the graph into two parts: the upper portion contains values of c (nonrelated DNA). These sequences have diverged to the point that, even at nonexacting conditions, they do not react. The DI for a given test DNA with respect to the reference DNA is the slope of the line drawn from its characteristic point through the origin. Any point on a given line has the same DI. The vertical distance from a point to the 45° diagonal gives the percentage of distantly related sequences, while the vertical distance from the diagonal to 100% above the point gives the percentage of unrelated sequences. Insight into the mechanisms of evolutionary divergence can be gained by considering the nature of nucleotide reassociation between DNA preparations containing different degrees of partially matched pairing but essentially no totally unmatched sequences, that is, as $c \rightarrow 0$, $a + b \rightarrow 100$.

Graphically maximum divergence is indicated by the vertical line from the 100% point on the abscissa, that is, the upward projection from the point of 100% binding under nonexacting conditions. Any point on or near this line of *maximum divergence* indicates that sequences in the test DNA have diverged from the reference DNA to the extent that any more changes in these nucleotide sequences will result in loss of their ability to reassociate to a detectable degree. Thus organisms whose DNA has undergone essentially all dispersed divergence with respect to the reference DNA fall on or near the line of maximum divergence; DNA samples exhibiting localized divergence or sequence conservation are found near the 45° line. A large number of organisms can be surveyed for biological relatedness by collecting and analyzing data according to these guidelines. In our hands, the graphical interpretations have been corroborated by subsequent $\Delta T_{m,e}$ determinations.

It is imperative that the limitations placed upon $a + b$ and b be observed. Misleading data and erroneous interpretations arise when inappropriate *exacting* and *nonexacting* conditions are selected. In addition, spurious data are generated if the sheared DNA fragments are too large. For a survey to have taxonomic significance, many reference cultures must be studied, including the type cultures of the species and genera involved.

H. Concluding Remarks

Although it is experimentally possible to estimate the degree of divergence of DNA sequences with respect to a given reference DNA, a major factor is missing from the analysis. At present, we have no reference point in evolutionary time for microbial divergence. In fact, we cannot readily distinguish recent gene transfer from conserved ancestral sequences. Jones and Sneath (37) suggest that bacterial evolution could follow reticulate modes of change with numerous partial fusions of phyletic lines. This concept presents severe conceptual and practical problems, however, for analysis of DNA reassociation data. On the other hand, Mandel (53) suggests that such reticulation of diverging lines is a minor problem because of the relative rarity of genetic interaction in nature. This issue still remains to be resolved.

From a deterministic point of view, no formal guidelines exist by which to correlate reassociation data with taxonomic groupings. It does seem clear that, when the DNAs from two organisms cannot reassociate, these organisms are not presently related. Brenner (8) suggests that, when the extent of interspecific DNA reassociation is virtually 100% and the thermal stability of the duplexes is identical to that of the reference reaction, the organisms in question are members of the same species. In this regard, a unit of classification applicable to DNA reassociation data is the "genospecies" (69). This concept involves a group of strains potentially able to contribute to or share in a common gene pool. Phenotypically, this results in a cluster of satellite strains around a central core. Probably even more useful in the present framework of taxonomy would be recognition of the type culture concept in the arbitrary definition of species. If this were done, reassociation data using the type cultures as reference DNA preparations would establish meaningful and comparable relationships (65).

REFERENCES

1. Baltimore, D., *Trans. N. Y. Acad. Sci.* **33,** 327 (1971).
2. Bautz, E. K. F., and B. D. Hall, *Proc. Natl. Acad. Sci. U.S.* **48,** 400 (1962).
3. Bendich, A., in *Methods in Enzymology*, S. P. Colowick and N. O. Kaplan (eds.), Academic Press, New York, 1957.
4. Bernardi, G., *Nature* **206,** 779 (1965).
5. Bolton, E. T., and B. J. McCarthy, *Proc. Natl. Acad. Sci. U.S.* **48,** 1390 (1962).
6. Bonner, J., G. Kung, and J. Bekhor, *Biochemistry* **6,** 3650 (1967).
7. Bradley, S. G., *Amer. Rev. Resp. Dis.* **106,** 122 (1972).
8. Brenner, D. J., *Develop. Ind. Microbiol.* **11,** 139 (1970).
9. Brenner, D. J., and D. B. Cowie, *J. Bacteriol.* **95,** 2258 (1968).
10. Brenner, D. J., G. R. Fanning, K. E. Johnson, R. V. Citarella, and S. Falkow, *J. Bacteriol.* **98,** 637 (1969).
11. Brenner, D. J., G. R. Fanning, A. Rake, and K. E. Johnson, *Anal. Biochem.* **28,** 447 (1969).
12. Brenner, D. J., M. J. Fournier, and B. P. Doctor, *Nature* **227,** 448 (1970).
13. Britten, R. J., and D. E. Kohne, *Carnegie Inst. Yearbook* **65,** 78 (1966).
14. Britten, R. J., and D. E. Kohne, *Carnegie Inst. Yearbook* **66,** 73 (1967).
15. Britten, R. J., and D. E. Kohne, *Science* **161,** 529 (1968).
16. Britten, R. J., M. Pavich, and J. Smith, *Carnegie Inst. Yearbook* **68,** 400 (1970).
17. Cairns, J., *Cold Spring Harbor Symp. Quant. Biol.* **28,** 43 (1963).
18. Clark, A. J., *Ann. Rev. Microbiol.* **25,** 437 (1971).
19. DeLey, J., *J. Theor. Biol.* **22,** 89 (1969).
20. DeLey, J., *J. Bacteriol.* **101,** 738 (1970).
21. DeLey, J., H. Cattoir, and A. Reynaerts, *Eur. J. Biochem.* **12,** 133 (1970).
22. DeLucia, P., and J. Cairns, *Nature* **224,** 1164 (1969).
23. Denhardt, D. T., *Biophys. Biochem. Res. Commun.* **23,** 641 (1966).
24. Enquist, L. W., and S. G. Bradley, *Advan. Frontiers Plant Sci.* **25,** 53 (1970).
25. Flamm, W. G., H. E. Bond, and H. E. Burr, *Biochim. Biophys. Acta* **129,** 310 (1966).
26. Fredericq, E., A. Oth, and F. Fontaine, *J. Mol. Biol.* **3,** 11 (1961).
27. Garen, A., *Science* **160,** 149 (1968).
28. Gasser, F., and M. Mandel, *J. Bacteriol.* **96,** 580 (1968).
29. Gehrke, C. W., D. L. Stalling, and C. D. Ruyle, *Biochem. Biophys. Res. Commun.* **28,** 869 (1967).

30. Gillespie, D., and S. Spiegelman, *J. Mol. Biol.* **12,** 829 (1965).
31. Greer, S., and S. Zamenhof, *J. Mol. Biol.* **4,** 123 (1962).
32. Gross, J. D., *Curr. Top. Microbiol. Immunol.* **57,** 39 (1972).
33. Hill, L. R., *J. Gen. Microbiol.* **44,** 419 (1966).
34. Hoyer, B. H., and R. B. Roberts, in *Molecular Genetics*, Part II, J. H. Taylor (ed.), Academic Press, New York, 1967.
35. Huang, P. C., and E. Rosenberg, *Anal. Biochem.* **16,** 107 (1966).
36. Johnson, J. L., and E. J. Ordal, *J. Bacteriol.* **95,** 893 (1968).
37. Jones, D., and P. H. A. Sneath, *Bacteriol. Rev.* **34,** 40 (1970).
38. Jost, J.-P., and H. V. Rickenberg, *Ann. Rev. Biochem.* **40,** 741 (1971).
39. Kennell, D. E., *Prog. Nucleic Acid Res. Mol. Biol.* **11,** 259 (1971).
40. Kersten, W., H. Kersten, and W. Szybalski, *Biochemistry* **5,** 236 (1966).
41. Kingsbury, D. T., G. R. Fanning, K. E. Johnson, and D. J. Brenner, *J. Gen. Microbiol.* **55,** 201 (1969).
42. Kirby, K. S., *Biochem. J.* **66,** 495 (1957).
43. Kirby, K. S., *Prog. Nucleic Acid Res.* **3,** 1 (1964).
44. Kit, S., *Ann. Rev. Biochem.* **32,** 43 (1963).
45. Klein, A., and F. Bonhoeffer, *Ann. Rev. Biochem.* **41,** 301 (1972).
46. Kohne, D. E., *Biophys. J.* **8,** 1104 (1968).
47. Kornberg, A., *Science* **163,** 1410 (1969).
48. Laird, C. D., B. L. McConaughy, and B. J. McCarthy, *Nature* **224,** 149 (1969).
49. Legault-De'mare, J., D. Desseaus, T. Heyman, S. Seror, and G. P. Ress, *Biochem. Biophys. Res. Commun.* **28,** 550 (1967).
50. Lengyel, P., *Cold Spring Harbor Symp. Quant. Biol.* **34,** 828 (1968).
51. Lengyel, P., and D. Söll, *Bacteriol. Rev.* **33,** 264 (1969).
52. Losick, R., *Ann. Rev. Biochem.* **41,** 409 (1972).
53. Mandel, M., *Ann. Rev. Microbiol.* **23,** 239 (1969).
54. Mandel, M., L. Igambi, J. Bergendahl, M. L. Dodson, Jr., and E. Scheltgen, *J. Bacteriol.* **101,** 333 (1970).
55. Marmur, J., *J. Mol. Biol.* **3,** 208 (1961).
56. Marmur, J., and P. Doty, *J. Mol. Biol.* **5,** 109 (1962).
57. Marmur, J., R. Rownd, and C. L. Schildkraut, *Prog. Nucleic Acid Res.* **1,** 231 (1963).
58. Martin, M. A., and B. H. Hoyer, *Biochemistry* **5,** 2706 (1966).
59. McCarthy, B. J., *Bacteriol. Rev.* **31,** 215 (1967).
60. McCarthy, B. J., and R. B. Church, *Ann. Rev. Biochem.* **39,** 131 (1970).
61. McCarthy, B. J., and B. L. McConaughy, *Biochem. Genet.* **2,** 37 (1968).
62. McConaughy, B. L., C. D. Laird, and B. J. McCarthy, *Biochemistry* **8,** 3289 (1969).
63. McConaughy, B. L., and B. J. McCarthy, *Biochim. Biophys. Acta* **149,** 180 (1967).
64. Miyazawa, Y., and C. A. Thomas, *J. Mol. Biol.* **11,** 223 (1965).
65. Monson, A. M., S. G. Bradley, L. W. Enquist, and G. Cruces, *J. Bacteriol.* **99,** 702 (1969).

66. Nomura, M., *Bacteriol. Rev.* **34,** 228 (1970).
67. Okanishi, M., and K. F. Gregory, *J. Bacteriol.* **104,** 1086 (1970).
68. Pastan, I., and R. Perlman, *Science* **169,** 339 (1970).
69. Ravin, A. W., *Amer. Natur.* **97,** 307 (1963).
70. Rogul, M., J. J. Brendle, D. K. Haapala, and A. D. Alexander, *J. Bacteriol.* **101,** 827 (1970).
71. Rolfe, R., *Proc. Natl. Acad. Sci. U.S.* **49,** 386 (1963).
72. Schildkraut, C. L., J. Marmur, and P. Doty, *J. Mol. Biol.* **4,** 430 (1962).
73. Shapiro, R., and R. S. Klein, *Biochemistry* **5,** 2358 (1966).
74. Skinner, D. M., and L. L. Triplett, *Biochem. Biophys. Res. Commun.* **28,** 892 (1967).
75. Stavis, R. L., and J. T. August, *Ann. Rev. Biochem.* **39,** 527 (1970).
76. Storck, R., and C. J. Alexopoulos, *Bacteriol. Rev.* **34,** 126 (1970).
77. Szybalski, W., *Fractions,* **1,** 1 (1968).
78. Walker, P. M. B., *Prog. Nucleic Acid Res. Mol. Biol.* **9,** 301 (1969).
79. Warnaar, S. O., and J. A. Cohen, *Biophys. Biochem. Res. Commun.* **24,** 554 (1966).
80. Wetmur, J. G., and N. Davidson, *J. Mol. Biol.* **31,** 349 (1968).
81. Jacobson, M. and K. Lark, *J. Mol. Biol.* **73,** 371 (1973).
82. Okazaki, R., T. Okazaki, K. Sakabe, K. Sugimoto, and A. Sugino, *Proc. Natl. Acad. Sci. U.S.* **59,** 598 (1968).
83. Sugino, A. and R. Okazaki, *Proc. Natl. Acad. Sci. U.S.* **70,** 88 (1973).

CHAPTER

FOUR

Microbial Proteins

F. J. REITHEL

1. INTRODUCTION

The spectrum of activity exhibited by the prokaryotic cell is essentially restricted to the mechanics of multiplication and its ancillary processes. There is no widespread ability to transduce one type of energy into another, and mechanical work is rarely observed, but there is an astonishing versatility for chemical interconversions. One view of this phenomenon is that evolutionary processes have resulted in every possible variety of environmental exploitation and that the hallmark of the enduring pattern called "life" is the synthesis of protein. In order that the pattern endure, there must be a component devoted to control as well as a mechanism for decreasing entropy. Whereas the nucleic acids supply a plan, a code, a message, it is the proteins that have unparalleled potentialities for catalysis and control. In this respect there is a unity in all living things.

At this time there is no justification for the view that protein biosynthesis in all organisms is nondivergent. There may be more than one mode of polypeptide synthesis or many modifications. It cannot be maintained that the control of protein synthesis in bacterial cells is identical to that in more differentiated organisms. Since prokaryotic cells do not possess the subcellular structures characteristic of the eukaryotic cells, a different control system is to be expected. Obviously the spatial segregation of functions in the two types of cells differs, and different modes of interaction should be anticipated. The rather surprising observation is that the molecular architecture of proteins is so similar in all cells. Small cells do not contain small proteins, nor are large proteins restricted to large cells. Indeed, the concept of the cell has become less useful and almost an impediment to the view of living structures at a molecular level. It has now been shown that mitochondria possess autonomy, just as virus DNA has an autonomy. The cell need no longer be considered the unit in biology but may be regarded as only one stage in a complicated pattern of transformations.

One of the measures of life is the protein pattern—the asymmetries, the sizes, the lengths of chains permitted by the linear translations of the nucleic acids. This pattern allows for the production of proteins with molecular weights, or aggregate weights, up to several million, and such proteins exist in bacteria.

2. THE PROTEIN DOMAIN

Fresh views and new concepts are often obscured by outmoded nomenclature and attendant semantic distractions. The term "protein molecule," as an example, is usually freighted with a complex of concepts directly derived from small-molecule chemistry. For the biologist it is probably more appropriate to think of proteins as microsystems rather than "molecules." Protein size can be conceived of more aptly as a uniquely monodisperse domain containing not only more than one molecule but more than one kind of molecule. Protein weight might better be expressed in daltons than by the term "molecular weight." As an example, the multimeric* proteins that possess glutamate dehydrogenase activity may exist in a monodisperse system comprising several polypeptide chains, a variety of nucleotides, and a substantial number of ions, as well as appreciable tightly bonded water. This complex participates in several equilibria and may respond to changes in the environment in a variety of ways. To speak of such a complex as a "molecule" subverts the essence of the word–idea relationship.

3. LEVELS AND ORIGINS OF ORDER IN PROTEINS

Despite the complexity that must be recognized it is possible to discern inherent limitations on variability. At present there is no evidence that the process of assembling a linear polypeptide polymerizes any molecules except amino acids. Hence there is general agreement with Anfinsen's proposition (3) that the linear polypeptides characteristic of a protein have a preferred three-dimensional ordering. Indeed, sufficient detailed knowledge has been accumulated in regard to bond angles and lengths so that it is possible to make reasonable predictions about the three-dimensional forms of a linear polypeptide of a given sequence (48, 13). Peptide bond sequences must show a definite relation expressable in two dihedral angles (Fig. 1). Peptide bonds are planar ($\omega = 0°$) or show slight angular twist. The relation of the two planes is a function of the two angles ψ and ϕ, which can have only certain values because of steric restrictions. Thus knowlcdgc of the range of permissible values of these dihedral angles allows prediction of the conformation of polypeptide sequences. We are thus on the threshold of an active period in which we will see speculative prediction of secondary and tertiary structure in linear polypeptides. The high probability of certain interactions has been confirmed by the interpretation of X-ray diffraction data.

Although it seems probable that current working hypotheses will result in a number of satisfying predictions, it is equally likely that surprising subtleties will be encountered. As an example, *Neurospora* glutamate dehydrogenase seems to exist as two species, having polypeptide chains that appear to be chemically identical and to differ only in conformation or details of tertiary structure (52). The two forms can be distinguished because they possess unlike antigenic sites. Both are catalytically active (see Section 9). At present these experimental observations run counter to prediction and suggest that tertiary folding patterns constitute a small group of structures rather than a unique structure.

* For nomenclature see F. J. Reithel, *Advan. Protein Chem.* **18,** 123 (1963).

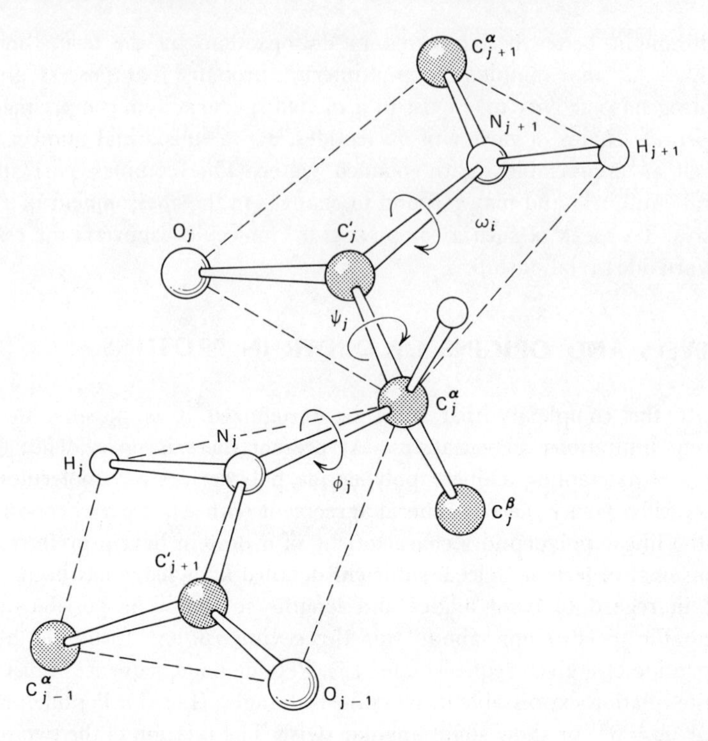

Fig. 1 Dipeptide assembly with dihedral angles phi and psi equal to 0° as shown. Peptide bond is planar since omega is usually 0°. Other values of ϕ and ψ result in a change of orientation of the peptide bond planes.

The next higher level of order has a somewhat different character. This level of order is imposed by the probability of association between polypeptide conformations. This probability is a rather sensitive function of the environment and is a kinetic relationship. Often the term "quaternary structure" is applied to this metastable association, but it is preferable to speak of the "state of association." It seems misleading to emphasize structure, which is, in concept, static rather than a process that preserves form. Emphasis on static structure tends to call attention to the symmetries that prevail, whereas emphasis on kinetic function recognizes the asymmetries that accompany function. Unlike the former intrachain relations, this level of order involves interchain relations. Because association and dissociation are equilibrium processes, it follows that at high dilution all proteins, in principle, dissociate. In practice, however, this dissociation is not always realized, particularly if the experimental conditions potentiate unfavorable conformational changes.

As the three-dimensional intrachain relationships of myoglobin became more and more certainly known from the X-ray diffraction work of Kendrew and his group, a generalization arose that has gained wide currency. This generalization is that most of the ionizable groups on the side chains are disposed outward and toward the solvent, and at the same time the hydrophobic residues form an inner protein domain virtually free of solvent. This is a useful view and emphasizes that proteins in the crystalline state enjoy conformations with a maximum of group interaction, resulting in stabilization. But this must be considered a norm from which there are frequent (and often marked) departures. Indeed, if proteins are to respond to the environment, if proteins are domains consisting of components other than polypeptide chains, and if one is permitted to envisage the changes termed "allosteric" by Monod, Changeaux, and Wyman (36), then conformational changes must occur. Thus X-ray diffraction interpretations may indicate the most probable conformation, but many others, almost as likely, must exist. Such conformational variations may or may not perturb interchain interactions. But it may be assumed that the protein domain is altered by changes in the concentration of any component and by the absorption of energy. This variation accompanying the preservation of the essence of form characterizes the living form.

The correlation of salting-out and salting-in phenomena with the interpretation of some reagent effects has been achieved by considering the free energy of transfer of amino acid residues into hydrophobic areas, as contrasted to aqueous solvent (64). It may be assumed that even the tertiary structure of a polypeptide is a steady state in which a few molecules in an assembly are somewhat unfolded. In the absence of any molecules that can bind specifically, these partially unfolded linear arrays assume the most stable folded form. However, if solutes are present that preferentially bind to the transiently exposed hydrophobic residues of the amino acids, the entire balance of forces is altered and a change of conformation is effected. In addition, some solutes may alter the solvent structure around residues that are exposed, again perturbing the fine thermodynamic balance so that residues formerly oriented toward the interior may now be oriented outward. Thus ions that increase the free energy of transfer of residues from the interior milieu to the solvent also increase the chemical potential of the exterior groups. Such ions are good salting-out agents, since they promote interaction of polypeptides with each other rather than the solvent. Alternatively, substances that bind preferentially to hydrophobic areas obviously change the balance of forces so that tertiary folding is altered and association–dissociation relations are changed accordingly. Although we do not possess enough detailed information to allow prediction of behavior as yet, the above discussion indicates how results may be rationalized.

It is pertinent here to discuss two items in response to the following statement (65): "The important point about the point groups is that they lead to structures of finite extent, that is, they are the closed type of design that would be used for an aggregate of well-determined size, such as an oligomeric enzyme. In such a structure the design is necessarily one in which the specific bonding potential of the parts is saturated in some closed configuration. That is, a fixed number of molecules come together, and no more can be added using the same kind of bonds." This presents a view of aggregation that does not account for the commonly observed phenomenon of protein polymerization, and does not take into consideration the full consequences of stepwise association.

The term "polymerization" refers to the production of protein species that seem to form from j units of the preparative weight, where $j = 2$ to 10. Suppose, as in the case of β-galactosidase, that the preparative species is about 500,000 daltons and can be shown to dissociate easily (as we have shown) in low ionic strength solutions to a species of 125,000 daltons. The preparative species can be denoted as $(4n)$ and the dissociation as $(4n) \rightarrow 4(n)$. If preparations of β-galactosidase are allowed to remain in storage at high protein concentration, higher aggregates form. These aggregates are $(8n)$, $(12n)$, $(16n)$, and larger. In this case the enzyme activity of these great aggregates is not far different from that of the $(4n)$ species. Generally it is considered that such aggregation results from disulfide bond formation after air oxidation, but there is reason to believe that this is not an accurate representation. Our experience indicates that aggregation occurs and that disulfide cross linking may or may not follow. This kind of protein polymerization is not unusual. It has been remarked in many proteins, but the process has not been systematically studied.

For proteins that normally exist as dimers, trimers, tetramers, or hexamers, relatively simple association mechanisms can be imagined. Such small oligomers may be stabilized by only one or two bonding modes, and the symmetry of the product precludes further subunit additions. In contrast to this class of proteins there is another that seems to form stable aggregates of 8 or 12 or 16 structural subunits. Studies of models and symmetries suggest that the structural subunits must have several specific intersubunit bonding modes. It is likely that assembly occurs stepwise, probably with conformational change at each step, and results in a product that is unsaturated in its intersubunit bonding sites. These may well be the groups involved in the polymerizations just referred to. As Bernal (6) has averred, symmetry is a product of rates of growth rather than the shapes of the identical aggregating units.

4. INTERACTION OF PROTEINS WITH OTHER CHEMICAL SPECIES

It was noted above that there exists no evidence for the incorporation of any components into a polypeptide chain except amino acids. And yet a high percentage of the proteins occurring in eukaryotic cells or organisms has been found to contain carbohydrate. It has been demonstrated in the recent past that the addition of covalently bonded carbohydrate is a postribosomal event and is associated with the smooth endoplasmic reticulum. It has been suggested, but by no means proved, that the addition of carbohydrate is related to the process of protein transfer out of the cell. So far very few bacterial proteins have been characterized as glycoproteins. They have been recovered from yeasts and molds, but it may be that the addition of carbohydrate is not a common feature of protein synthesis in bacteria. This is one of the little mysteries that await an early solution.

"Protein," then, can be considered as a pattern of interactions characteristic of linear polypeptides ordinarily containing between 100 and 1000 amino acid residues. (Myoglobin, 17,000 daltons, consists of 153 amino acid residues.) Given the rotational restrictions of the dihedral angles, ψ and ϕ, the possibilities of ionic and hydrogen bonding, and the interaction with the solvent that results in hydrophobic bonding, a compromise is reached that achieves a minimum energy state for the system. In the majority of proteins that have been investigated successfully, the pattern reflects a high intrachain interaction with a high degree of solvent exclusion. However, the exposed areas still contain many groups that can form reasonably stable complexes with similar groups on other structured polypeptides. This is often referred to (rather inaccurately) as "hydrophobic interaction between subunits." In fact, it would be much better to speak of "protomeric association." If small molecules bind to an amino acid residue that normally would associate protomers in the absence of a competitor, we will observe one variety of allosteric effect accompanied by dissociation.

Massive interactions of polypeptide chains with neutral carbohydrates are uncommon. This is not surprising since carbohydrates are saturated with hydrogen-bonded water and possess relatively few structures that would interact readily with predominantly hydrocarbon areas. Thus proteins having a carbohydrate content will be found to have covalently bonded chains. Interaction with lipid is obviously another matter. There seems no reason to doubt that lipids offer manifold modes of association with the so-called hydrophobic areas of polypeptide structures. Our notions about lipoproteins, and membranes that are largely lipoprotein, are vague or controversial.

Proteins react with nucleotides in a variety of ways, including covalent bonding. Since so many nucleotides are coenzymes or prosthetic groups, it is clear that sterospecific binding patterns are common. However, when nucleotides are polymerized into nucleic acids, the possibilities of binding are strongly limited. The Watson–Crick helix is a highly charged and relatively rigid structure.

Yet another kind of stereospecificity involves specific charge patterns such as those observed in hemagglutination. The sialic acid residues of the linear oligosaccharide side chains forming part of the erythrocyte surface presumably form charge patterns. Viral proteins, among others, bind specifically to these receptor mosaics and cause agglutinated aggregates of cells to form. These mosaic patterns are occasionally found in other carbohydrate materials. The specificity is underlined by the observation that antiserum is also bound readily.

Again in the antigen–antibody reaction we may observe an exquisite specificity that can characterize the interaction between polypeptide conformations and a variety of polymeric substances. Under special conditions a protein may arise, an antibody that expresses the ultimate in the potential for binding specificity.

5. REFOLDING AND REASSEMBLY OF PROTEIN STRUCTURES

The experimentation in Anfinsen's laboratory about 10 years ago stimulated a wide variety of refolding, renaturation, and reconstitution experiments. Proteins, both simple and complex, have been transformed into random linear polypeptides and then demonstrated to reform structures similar to those of the initial state. Step by step, larger and larger structures have been shown to reassemble (28), such as ribosomes, complete phage structures, and recently whole cells (24). We are currently optimistic about the prospect of demonstrating structural subunits derived from all kinds of biological assemblies. Such subunits may not be particularly simple chemically but will represent the unit elements of biological form. Of course there will be some structures, such as cell walls, that are cross linked into a continuum, but even here we may hope to find an element of structure. There is probably no one who has written more incisively on this subject than the late Professor Bernal, who provocatively suggested a whole new generalized crystallography (5). Classical crystallography derives from a study of the close packing patterns of identical units. As these relations evolved, it was found that all crystals could be classified into 32 classes involving symmetry rotations of 2, 3, 4, and 6. Indeed a review of the types of protein association that have been carefully characterized reveals, correspondingly, dimers, some trimers, tetramers, and hexamers. One of the earlier association models proposed (19) involved just such symmetries. But one sees about him in a variety of natural forms structures of 5-, 7-, and 9-fold symmetry. Indeed, even nonintegral rotations became respectable after Pauling developed the serious consideration of the α helix with 3.6 residues per turn.

It has been emphasized that symmetrical structures can evolve from quasi-identical structures with a wide choice of symmetries and that often topological rather than crystallographic principles are evident in biological structures. Examples are the open cylinder and the closed sphere, both of them capable of being two-dimensional crystals of quasi-identical units (65). Both of these structures are found among the viruses. But earlier we considered thermodynamic stability. Two-dimensional crystals do not have thermodynamic stability, but three-dimensional arrays are minimum-energy structures. As far as this kind of symmetry and order is concerned, we can think of proteins as kinetic structures formed from structural subunits. These structural subunits, because they can exist in several conformations of similar energy, must be considered as quasi-identical units with characteristic patterns of bonding potentialities. Bernal has suggested considering globular protein subunits as polyhedra whose faces have rather different electronic states. This reduces the study of association–dissociation of proteins to two main topics: chemical topology and equilibrium.

But, in fact, many association experiments are orderly, just as many self-assembly experiments are successful, because only limited possibilities of interaction are present. For example, completely unfolded and randomized linear polypeptides do not reassociate into neat three-dimensional arrays unless the process is slowed so that stepwise association can occur, the groups with the largest binding constants reacting first. With a careful selection of such experimental conditions, all kinds of parasitic side reactions occur and an intractable, insoluble network forms such as we observe in the heat denaturation of proteins. This necessity to select conditions that allow stepwise association has been recognized by those experimenting with self-assembly, but not by those studying protein association. Indeed, in the case of self-assembly, the structures being assembled are already rather complicated and highly asymmetric in many cases, so that the structure resulting forms a one- or two-dimensional crystal rather than a three-dimensional one. In the case of protein dissociation we are only very recently finding ways to dissociate stepwise. As a consequence the use of guanidine HCl and sodium dodecyl sulfate will become much less important, since they cause unfolding as well as dissociation and involve several associaton patterns simultaneously.

Cook and Koshland (9) describe empirical experiments to determine whether mixtures of structural subunits from various enzymes will form hybrids upon reassociation. To a limited degree the results confirm that association is a specific process and does not readily accommodate a variety of polypeptide chains.

One of the more conspicuous successes in reassembly has involved *Escherichia coli* ribosomes (63). The 70S ribosomes (2.3×10^6 daltons) can be resolved into 30S (0.7×10^6 daltons) and 50S (1.8×10^6 daltons) subunits, each of which has distinguishable functions in linear polypeptide synthesis. The 30S subunit combines 5S and 23S RNA with at least 30 protein species. The proteins can be dissociated from the RNA in high concentrations of CsCl (often used for density gradient centrifugation) at low concentrations of Mg^{2+}. Reversal of such dissociations resulted in structures that were effective in the synthesis of peptide bonds. The proteins of each subunit were characteristic for the subunit and were not substitutive for one other. Fractionation of these proteins (63) allowed experimentation with subunits from which protein components were missing and has permitted the following conclusions. Only the 30S subunit is structured to become active in initiating a polypeptide chain. It is the subunit that specifically binds formylmethionyl–tRNA. [At present formylmethionine is believed to form the initial N-terminal position in all bacterial proteins (35). This has not been found to be so for eukaryotes.] Not until this process has occurred will 30S and 50S subunits self-assemble on mRNA strands. It is the 50S subunit that appears to be involved in the mechanism of the successive addition of amino acid units to the linear polypeptide being synthesized. When this elongation process is terminated, the subunits dissociate.

Pools of ribosomal proteins appear to exist in the cell, and there is evidence that they associate into specific aggregates in a stepwise manner. Kinetically, the orderly association of an RNA structure with 20 different proteins represents a nightmare for the physical chemist and can be contemplated only as a stepwise process. All proteins were not found to be indispensable, and several different functions were indicated by the data. Reassembly occurred at temperatures near 40°, but not in the cold, a fact for which no explanation is at hand. It should be emphasized that the assembly of prokaryotic ribosomes differs from that of eukarytic ribosomes.

The association and dissociation of protein structure are equilibrium processes, and the time required to reach equilibrium may be very short (less than a minute) or very long (several days). Since these times are related to rates of reaction, it follows that there is a wide range of reaction probability. Attempts to formulate models for simple linear polymerization (42) encounter the same problem that arises in describing crystallization. Some type of nucleation process must occur, and some molecules must achieve an activated state, but obviously not all at the same time. There is speculation that interaction with a small molecule, an initiator, results in a conformation that is just right for polymerization. The same idea is inherent in the suggestion of a morphopoietic factor in self-assembly considerations. (Such a factor is any substance that interacts with a structural subunit and is an obligatory participant in the assembly to the final structure. It may or may not be part of the final product.) Indeed, both ideas are identical with the concept of an allosteric effect in which a small molecule binds to a macromolecule and thereby changes its behavior either toward small molecules or toward other macromolecules. Of course the number of possibilities is inconceivably large and suggests that this type of research is likely to remain empirical.

6. INTERMOLECULAR FORCES

In the foregoing discussion it is evident that noncovalent bonds preponderate in the interactions between proteins and other molecules, large and small. For the biologist (and indeed for most experimentalists) a detailed discussion of forces is of little pragmatic value. Yet one attempts to understand, to comprehend, how it is that molecules adhere. Even such ideal bodies as helium atoms can interact, since liquid helium is known, and hence any type of dissymmetry, inherent or induced, can give rise to attractive forces. On the other hand, electron clouds must not overlap, and hence repulsion develops when this condition is approached by two atomic nuclei surrounded by electrons. In general, for biological macromolecules the potential energy equals a sum of values. One of these values is a repulsive force that is inversely proportional to the twelfth power of the distance multiplied by an empirical factor *a*. Obviously this term can be comprehended only if the distance is extremely small. Another of the values in the sum is an attractive force, inversely proportional to the sixth power of the distance multiplied by an empirical factor *b*. This force was the first described by Fritz London. A third value derives from charge–charge interaction. The total may be expressed as follows:

$$\text{Potential energy} = \sum \frac{a}{r^{12}} - \sum \frac{b}{r_6} + \sum \frac{qq'}{r\epsilon}$$

$$= -\ \text{energy of attraction}$$

Classical charge–charge interactions are relatively easy to conceptualize. The interaction between two H^+ at 562 Å is 600 cal/mole. The same value obtains for H^+ and H_2O at a distance of 14.6 Å and for two H_2O at 5.5 Å. But in water simple charge-charge interaction is less likely because the formation of a hydrated complex for each ion has a larger negative free energy than the formation of a salt link. As noted above, the electrostatic force is inversely proportional to the dielectric constant. Thus a bond will have a lower potential energy in an aqueous environment than in a nonaqueous environment for any given distance.

In any solvent situation unsymmetrical molecules will exhibit dipole interactions and will interact even with symmetrical molecules by induction. The sum of these interactions is referred to as the Lennard–Jones potential energy diagram (see Fig. 2).

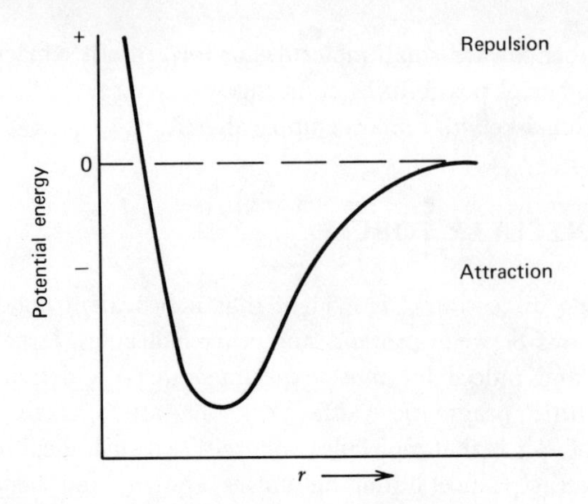

Fig. 2 Relation of dipole interaction to attraction or repulsion.

It is evident that there is a small value of r for a potential energy of zero. This is the van der Waals radius, which can be diminished only by formation of a covalent bond. At the most negative value of the curve the potential energy is 5 to 10 kcal/mole. In view of the value for RT at room temperature, 600 cal, such a bond would not be easily altered by thermal changes. In large molecules where many such bonds are formed, stability may be marked.

The forces involved in forming hydrogen bonds of various kinds, as well as hydrophobic bonds, are those just described. However, in hydrophobic bonding it is the thermodynamic properties of the solvent that provide stabilization.

In sum, there are classical interactions involving both charge and polarizability effects, quantum-mechanical forces that contribute to the Lennard–Jones potential and to the formation of hydrogen bonds, and thermodynamic forces contributing to hydrophobic bonds. In most protein behavior it is likely that a blend of these various forces is involved, and only rarely a single type. Even in a diversified biological structure such as T-even phage there is evidence (61) that structure is not dependent on covalent bonding, or even on the presence of disulfide bonds. In the reference given, T-even phage preparations were subjected to the reagents commonly used for protein dissociation: extremes of H^+ concentration, urea, guanidine hydrochloride, acetic acid, and detergents. The action of each reagent was noted with reference to the following specific interactions or attachments: sheath contraction; tube, sheath, or fiber to base plate; head to tail; polysheath, base plate, tube, capsid, fiber, and contracted sheath. Independently of the reagent used, the various structural parts were ranked in the same order of increasing resistance to dissociation: base plates, tubes, capsids, fibers, and contracted sheaths. It may be presumed that the forces maintaining the various assemblies are similar but that there are gradations of strength.

As noted, disulfide bonds seem not to be necessary (see also p. 106) to stabilize structures. Often reducing agents are added to urea and guanidine hydrochloride in order to reduce disulfide linkages. In actuality they are often unnecessary for reducing disulfide bonds (although β-mercaptoethanol does alter hydrophobic binding) but are effective in preventing rapid disulfide bond formation after dissociation has been accomplished.

7. PROTEIN-MEDIATED CONTROLS

There has been keen curiosity about the relation between catalytic activity and protein association, particularly in regard to the globular enzymes. No fruitful generalization has yet emerged. Postulation of an active site for every structural subunit is certainly incorrect. And this speculation leads into a consideration of control. In bacteria there have been described an abundance of mechanisms: repressor action, feedback mechanisms, allosteric responses, the operation of regulatory subunits, isoenzyme formation, cyclic AMP effects. So far most of the attention has been focused on control of intermediary metabolism and enzyme action, but there must also be mechanisms operating to control the assembly of structures. Presumably the evolutionary process has resulted in genetic patterns which direct the synthesis of macromolecules that not only associate into specific structures but also are subject to fine control in rate. In turn, it is not altogether unexpected that control of genetic expression seems to be mediated by repressor proteins. The synthesis of the *lac* repressor is coded by the i gene situated close to the *lac* operon. It was proposed by Jacob and Monod that the operator site adjacent to the z gene of the *lac* operon, and responsible for expression of the operon, was blocked by a repressor of unspecified character. Expression of the operon was thought to be secured by exposing the cell to an inducer that could bind specifically to the repressor and, competing, liberate the o gene to mediate operon expression.

Only protein is likely to possess the potentiality for the specific kind of binding envisaged. By utilizing a suitable mutant (20) it was found possible to concentrate enough *lac* repressor to demonstrate its protein nature. It is believed to be a tetramer of 150,000 daltons that normally forms only 0.002 to 0.02% of the cell protein. It has been shown to bind to the *lac* operon carried by a hybrid bacteriophage, but it binds only to duplex DNA. The kinetics of the interaction between DNA and the repressor pose problems, and the mechanism is obscure. Determination of the amino acid sequence of the linear polypeptide involved has been made.

Solution to the problem of interaction has a more general aspect as well. Several enzymes are now known that require a macromolecular substrate. We do not know how to rationalize this requirement as yet, although some progress has been made in the study of lysozyme action. In the present instance the gratuitous (nonmetabolized) inducer isopropyl thiogalactoside clearly competes for binding sites or alters the conformation of the repressor so that it no longer can interact with DNA. Moreover, the outer regions of DNA do not offer much except phosphate groups for bonding, and it is difficult to envisage the kind of specificity assumed to exist.

In addition to operon regulation there is also control at the metabolic level. It has been known for some years that glucose suppresses the synthesis of *lac* operon enzymes as well as several others. Recent results (11) suggest that this suppression is a result of lowering cyclic AMP levels in the organism and that cyclic AMP is directly involved in enzyme synthesis.

The metabolic controls in bacteria are numerous enough to merit their own monograph. It has been suggested (53) that the complexity is necessitated by the minimal compartmentation of the bacterial cell. However convenient this is as a working hypothesis, limited cell compartmentation must be only one of several contributing factors.

8. ISOENZYMES AND MULTIPROTEIN ENZYME COMPLEXES

In some respects this section is a subdivision of the preceding one, but it is not yet clear that attention should be restricted to the aspect of control. As noted (30), it is difficult to state succinctly the nature of all isoenzyme systems. When we refer to lactic dehydrogenase, for instance, we usually use the singular form of the noun and a specific protein is implied. But both grammar and usage are misleading. Lactic dehydrogenase is singular only in activity, and it is this *activity* that bears the designation (1.1.1.27) of the International Union of Biochemistry (15). A large number of proteins are isolable—in principle, a different protein from each type of cell—that can have this activity as an attribute. Just what minimum structure must exist for such an activity to appear is still a mystery. Whatever it is, this minimum structure must be modifiable, since both the specificity and the kinetic constants may vary substantially, and presumably both binding and catalytic functions may be involved.

It is now recognized that the term "pure enzyme" is a remnant of the romantic period of enzymology. If one aspires to the quintessence of singularity, it is necessary to isolate a single isoenzyme fraction. This is not separable from others by criteria based only on mass, but is likely to be distinguishable by electrophoresis and found to differ in detail kinetically from its other isoenzyme counterparts. In the case of the lactic dehydrogenase isoenzymes there is evidence that each differs slightly in amino acid composition, in inhibition by pyruvate,* and in stability at elevated temperatures. There seems to be a good possibility that these isoenzymes can be mediators of a fine control system, but this cannot be assumed to be a general role. In mammalian muscle, a highly specialized tissue, it seems to be established that the five isoenzymes of the tetrameric enzyme ($4n$) reflect random assortment of two distinct types of structural subunit. Evidently these subunits are similar enough so that all possible combinations can be realized. In contrast, the five isoenzymes of ($2n$) alcohol dehydrogenase in *Drosophila* seem to be derived from the single locus (23). Two stable forms of this enzyme protein exist having the same amino acid sequence but, presumably, differing tertiary structures.

The importance of isoenzymes in control mechanisms is well exemplified by aspartokinase in *E. coli* and in *Salmonella typhimurium* (10). A number of intermediates form alternative reaction sequences by which the carbon structure of aspartate is transformed into lysine, threonine, or methionine. If aspartokinase, the first catalyst of this metabolic complex, were inhibited by the end products, regulations of the syntheses could not be independent. In fact, however, they are, because of the existence of three isoenzymic forms of aspartokinase: threonine sensitive, methionine repressible, and lysine inhibited, each of which has been isolated.

Two-protein enzyme complexes have been recognized often enough so that they must be classified as a separate category (14). The most intensely studied of these is the tryptophan synthetase complex from *E. coli*. The enzymatic activities embraced by the systematic designation (4.2.1.20) include:

1. Indole + L-Ser → L-Try
2. Indole glycerol P ⇄ indole + glyceraldehyde-3-P
3. Indole glycerol P + L-Ser → L-Try + glyceraldehyde-3-P

It has been shown (22) that the enzyme–protein complex may be designated as $\alpha_2\beta_2$, consisting of two α polypeptides and a dimer of β polypeptides. Reaction 1 is catalyzed (but poorly) by the β_2 dimer; reaction 2 is catalyzed by the α monomer. Only the associated complex of the two dissimilar polypeptides, $\alpha_2\beta_2$, catalyzes reaction 3. Two equivalents of pyridoxal P are bound to the β_2 dimer, and it is presumed that there are two catalytic regions. It seems entirely

* This is currently in dispute (66) and serves to focus on the problem of devising enzyme assays appropriate for the concentrations of catalytic protein found in tissues.

likely that the changes in catalytic efficiency are due to conformational changes attendant upon aggregation of the polypeptide chains, and in this sense both kinds of chains are necessary for the formation of the catalytic array of groups. It is not yet evident how the activity of this enzyme aggregate is controlled *in vivo*.

Another pyridoxal P-containing enzyme protein seems to be rather similarly constituted (26). The glycine decarboxylase complex from *Peptococcus glycinophilus* also consists of two proteins, only one of them binding pyridoxal P. Alone, each is inactive; combined, each contributes to catalyzing the decarboxylation of glycine.

It has also been shown that RNA polymerase complex from *E. coli* is comprised of three polypeptide sequences in an assembly denoted as $\alpha_2\beta\beta'$ (7). This assembly catalyzes the transcription of RNA on a DNA template, but its activity is greatly increased by another protein of 90,000 daltons, labeled the σ factor (8). This protein evidently interacts with the DNA template at the appropriate sequence for initiation. Yet another protein, termed ω, of 9000 daltons has been detected to be associated with $\alpha_2\beta\beta'$ (7). Finally, a termination factor, ρ, of about 50,000 dalton subunits may function as a part of the complex. There is evidence that σ factors from both *E. coli* and T4 phage stimulate transcription and hence have rather specific binding to DNA and a somewhat less specific interaction with polymerase protein.

Such complexes are not unlike those of the α-keto acid dehydrogenases or the enzyme group that, associated in a single complex, catalyzes the oxidation of fatty acids. In each of these three complexes some type of physical movement of the substrate with respect to the enzyme component has been postulated. This movement is suggested because the substrate undergoes a series of reactions with a succession of proteins and yet does not dissociate until the end of the reaction series. It is not easy to build or envisage a model that expresses a mechanical analogy in strictly molecular terms. However, in the case of RNA polymerase it may be that the RNA–DNA complex formed during RNA elongation is thermodynamically less stable than the DNA duplex and that the synthetase momentarily labilizes the DNA at the position where nucleotide is being added. Thus a wave of instability would flow along the template, accompanied by synthesis of an RNA structure and reformation of DNA duplex posterior to the direction of propagation.

9. COMPLEMENTATION AND ENZYME ACTIVITY

There has developed a kind of experimentation in which genetic information serves to provide data on the structure of the multimeric proteins. This technique, referred to as genetic complementation (16), involves the restoration, at least partially, of the normal phenotype in heterokaryons or heterozygotes from different allelic mutants. It has been proposed that the qualitative demonstration of such complementation indicates a causal relation between gene alteration and changes in the structural features of a multimeric protein composed of quasi-identical structural subunits. For eukaryotes this kind of experimentation is straightforward, but for prokaryotes a few refinements must be added.

The use of complementation techniques in protein structure research really deserves a special name because complementation is fundamentally a genetic technique applicable to a variety of problems. In any organism unlinked mutations may be expected to show complementation if two mutant chromosomes, or relevant parts, are somehow juxtaposed in the same cell. Some closely linked mutants may also show complementation, although noncomplementary mutants are usually those closely linked.

The relation of observations on complementation to protein function has been well exploited by studies on glutamate dehydrogenase in *Neurospora* mutants. In a few heterokaryons derived from pairs of *am* mutants, each of which produces either no enzyme or an inactive form, there appeared enzyme activity that was definite but at a low level. It has been proposed (17) that hybrid enzyme molecules are formed, that defective protomers, unable to self-associate to enzymically competent assemblies, can form hybrids in which critical deficiencies are eliminated.

Enzymes formed in such complementation experiments are subactive, and we can only speculate that the molecular fitting, a preoccupation since the days of Paul Ehrlich, is imprecise. But little enough is known about the relation between enzymic activity and the essential structural features of multimeric enzymes so that the working hypothesis above is acceptable.

Glutamate dehydrogenase has been a fruitful experimental system but an activity full of complexities. Although the enzyme from beef liver has been most adequately studied, that from other sources does not differ markedly. The polypeptide units have a molecular weight of about 50,000, and six of them associate to the smallest catalytically active protomer. In the proper environment, these 300,000-dalton aggregates associate further, up to the usual preparative size of 2.2×10^6 daltons (59). The association–dissociation equilibrium doubtless is a composite of eight steps, all involving species with catalytic activity. The protein can engage, in various modes, substrates, nucleotides, steroids, and divalent metals, each of which may affect the protein in such a way that either association or dissociation is promoted (51). As noted previously, such a protein should dissociate upon high dilution, and it does.

Why catalytic activity should disappear below the 300,000-dalton level cannot be rationalized at present. This is a well-documented, but not a unique, case. One obvious explanation is that the reagents used to dissociate the catalytic protomer into its constituent chains cause extensive unfolding. β-Galactosidase has polypeptide chains of about 35,000, a catalytic protomer of about 130,000, and a usual preparative size of about 520,000 daltons. At 130,000 daltons catalytic activity disappears without measurable unfolding (1), and the presumption is that the conformation requisite for catalytic activity is metastable. Although stabilization is achieved by further association, it is highly likely that additional search will reveal a surrogate stabilizer. It has been proposed that association is responsible for the catalytic activity of glutamate dehydrogenase because of an allosteric effect (41). This is useful in juxtaposing ideas and emphasizes the possible relation between association and any conformational change that may develop. Indeed good evidence has been presented to demonstrate the allosteric effects of association (40). Nevertheless, the fine detail of the catalytic assembly is not available to us, nor have mechanistic details been related causally to structural changes.

One attempt to characterize the active site is based on interpretation of genetic evidence (29). Proteins with β-galactosidase activity were extracted from a large number of mutants (*E. coli* K12). The kinetic constants for several substrates were correlated with the genetic mapping. Interpretation of the data was that binding properties are not distributed at random along the gene that codes the polypeptide chain but are restricted to a small number of regions. Presumably the linear polypeptide is folded in such a manner that the binding region is assembled from several disparate amino acid sequences. It was noted that at least two such regions were close to the polypeptide sections involved in structural association–dissociation.

It has been recognized for several years that experimentation relating to these problems suffers from the strong concentration dependence of protein behavior. Enzymatic activity is usually measured under conditions where the protein concentration is less than 10^{-3} mg/ml, whereas physical measurements ordinarily demand concentrations of 10^{-1} mg/ml or more. Vexingly it is just in this nonoverlapping range that many proteins dissociate. Experimentally this difficulty is being overcome, but much of the literature antedating 1967 should be read with this cautionary observation in mind.

In vitro hybridization of glutamate dehydrogenases has been achieved by the repeated freezing and thawing of solutions of mixed mutant varieties. This technique has been employed in studying isoenzymes such as those of lactate dehydrogenase and results in marked dissociation and reassociation. If mutant proteins are present, reassociation may produce hybrids, as observed in isoenzyme mixtures. The mechanism of this process is not known, but a recent investigation (60) has suggested the following. As ice forms, small molecules are excluded to some degree. This could produce a moving zone of a strong concentration gradient in which a protein conformational change may be induced. A return to the original concentration upon thawing would predispose to reformation of the starting product or its equivalent.

Complementation studies (38) led to the conclusion that, in *Neurospora* changes in the genetic loci *ma*-1 and *ma*-2 affect simultaneously malate dehydrogenase and aspartate aminotransferase. It was subsequently shown (37) that these enzymatic activities were attributes of the same protein.

A general test for complementation in bacteria is that of abortive transduction, clearly described by Fincham (17) along with related matters. The work on *Salmonella* carried on by Hartman and his colleagues has led to data supporting the operon hypothesis, the concept of polycistronic messenger RNA, as well as polarity in the operon. Whether the operon function occurs commonly in eukaryotes has not been determined. In the case of *E. coli*, λ phage has been widely used to carry, into the bacterial cell, fragments of DNA from another bacterial cell. Because λ phage attaches to the *gal* region of the bacterial chromosome, it is uniquely useful for studying complementation in this region. For biochemical purposes, however, the partial heterozygotes obtained are not particularly satisfactory since they tend to be unstable.

In vitro hybridization has been used for studying the alkaline phosphatase of *E. coli*. This enzyme protein is a dimer of two polypeptide chains associated in a complex involving three zinc ions. Since it can be dissociated easily, it could be subjected to techniques developed for obtaining hybrid hemoglobins.Thus protomers prepared from different strains of *E. coli* could be prepared and mixed in the presence of zinc ion (55) to produce dimers corresponding to those of each mutant as well as a hybrid.

Experimentation with this enzyme has suggested an answer to another general question as well, that concerning the *in vivo* process of protein association. It is currently assumed that polypeptide chains attain a stable conformation, based on intrachain features, on or near the site of synthesis, presumably the ribosome. It is not known whether conditions are favorable for association at the same site, and in the case of proteins that contain at least two species of polypeptide chains additional difficulties can be anticipated. Since alkaline phosphatase normally accumulates between the cell membrane and the cell wall, any active enzyme can be removed by treating cells with EDTA and lysozyme. By decreasing the concentration of Zn^{2+} in the medium as well, it has been possible to demonstrate that *E. coli* can produce a pool of alkaline phosphatase protomers (62) bound to a membrane structure (containing some RNA and heavier than 70S) that was not ribosomal and was in the endoplasm. The data obtained pointed to a dimerization process occurring after the release of protomers from the ribosome and not closely coupled, since synthesis was not inhibited by lack of dimerization. This has been confirmed (54) by a study showing that spheroplasts could not form an active dimer but could synthesize protein antigenically identical to the enzyme subunit. It was concluded that dimerization occurs only in the periplasmic space.

10. PROTEIN COMPONENTS OF MEMBRANES

Much of the work now in progress on bacterial membranes was made possible by the seminal discovery of Weibull in 1953 that lysozyme can effect the removal of the cell wall without destroying the cell membrane if a medium of the proper osmotic potential is used. The complexity of the membrane is suggested by the number of specific binding proteins that have been isolated.

A sulfate-binding protein has been demonstrated (45) in *S. typhimurium* by a combination of techniques. One device was to use diazo-7-amino-1,3-

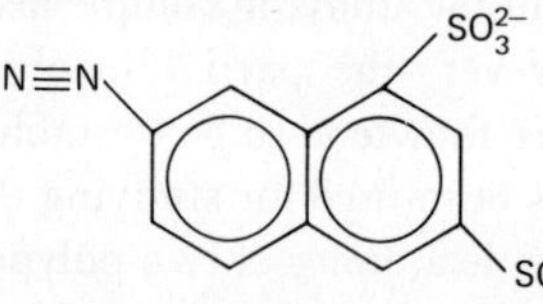

naphthalene disulfonate, which can combine with histidine and tyrosine. Since the reagent cannot penetrate the cell, inhibition of binding indicated location of the receptor on the outside of the cell. Release of binding ability on spheroplast formation allowed verification that binding is associated with a protein, permitted preparation of an antibody that specifically inhibited the binding of SO_4^{2-} by purified protein, and, finally, led to isolation of the protein itself (44).

Recently a membrane ATPase has been isolated from *Streptococcus faecalis* (57) and shown to have a size of 385,000 daltons; it is assembled from six protomers, each containing an α and a β monomer. Reconstitution studies indicate that the isolated enzyme can be reinserted into specific sites on the protoplast* membrane and held there by Mg^{2+} complexing.

The work of Kennedy and his colleagues on the lactose permease protein, or M protein, of *E. coli* exemplifies some of the problems that may be encountered in investigating enzymes which are membrane components. This protein, the product of the y gene of the *lac* operon, facilitates the transport of β-galactosides into the intact cells. It has been demonstrated to be a component of the protoplast membrane and is inhibited by *N*-ethylmaleimide (NEM) reaction with cysteine groups of the protein. It was isolated by the following

* The term "spheroplast" is used to designate a cell whose cell wall has been damaged but not removed. A "protoplast" is a cell whose cell wall has been totally removed.

strategy. The protein was saturated with galactosyl thiogalactoside (which protects against NEM inactivation), and the other SH groups were allowed to react with NEM. Then, after removal of the galactoside, the SH groups of M protein were allowed to react with NEM–^{3}H. The labeled protein was successfully extracted from the membrane structure with the neutral detergent Triton* X-100 or with sodium dodecyl sulfate (SDS) (25). By comparing its electrophoretic velocity in acrylamide gel containing SDS, or its migration on a Sephadex column equilibrated with SDS, against protein standards, an estimate of molecular size could be made. The estimated size under these conditions was 30,000 daltons, perhaps the size of the monomer and similar to the value obtained by several other groups for binding proteins released by osmotic stress (43). But proteins extracted in this way must be studied in the presence of detergent, since removal of the reagent is accompanied by disordered aggregation. Thus it is not known what properties such proteins have in isolation or when separated from the phospholipid of the membrane. Improvements in techniques of preparing such proteins are much needed.

It is also possible to release proteins from bacteria by cold osmotic stress after exposure to Tris-EDTA (39). Some of these proteins are being referred to as binding proteins, since they bind various low-molecular-weight substances such as galactose, leucine, and SO_4^{2-}. From a chemical standpoint, however, this is a rather unsatisfactory designation because it is not obvious what sort of binding process is involved, although dissociation constants can be measured by equilibrium dialysis in many cases. The specific binding of leucine by a protein isolated from osmotic shock fluid (*E. coli*) disappeared in urea or guanidine and reappeared when these reagents were removed (46). Efforts to demonstrate conformational changes were unrewarded. Moreover it has been proposed that such proteins participate as stereospecific reagents in an active transport process. It is hoped that this working hypothesis will help to clarify the mechanisms involved, but so far no direct evidence is available. When isolated binding protein was restored to shocked cells (4), behavior suggesting a simple correlation was not observed.

At least two general classes of proteins have been related to membrane transport phenomena. One group consists of those cited above, so-called binding proteins. The binding mechanism may be very different from protein to protein. The protein may act as an enzyme, or it may act in concert with other proteins, functioning as a recognition site. In either case some type of conformational change might be expected, but so far none has been clearly demonstrated. The second group consists of proteins that might furnish energy for transport, such as the HPr protein (27). In the presence of a soluble

* Triton X-100 is alkylphenoxypolyethoxyethanol.

enzyme, HPr (also soluble) is modified with a labile phosphate group by interaction with phosphoenolpyruvate. This modified HPr then acts in concert with an enzyme group, whose specificity is directed to certain sugars and located in the membrane, to produce sugar phosphates on the inside of the membrane. Many possibilities have been suggested by the above investigations, and it seems certain that the plasma membrane will be found to be a complex assembly of functional proteins.

Study of bacterial chemotaxis (2) has revealed that *E. coli* possesses at least five chemoreceptors. In all likelihood these receptors form part of the membrane. Since they respond to galactose, glucose, ribose, aspartate, and serine, it is probable that they are protein, but they do not seem to be any of the proteins discussed above.

11. PROTEIN TURNOVER

There is now experimental evidence that, in all types of cells, the processes of protein synthesis and of protein degradation are constantly active. The ratio of activities may vary widely, and the nature of the coupling between these processes is not known. Although it has been accepted for about 25 years that this relationship, usually referred to as turnover, exists in mammalian cells, there has been less certainty about this aspect of metabolism in bacteria. Indeed, several papers supported the view that some bacterial proteins were stable. Further careful work, however, revealed that viable, but nonproliferating, *E. coli* and yeast did exhibit turnover. Degradation of protein was found to occur in *E. coli* during logarithmic growth at a rate of 2.7% per generation (18). It has also been reported (47) that degradation rates are insensitive to the physiological state of the cell. If this is verified, there is no coupling between synthesis and degradation and the degradation process must reflect some other characteristic of living systems.

It has been noted that in bacteria, growing exponentially, the stability of protein tends to exceed the dividing time. In mammals, a tissue protein has a half-life smaller than the half-life of the cell from which it derives. Moreover, the rate of protein degradation in mammals (21) is demonstrably modified by alterations in the physiological state. It seems safe to generalize that the level of protein in bacteria is much more a function of the synthetic processes and is very probably controlled primarily by alterations in their rate, although a measurable degradation obtains under all conditions. Since the level of protein in mammals seems just as clearly the result of two processes of more comparable rates, it must not be assumed that either synthetic or degradative processes are identical in prokaryotes and eukaryotes.

Since there seems to be no mandatory coupling of degradation to any other process, it is at least possible that degradation is a reflection of entropy increase. Proteins are very mobile systems and respond to changes in the environment. They participate, many of them, in equilibria involving association–dissociation and conformational changes. It may be assumed that all of these equilibria are not perfectly reversible and that occasionally there is sufficient unfolding so that proteolysis occurs. Presumably the hydrolyzing of a few peptide bonds would markedly alter the stability of the molecule. The result would be a low but constant background of protein "wear-and-tear." The concept of protein wear-and-tear was once widely held but was never clearly formulated. It was abandoned for the concept of the steady state, but the latter never carried an explicit description of protein degradation. Proteins, as such, may be very stable and resist degradation in storage for long periods. However, the rigors of the catalytic or transduction process might predispose toward an increase in entropy and dissolution.

12. PROTEIN SYNTHESIS

Bacteria—*E. coli* in particular—have provided biologists, molecular and otherwise, and biochemists with excellent experimental material for studying protein syntheses. As noted at the beginning of this chapter, protein synthesis is a major activity of these organisms. Physicists have been intrigued with the processes of information transfer that must be part of gene-directed protein synthesis, and these DNA–RNA transcription processes will be considered in another chapter. Chemists have become impressed with the ability of bacteria to lower activation energy and to reduce the energy barrier to reactions of the most varied kinds. Since this reduction is achieved by enzyme catalysts elaborated by the cell, attention has been given to protein synthesis by pragmatists as well as theoreticians. The number of papers cited in a recent review of protein synthesis (33) totals 426. It would be presumptuous to offer anything but selected remarks on this subject.

A bacterial cell-free system has been devised (31) that converts appropriate amino acid derivations into specific linear peptide sequences. Ever since Nirenberg and Matthaei demonstrated that a synthetic polynucleotide could serve as a directive for polymerization of an amino acid, there has been wide agreement that mRNA served as a guide to the sequence of amino acid addition. But the proof would be addition of a specific mRNA to a synthesizing system with the subsequent isolation of a specific protein. In the investigation cited (31), an 8S RNA from rabbit reticulocytes served as a guide to hemoglobin synthesis. As activator for the mixture of amino acids necessary, a commercially available transfer RNA from *E. coli* was added. The catalytic structure for translation and creation of a linear polypeptide was provided by ribosomes from *E. coli* strain Q13. Previous attempts to demonstrate a specific synthesis in such a system had failed, presumably because initiation of polymerization required that the amino group of the N-terminal amino acid be substituted. As noted in a previous section, bacterial polypeptide arrays initially have formylmethionine at the N-terminal end, although this amino acid may be removed later. In the present experiments *N*-acetylvalyl–tRNA was used because both the alpha and the beta subunits of rabbit globin have valine in the N-terminal position. The product of the experiments cited (31) appeared to be identical to rabbit globin by several criteria.

Although the preparations used in the experiments just cited were effective in forming a specific polypeptide sequence, much remains unexplained. Initiation of polypeptide synthesis requires not only a specific derivative of an activated amino acid but also several protein factors involved in the mRNA–ribosome complex (34). Attempts to purify the ribosomal preparations have made evident the role of protein in polypeptide elongation and termination as well (56).

Thus, here again, protein interactions provide a hierarchy of controls. For example, in the early stages of T4 phage infection of *E. coli* B, protein synthesis can be inhibited by chloramphenicol, and during this interval mRNA is presumed to accumulate (32). This mRNA is believed to transcribe for enzymes needed to initiate the synthesis of viral DNA. When chloramphenicol was diluted out and further RNA synthesis inhibited by rifamycin, translation of the mRNA occurred and "early enzymes" appeared.

In yet another instance a synthetase for transfer RNA has been postulated as a regulatory substance (12). The enzyme catalyzing the synthesis of histidyl–tRNA has been isolated from *S. typhimurium* and has been found to form a very stable complex with $tRNA^{His}$—so stable, in fact, that the suggestion has been made that the synthetase may be a repressor of the histidine operon.

13. CONCLUSIONS

From the foregoing discussion one cannot draw the conclusion that the proteins found in bacteria differ markedly from those occurring in other forms of life. In attempting to conceptualize protein one is forced to oscillate between structure and function, symmetry and asymmetry, and to consider now one, now the other. Proteins offer the most varied types of interactions and are omnipresent in all living forms, exhibiting simultaneously structural and functional purposes. This simultaneity is evident in membrane proteins. The lack of subcellular structures in bacteria does not mean, however, a lack of organization. It is more than likely that most of the enzymic proteins within bacteria are present in massive complexes that merge in a continuum of structure.

It has often been remarked that DNA is immortal since it is self-perpetuating. One might add that the unique and precious processes that sum to life, that decrease entropy, reach fulfillment in the structures of protein that confer upon the living substance its form, as well as providing controls in bewildering variety.

REFERENCES

1. Contaxis, C. C., and F. J. Reithel, *Biochem. J.* **124,** 623 (1971).
2. Adler, J., *Science* **166,** 1588 (1969).
3. Anfinsen, C. B., *J. Polymer Sci.* **49,** 31 (1961).
4. Anraku, Y., *J. Biol. Chem.* **243,** 3116, 3123, 3128 (1968).
5. Bernal, J. D., *The Origin of Life*, Appendix 3, World Publishing Co., Cleveland, Ohio, 1967.
6. Bernal, J. D., *J. Mol. Biol.* **24,** 379 (1967).
7. Burgess, R. R., *J. Biol. Chem.* **244,** 6168 (1969).
8. Burgess, R. R., A. A. Travers, J. J. Dunn, and E. K. F. Bautz, *Nature* **221,** 43 (1969).
9. Cook, R. A., and D. E. Koshland, *Proc. Natl. Acad. Sci. U.S.* **64,** 247 (1969).
10. Datta, P., *Science* **165,** 556 (1969).
11. De Crombrugghe, B., R. L. Perlman, H. E. Varmus, and I. Pastan, *J. Biol. Chem.* **244,** 5228 (1969).
12. De Lorenzo, F., and B. N. Ames, *J. Biol. Chem.* **245,** 1710 (1970).
13. Dickerson, R. E., and I. Geis, *Structure and Action of Proteins*, Harper and Row, New York, 1969.
14. Ebner, K. E., *Acc. Chem. Res.* **3,** 41 (1970).
15. *Enzyme Nomenclature: Recommendations of the International Union of Biochemistry*, Elsevier, Amsterdam, 1965.
16. Fincham, J. R. S., *Sci. Prog.* **56,** 165 (1968).
17. Fincham, J. R. S., *Genetic Complementation*, Benjamin, New York, 1966.
18. Fox, G., and J. W. Brown, *Biochim. Biophys. Acta* **46,** 387 (1961).
19. Gilbert, G. A., *Proc. Roy. Soc.* **A250,** 377 (1959).
20. Gilbert, W., and B. Müller-Hill, *Proc. Natl. Acad. Sci. U.S.* **56,** 1891 (1966).
21. Goldberg, A. L., *J. Biol. Chem.* **244,** 3217 (1969).
22. Goldberg, M. E., T. E. Creighton, R. L. Baldwin, and C. Yanofsky, *J. Mol. Biol.* **21,** 71 (1966); T. E. Creighton and C. Yanofsky, *J. Biol. Chem.* **241,** 980 (1966).
23. Jacobson, K. B., J. B. Murphy, and F. C. Hartman, *J. Biol. Chem.* **245,** 1075 (1970).
24. Jeon, K. W., I. J. Lorch, and J. F. Danielli, *Science* **167,** 1626 (1970).

25. Jones, T. H. D., and E. P. Kennedy, *J. Biol. Chem.* **244,** 5981 (1969).
26. Klein, S. M., and R. K. Sagers, *J. Biol. Chem.* **241,** 197, 206 (1966).
27. Kundig, W., et al., *J. Biol. Chem.* **241,** 3243 (1966).
28. Kushner, D. J., *Bacteriol. Rev.* **33,** 302 (1969).
29. Langridge, J., *Proc. Natl. Acad. Sci. U.S.* **60,** 1260 (1968).
30. Latner, A. L., and A. W. Skillen, *Isoenzymes in Biology and Medicine*, Academic Press, New York, 1968.
31. Laycock, D. G., and J. A. Hunt, *Nature* **221,** 1118 (1969).
32. Lembach, K. J., and J. M. Buchanan, *J. Biol. Chem.* **245,** 1571 (1970).
33. Lengyel, P., and D. Söll, *Bacteriol. Rev.* **33,** 264 (1969).
34. Mechanism of Protein Synthesis, *Cold Spring Harbor Symp. Quant. Biol.* **34** (1969).
35. Migita, L. K., and R. H. Doi, *J. Biol. Chem.* **245,** 2005 (1970).
36. Monod, J., J.-P. Changeaux, and J. Wyman, *J. Mol. Biol.* **12,** 88 (1965).
37. Munkres, K. D., *Arch. Biochem. Biophys.* **112,** 347 (1965).
38. Munkres, K. D., N. H. Giles, and M. E. Case, *Arch. Biochem. Biophys.* **109,** 397 (1965).
39. Neu, H. C., and L. A. Heppel, *J. Biol. Chem.* **240,** 3685 (1965).
40. Nichol. L. W., G. D. Smith, and A. G. Ogston, *Biochim. Biophys. Acta* **184,** 1 (1969).
41. Nichol, L. W., G. D. Smith, and D. J. Winzor, *Nature* **222,** 174 (1969).
42. Oosawa, F., and S. Higashi, *Theor. Biol.* **1,** 79 (1967).
43. Pardee, A. B., *Science* **162,** 632 (1968).
44. Pardee, A. B., *J. Biol. Chem.* **241,** 5886 (1966).
45. Pardee, A. B., and K. Watanabe, *J. Bacteriol.* **96,** 1049 (1968).
46. Penrose, W. R., R. Zand, and D. L. Oxender, *J. Biol. Chem.* **245,** 1432 (1970).
47. Pine, M. J. *J. Bacteriol.* **92,** 847 (1966).
48. Ramachandran, G. N., and V. Sasisekharan, *Advan. Protein Chem.* **23,** 283 (1968).
49. Rechcigl, M. Jr., *Texas Rept. Biol. Med.* **26,** 147 (1968).
50. Reithel, F. J., *Concepts in Biochemistry*, McGraw-Hill, New York, 1967.
51. Reithel, F. J., *Advan. Protein Chem.* **18,** 148 (1963).
52. Roberts, D. B., *J. Mol. Biol.* **45,** 221 (1969).
53. Sanwal, B. D., and R. Smando, *J. Biol. Chem.* **244,** 1817, 1824 (1969).
54. Schlesinger, M. J., *J. Bacteriol* **96,** 727 (1968); M. J. Schlesinger, J. A. Reynolds, and S. Schlesinger, *Ann. N.Y. Acad. Sci.* **166,** 368 (1969).
55. Schlesinger, M. J., and C. Levinthal, *J. Mol. Biol.* **7,** 1 (1963).
56. Schlessinger, D., and D. Apirion, *Ann. Rev. Microbiol.* **23,** 387 (1969).
57. Schnebli, H. P., A. E. Vatter, and A. Abrams, *J. Biol. Chem.* **245,** 1122 (1970).
58. *Study Week on Molecular Forces*, Pontificiae Scientiarum Scripta Varia 31, Wiley, New York, 1967.
59. Sund, H., and W. Burchard, *Eur. J. Biochem.* **6,** 202 (1968).
60. Taborsky, J., *J. Biol. Chem.* **245,** 1054 (1970).
61. To, C. M., E. Kellenberger, and A. Eisenstartz, *J. Mol. Biol.* **46,** 492 (1969).
62. Torriani, A., *J. Bacteriol. 96,* 1200 (1968).

63. Traub, P., and M. Nomura, *J. Mol. Biol.* **34,** 575 (1968).

64. Von Hippel, P. H., and T. Schleich, *Acc. Chem. Res.* **2,** 257 (1969).

65. Warren, K. B. (ed.), *Symposium of the International Society of Cell Biology*, Vol. 6, Academic Press, New York, 1967.

66. Wuntch, T., R. F. Chen, and E. S. Vesell, *Science* **167,** 63 (1970).

CHAPTER

FIVE

Microbial Polysaccharides

E. E. WOODSIDE and J. B. G. KWAPINSKI

1. CLASSIFICATION AND PHYLOGENETIC RELATIONSHIPS OF POLYSACCHARIDES

Microbial polysaccharides represent a highly diversified group of biopolymers which are characteristic not only of organisms in the Protista kingdom but also of organisms in both the plant and animal kingdoms. For example, bacterial cell wall lipopolysaccharides represent exceptionally complex polyhydroxy macromolecules which may contain monomeric constituents found only in bacterial genera. On the other hand, structural and intracellular polysaccharides of both animal (glycogen and/or sulfated heteropolysaccharides) and plant (cellulose, amylopectin, and/or starch) origin are also known to be present in bacterial and fungal species; consequently, any discourse on the synthesis, structure, and function of bacterial polysaccharides transgresses the physicochemical and biosynthetic mechanisms that are involved in the synthesis and accumulation of polysaccharides in both the plant and the animal kindgoms.

Since numerous cellular macromolecules, such as DNA, RNA, protein, and lipids, which persist in all forms of life (viruses excluded), are synthesized via the same general biochemical mechanisms, the homogeneity of biochemical processes in living cells of diverse phylogenetic origin is a reflection of their common evolutionary origin. With respect to chemotypic interrelationships between the major cellular biopolymers, it is of interest to note that both DNA and RNA are, in essence, substituted polysaccharides. In addition, a large number of cellular proteins and lipids are known to contain covalently—linked monosacchardie and/or oligosaccharide moieties within their primary structures, and consequently they have been classified as glycoproteins (122) and glycolipids (114), respectively. Although any discussion of polysaccharides could well include the carbohydrate-containing biopolymers mentioned above, the scope of this chapter will be limited to polysaccharides per se, the substituted polysaccharides (lipopolysaccharides and peptidoglycans), and carbohydrate-containing polyols (teichoic acids) of microbial origin, with special reference to their interactions with each other and with other cellular macromolecules.

Polysaccharides have been classified into numerous groups on the basis of their source, biological function, cellular distribution, stereochemotype, or phylogenetic origin. The choice of one or another of these taxonomic groupings has invariably been based on utilitarian concepts; however, the chemotypic classification of polysaccharides appears to be the most desirable form of taxonomy. Since variations on theme with respect to the primary structure of specific polysaccharides have been utilized as one of the criteria for differentiating the major forms of life, a chemotypic tabulation of polysaccharides with reference to phylogenetic interrelationships is presented in Tables 1 through 3. On the basis of their diverse polysaccharide contents, bacteria and the other transitional forms of life (Protista) represent a primitive group of living organisms which infringe on the physicochemical properties of polysaccharides that have been generally attributed to organisms in either the plant or the animal kingdom; consequently, in addition to the prokaryotic nature of the bacterial cell, the diversity in their types of polysaccharides, which overlap into both the plant and the animal kingdoms, gives further credence to the postulate that bacteria and their related forms are primitive anchestial species which merit an independent kingdom.

With the exception of pustulan, all D-glucans that have been found in either the plant or the animal kingdom have also been isolated from one or another

Table 1. Distribution of Neutral D-Glucans in Various Biological Kingdoms[a]

		Protista[b]			
D-Glucan Anomers	Predominant Linkages	Lower Protists	Higher Protists	Plant	Animal
α-D-Glucans					
Glycogen	α-(1 → 4), α-(1 → 6)	+ (82)	+ (225)	+ (92)	+ (92)
Amylopectin	α-(1 → 4), α-(1 → 6)	+ (265)	+ (172)	+ (230)	−
Amylose (starch)	α-(1 → 4)	+ (102)	+ (169)	+ (230)	−
"Mycodextran"	α-(1 → 2)	+ (189)	−	−	−
Dextrans	α-(1 → 6), α-(1 → 4), α-(1 → 3)	+ (234)	−	−	−
Nigeran	α-(1 → 3), α-(1 → 4)	−	+ (81)	−	−
β-D-Glucans					
Crown gall glucan	β-(1 → 2)	+ (214)	−	−	−
Laminaran	β-(1 → 3)	−	+ (203)	+ (50)	−
Lentinan	β-(1 → 3)	−	+ (173)	−	−
Cellulose	β-(1 → 4)	+ (79)	+ (31)	+ (230)	+ (95)
Luteose	β-(1 → 6)	−	+ (7)	−	−
Pustulan	β-(1 → 6)	−	−	+ (119)	−
Lichenans	β-(1 → 3), β-(1 → 4)	−	+(203)	+ (205)	−
Pachyman	β-(1 → 3), β-(1 → 6)	−	+ (47)	−	−
Yeast D-glucan	β-(1 → 3), β-(1 → 6)	−	+ (99)	−	−
Pullulan	β-(1 → 4), β-(1 → 6)	−	+ (44)	+	−

[a] Numbers (after +) in parentheses refer to items in the reference list at the end of the chapter; the reported presence or absence of D-glucans in the various kingdoms is noted by a + or − symbol, respectively.
[b] Organisms in the Protista kingdom are subdivided into the lower protists (mycoplasma, bacteria, and blue-green algae) and the higher protists (slime molds, fungi imperfecti, fungi, algae, and protozoa).

Table 2. Distribution of Polysaccharides in Various Kingdoms[a]

Polysaccharide Chemotype	Protista: Lower Protists	Protista: Higher Protists	Plant	Animal
Homoglycans				
Galactans	+ (207)	+ (51)	+ (11)	+ (119)
Mannans	+ (192)	+ (83)	+ (188)	−
Ketoglycans	+ (74)	+ (182)	+ (119)	−
Xylans	−	+ (204)	+ (230)	−
Colominic acid (poly-*N*-acetylneuraminic acid)	+ (179)	−	−	−
Chitin (glucosaminoglycan)	−	+ (218)	−	+ (128)
Galactosaminoglycan	−	+ (60)	−	−
Vi-antigen (poly-*N*-acetylgalactosaminuronic acid)	+ (114)	−	−	−
Pectic acid (polygalacturonic acid)	−	+ (8)	+ (30)	−
Polyglucuronic acid	−	+ (26)	−	−
Polymannuronic acid	+ (42)	−	−	−
Heteroglycans				
Neutral heteroglycans	+ (196)	+ (233)	+ (17)	−
Complex heteroglycans with uronic acid moieties	+ (132)	+ (26)	+ (230)	−
Dermatan sulfate	+ (56)	−	−	+ (129)
Hyaluronic acid	+ (140)	−	−	+ (129)

[a] See the footnotes to Table 1; homoglucan polymers are listed in Table 1 and therefore are not included above.

of the organisms in the Protista kingdom (Table 1). Since both pustulan and luteose are β-(1→6)-linked D-glucans, it is suggested that these D-glucans may possess similar stereochemical structures. In any event, the plant polysaccharide pustulan (119) may well have its origin in the polysaccharide luteose (7), which has been isolated from *Penicillium* species.

Of particular interest is the fact that lower protists contain a greater number of α-D-glucans than the higher protists, whereas a greater number of β-D-glucans are prevalent in the higher protists (Table 1). Numerous other β-D-glucans, which possess mainly β-(1→3) linkages, with minor amounts of β-(1→4), β-(1→2), and/or β-(1→6) linkages, have been partially identified in fungi and fungi imperfecti (181). Although β-D-glucans are generally classified as structural or extracellular polysaccharides, with limited or undefined functions, it has been shown that the algal polysaccharide laminaran (203) and numerous extracellular -D-glucans of fungi (181) serve as food-reserve polymers; consequently, in addition to the well-established roles of α-D-glucans (glycogen, amylopectin, and amylose) as intracellular food-reserve polymers, intracellular (laminaran) and extracellular β-D-glucans can also serve as readily available carbon and energy sources. Since other glycans (Table 2), such as ketoglycans [levans (74) and inulin (182)], and xylan (204) also function as food-reserve polysaccharides, it is obvious that a classification on the basis of biochemical function (storage polysaccharides) is heterogeneous with respect to the chemotypic classification (Tables 1 and 2).

The β-D-glucan cellulose, which is the most abundant organic compound in nature and is the most prevalent in the plant kingdom, has been reported as a structural or extracellular polysaccharide in all major forms of life. Similarly, galactans have been found in all major forms of life; however, the anomeric configurations of these homoglycans vary from one kingdom to another. For example, galactocarolose in *Penicillium* species is a (1→5)-linked D-galactofuranosyl polymer (51), whereas the galactan in plants (lupin) is a β-(1→4)-linked D-galactopyranose polymer (11). Similarly, the homoglycans (mannans, ketoglycans, and xylans) and the neutral and complex heteroglycans (Table 2) in the various kingdoms can generally be differentiated from one another on the basis of variations in their anomeric configurations, the modes of linkage between the respective monomers, and/or the qualitative

Table 3. Distribution of Substituted Polysaccharides and Carbohydrate-Containing Biopolymers in Various Kingdoms[a]

	Protista			
Biopolymeric Class	Lower Protists	Higher Protists	Plant	Animal
Substituted polysaccharides				
DNA (57)	+	+	+	+
RNA (57)	+	+	+	+
Lipopolysaccharides	+ (170)	± (240)	−	−
Peptidoglycans	+ (78)	−	−	−
Polyols				
Teichoic acids	+ (19)	−	−	−
Glycolipids	+ (22)	+ (158)	+ (3)	+ (180)
Glycoproteins	+ (134)	+ (23)	+ (209)	+ (189)

[a] See the footnotes to Table 1.

and quantitative distribution of the monomers within the respective polymers. Consequently, of the numerous homo- and heteroglycans, cellulose and glycogen represent the universal polysaccharides that persist in all three major forms of life.

On the basis of the binary kingdom classification, the homoglycan chitin would be present in both the plant (fungi and primodial algae) and animal kingdoms. In contrast, on the basis of the ternary kingdom classification, chitin would persist in only the higher protists (72) and the animal kingdoms (126, 228).

Closely related chemotypic [substituted and unsubstituted β-(1→4)-linked D-glucans] analogs of chitin [β-D-(1→4)-linked 2-acetamido-2-deoxy-D-glucan] in the lower protists (bacteria) and the plant kingdoms would be the 3-*O*-lactyl-"substituted chitin" moiety in the bacterial peptidoglycan [2-acetamido-2-deoxy-D-glucose in β-(1→4) linkage with its 3-*O*-lactyl muramic acid derivative], and cellulose [unsubstituted β-(1→4)-linked D-glucan], respectively. The stereochemical similarities between chitin and the glycan moiety of the bacterial peptidoglycan were further delineated by the early finding that both chitin and peptidoglycan serve as substrates for muramidase (lysozyme). These correlations further emphasize the fact that, in addition to taxonomy by chemotype, the more utilitarian classifications of polysaccharides, such as those based on biological function and cellular distribution, serve as aids in clarifying the phylogenetic relationships between organisms in the various kingdoms.

There exist some unique phylogenetic relationships between sulfated polysaccharides in the various kingdoms (Tables 1 and 2). Sulfate-containing polysaccharides in the plant kingdom are limited to sulfonic acid derivatives (213). In contrast, dermatan sulfate has been found in both bacterial (56) and animal (129) genera. Furthermore, although chondroitin sulfates have been reported in the animal kingdom (128), the nonsulfated heteroglycan, 3-*O*-α-glucuronyl-*N*-acetylgalactosamine, which can be considered as an analog of mammalian chondroitin sulfate, has been isolated from *Bacillus subtilis* (125). Consequently, since plant polysaccharides are known to be void of amino sugars [D-glucosamine has been reported in glycolipids (43) and glycoproteins (256) of plants], it is suggested that the glycosaminoglycans in the animal kingdom have their origin in identical or analogous polymers that have been found in Protista (26, 56, 132, 140, 227).

In addition to the chemotypic similarities between bacterial polysaccharides and the glycans in higher protists, as well as glycans in both the plant and animal kingdoms, numerous polysaccharides have been isolated only from organisms in the lower protists. These include homoglycans (dextran, myucodextran, crown gall glucan, colominic acid, Vi-antigen, and polymannuronic acid), substituted polysaccharides (lipopolysaccharides and peptidoglycans), and the polyol phosphates (teichoic acids); consequently, the isolation of one or more of these specific polyhydroxy macromolecules from an organism would tentatively characterize the phylogenetic origin of the species as a lower protist.

The above classification of polysaccharides represents a minimal chemotypic grouping of these biopolymers. For example, the D-glucans in Table 1 can be further subdivided as branched or linear homoglucans; furthermore, homoglycans, such as galactans, mannans, xylans, and ketoglycans, can be similarly subdivided (57). Since uronic acids have been reported in approximately half of the more than 200 polysaccharides of biological origin (263), the single taxonomic group consisting of heteroglycans containing uronic acids comprises a major portion of the chemotypically defined polysaccharides. Bacterial heteroglycans in this group would include not only such polyuronides as alginic acid (86), which contains only D-mannuronic acid and L-guluronic acid residues, but also numerous extracellular and capsular polysaccharides that may contain relatively small quantities of uronic acid (1 mole uronic acid per 4 moles of neutral sugar monomers) residues within the repeating oligosaccharide subunit of the polymer (170).

Polysaccharides have also been chemotypically classified as acidic polysaccharides (28). Specific bacterial polysaccharides in the various chemotypic classes (Tables 2 and 3) contain a host of anionic functional groups, such as hexuronic acids, neuraminic acid, aminohexuronic acids, acyl substituents, phosphate, and/or sulfate, within their primary structures, and consequently can also be classified as anionic polysaccharides. For example, the type-specific capsular antigens of *Hemophilus* species (284), which contain polyribose phosphate, polyglucose phosphate, or polygalactosamine phosphate polymers, as well as colominic acid, Vi-antigen, and polymannuronic acid (Table 2), can be concomitantly classified as homoglycans and anionic polysaccharides. Similarly, the bacterial heteroglycans, uronic-acid-containing heteroglycans, dermatan sulfate, and hyaluronic acid (Table 2), and the substituted polysaccharides, lipopolysaccharides (Table 3), are, in essence, anionic polysaccharides. Numerous bacterial anionic polysaccharides can also be classified as exopolysaccharides, since they persist either as cell wall biopolymers, capsules, or secreted extracellular slime (28, 170).

Since nucleic acids contain a backbone region of ribose (RNA) or deoxyribose (DNA) units linked together via phosphate ester linkages, and since all the sugar residues are substituted with either purine or pyrimidine bases at the C-1 position of the respective sugar moieties via *N*-glycosidic linkages, RNA and DNA macromolecules have been classified as substituted polysaccharides (57, 275). The carbohydrate nature of specific bacteriophage DNA is further exemplified by the covalently bound mono- and diglucosyl residues in the hydroxymethylcytosine moieties of T-even coliphage DNA (159). With respect to analogies between RNA and substituted bacterial polysaccharides, structural similarities between type b *Hemophilus influenzae* capsular polysaccharide (a double-stranded polyribophosphate polymer) and RNA have been noted (220). The ability of pancreatic ribonuclease to hydrolyze the capsular polyribophosphate not only aided in the elucidation of the steric similarities between the type b antigen and RNA but also gives further credence to the classification of nucleic acids as substituted polysaccharides.

The functional classification of intracellular glycogen and amylopectin as storage compounds that mutually exclude one another within a specific cell has been questioned in recent years (68, 92). For example, physicochemical distinctions between amylopectin and glycogen are not completely defined, in that glycogenlike polymers are known to coexist with amylopectin in sweet corn (92). Although glycogenlike polymers are more prevalent in bacterial species than amylopectin, it has been recently noted that *Clostridium botulinum* type E synthesizes an intracellular amylopectin (storage α-D-glucan) that is rapidly depleted upon sporulation (265). Furthermore, the supposition that glycogen exists only as an intracellular storage polymer which functions solely as readily available carbon and energy sources appears to be no longer tenable. In bacterial cells it has been shown that, during the logarithmic growth phase, a major portion of the newly synthesized glycogen exists in a bound form with other unidentified biopolymers (274, 277). Furthermore, glycogen–DNA and glycogen–RNA complexes have been found in purified mammalian, plant, and bacterial DNA and RNA preparations (275). In addition, repartitioning of oyster glycogens by a phenol–water extraction procedure resulted in the isolation of small quantities of glycogen that were associated with protein and/or DNA (276). These findings suggest that bound forms of intracellular glycogen may possess still undefined cellular functions in addition to serving as carbon and energy sources.

2. GENERAL PHYSICOCHEMICAL CHARACTERIZATION OF MICROBIAL POLYSACCHARIDES

Microbial polysaccharides are physicochemically and immunochemically highly active compounds and interact readily with other macromolecules, as well as with small molecules, possessing any kind and amount of easily available ionic, hydrogen, or hydrophobic bonding sites. Since the net charge on the saccharide's own molecules is negligible and variable, it is suggested that polysaccharide complex formation is mediated via hydrogen bonding, which may vary in strength from weak van der Waals forces to a covalent bond.

The types of sugars in microbial polysaccharides are as numerous as the types of amino acids in protein macromolecules, but the number of different sugars occurring in a specific microbial polysaccharide is smaller, partly because it consists of repeating units containing generally not more than three to five different sugars. The molecular structures of most microbial polysaccharides are more expanded than those of globular proteins, but some polysaccharides are more complicated in their molecular structures. Glucose, galactose, and rhamnose are the most common sugars occurring in bacterial polysaccharides.

With respect to the physicochemical characterization of microbial polysaccharides and their monomeric constituents, a series of review articles has recently appeared in the literature (16, 24, 28, 78, 84, 101, 170, 186, 233). The studies have shown the applicability of laser-Raman (253), attenuated reflectance-infrared, (ATR–IR) (248), polarized ATR–IR (247), near-infrared (NIR) (249), electron spin resonance (ESR) (280), and nuclear magnetic resonance (NMR) (48, 73) spectroscopic methods for the characterization of the functional groups in carbohydrates and/or the detection of carbohydrate–biopolymeric complex formation.

An area of primary concern to the contemporary microbiologist is the structural interrelationships between cellular macromolecules and the physicochemical forces involved in defining the architectural organization of biopolymers at the cellular level. In bacterial cells, intracellular polysaccharides, which may serve as carbon and energy sources (unbound glycans) and fulfill other undefined functions (such as glycan–DNA and glycan–RNA complexes), are relatively limited when compared to the multitude of extracellular glycans (lipopolysaccharides, D-glycans, capsular homo-and heteroglycans, peptidoglycans, teichoic acids, and/or exopolysaccharides that are prevalent in bacteria.

Extracellular polysaccharides of bacteria are either covalently or noncovalently linked to each other and/or other extracellular biopolymers (lipoprotein, protein), as well as to the macromolecules (lipid, polysaccharides, and protein), that are prevalent in the bacterial cytoplasmic membrane (275). These substituted polysaccharides often form a backbone in a heteropolymer, such as lipopolysaccharide (LPS) or peptidoglycan, found in microbial cell walls; but a terminal saccharide molecule exerts a profound immunochemical influence on the conformational subunit of a heteropolymer or a homopolymer, expressing its fine antigenic specificity.

Many of the physiological, biochemical, and immunological, as well as the physicochemical, properties of microbial polysaccharides have been elucidated,

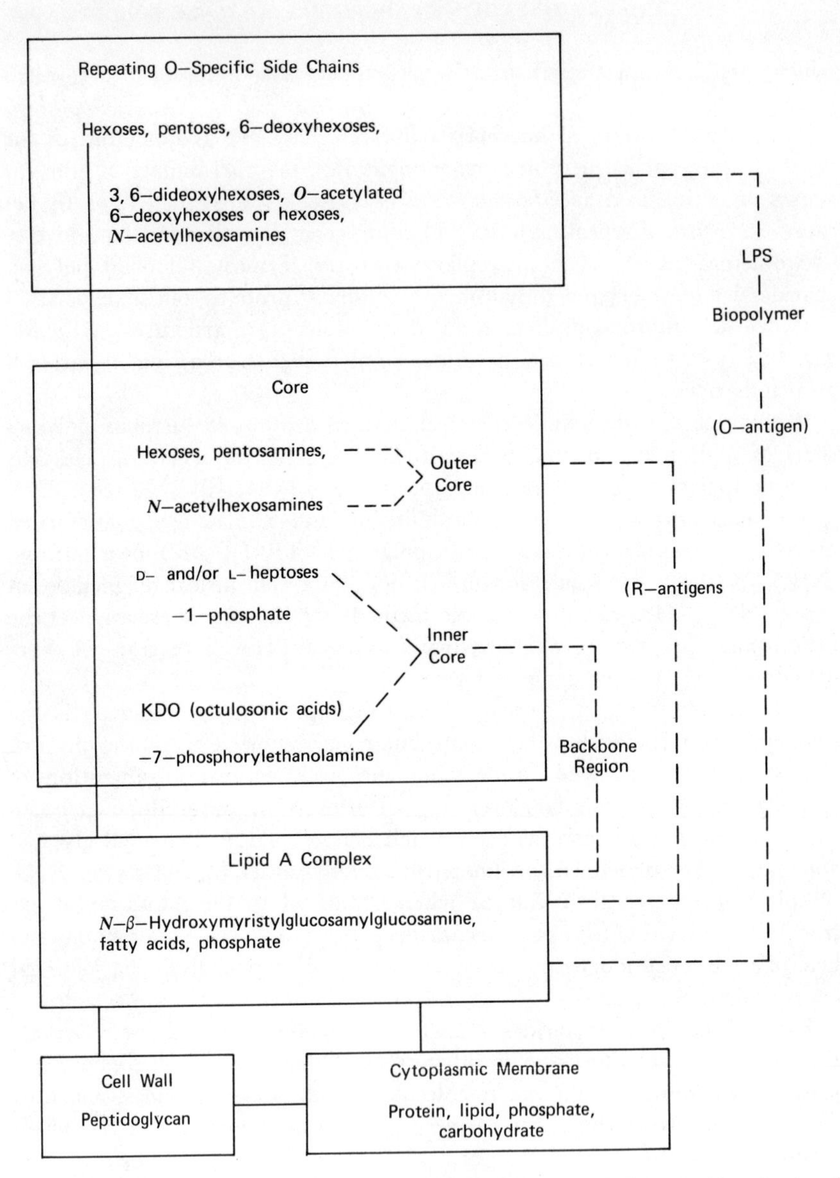

Fig. 1. Tentative distribution of monomeric constituents in lipopolysaccharide subunits of Enterobacteriaceae.

and recent monographs and review articles attest to the continued interest in both delineating the multiplicity of their biological functions and characterizing the architectural structures of carbohydrate-containing biopolymers (5, 10, 24, 25, 28, 68, 78, 113, 126, 127, 135, 152, 181, 212, 235, 242, 262, 263, 271). Another area of recent interest has been the characterization of covalently and noncovalently linked noncarbohydrate residues that persist in either partially or highly purified glycans or their substituted derivatives (19, 22, 28, 56, 61, 62, 89, 90, 101, 134, 162, 171, 178, 180, 228, 247, 275, 276, 281). For example, the immunogenic specificity of a synthetic gum arabic antigen against *Salmonella typhimurium* infections in mice is dependent on the presence of covalently linked acetyl or propionyl groups within the primary structure of the plant galactan (124). Since chemically labile acyl groups have also been shown to be covalently attached to the carbohydrate moieties of microbial exopolysaccharides (66, 239), and since such exopolysaccharides are known to function as both phage receptor sites and substrates for phage-induced polysaccharide depolymerases (210, 241), it is suggested that such noncarbohydrate residues within the primary structure of exopolysaccharides may also possess undefined biological functions related to either phage absorption and/or depolymerase activities. With respect to bacterial homoglycans, it is noteworthy that all highly purified *Escherichia coli* B glycogen fractions inadvertently contained either bound or coprecipitated form(s) of nondialyzable phosphate (275).

The substituted heteroglycan, LPS (Fig. 1), represents one of the most chemically heterogeneous biopolymers; in addition to its complex carbohydrate moiety (16), it may contain numerous noncarbohydrate substituents such as low-molecular-weight acyl groups, "lipid A," ethanolamine, and phosphate within its primary structure (197). In some instances, these noncarbohydrate residues have been shown to be essential for the elicitation of specific physiological, biochemical, immunological, and/or biological functions of the glycan (120, 121, 164).

Functionally, polysaccharides in microorganisms have been conveniently classified into two principal biological groups, namely, extracellular and intracellular polysaccharides. In lower protists, the intracellular polysaccharides, such as glycogen, starch, amylopectin, and mannans, have generally been considered as food-reserve polysaccharides; however, the ability of these intracellular glycans to complex (63, 275) with other intracellular polymers (DNA, RNA, and/or proteins) suggests that intracellular glycans may exist as physiologically and/or biochemically active complexes which may participate in a host of cellular regulatory mechanisms governing cellular growth and division. Furthermore, since glycogen has been found in cell envelope fractions of *E. coli* (260, 275), it is evident that glycogen can be defined as both an intracellular and an extracellular glycan. These findings also emphasize that, on the basis of biopolymeric complex formation, intracellular homoglycans may possess other biological functions, in addition to serving as food-reserve polysaccharides.

Glycogen in bacteria appears to exist in essentially two discernible forms, namely, a "free" unbound form and a firmly, but noncovalently, bound form. Glycogen forms *in vitro* complexes with numerous other cellular biopolymers, such as anionic polysaccharides, nucleic acids, lipopolysaccharides, and proteins. Most of these polymers possess the potential of interacting with cell wall glycogen. In a concentrated solution, the conformations of polysaccharides and their secondary structures are influenced by the large number of inter- and intramolecular hydrogen-bonded interactions occurring between similar and dissimilar macromolecules. The potential polar sites in the neutral glycogen molecules that may be involved in such hydrogen-bonded interactions are the —OH, —C—O—C, and —C=O functional groups of the glycogen molecule and the same and other polar sites, such as NH_2, COO—, —SO_3—, and —PO_3—, in the above cell wall polymers. It may be expected that *in vivo* hydrogen-bonded interactions between the cell wall polymers may alter the conformations of the respective biopolymers so that the immunological classification of bacteria based on chemotype may not be entirely valid. In this respect, stereochemotypes are probably more accurate criteria for immunochemical classification.

The major substituted polysaccharides and polyols in bacteria are peptidoglycans, teichoic acids, low-molecular-weight glycolipids, LPSs, ribonucleic acid, and deoxyribonucleic acid. In the specific aperiodic polyhydroxy macromolecules, RNA and DNA, the ribose and deoxyribose moieties are linked through phosphate ester bridges, whereas their *N*-glycosidic linkages occur as pyrimidine and purine *N*-glucosides. These highly informational macromolecules are distributed mainly as either intracellular or membrane-associated biopolymers. The sulfonated bacterial polysaccharides, possessing —SO_3—functional groups, are found in bacterial capsules of some Gram-negative and Gram-positive bacteria, for example, in clostridia.

The teichoic acids represent diverse types of multifunctional polyhydroxy macromolecules in which the repeating polyol (glycerol or ribitol) may or may not contain a monosaccharide substituent. For example, the repeating unit of the cell wall teichoic acid of a *Bacillus* species was found to be poly(glycerol)phosphate-*N*-acetylglucosamine, whereas in a *Staphylococcus* or *Micrococcus* species it was a poly(glycerol)phosphate-*N*-acetylglucosamine (13). In contrast, other numerous poly(glycerol) phosphate and poly(ribitol) phosphate macromolecules occurring either in the cell wall or membrane teichoic acids contain glucosyl substituents on the glycerol or ribitol moieties, but not within the repeating polyol subunits (19, 21). The teichoic acids possess an overall negative charge (anionic phosphate groups) and have a high affinity for monovalent and bivalent cations, and therefore it is believed that the major function of teichoic acids is to serve as cation-binding polymers (121, 123). The molecular arrangement and the manner in which cell wall and membrane teichoic acids are linked with other cellular biopolymers have been extensively investigated (53, 229, 267). It has been found, for example, that a covalent linkage between teichoic acid and peptidoglycan occurs between a terminal phosphate of the teichoic and peptidoglycan. Similarly, the teichoic acid may be attached to muramic acid through a phosphodiester linkage. In addition to functioning as cation-binding polymers, teichoic acids are known to possess immunochemical properties (184, 229).

Although the physicochemical properties of bacterial lipopolysaccharides have been extensively investigated, a LPS macromolecular model for a given bacterial species has yet to be fully elucidated. For example, the physical structure of *E. coli* LPS has been shown to vary with the physicochemical methods employed for the isolation of the LPS moiety. In some instances, physicochemical analysis revealed celluloselike micellar structures, consisting of aggregated bundles of polysaccharides (6) or a phospholipidlike, ordered "leaflet structure" (221). In other studies, lipopolysaccharides of *E. coli* appeared in the electron microscope in the form of droplets and long ribbons (224). The Boivin-type LPS from *E. coli* seems to occur in the shape of a doughnut, flat sheet, slender rodlike filament, or snake (27). Such LPS complexes with varied physical structures still possess endotoxic activity.

A LPS of *S. typhimurium* has the form of a homogeneous ribbon consisting of two monolayers of polysaccharide and covalently bound lipid so assembled that the two are bound together by their hydrophobic lipid moieties, whereas the polysaccharides constitute the exposed, hydrophilic surfaces of the particle (227). This assembly separates into monolayers when sonicated, heated, or exposed to ether or alkali to form homogeneously stained discs. Both the ribbon and the disc structure have a trilaminar appearance resembling cross sections of arrays of biomolecular leaflets. The trilaminar structure is determined by the whole polysaccharide molecule, although the lipid moiety is the major determinant.

With respect to the monomeric carbohydrates, the LPS molecules are built from repeating units synthesized as oligosaccharides on a lipid carrier and incorporated into the LPS macromolecules. The polysaccharide moiety combined to lipid in rhizobia and agrobacteria consists of glucose and rhamnose, as well as glucosamine, 4-OMe-glcA, galactose, and fucose (88). A LPS of *Proteus vulgaris* OX19 contains D-galactose, glucose, galNAc, fucose, heptose, and 2-keto-3-deoxyoctanoic glcNAc acid (200). The saccharide parts of all *Neisseria perflava* and *Neisseria catarrhalis* cell wall LPSs contain, respectively, D-glucose and L-rhamnose, both linked 1→3 in a chain, and D-galactose together with OAc (2). In general, the terminal O-antigenic side chains of the LPS may contain from two (257) to five (111, 171) different monomers within the repeating subunit; however, the multiunit LPS biopolymer may contain as many as ten specific monosaccharides and/or substituted monosaccharides within its primary structure. From the results of recent studies (111, 160, 164, 165, 171, 271, 275), a tentative model of the multiunit LPS macromolecule is depicted in Fig. 1. The three major subunits consist of the O-specific side chain, which elicits the O-antigenic specificity of an organism, a core region, and the innermost lipid A glucosaminyl–glucosamine complex, which is involved in linkages to both the cytoplasmic membrane and the peptidoglycan moieties of the bacterial cell.

Recent physicochemical studies on purified LPS preparations have shown that *O*-acyl-substituted glucose (109, 111), galactose (163), abequose (110), or rhamnose (164) moieties persist within the serologically specific O-antigenic side chains of *Shigella, Salmonella*, and *Klebsiella* species. In certain instances, both an *O*-acetylated hexose, deoxyhexose, or dideoxyhexose and *N*-acetylated glucosamine have been found in the repeating side chain. With respect to the O-antigenic side chains of *Shigella flexneri*, types 3C and 4B, which contain *N*-acetylglucosamine, 2-*O*-acetyl-L-rhamnose, D-glucose, and L-rhamnose residues, the immunodominant monomer of their common group 6 antigen was found to be the substituted monosaccharide, 2-*O*-acetyl-L-rhamnose (164). In another study, it was shown that removal of acetyl-linked pyruvic acid from the capsular polysaccharide of *Pneumoccus* type IV results in the unmasking of its group-specific antigen (116). Although the immunological specificity of complex bacterial polysaccharides has been thought to reside in specific unsubstituted monosaccharides, the above studies have shown that acyl-substituted monosaccharides also function as type- and/or group-specific immunodominant monomers.

In the lipopolysaccharides of shigellae, the core region also contains *N*-acetylated hexosamines and *O*-acetylated glucosyl or rhamnosyl substituents (219, 231). Although originally each genus of Enterobacteriaceae was assumed to contain specific invariable monomers in its core LPS subunit, it has been revealed more recently that the core regions of the *E. coli* (223) and *S. typhimurium* (160) LPS polymers contain more than one polysaccharide subunit. Probably the core LPS subunits in other genera are also heterogeneous, as indicated by the presence or absence of functional monomeric constitutents, such as arabinosamine, D- and L-heptose, and variable amounts of 2-keto-3-deoxyoctonic acid (KDO), in S- and R-polysaccharide moieties of *Salmonella* and *Escherichia* and *O*-acyl substituents in S- and R-polysaccharide moieties of *Salmonella* and *Shigella*. Indeed a single strain of *Salmonella minnesota* was found to contain two distinct LPS polymers (75). Studies on the molecular weight, polydispersity, and chemical composition of different polysaccharide fractions obtained from Gram-negative bacteria have further suggested that LPS macromolecules within bacterial cell walls do not exist as single, homogeneous entities. Actually, the LPS polymers seem to persist not only as multiple, discrete biopolymeric entities, but also as biopolymeric complexes with other cell wall polymers.

In Enterobacteriaceae the LPS macromolecule may also form complexes with other skeletal heteropolysaccharides, such as peptidoglycan, acidic capsular polysaccharides, and the common acidic exopolysaccharide, colanic acid. A tentative representation of the spatial distribution and macromolecular organization of the cell wall polymers in Enterobacteriaceae is shown in Fig. 2. In

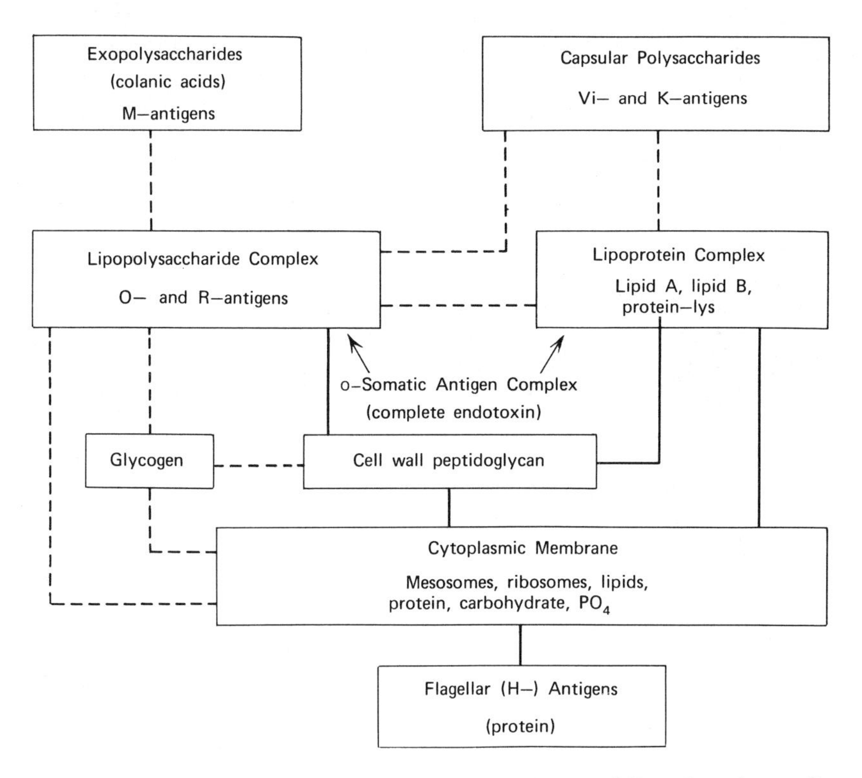

Fig. 2. Tentative interactions between cell wall biopolymers of Enterobacteriaceae. Complex formation mediated through covalent (—) and noncovalent (ionic and/or hydrogen: ———) bonding forces.

this model, a multiplicity of intermacromolecular interactions that are likely to prevail at the cell surfaces of Enterobacteriaceae is shown. For example, the most reactive cell wall biopolymer, LPS, is capable of forming complexes with all other cell wall polymers, such as exopolysaccharides, capsular polysaccharides, peptidoglycan, and glycogen, as well as with the cytoplasmic membrane. The K- and O-heteropolysaccharides are covalently linked in some undefined manner.

The peptidoglycan is covalently bound to the cell wall lipoprotein. On the average, one lipoprotein molecule is covalently linked to every tenth peptidoglycan residue (35). The linkage between these biopolymers is in the site between a lysyl group of the lipoprotein and the carboxyl group of the optical L-center of the diaminopimelic acid residue in the peptidoglycan. It has been postulated that the lipoprotein moiety of the lipoprotein–peptidoglycan (murein–lipoprotein) complex stabilizes the total suprastructure of the rigid cell wall layer (34). With respect to lipid "A"–core LPS complexes, it has been proposed that the terminal KDO residue of the core LPS is ketosidically linked by an extremely labile covalent bond to either one of the hydroxyl groups of the glucosamine residues or to the β-hydroxymyristic acid residue of the lipid A complex (174, 198). The above studies exemplify the use of physicochemical methods for the characterization of the bonding forces involved in the orientation of macromolecular complexes at the cellular level.

3. INTRACELLULAR MICROBIAL POLYSACCHARIDES

In Protista, intracellular homoglycans that serve as food-reserve polymers include mannans, starch, amylopectin, glycogen, levans, xylans, and laminaran (Tables 1, 2, and 5). With respect to the multiple dispersion of chemotypically related α-D-glucan polymers in protists, it has recently been noted that the higher protist, *Cryptococcus laurentii,* synthesizes intracellular glycogen (branched α-1→4, α-1→6-D-glucan) and extracellular starch (linear α-1→4-D-glucan) biopolymer (225). In contrast, the lower protist, *Mycobacterium smegmatis*, synthesizes intracellulary located lipid-bound trehalose (1-α-glucosido-1-α-glucoside) and glycogen (67). Other intracellular carbohydrate-containing polymers include the substituted polysaccharides (DNA and RNA), the membrane-associated teichoic acids and lipoteichoic acids (14, 53, 229, 267), and the covalently bound monosaccharides and oligosaccharides in bacterial glycoproteins and glycolipids (Table 3).

Although heteroglycans are listed in Table 4 as intracellularly located, such a classification has only extremely limited validity. A cursory review of these intracellularly located heteroglycan fragments suggests that such complex glycans probably represent intracellularly synthesized extracellular polysaccharides that have not yet been assembled and transported across the cytoplasmic membrane. In support of this interpretation are the well-documented biosynthetic mechanisms involved in the intracellular synthesis of both the core and O-antigenic side chains of LPS polymers before their assembly and transport across the cytoplasmic membrane to their extracellular location (198, 217).

Actinomyces species have been reported to contain a glycogenlike polymer as well as a glycosaminoglycan (149, 150). Intracellular polysaccharides in *Mycobacterium tuberculosis* have been identified as consisting of moieties of arabinose and mannose. A polysaccharide isolated from *M. avium* consisted of mannose and inositol occurring in 2:1 ratio. Other polysaccharides obtained from *M. tuberculosis* were described as compounds constituted by D-arabinose or glucose, or by D-arabofuranose, glucosamine, and inositol (100). Other polysaccharides isolated from *M. tuberculosis* were identified as 3,5-dimethyl-D-arabinoside, 2,3,4-trimethyl-L-rhamnoside, 3,4,6-trimethyl-D-mannoside, 3,-4-dimethyl-D-mannoside, and dimethyl-2-acetamide glucoside, combined in the molecular ratio of 5:5:5:4:1. Galactose, mannose, arabinose, rhamnose, glucose, xylose, and ribose were detected in the polysaccharide–nucleic acid preparations obtained from *M. ulcerans, M. phlei*, and *M. smegmatis* (39). These studies suggest that heteroglycans may be prevalent as intracellular moieties in *Actinomyces* and *Mycobacterium* species; however, their intracellular location may only be transient and may reflect the intracellular synthesis of polysaccharide subunits that function primarily as skeletal (extracellular) biopolymers (20, 236).

The ability of intracellular glycogen to form complexes with intracellularly located substituted polysaccharides (DNA and RNA) has been amply mentioned in Sections 1 and 2 of this chapter. It is also well established that purified preparations of intracellular homoglucans, such as amylopectin, soluble starch, glycogen, and lichenan, may contain trace amounts of other nondialyzable moieties, such as phosphate, protein, lipid, and/or galactose (93, 94, 195, 208, 246, 276, 277). In amylopectin preparations, the bound phosphate is ester-linked through the primary hydroxyl groups of the glucose residues (208), and the ratio of phosphate to D-glucose residues has been reported to be 1:400 (92). With respect to galactose residues in purified liver glycogen from galactose-toxic chicks, the ratio of galactose to glucose residues was found to be 1:500 (141, 195). Partitioning of glycogens by a phenol–water solvent system has been shown to yield highly purified glycogen preparations in the water-soluble phase that still contain minimal amounts of both phosphate and protein (157, 276). In contrast, the phenol-soluble and phenol–water-insoluble glycogen fractions contained increasing amounts of

Table 4. Molecular Compositions of Group-Specific Polysaccharides of Some Protists

Monomers in Polysaccharide Units[a]	Organism
Extracellular Glycans	
Glc, glcNAc	*Pasteurella multocida*
Glc, glcNAc, or 4-methoxyglcNAc	*Rhizobium*
D-Glc, D-glcNAc, D-gal, L-fucose	*Aerobacter*
Glc, gal, D-glcNhc, L-fucose	*Azotobacter chroccoccum*
GalNAc, D-glc, rham (43:2:1); hexuronic acid, lactone	*Azotobacter vinelandii*
Glc, glcN, gal, man, galN, glcNAc, manNAc, rib	*Cl. perfringens*
Rham (1 → 4)glc, glcNAc, man, hept	*S. marcescens*
GlcNAc, xylose, gal, man	*Crypt. neoformans*
GlcNAc, gal, man, L-rham (1:1.5:1.1:1.5)	*Ps. aeruginosa*
L-Rham, D-glc (4:1)	*Acinetobacter calcoaceticus*
Acetylglc	*Mycococcus*
Gal, glc, glcN, heptose (rham, man, gaIN, manN, deoxyhexose)	*Arizona*
Aldobionic acid (ara)	*Vibrio comma*
Intracellular, Skeletal, and or Membrane-Associated Glycans	
Glc, glcNAc, gal, man/ara, muramic acid	*Nocardia*
Glc, glcN, man, ara	*Actinomyces*
Glc, glcN, rham	*Streptomyces* spp.
D-Gal, D-man, D-ara (2:1:3); glc, gal, ara, glcN; gal, glcN, man, ara	*C. diphtheriae*
Glc. glcN, gal, man, ara	*C. hoffmannii*
Glc, glcN, man, rham, 2-keto-3-deoxyoctonate, fucose	*Xanthomonas*
Glc/gal, rham, glcN, D-fucoseN, ara/xyl	*Chr. visolaceum*
GlcN; gal, rham, glcN; gal, glc, rham	*Listeria monocytogenes*
Glc, man, heptose, xyl, 3,6-dideoxyaldoxexoses	*Past. pseudotuberculosis*
Glc, ara	*Micromonospora*
6-Phosphoryl-α-D-glc-(1 → 3)-galNAc	*Micrococcus*
GlcA-(1 → 3)-galNAc	*B. licheniformis*, *B. megaterium*
Glc, galNAc, glcNAc, P	*B. cereus*
Gal, glcNAc, galNAc (2:1:1)	*B. subtilis*
Gal, glcN, xyl	*B. anthracis*
Gal, glc, ara	*Mycococcus*
Gal, glcN, ara, ribitol	*Dermatophilus*
Gal, glc, rham, glcN, P (1:1:6:1:1)	*Lactobacillus planterium*
Gal, glc, rham (1:1:1)	*Lactobacillus acidophilus*
L-Rham (1 → 3)-D-galNAc, *O*-α-D-galpyranosyl (1 → 6)-D-glcNAc	*Lactobacillus* spp.
Gal, glc, man (2:2.5:3), pyruvic acid, lactic acid	*Brevibacterium oleocaptus*
D-Gal, D-glc, fructose, ribose	*Rothia dentocariosa*

Table 4. (*Continued*)

Monomers in Polysaccharide Units[a]	Organism
Man:inositol (2:1)	*M. avium*
Man, gal, glc, rham, ara, xyl	*M. ulcerans, M. phlei, M. smegmatis*
3,5-Me-D-ara:2,3,4-Me-L-rham:3,4,6-Me-D-man: 3,4-D-man:Me-2-glcNAc (5:5:5:4:1)	*M. tuberculosis*
Polysialic acid	*N. meningitidis* group B and C
Acetyl (1 → 6) mannosamine phosphate	*N. meningitidis* group A
Exocellular Glycans	
D-galactan	*Pencillium charlesii*
α-(1 → 6)-or α-(1 → 3)-Glucans	*Leuconostoc, Pediococcus*
α-D-Glucopyranose (1 → 4, 1 → 6)	*Pullularia pullulans*
Cellulose	*Acetobacter*
Phosphomannan	*Hansenula, Hist.*
Glc, xyl	*Chlamydomones ulvaensis*
D-Glc, D-gal, D-man, L-rham, D-glcNAc	*Ps. aeruginosa, Mycococcus*
D-Man, arab	*M. tuberculosis*

[a] Abbreviations: gal, galactose; galA, galacturonic acid; glc, glucose; glcA glucuronic acid; man, mannose; rham, rhamnose; fuc, fucose; galN, galactosamine; galNAc, *N*-acetylgalactosamine; glcN, glucosamine; glcNAc, *N*-acetylglucosamine; manN, mannosamine; manNAc, *N*-acetylmannosamine; P, Phosphorus, arab, arabinose, xyl, xylose.

both bound protein and phosphate, respectively; furthermore, the pheonol-water-insoluble glycogen also contained 2% DNA (276), whereas HeLa cell glycogen partially existed as RNA-glycogen complexes (176). Consequently, on the basis of the persistence of trace amounts of bound residues other then glucose in purified homoglucan preparations, these homopolymers could also be chemotypically classified as substituted polysaccharides (Tables 1 and 3).

Trace amounts of noncarbohydrate residues in intracellularly located homoglucans are known to affect the physicochemical characteristics of the respective homoglucans. For example, the bound phosphate in amylopectin alters the physicochemical behavior of amylopectin (93). A correlation has also been shown between increasing innate turbidities of glycogen–water suspensions and increasing amounts of bound phosphate and protein in the glycogen preparations (276). In the latter instance (increased turbidity of glycogen suspensions as a function of bound protein, phosphorus, and/or DNA), weight average molecular weight determinations of glycogen by light-scattering or sedimentation techniques (92) may yield erroneously high values as a consequence of bound microions and/or macroions (190, 275).

Although polysaccharide interactions with other macroions or microions can be detected in numerous solvent systems and in the solid state, it is not to be inferred that the binding sites and bonding forces responsible for complex formation are identical under the various environmental conditions. For example, in studies on the conformations and secondary structures of carbohydrates or their complexes in the solid state (X-ray diffraction, ATR–IR, ESR, laser-Raman, or NIR studies), both inter- and intramolecular hydrogen bonding forces between the polar sites (—OH, —C—OC—, —C=O, NH_2—, —COO—, —PO_3—, and —SO_3—) on the respective carbohydrate and/or ligand moieties were enhanced (28). In this respect, the involvement of hydrogen bonding forces in carbohydrate–ligand complex formation in the solid state has been amply verified (246–248, 275, 276). Similarly, it has also been shown that polysaccharide–ligand complex formation in water (62, 247, 281) or organic solvent systems (61, 63–65, 276, 277) is partially dependent on both hydrogen-bonded and electrostatic interactions. Although the functional groups within the respective biopolymers that are involved in complex formation in solvent systems still remain to be defined, the above studies demonstrate the complex nature of the binding forces, which are partially dependent on solute–solvent interactions.

Variations in the conformations of intracellular polysaccharides have been noted in both aqueous solution and in the solid state. X-ray diffraction studies have shown that starch (273) posseses a helical structure, whereas numerous glycogen preparations are amorphous (247) in the solid state. Varations in the conformations of glycogens from different biological sources could be detected by ATR and polarized ATR–IR techniques (247). It has also been inferred that both helical and random-coil conformations of glycogen and amylopectin occur in aqueous solution (69, 70). The solution properties of starch are numerous, and it has been shown that starch can exist as random coils (59) or helices (70), which occur primarily in the presence of complexing agents such as iodine, abosrbed fats, or polar organic molecules that are complexed within the starch helix (92). Slow cooling of randomly coiled starch solutions results in the linear alignment of intermolecularly hydrogen-bonded starch molecules with the concomitant formation of insoluble micellar starch bundles; in contrast, rapid cooling of starch solutions results in intermolecular hydrogen bonding between random coils, with the simulataneous formation of starch gels (92). These studies reveal that both starch and glycogen preparations can assume numerous conformations under the influence of varied physicochemical environments and in the presence of specific complexing agents.

The organic solvent systems used for the isolation of polysaccharides from cellular homogenates are also known to influence the conformation of the specific glycans. For example, on the basis of differences in ATR–IR absorption spectra, it was shown that the conformations of *E. coli* O111:B4 LPS preparations which had been extracted by three different solvent systems were significantly different from one another (275). Similarly, extraction of intracellular glycogen by a combination of the TCA–ethanol–water, phenol–water, and 30% KOH–solvent systems resulted in the isolation of glycogens with different ATR–IR absorption spectra and, consequently, different conformations (247). These results further demonstrate that the conformation of a specific polysaccharide is at least partially dependent on the primary organic solvent system used to isolate the polysaccharide.

Although *in vitro* model system studies have demonstrated that numerous glycan–biopolymeric complexes can be detected in a host of physicochemical environments, information as to the existence of such noncovalently linked complexes *in vivo*, or their potential physiological, immunological, or biochemical functions, remains to be elucidated. It has been noted that glycogen–gelatin complex formation enhances the rate of glycogenolysis by α-amylase (278). Since gelatin is known to induce conformational changes of glycogen in the solid state (247, 248, 280), it is suggested that, via complex formation, a more favorable conformation of glycogen was induced that permitted α-amylolysis to proceed at an accelerated rate. It has also been observed that 30% KOH-stable RNA from *E. coli* B cells probably existed as glycogen–RNA complexes; consequently, it was suggested that glycogen may function as a stabilizing or protective agent inhibiting the alkaline destruction of the RNA moiety (275). The potential functions of noncovalently linked glycan–RNA (89, 90, 176, 275, 276), LPS–RNA or LPS-polynucleotide (275), glycogen-DNA (176, 275, 276), glycogen-LPS (275), and glycan-protein (48, 49, 247, 275, 276) complexes, which have been partitioned by organic solvent systems from subcellular fractions of bacteria, have not been clearly defined and may exhibit a multiplicity even greater than that of the respective complexes.

Partitioning of CP-grade substituted polysaccharides (nucleic acid preparations) of intracellular origin from organisms in the Protista, plant, and animal kingdoms by a TCA–ethanol–water solvent system revealed the presence of both glycogen–DNA and glycogen-RNA complexes in all preparations (275). Furthermore, although none of the DNA preparations bound additional quantities of glycans, RNA preparations bound additional amounts of glycogen, dextran, lichenan, *E. coli* LPS, or *S. enteriditis* LPS, either as ternary (RNA-glycogen-lichenan and RNA-glycogen-*S. enteriditis* LPS) or as binary (RNA-glycogen, RNA-dextran, RNA-lichenan, and RNA-*E. coli* LPS) complexes (275). These studies revealed that RNA was more amenable to polysaccharide complex formation than DNA and suggest that differences in the three-dimensional conformations of the respective substituted polysaccharides,

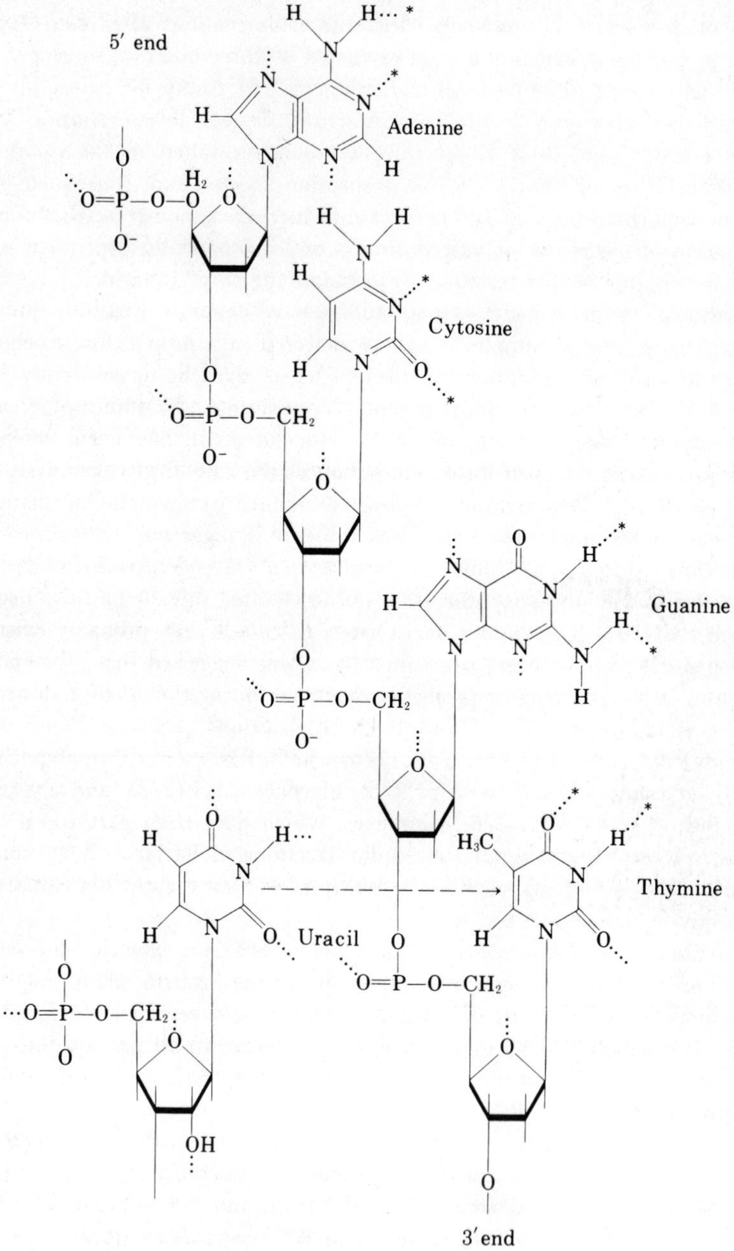

5′ end

Adenine

Cytosine

Guanine

Uracil

Thymine

3′ end

that is, the single- and double-stranded helical tertiary structures of native RNA and DNA, respectively (259), are, in part, limiting factors that govern the permissiveness of nucleic acid-glycan complex formation.

The functional groups and bonding forces involved in glycan–nucleic acid complex formation have not been defined; however, on the basis of other physicochemical studies, which have demonstrated the involvement of hydrogen bonding forces in glycogen–polyelectrolyte interacctions (61, 63, 65, 247, 248, 276, 280), numerous functional groups within the nucleic acids can be tentatively implicated in complex formation. For example, the following functional groups in the respective macromolecules have been reported to be involved in glycogen–LPS (247, 275) and glycan-gelatin (61, 247, 248, 276) complex formation: C=O, $—SO_3—$, $—PO_3—$, peptide bond, and the heteroring oxygen and primary and secondary OH groups of the sugar residues. The functional groups within the glycogen macromolecule that would be concomitantly involved in hydrogen-bonded interactions with other cellular polymers are the primary and secondary OH and heteroring oxygen groups (61, 246, 247). With respect to the primary and secondary structures of RNA and DNA (Fig. 3), it is proposed that complex formation may involve hydrogen-bonded interactions between the secondary OH and heteroring oxygen groups in the respective pentoses and/or the P=O groups of the sugar–phosphate backbones of the RNA and DNA polymers and the aforementioned functional groups of glycogen. Furthermore, the same functional groups in glycogen could also form hydrogen bonds between specific polar (NH, NH_2, and/or C=O) groups within the purine and pyrimidine moieties of the RNA and DNA polynucleotide chains (Fig. 3).

The bonding forces [van der Waals (hydrophobic), ionic, and hydrogen bonds] involved in the stabilization of the secondary and tertiary structures of nucleic acid polymers (259) are identical to the noncovalent bonding forces that have been implicated in glycan–biopolymeric complex formation (62, 65, 275, 276, 281). Since RNA, in most instances, exists as a single-stranded helix that involves a conformation with less intermolecular hydrogen bonding than the double-stranded helix of DNA (259), it is obvious that RNA polynucleotides contain more readily available hydrogen bonding sites which can preferentially form complexes with glycans than the intermolecularly hydrogen-bonded (hydrogen bonding between base pairs; cf. Fig. 3) double

Fig. 3. Some potentially strong hydrogen bonding sites within the primary structures of DNA and RNA tetranucleotide models. Polar groups capable of intra- and intermolecular hydrogen bonding with other cellular biopolymers are represented by . . . whereas hydrogen bonding sites involved in base pairing are denoted by . . . *. The RNA tetranucleotide can be visualized by substituting uridine (--→) for the thymidine residue and ribose for the deoxyribose residues in the other bases of the DNA tetranucleotide.

strands of DNA polynucleotides. This interpretation is in agreement with the experimental findings that RNA formed complexes with numerous glycans (275) whereas DNA did not form complexes with glycans, and only inherent glycogen–DNA complexes could be ascertained in the presence of TCA–ethanol–water (275) or phenol–water (276) solvent systems.

The biological functions of intracellular homoglucan–biopolymeric complexes have not been elucidated; however, it is probable that specific functions may well be a multiple factor of the respective complexes. For example, it has been postulated that bound glycogen in RNA- and/or DNA-glycogen complexes may function in numerous direct and indirect ways as a radioprotective agent for the respective DNA and RNA moieties (274, 275, 279). In addition, such complexes may be involved in cellular regulatory mechanisms that govern the synthesis and/or release of biological information from such informational macromolecules as DNA or messenger RNA. Furthermore, since bound forms of glycogen have been found to be associated with 70S ribosomal fractions of *Sarcina lutea* protoplasts (275), it is conceivable that ribosomal–glycogen complexes may also be involved in the regulatory mechanisms that govern the synthesis of cellular proteins. In any event, the ability of glycogen to bind with a host of other cellular polymers suggests that intracellular homoglycans may possess numerous undefined functions in addition to serving as carbon and energy sources for cellular metabolism.

4. EXTRACELLULAR MICROBIAL POLYSACCHARIDES

The extracellular polysaccharides are defined as the glycans that are exterior to the eytoplasmic membrane and can be conveniently subdivided into skeletal (cell envelope) glycans and exoglycans (glycans exuded into the growth medium). Extracellular glycans of Gram-positive and Gram-negative bacteria, as well as those in some higher protists, such as *Candida* (85), represent the majority of the well-characterized microbial glycans (Tables 1 to 5). Although glycans of the eukaryotic higher protists have been partially listed in Tables 1 to 5 for the purpose of elucidating phylogenetic interrelationships, the scope of this chapter is primarily limited to polysaccharides that are prevalent in prokaryotic microorganisms. With respect to mannose-containing glycans in numerous eukaryotic yeasts, it has recently been reported that 410 of 450 yeast species contained 180 different glycans having mannose as the predominant monomeric constituent (233, 243). The presistence of a multitude of extracellular α- and β-D-glucans, as well as heteroglycans (Tables 1 to 5), in higher protists (slime molds, fungi imperfecti, and fungi) is well documented,

Table 5. Intracellular and Extracellular Homopolysaccharides in Protista

Homopolysaccharide	Predominant Polymeric linkages	Microorganisms
Starch (amylose)	α-(1 → 4)	*C. diphtheriae, Cl. butyricum, E. coli, T. innocua, T. liquefaciens, T. mucorugosa, A. niger, Polytomella coeca, A. pasteurianum, A. mucosum, T. neoformans*
Glycogen and amylopectin	α-(1 → 4), α-(1 → 6)	*D. pneumoniae* (types I, II, III), *B. megatherium, M. tuberculosis, N. perflava, S. cerevisiae, Cycloposthium, Tetrahymena pyriformis, E. coli, Cl. botulinum*
Dextran	α-(1 → 6), α-(1 → 4), α-(1 → 3)	*S. bovis, S. viridans, S. viscosum, Streptobact. dextranicum, L. mesenteroides, L. dextranicum, Betabacterium vermiformis, Betacoccus arabinosaceous, A. capsulatum*
Cellulose	β-(1 → 4)	*A. pasteurianum, A. xylinum, A. ranceus, A. kutzengianum, A. acetigenum*
Polyglucosan	β-(1 → 2)	*Agrobacterium tumefaciens*
Luteose	β-(1 → 6)	*P. luteum, P. islandicum*
Glucan	β-(1 → 3)	*S. cerevisiae*
Polygalactan		*M. mycoides, P. charlesii, A. niger*
Polymannan		*B. polymyxa, Desulfovibrio desulfuricans, S. cerevisiae, P. charlesii*
Polyfructosan (levan)	β-(2 → 6)	*B. subtilis, B. cereus, B. mesentericus, B. megathericum, B. polymyxa, B. pumilus, S. salivarius, B. fluorescens, Ps. chlororaphis, Ps. aureofaciens, Acet. levanicum, Alc. viscosus*
Chitin (poly-2-*N*-acetamido-2-deoxyglucosan)	β-(1 → 4)	*S. cerevisiae, Rhodotorula glutins, Asp. niger, Neurospora crossa, Nadsonia fulvescens, Endomyces decipiens, Endomycosis capsularis*

and the reader is referred to recent reviews and articles (40, 46, 85, 96, 120, 145, 233, 243, 244) for a more detailed discussion of glycans in higher protists.

Representative exoglycans of the lower protists include the dextrans, levans (ketoglycans), cellulose, hyaluronic acid, and a limited number of uronic-acid-containing heteroglycans (Tables 1 to 5). The remaining glycans (excluding DNA and RNA), which have not previously been classified as either intracellular or exoglycans, persist as skeletal (envelope) glycans. In contrast to the aforementioned catabolically functional role of intracellular polysaccharides as food-reserve biopolymers, the extracellular polysaccharides generally cannot be metabolically utilized as potential energy and carbon sources by the microorganism that synthesized the biopolymer.

The subdivision of extracellular polysaccharides into skeletal glycans and exoglycans is a convenient but not absolute system of classification. For example, in certain bacterial species which contain concomitantly both a capsular (skeletal) glycan and an exoglycan, it has been found that both glycans may possess the same chemical composition (269). Such chemical homogeneity between capsular glycans and exoglycans has been reported in *Streptococcus* species (147), where hyaluronic acid was the common polymer, and in *Azotobacter vinelandii* (52), *Pseudomonas* species (33, 241), and *Klebsiella* species (238, 270), where genus-specific anionic heteroglycans were implicated as the common biopolymers. In the above instances, the exoglycans can be considered as excreted capsular glycans. In other cases, capsular glycans persist primarily as skeletal glycans. Well-defined capsular glycans include the common M-antigens (9, 80) and colanic acid glycans (237) in mucoid strains of Enterobacteriaceae, Vi-antigens of *Salmonella, Arizona*, and *Escherichia* species (112), and the type-specific capsular antigens (269) of *E. coli* (K-antigens), *Streptococcus* (*Diplococcus*) *pneumoniae* (222, 234), *Neisseria meningitidis* (134, 258), and *Hemophilus influenzae* (161, 220, 284).

Most microbial exocellular polysaccharides are heteropolysaccharides, whereas homopolysaccharides are produced by a few species, for example, the glucans found in culture filtrates of *Leuconostoc* and *Pediococcus*. Dextrans are branched molecules with chiefly α-(1→6) [but occasionally α-(1→3) or α-(1→4)] linkages. Extracellular cellulose is produced by *Acetobacter*. Other well-identified types of extracellular homopolysaccharides of protists include (*a*) pullulans, synthesized by *Pullularia pullulans*, a linear polymer of α-D-glucopyranose with 1→4 and 1→6 linkages, (*b*) phosphomannans, secreted by *Hansenula* and *Histoplasma capsulata*, (*c*) galactans composed of D-galactofuronosyl residues, produced by *Penicillium charlesii*, (*d*) starch [linear α-(1→4)-linked D-glucan] in *Cryptococcus laurentii* culture filtrate (225) and α-galacto homopolymer, secreted by mycococci (152). Well-identified, extracellular microbial heteropolysaccharides are as follows: acidic polysaccharide composed of D-glucose, D-galactose, D-mannose, L-rhamnose, and D-glucuronic acid residues, and produced by *Ps. aeruginosa* (37); a polysaccharide containing glucose and xylose, secreted by *Chlamydomonas ulvaensis*; and a polysaccharide consisting of D-mannose and arabinose, occurring in the culture filtrates of *M. tuberculosis* (151).

On a physicochemical basis, capsular heteroglycans (with a few exceptions), lipopolysaccharides, and teichoic acids can all be classified as macromolecular polyanions. The overall negative charges in these polymers are attributable to the presence within their primary structures of such functional groups as sulfate, phosphate, hexuronic acids, sialic acids, and/or aminohexuronic acids. It has been postulated that one of the major physicochemical functions of teichoic acids is to serve as potential cationic-binding polymers (19, 121). In an analogous manner, both LPS and most capsular heteroglycans in Gram-positive and Gram-negative bacteria could also function as cation-binding polymers. The physiological significance of skeletal cation-binding polymers would be the provision of a physicochemical mechanism for the concentration of cations from the external environment, thereby ensuring the cation concentrations that are essential for cellular metabalism.

Correlations between encapsulation of bacteria and increased virulence and pathogenicity of the respective microorganism have been well documented. The increase in virulence has been attributed to the ability of capsular glycans to inhibit the phagocytic activity of the host macrophages and leukocytes (29, 252). Capsular antigens, which form a thick layer over the skeletal O-somatic antigens in Gram-negative bacteria, may also function as protective barriers against O-phage absorption in that the O-antigen receptor sites, which are requisite for the initial physicochemical absorption process, are not accessible to the virus (139). Other biological functions that have been proposed for the capsular glycans include the potential ability to act as osmotic barriers and to protect the cell against dessication (9).

With a few exceptions (glycogen, hyaluronic acid, and possibly peptidoglycans), extracellular glycans of the lower protists are immunologically specific biopolymers that permit immunological differentiation between genera, species, and subtypes of a single species. For example, over 80 different subtypes can be identified in encapsulated *E. coli* (139) by their specific K-antigens (A-, B-, and L-antigens). Similarly, 6 subtypes of *Hemophilus influenzae* (272, 284), 5 subtypes of *Neisseria meningiditis*, (134, 258), and more than 75 immunologically distinct capsular serotypes of pneumococci have been defined. Intrageneric [pneumococcal glycans of types III and VIII, Table 6 (45, 104)] and intergeneric [*Lipomyces* (107) and *Klebsiella* (105, 165)] cross reactions with pneumococcal polysaccharides have been shown to be dependent on the stereochemical similarities between the oligosaccharide repeating units of the capsular polysaccharides in question. The chemical compositions of the monomeric constitutents and limited physicochemical characterizations of the respective polymers have been reported. For detailed immunochemical information on bacterial capsules, the reader is referred to the numerous review articles (25, 104, 139, 147, 170, 234, 242, 264, 269).

The capsular glycans of pneumococci were the first to be extensively investigated (104, 108, 185, 234, 264). Although the monomeric carbohydrate moieties in many of the type-specific capsular antigens have been defined (Table 6), the primary, secondary, and tertiary structures of only a few types have been definitely established. For example, the monomeric constituents in the pneumococcal type I anionic heteroglycan are galactose, fucose, glucosamine, and galacturonic acid, whereas the monomeric residues of type XVII capsular material (a neutral heteroglycan) are glucose and rhamnose (175); however, the fine structures of both polymers remain to be elucidated.

The tentative fine structures of some of the fully investigated capsular heteroglycans of pneumococci are depicted in Table 5. The ratio of L-rhamnose, D-glucose, and D-glucuronic acid monomers in the type II heteroglycan was previously thought to be 3:1:2, respectively (76, 103); however, later studies have shown the ratio to be 3:2:1 (106, 156). With the exception of the branching D-glucosiduronic acid in the repeating hexasaccharide of the type II heteroglycan, the anomeric configurations of the linkages are not definitely known. The basic disaccharide unit in the type III heteroglycan is cellobiuronic acid, a β-(1→4) glycosidically linked dimer of glucose and glucuronic acid residues are joined together via β-(1→3) glycosidic linkages (215). In the polymeric form, the cellobiuronic acid residues are jointed together via β-(1→3) glycosidic linkages. The type VI (211) and type VIII (132) polymers are the only other pneumococcal heteroglycans in which the fine structure has been elucidated. The type VIII heteroglycan is known to be a linear polymer with a minimum molecular weight of 140,000 and to contain (1→4) glycosidic linkages, exclusively.

The type V capsular heteroglycan contains, in addition to D-glucose and D-glucuronic acid, two unusual aminodideoxyhexoses, namely, 2-amino-2,6-deoxy-L-talose and 2-amino-2,6-deoxy-L-galactose. The type VII heteroglycan of pneumococci consists mainly of D-galactose, D-glucose, and L-rhamnose residues with small amounts of glucosamine and galactosamine. The monomers in the type IX heteroglycan are glucose, glucuronic acid, glucosamine, and an unidentified amino sugar, whereas the type XII pneumococcal capsule contains glucose, galactose, and a small amount of a 2-acetamido-2-deoxyhexose.

The type-specific pneumococcal heteroglycans in 15 of the 34 capsular polysaccharides listed in Table 6 contain, in part, monomeric constituents (ribitol or glycerol and PO_4) that may also be concomitantly present in the skeletal teichoic acid polymer of pneumococci (191, 222, 275). In these instances, the capsular heteroglycan fraction may be heterogeneous and contain both the specific capsular heteroglycan and a partially degraded skeletal and/or membrane teichoic acid residue (19). Alternatively, since teichoic acids are known to contain highly labile phosphodiester linkages within the teichoic acid chain (19, 75), it is possible that the ribitol or glycerol and PO_4 moieties within the heteroglycan polymer may represent a labile teichoic acid fragment that is still attached to the capsular heteroglycan, and in this respect may represent the primary linkage between this capsular glycan and a cell envelope teichoic acid.

The above data emphasize the chemical heterogeneity of the pneumococcal capsular antigens which are responsible for the immunogenic specificity of the skeletal glycans. In this respect, the capsular antigens of both Gram-positive and Gram-negative bacteria (the immunologically specific determinant groups of cell wall antigens) share a common immunological function that is attributable also to such other skeletal polymers as the teichoic acids (21, 53, 112, 275) and the lipopolysaccharides (135, 170). The capsular carbohydrates of *Streptococcus bovis* are pure glycans containing approximately 4 and 5 μM/mg of glucoseand relatively free of contaminating sugars such as rhamnose and glucosamine (136). A marked immunologic relationship was revealed among the capsular glucans of *Strep. bovis*. Isomaltose inhibited 90% of the reaction between a glucan and its homologous antiserum, thus indicating that isomaltose was a common feature of the capsular glucans isolated from a few selected strains of *Strep. bovis*. The immunological relationship existing among the various glucans depends, in part, on the relative proportion of the α-(1→6) and α-(1→4) linkages in the D-glucan. In this respect, there seems to be a considerable chemical analogy between *Strep. bovis* glucans and dextran (136). In fact, the dextranlike capsule may, in essence, be identical to one of the numerous dextrans (Table 5) that are known to be exuded (exoglycans) into the growth medium by many Gram-positive organisms (127).

The type-specific capsular polysaccharides (K-antigens) of klebsiellae are all acidic polysaccharides (165) and usually contain galactose, mannose, and galacturonic acid. Whereas molecules of galactose are linked only in the 3 position, the mannose molecules are linked in both the 3 and the 4 positions. In some species of *Klebsiella*, the capsules possess aldobiuronic acid, galacturonosylmannose and a neutral disaccharide, galactosylmannose-4 (or -3), but many other capsular polysaccharides are composed of glucuronic acid, glucose, and fucose in a molar ratio of 1:2:1. Fucose molecules are linked through C-3 and half of the glucose is at branch points linked through both C-4 and C-6, and the remaining glucose is at the nonreducing end of the polysacca-

Table 6. Structures of Extracellular Polysaccharide Units in Pneumococci

Polysaccharide Unit Structure or Monomers Detected	Immunotype
D-GalA, glc, galN, glcN GalA-(1 → 3)-glcN-(1 → 3)galA, galA→galN, galA→glcN	I
[→3)L-Rham-(1 → 3)-L-rham-(1 → 4)D-glcA-(1 → 4)-D-glc-(1 → 3)-L-rham-(1→]$_n$, with -D-glcA-(1 → 6) branch on D-glc	II
[→3)-β-D-GlcA-(1 → 4)-β(?)-D-glc-(1→]$_n$	III
D-Gal, D-galNAc, manNAc, fucNAc, pyruvic acid	IV
D-Gal, D-glcA, "pneumosamine" a-AcNH-2,6-dideoxy-L-talose, 2AcNH-2,6-dideoxy-L-galactose (L-fucosamine). Oligosaccharides indicated: 2)-β-D-glcA-(1 → 3)-L-fucN, D-glcA-(1 → 3)-L-fucN-(1 → 4)-D-glc	V
[→2)-α-D-gal-(1 → 3)-α-D-glc-(1 → 3)-α-L-rham-(1 → 3)-ribitol-1 or 2, or 4 or 5-O·P(O)(ONa)·O]$_n$	VI
D-Gal, D-glc, L-rham, galNAc, D-glcNAc. Nonreducing end groups: β-D-gal and L-rham	VII
[→4)-β-D-GlcA-(1 → 4)-β-D-glc-(1 → 4)-α-D-glc-(1 → 4)-α-D-gal-(1→]$_n$	VIII
D-Glc, *D*-glcA, D-glcNAc, manNAc. Oligosaccharides indicated: α-D-glcA-(1 → 3)-D-glc, glcA-(1 → 3)-glcNAc, manNAc-(1 → 3)-glc-(1 → 3)-manNAc	IX
Gal(*f*), galN, glcN, ribitol, PO_4	X
Gal, glc, glcN, ribitol, PO_4	XI
D-Gal, D-glc, D-glaN, L-fucN	XII
Gal, glc, glcN, ribitol, PO_4	XIII

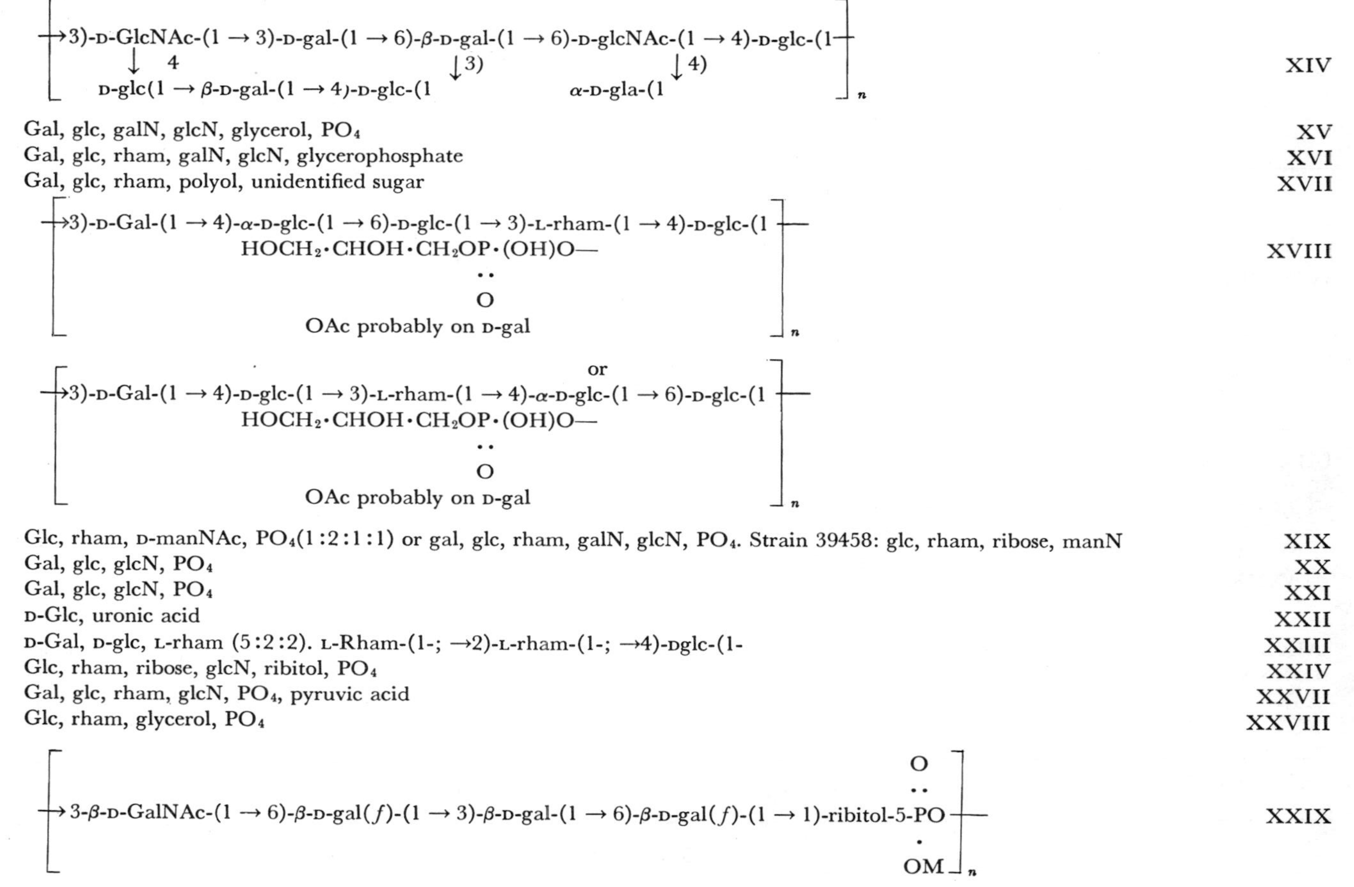

Composition / structure	No.
[→3)-D-GlcNAc-(1 → 3)-D-gal-(1 → 6)-β-D-gal-(1 → 6)-D-glcNAc-(1 → 4)-D-glc-(1→]$_n$; D-glc(1 → β-D-gal-(1 → 4)-D-glc-(1 → 4 of GlcNAc; ↓3) on β-D-gal; α-D-gla-(1 ↓4) on D-glcNAc	XIV
Gal, glc, galN, glcN, glycerol, PO_4	XV
Gal, glc, rham, galN, glcN, glycerophosphate	XVI
Gal, glc, rham, polyol, unidentified sugar	XVII
[→3)-D-Gal-(1 → 4)-α-D-glc-(1 → 6)-D-glc-(1 → 3)-L-rham-(1 → 4)-D-glc-(1 →]$_n$ HOCH$_2$·CHOH·CH$_2$OP·(OH)O— (with :O) OAc probably on D-gal; or [→3)-D-Gal-(1 → 4)-D-glc-(1 → 3)-L-rham-(1 → 4)-α-D-glc-(1 → 6)-D-glc-(1 →]$_n$ HOCH$_2$·CHOH·CH$_2$OP·(OH)O— (with :O) OAc probably on D-gal	XVIII
Glc, rham, D-manNAc, PO_4(1 : 2 : 1 : 1) or gal, glc, rham, galN, glcN, PO_4. Strain 39458: glc, rham, ribose, manN	XIX
Gal, glc, glcN, PO_4	XX
Gal, glc, glcN, PO_4	XXI
D-Glc, uronic acid	XXII
D-Gal, D-glc, L-rham (5 : 2 : 2). L-Rham-(1-; →2)-L-rham-(1-; →4)-Dglc-(1-	XXIII
Glc, rham, ribose, glcN, ribitol, PO_4	XXIV
Gal, glc, rham, glcN, PO_4, pyruvic acid	XXVII
Glc, rham, glycerol, PO_4	XXVIII
[→3-β-D-GalNAc-(1 → 6)-β-D-gal(*f*)-(1 → 3)-β-D-gal-(1 → 6)-β-D-gal(*f*)-(1 → 1)-ribitol-5-PO(:O)(·OM) →]$_n$	XXIX

Table 6. (Continued)

Polysaccharide Unit Structure or Monomers Detected	Immunotype
Gal, glc, galN, ribitol, PO_4	XXX
D-Gal, L-rham, glcA (2:2:1), also OAc, GlcA, gal	XXXI
Gal, glc, D-galA	XXXIII
Gal, glc, ribitol, PO_4	XXXIV
$[\rightarrow 3)$-D-Glc-$(1 \rightarrow 3)$-D-glc-$(1 \rightarrow]_n$, with -D-glc-$(1 \rightarrow 2)$-D-glc attached $1 \rightarrow 6$	XXXVII
$[\rightarrow 3)$-β-D-Gal(f)-$(1 \rightarrow 3)$-α-D-glc-$(1 \rightarrow 2)$-β-D-gal(f)-$(1 \rightarrow 3)$-α-D-gal-$(1 \rightarrow 2)$-ribitol-5-O·PO(OH)(:O)$]_n$	XLI(XA-XXXIV)
$[\rightarrow 3)$-β-D-Gal$(1 \rightarrow 4)$-α-D-glc$(1 \rightarrow 6)$-α-D-glc$(1 \rightarrow 4)$-α-D-gal-1$]_n$, with OAc $\downarrow$ and $HOCH_2 \cdot CHOH \cdot CH_2O \cdot P(OH)(:O) \cdot O \uparrow$ on the second glc	XLIII(XIA)

Table 7. Molecular Constitutions of Extracellular Polysaccharides of Klebsiellae

Molecular Constitution of Polysaccharide	Immunotype
D-Glc, D-glcA, L-fuc, mostly linked 1,3: pyruvic acid	1
$\left[\rightarrow 3)\text{-}\beta\text{-D-Glc-}(1 \rightarrow 4)\text{-}\beta\text{-D-man-}(1 \rightarrow 4)\text{-}\alpha\text{-D-glc-}(1 \rightarrow\right]_n$ with α-D-glcA linked $1 \rightarrow 3$ to β-D-man	2
D-Gal, D-man, galA : galA → D-man	3
$\left[\rightarrow 3\text{-}\beta\text{-D-gal-}(1 \rightarrow 3)\text{-}\alpha\text{-D-gal-}(1 \rightarrow 3)\text{-}\beta\text{-D-glc} \rightarrow\right]_n$ with α-D-glcA linked $1 \rightarrow 4$ to α-D-gal	8
$[\text{D-GlcA-}(1 \rightarrow]_n$ D-gal, L-rham	9
or	
$[\text{L-Rham-}(1 \rightarrow]_n \rightarrow 3)\text{-D-gal-}1 \rightarrow 4)\text{-D-glcA-}(1 \rightarrow$	47
$\left[6)\text{-}\beta\text{-D-Glc-}(1 \rightarrow 4)\text{-}\alpha\text{-D-glcA-}(1 \rightarrow 3)\text{-L-fuc-}(1\right]_n$ with β-D-glc linked $1 \rightarrow 4$ to α-D-glcA	54(A3S1)
D-Glc, L-rham, D-man, glcA → man	64

Table 8. Molecular Structure of Capsular Polysaccharides of *Escherichia coli*

Structure of Polysaccharide Unit	Polysaccharide Antigen
[gal 1 ↓ 3 —?)Glc-(1 → 3)-glcA-(1 → 3)-fuc—]$_n$	K27
[→6)Man-(1 → 3)-glc-(1 → 6)-man-(1 → 3)-glc-(1 → 3)-β-D-glcA-(1 → 3)-gal-α-(1 —]$_n$	K29
[→3)-Man-(1 → 2)-β-D-glcA-(1 → 3)-D-gal-(1 —]$_n$	K30
[→3)-D-Gal-(1 → 3)-D-galA-(1 → 2)-fuc(*f*)-(1 —]$_n$	K42
[rham (2 or ↓ ... 6) rham ↓ ... glcA 2- or 4-(GlcA-(1 → -man-(1 → 3)-man-(1 → 3)-glcNAc-(1 → ?)-man-(1 → 3)-man-(1 → 3)-glcNAc-(1—]$_n$	K85

rides. The structures proposed for capsular polysaccharides of klebsiellae (54) are listed in Table 7. It is apparent that the capsular polysaccharide is composed of tetrasaccharide repeating units. Only a few strains of klebsiellae possess capsules containing rhamnose as a nonreducing end group, linked through the 2- or 4-position, or fucose. Residues of glucose and/or galactose, or mannose and glucuronic acid, are frequently found in the capsular polysaccharides, either linked 1→3 or existing as branch points (Table 7). The type 9 capsular polysaccharide has been shown to consist of a pentasaccharide repeating unit containing D-glucuronic acid, D-galactose, and L-rhamnose in a ratio of 1:1:3, respectively (165).

The compositions of some capsular polysaccharides of *E. coli* and their characteristic repeating units are shown in Table 8. In the capsules of *E. coli*, galactose is the most common monomer, followed by glucose, mannose, and fucose, whereas rhamnose and neuraminic acid are found very rarely.

The capsular, group-specific polysaccharides of *Neisseria meningiditis* are acidic polysaccharides, composed mainly of amino sugars. The group-specific polysaccharide consists of D-glucosamine-6-phosphate, D-glucosamine, D-galactosamine, and D-glucose (12). Polysaccharide constituents of capsules of *Hemophilus influenzae* contain either glucose, ribose, galactosamine, and glucosamine, or actylglucosamine and acetylglucosamine and uronic acids. Capsular polysaccharides of *H. influenzae* types a, b, c, and f possess a common structural feature, consisting of (1→1)-linked disaccharide subunits that are bridged by phosphodiester groups. The type b capsular antigen is composed of two polyribose phosphate chains with phosphate diester bridges between C-3 and C-5, respectively (220). The two chains are linked through 1,1-(β,β)-glycosidic bonds between the ribose units of the chains. This feature resembles the double-stranded multilinked structure of ribonucleic acid (Fig 3).

A capsular polysaccharide of *Pseudomonas aeruginosa* has been found to contain glucuronic acid, galactose, mannose, and L-rhamnose in a molar ratio of 1:1.5:1.1:1.5 (33). Capsular antigens of *S. marcescens* contain one or more of the following constituents: a rhamnoglucan with the glucose linked mainly through 1,4 and occasionally through 1,6 bonds, a glucoheptan (probably a skeletal glycan) containing D-glycero-D-mannoheptose and L-glycero-D-mannoheptose, a rhamnoglucoheptose or a gluconic acid, glucose, and mannose, the mannose being bound through the 3 position. Capsular polysaccharides in *Aerobacter* consist of L-fucose, glucuronic acid, and D-glucose or D-galactose.

The capsules of *Rhizobium* contain glucose with a smaller amount of glucuronic acid in certain species; 4-methoxyglucuronic acid seems to occur in the polysaccharide of capsules of *R. trifolii*. The capsular polysaccharides of *Azotobacter* consist of galacturonic acid, glucose, and rhamnose in a proportional ratio of 43:2:1 (52), whereas the capsular polysaccharide of *A. chroococcum* contains both hexoses in addition to L-fucose and D-glucuronic acid. Capsules of *Cryptococcus neoformans* possess galactose, mannose, xylose, and glucuronic acid.

The capsular polysaccharide from type A of *Pasteurella multocida* has been identified as hyaluronic acid (49). A polysaccharide component of the polysaccharide–protein complex in capsules of *P. multocida* contains fructose, glucose, mannose, and glucosamine (144). The mucus antigen (M-antigen) of *Salmonella, Arizona*, and *E. coli* contains glucuronic acid, galactose, glucose, and fucose (8). The capsule of *Acinetobacter calco-aceticus* consists of L-rhamnose and D-glucose, combined in a ratio of 4:1 (133). In addition, capsules of mycococci have been reported to contain acetylhexose (152).

An area of current interest for molecular microbiology is the detection and elucidation of biopolymeric complexes and the definition of both their spatial distribution patterns within the cell and their potential biological functions. *In vitro* interactions between intracellular polysaccharides and both microions and other cellular biopolymers (DNA, RNA, and protein) have been amply described and referenced in the preceding discussion. The tentative linkages between extracellular polysaccharides (exopolysaccharides, capsular polysaccharides, LPS, and peptidoglycans) and other extracellular biopolymers (skeletal proteins, lipoproteins, lipids, teichoic acids, and membrane-associated teichoic acids) have been partially defined in numerous articles and reviews (14, 19, 22, 25, 34, 101, 135, 142, 180, 191, 194, 197, 216, 226, 229, 267, 271, 275). A tentative listing of bacteria in the order of increasing complexity of their skeletal glycans and other skeletal polymers would be as follows: Gram-positive micrococci and bacilli, diplococci, lactobacilli, streptococci, corynebacteria, and both Gram-negative organisms and microorganisms in the order Actinomycetales.

The substituted heteroglycan present in the walls of all bacteria is the rigid, water-insoluble peptidoglycan layer immediately surrounding the cellular cytoplasmic membrane. The peptidoglycan polymer not only is responsible for the rigidity of the bacterial cell wall, but also is the common site at which other nonpeptidoglycan skeletal biopolymers (LPS, teichoic acids, proteins, lipoproteins, and, in some instances, capsular biopolymers) are covalently and/or noncovalently linked to one another (19, 101, 275). The peptidoglycan also represents the cell wall biopolymer where synthesis is adversely affected by a host of cell wall antibiotics (32, 78, 236). Gram-positive bacteria generally contain considerably greater amounts (50–70%) of peptidoglycan in their cell walls than Gram-negative bacteria (5–15%). Conversely, Gram-negative bacteria (along with Gram-positive streptococcal organisms and bacteria in the order Actinomycetales) contain a greater array of chemically diverse biopolymers than Gram-positive bacteria. For example, the backbone of the group A polysaccharide of streptococcal cell walls is formed by an L-rhamnose polymer, with the rhamnose molecules joined by α-1,3-glycosidic linkages. The molecular compositions of the more thoroughly investigated streptococcal polysaccharides are shown in Table 9. A recent study has shown that the heteroglycan in *Strep. faecalis* cell walls is a diheteroglycan of glucose and galactose residues with approximately 18 of the following units per molecule (202):

$$\left[\overset{\beta\text{-}(1\rightarrow4)}{\rule{0pt}{0pt}}\text{Glu}\overset{\beta\text{-}(1\rightarrow6)}{\rule{0pt}{0pt}}\underset{\begin{array}{c}|\beta\text{-}(1\rightarrow4)\\ \text{Glu}\\ |\beta\text{-}(1\rightarrow4)\\ \text{Gal}\end{array}}{\text{Glu}}\overset{\beta\text{-}(1\rightarrow4)}{\rule{0pt}{0pt}}\text{Gal}\overset{\beta\text{-}(1\rightarrow4)}{\rule{0pt}{0pt}} \right]_{18}$$

The Gram-positive micrococci and aerobic bacilli represent the bacteria that contain the fewest skeletal glycan polymers. In addition to the basic peptidoglycan (41, 143, 201) moiety, micrococci generally contain either a wall teichoic acid polymer (14, 20, 53, 187) or a heteroglycan (1, 201, 206), as well as a common protein that is present in all recognized pathological immunotypes (78, 168). In some strains, the protein content may account for as much as 30% of the dry weight of the cell wall (283). The skeletal polymers that contain the immunologically group-specific determinants are the glucosyl- and/or *N*-acetylglucosaminyl-substituted teichoic acid and specific heteroglycans (2, 14, 19, 20, 53, 78, 177, 201, 206). On the basis of the studies

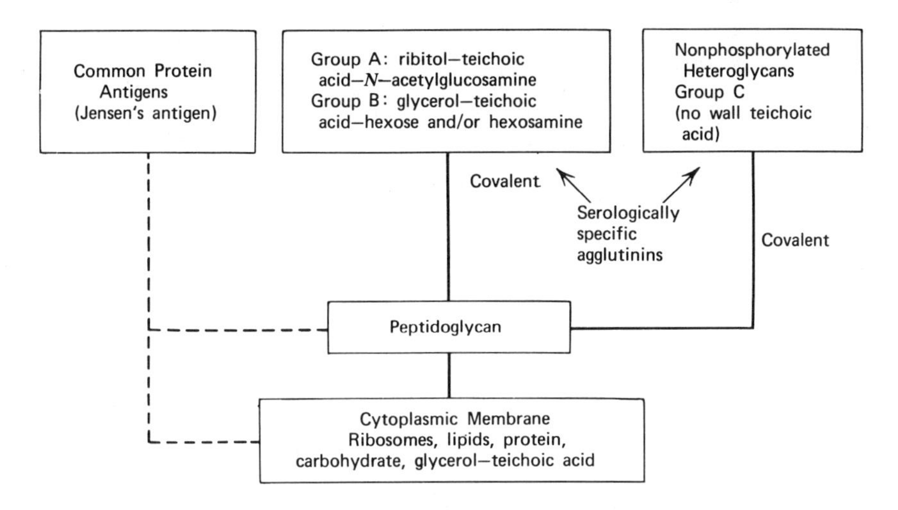

Fig. 4. Tentative orientation of cell wall biopolymers in micrococci.

Table 9. Molecular Constitutions of Streptococcal Polysaccharides

Molecular Constitution	Immunodominant Haptenic Group	Immunogroup	Immunotype
L-Rham, 60; glcNAc, 30	β-D-GlcNAc	A	
L-Rham, 70; glcNAc, 17	L-Rham, glcNAc	A-intermediate	
L-Rham, 85; glcNAc, 3	L-Rham-(1 → 3)-L-rham-(1 →	A–variant	
D-Gal, 10; L-Rham, 50; glcNAc, 12	$[$L-Rham-(1—$]_n$	B	
D-Gal, D-glc, glcNAc, sialic acid	$[$β-D-Gal-(1—$]_n$		II
L-rham, 42; D-galNAc, 40; glcNAc, 5	$[$GalNAc-(1 → ?)-L-rham-(1—$]_n$	C	
L-Rham, 59; D-galNAc, 22; glcNAc, 4		C–intermediate	
L-Rham, 88; D-galNAc, 2; glcNAc, 3	$[$L-Rham-(1—$]_n$	C-variant	
Teichoic acid, kojibiose, kojitriose		D	
D-Glc, rham, D-glcNAc, galN, ribitol, teichoic acid	α-D-Glc, D-glcNAc		I
Glc, rham, galN, glcN			XXVI
D-Glc, 22; L-rham, 44; glcNAc, 2	$[$β-D-Glc-(1 → ?)-1-rham—$]_n$	E	

D-Glc, 14.5; rham, 35; galNAc, 18.5	β-D-Glc-(1 → 3)-galNAc-rham	F	
Gal, 46.5; glc, 21.5; rham, 23; galNAc, 16.5			I
Gal, 12; glc, 26; rham, 24.5; galNAc, 19.5			II
[β-D-GalNAc-(1 → 2)-α-D-gal-(1 → 2)-α-rham-(1 → 4)-β-D-glc-(1 → 4)-D-glc-]$_n$			
Gal, 31; glc, 44; rham, 9	β-D-Glc, α-D-gal?		III
Gal, 31; glc, 35; rham, 26; glcN, 8	β-D-Gal		IV
Gal, 19; glc, 9; rham, 50; galN, 14.5	β-D-Glc?		V
D-Gal, 23.5; L-rham, 40.5; galNAc, 25.5	[L-Rham-(1—]$_n$	G	
Gal, glc, rham; occasionally galN, glcN, man in group K		H–K, M–Q, S	
GlcNAc, 6.0; galNAc, 6.4; L-rham, 12.9; D-gal, 12.1	β-GlcNAc-rham	L	
D-Gal, D-glc, rham, glcN	D-Gal	R	
Glc, 31; rham, 47; glcN, 6	L-Rham-(1-	z1	
Gal, 11; rham, 30; glcN, 30.5		z2	
Rham, 50; glcNAc, 17.5; galNAc, 16.5	3-O-α-GlcNAc-galNAc-	z3	
Gal, 12; rham, 48; glcN, 38.5	[D-Gal-(1—]$_n$	z4	
Gal, 11.5; glc, 15.5; rham, 47; glcN, 23		z6	

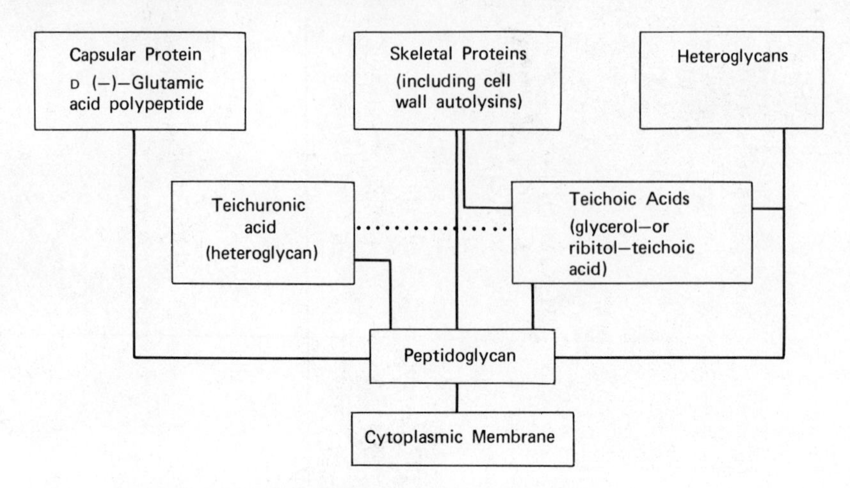

Fig. 5. Tentative orientation of cell wall biopolymers in bacilli.

cited above, a tentative model for the sequential macromolecular organization of cell wall biopolymers in *Micrococcus* is depicted in Fig. 4.

Similarly the bacilli generally contain only a limited number of cell wall glycans. The nonglycan polymers include a capsular polypeptide, skeletal proteins (38, 78, 122) that may account for as much as 10% of the cell wall weight (122), and either glycerol- (19, 77, 266) or ribitol-(15, 19) teichoic acid. The glycan polymers include peptidoglycan and the immunolgoically specific heteroglycans such as those found in *Bacillus licheniformis* [teichuronic acid (122, 123)], *B. megaterium* (77), and *B. anthracis* (232). In some cases, the heteroglycans have been shown to be covalently linked to the peptidoglycan moiety (15, 98, 123). In contrast, the *B. megaterium* heteroglycan appears to be linked to both the glycerol-teichoic acid and the peptidoglycan biopolymers (77). The above relationships between the skeletal biopolymers in *Bacillus* species are diagrammatically represented in Fig. 5.

The interrelationships between cell wall and eytoplasmic membrane biopolymers in Streptococcaceae and Lactobacillaceae have been similarly discussed in a recent publication (275). The chemical compositions and intramolecular arrangements of the monomeric constitutents of numerous microbial polysaccharides that are involved in extracellular macromolecular complex formation have been presented in Tables 4 to 9. In the class Schizomycetaceae, the microorganisms that possess the most complex array of skeletal biopolymers include bacteria in the family Enterobacteriaceae (275), and organisms in the order Actinomycetales. Organisms in the Actinomycetales have been of primary interest from two major viewpoints, namely, their ability

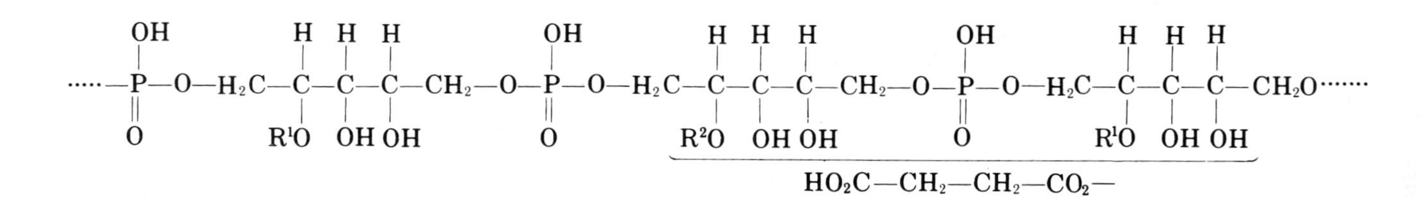

Fig. 6. Tentative formula for the structure of the ribitol-teichoic acid polymer of *Streptomyces streptomycini* (193).

to cause disease (*Mycobacterium, Nocardia*, and *Actinomyces* species) and their ability to produce over 500 specific antimicrobial agents, some of which have proved to be clinically useful antibiotics (*Streptomyces* species). The usual variety of carbohydrate monomers in specific antimicrobial agents from *Streptomyces* species (97, 98, 118) is analogous to the host of monomeric sugars found in the extracellular polysaccharides of microorganisms in the order Eubacteriales; consequently, it has been suggested that the carbohydrate-containing antimicrobial agents from *Streptomyces* species may, in essence, be cell wall degradation products that have been released into the culture fluid.

Although corynebacteria are not presently classified in the same phylogenetic order as mycobacteria, recently noted similarities between the complex extracellular polysaccharides and lipids (as well as extracellular macromolecular complexes) of these two genera have resulted in a tentative grouping of these organisms within the same order (25, 275), namely, Actinomycetales. The following complex polysaccharides have been identified in cells of *Corynebacterium diphtheriae*: (*a*) a polysaccharide composed of D-galactose, D-mannose, and D-arabinose in a molar ratio of 2:1:3, with a smaller amount of glucosamine and muramic acid (117); (*b*) an arabinose–glucose–galactose–glucosamine polymer; (*c*) an arabinose–glucose–galactose–glucosamine–mannan; (*d*) an arabinose–glucose–galactose–mannan; and (*e*) an arabinose polymer (112, 119). A mannan and a galactose–arabinose–mannan were also identified (25). Polysaccharides extracted from cells of *C. pseudodiphtheriae* (150, 153) were identified as a glucose–arabinose–mannan, a glucosamine–glucose–galactose–arabinose–mannan, and a glucosamine–glucose–galactose–arabinose–rhamnose–mannan.

Only limited studies have been reported on the cell wall constituents of streptomycetes (148, 150, 154, 193). Both peptidoglycan (150) and teichoic acid (138) have been reported to be present in these organisms. The tentative structure of the ribitol-teichoic acid polymer in *Streptomyces* is shown in Fig. 6 (193). The carbohydrate monomers that have been identified in the somatic polysaccharides of *Streptomyces* species include glucose, arabinose, rhamnose, and glucosamine (148, 150, 154, 193). The carbohydrate component of cell walls in *Micromonospora* contains glucose and arabinose (151), and the cytoplasm polysaccharide of dermatophili consists of glucosamine, galactose, arabinose, and ribitol (155).

Actinomyces species have been reported to contain numerous types of polysaccharides (138, 149, 150). Both peptidoglycan (150) and a teichoic acid polymer (138) are present, as well as complex heteroglycan(s) whose monomeric constituents include glucosamine, glucose, arabinose, mannose, and/or rhamnose (149, 150).

With respect to extracellular lipids, a *Nocardia* species was shown to contain both free (17.5% of the dry weight) and bound (20% of the dry weight) lipids as major constituents of the cell walls; in addition, an unusual peptidoglycan moiety that contains *N*-glycolmuramic acid residues constitutes approximately 40% of the dry weight (251). Although the bound lipids (mainly nocardic acids) are thought to be ester-linked to an arabinogalactan polymer (25, 275), other carbohydrate monomers (glucose, mannose, and galactosamine (151) may persist as non-lipid-bound polysaccharides.

The cell walls of most strains of nocardiae were found to contain the following saccharide monomers: muramic acid, hexosamine, glucose, galactose, and either mannose or arabinose (55, 154). The cell walls of *Nocardia-pelletieri* (*S. pelletieri*) contain two kinds of saccharides: muramic acid and hexosamine, whereas the cell walls of *N. polychromogenes* contain hexosamine exclusively (151). Polysaccharides extracted from cells of *N. asteroides* were found to contain between two and seven saccharides, depending on the strain, identified as hexosamine, galactose, glucose, xylose, rhamnose, sorbose, or ribose. A polysaccharide isolated from *N. brasiliensis* cells was described as a polymer of D-arabinose and D-galactose, combined in a 3:1 molar ratio, with a main chain of galactose, linked 1,3 and of arabinose linked 1,2 or 1,3, or both (71).

The isolation and physicochemical characterization of the numerous cell wall biopolymers in Mycobacteriaceae have been under detailed study in recent years. Mycobacterial cell wall preparations are composed mainly of lipids (approximately 70% of the dry weight as phospholipids, sulfolipids, glycolipids, waxes, low-molecular-weight fatty acid–hexose complexes) and proteins (approximately 25% of the dry weight as lipoproteins, proteins, poly-D-glutamic acid, and peptidoglycans), which are either covalently and/or noncovalently linked to monomeric, oligomeric, or polymeric forms of numerous carbohydrates (4, 18, 25, 36, 67, 91, 131, 137, 146, 183, 268, 282). Some of the polysaccharide subunits, which ultimately are oriented as cell wall biopolymers, were discussed as intracellular polysaccharides in Section 3 (also Tables 4 and 5) of this chapter (22, 39, 91, 100, 151). On the basis of the above studies, a tentative model depicting macromolecular complex formation between the numerous cell wall polymers is shown in Fig. 7.

The simplest polymeric complex within the mycobacterial cell wall that is responsible for "adjuvant activity" has not yet been fully elucidated (Fig. 7). For example, it was previously thought that a peptidoglycan–LPS–wax D complex was the ternary macromolecular complex responsible for adjuvant activity (131); however, a more recent study suggests that a water-soluble peptidoglycan–LPS moiety (Fig. 7) elicits adjuvant activity (183). Current

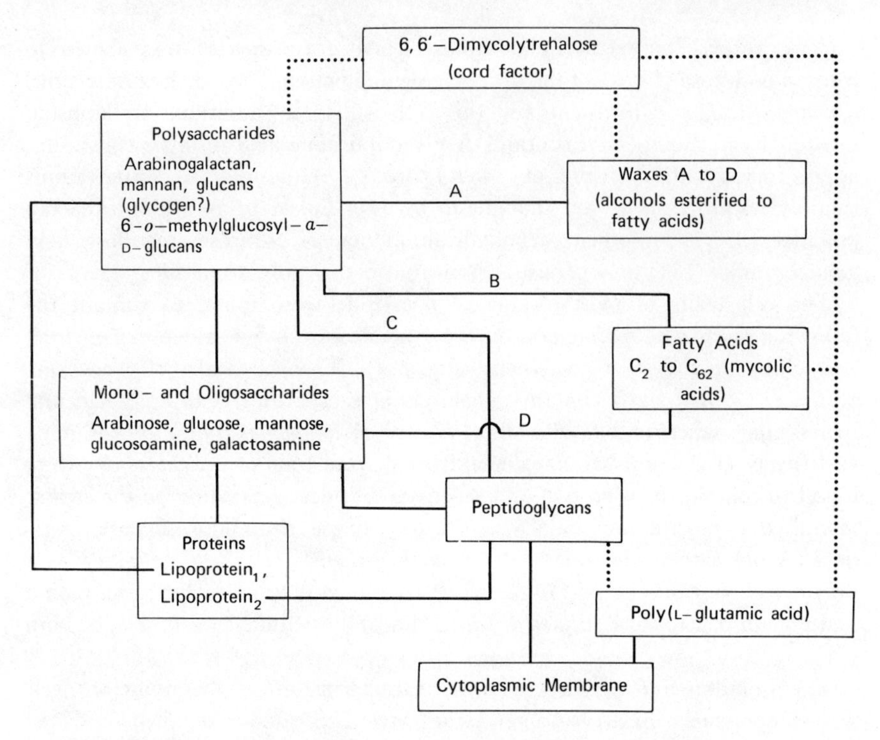

Fig. 7. Tentative interactions between cell wall constituents of Mycobacteriaceae. Complex formation mediated through covalent and/or noncovalent bonding forces (—); bonding forces unknown (.....). A + B + C: peptidoglycolipid exhibiting adjuvant activity; B: Lipopolysaccharides (heat-stable O-antigen); C: water-soluble complex exhibiting adjuvant activity; D: low-molecular-weight acylated sugars.

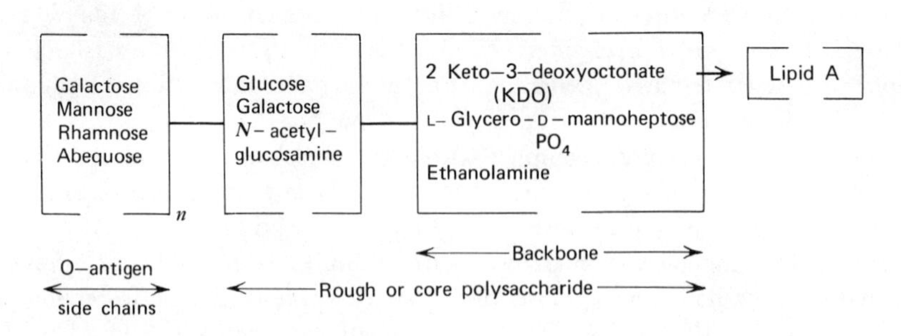

Fig. 8. Postulated structure of the lipopolysaccharide of *Salmonella typhimurium*.

studies are being conducted to determine whether the recently characterized poly(L-glutamic acid), in combination with one or more of the multitude of cell wall constituents in mycobacteria (Fig. 7), can exhibit adjuvant activity (268).

Although it has been suggested that the presence of either cell wall or membrane teichoic acids is a characteristic component that is common for Gram-positive bacteria (19), no evidence has been presented for the occurrence of these substituted polysaccharides in either corynebacteriaceae or mycobacteriaceae. The apparent absence of teichoic acids in these Gram-positive organisms coincides with the absence of such substituted polysaccharides in most Gram-negative bacteria. It is probable that the lipopolysaccharides of Gram-negative organisms possess physiological, biochemical, and immunological functions similar to those attributed to the teichoic acids of Gram-positive organisms (121). It would appear that the lipopolysaccharides and glycolipids, in conjunction with the phosphatides, are the macromolecular entities in mycobacteria and corynebacteria that perform cellular functions similar to those of the heteropolysaccharides and teichoic acids of other Gram-positive organisms.

The skeletal polysaccharides and lipopolysaccharides of Gram-negative bacteria have been extensively investigated, and their complex structures (Fig. 1), as well as their interactions with other cell wall biopolymers (Fig. 2), were discussed in Section 2 of this chapter. The tentative LPS structures of a *Salmonella* (199) and a *Shigella* (130) species are shown in Figs. 8 and 9, respectively. Although unsubstituted monomeric sugars within the O-antigenic

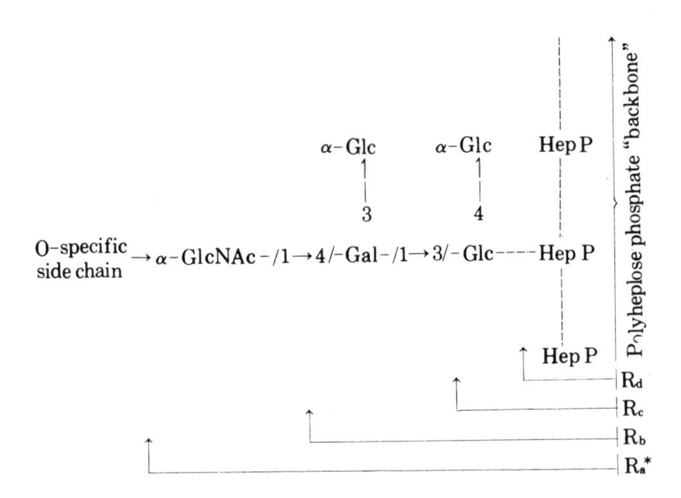

Fig. 9. Structure of *Shigella* lipopolysaccharide (130). * R_a through R_d representative of LPS-deficient mutants.

side chains of the LPS were formerly thought to elicit the immunogenic specificity of these Gram-negative organisms, recent studies have shown that O-actylated monomeric sugars are, in essence, the immunodominant "substituted" monomers in both *Shigella flexneri* (164) and *Salmonella typhimurium* (110); furthermore, either acetylated or propionated galactans from gum arabic have been found to function as artificial *S. typhimurium* antigens (124). These studies stress the need for utilizing mild extraction procedures in order to define the presence of labile acyl substituents within LPS macromolecules.

Other Gram-negative bacteria known to possess labile O-acetyl groups in the monomeric sugars of their O-antigenic side chains include *Klebiella* (163), *Pseudomonas* (271), *Salmonella newport*, and *Salmonella kentucky* (109, 111). In Gram-negative meningococci, the group-A-specific substituted glycan is mainly 1→6-linked mannosamine phosphate, which is partially *O*- and *N*-acetylated (167), whereas the group-B- and group-C-specific carbohydrates are immunologically different polysialic acids (87). In an early study (166), three polysaccharides occurring in an acetylated form were found in cells of *Vibrio comma*: (*a*) a polysaccharide consisting of galactose and aldobionic acid (composed of galactose and glucuronic acid), (*b*) a polysaccharide containing arabinose and aldobionic acid, and (*c*) a glucose polymer. In retrospect, it appears from the above studies that low-molecular-weight substituents (PO_4 and NH_2 groups in amino sugars), as well as undefined chemically labile acyl groups, may be covalently linked to the sugar monomers in O-antigenic side chains and may also be involved as noncarbohydrate antigenic functional groups of native LPS preparations.

The physicochemical characterization of chemically homogeneous LPS preparations has resulted in the chemotypic classification of Enterobacteriaceae (139, 170, 171) on the basis of qualitative and quantitative differences in their monomeric sugars. In *Arizona* species, the invariable monomers were glucosamine, heptose, galactose, and glucose, whereas monomers that varied from species to species were rhamnose, colitose, mannose, galatosamine, D-mannosamine, and deoxyhexose; in contrast, *Citrobacter* LPS invariably contained glucosamine, KDO, heptose, galactose, and glucose, with variable amounts of rhamnose, mannose, and/or xylose in four different strains (261). The glycan component of the LPS of *Xanthomonas* consists of glucosamine, KDO, glucose, mannose, rhamnose, and uronic acid, with fucose found in only a few strains (115, 254). The polysaccharides of *Chromobacterium violaceum* contain glucose and/or galactose, and rhamnose, with variable amounts of glucosamine, D-fucosamine, fucose, arabinose, and/or xylose in some strains (58, 250). Although differentiation between the O-antigenic side-chain monomers and the core LPS monomers (Fig. 1) has not been elucidated in some of the above studies, it has been inferred that the nonvariable monomers persist in the core region of the LPS, whereas the variable monomers represent the specific O-antigenic side-chain sugar residues.

Recent chemical analyses of organic-solvent-partitioned LPS preparations (75, 160, 255) have noted that the core LPS subunits also exhibit considerable variations in the carbohydrate monomers within an LPS of a particular bacterial species (the variable monomeric constituents were discussed in Section 2 of this chapter). In this respect, both the core and the O-antigenic side-chain subunits of an LPS macromolecule have been found to be heterogeneous, in certain instances, within an Enterobacteriaceae genus.

The heterogeneity of O-antigenic side chains has been amply related to immunospecificity between enterobacterial strains and genera. In general, the O-antigenic subunits of LPS in Enterobacteriaceae, (Figs. 8 and 9) contain three to five different monosaccharides, and in some instances the subunits are highly branched (109, 111, 163). A recent study has shown that the O-specific side chain of *Serratia marcescens* is composed of approximately 43 repeating disaccharide units of the following composition (257):

$$\left[\rightarrow 6)\text{-}\beta\text{-D-Glc-}(1-2)\text{-}\beta\text{-L-Rha-}(1\right]_{43}\rightarrow$$

These studies reveal that O-antigenic side chains vary in complexity from a simple linear disaccharide repeating unit to highly branched repeating units composed of as many as five monomeric sugar residues.

The preceding discussion not only has stressed the physicochemical nature of intracellular and extracellular microbial polysaccharides per se, but also has been oriented toward reviewing the recent literature which has shown the presence of *in vitro* polysaccharide complex formation with a host of other micro- and macroions. In some instances, the biological, immunological, and physicochemical properties of either intracellular or extracellular polysaccharides were found to be dependent on the presence of noncarbohydrate residues within the primary structure of the glycans. The probable cellular functions of numerous extracellular glycan–biopolymeric complexes (interactions between lipopolysaccharides, capsular glycans, peptidoglycans, teichoic acids, proteins, and/or lipids) have been extensively reviewed; however, the biological functions of intracellular glycan–biopolymeric (–DNA,–RNA, and/or –protein) complexes have yet to be defined.

REFERENCES

1. Adams, G. A., C. Quadling, and M. Yaguchi, *Can. J. Microbiol.* **16,** 1 (1970).
2. Adams, G. A., T. G. Tornabene, and M. Yaguchi, *Can. J. Microbiol.* **15,** 365 (1969).
3. Allen, C. F., and P. Good, *J. Amer. Oil Chem. Soc.* **42,** 610 (1965).
4. Amar-Nacasch, C., and E. Vilkas, *Bull. Soc. Chem. Biol.* **52,** 145 (1970).

5. Aminoff, D., W. Binkley, R. Schaffer, and R. W. Mowry, in *The Carbohydrates, Chemistry and Biochemistry*, Vol. 2-B, W. Pitman and D. Horton (eds.), Chap. 45, Academic Press, New York 1970, pp. 740–807.
6. Anacker, R. L., W. D. Bickel, W. T. Haskins, K. C. Milner, E. Ribi, and J. A. Rudbach, *J. Bacteriol.* **91,** 1427 (1966).
7. Anderson, C. G., W. H. Haworth, H. Raistrick, and M. Stacey, *Biochem. J.* **33,** 272 (1939).
8. Anderson, D. M. W., and N. J., King, *J. Chem. Soc.*, p. 5333, 1961.
9. Anderson, E. S., and A. H. Rodgers, *Nature* **198,** 714 (1963).
10. Anderson, R. G., H. Hussey, and J. Baddiley, *Biochem. J.* **127,** 11 (1972).
11. Andrews, P., L. Hough, and J. K. N. Jones, *J. Chem. Soc.*, p. 806, 1954.
12. Apicella, M. A., and J. A. Robinson, *Infect. Immunol.* **6,** 773 (1972).
13. Archibald, A. R., and J. Baddiley, *Advan. Carbohydr. Res.* **21,** 323 (1966).
14. Archibald, A. R., J. Baddiley, and J. E. Heckels, *Nature New Biol.* **241,** 29 (1973).
15. Armstrong, J. J., J. Baddiley, and J. Buchanan, *Biochem. J.* **80,** 254 (1961).
16. Ashwell, G., and J. Hickman, in *Microbial Toxins*, Vol. 4, G. Weinbaum, S. Kadis, and S. J. Ajl (eds.), Chap. 5, Academic Press, New York, 1971, pp. 235–266.
17. Aspinall, G. O., R. Begbie, A. Hamilton, and J. N. C. White, *J. Chem. Soc.* (C), p. 1065, 1967.
18. Azuma, I., Y. Yamamura, and A. Misaki, *J. Bacteriol.* **98,** 331 (1969).
19. Baddiley, J., *Acc. Chem. Res.* **3,** 98 (1970).
20. Baddiley, J., in *Biochemistry of the Glycosidic Linkage*, PAABS Symposium, Vol. 2, R. Piras and H. G. Pontis (eds.), Academic Press, New York, 1972, pp. 337–343.
21. Baddiley, J., J. H. Brock, A. L. Davison, and M. D. Partridge, *J. Gen. Microbiol.* **54,** 393 (1968).
22. Ballou, C. E., *Acc. Chem. Res.* **1,** 366 (1968).
23. Barker, S. A., C. N. D. Cruickshank, and J. H. Morris, *Biochim. Biophys. Acta* **74,** 239 (1963).
24. Barker, S. A., and P. J. Somers, in *The Carbohydrates, Chemistry and Biochemistry*, Vol. 2–B, W. Pigman and D. Horton (eds.), Chap. 41, Academic Press, New York, 1970, pp. 569–587.
25. Barksdale, L., *Bacteriol. Rev.* **34,** 378 (1970).
26. Bartnicki-Garcia, S., *Ann. Rev. Microbiol.* **23,** 87 (1969).
27. Beer, H., A. Braude, and C. Brinton, Jr., *Ann. N. Y. Acad. Sci.* **113,** 450 (1966).
28. Bettelheim, F. A., in *Biological Macromolecules*, Vol. 3, A. Veis (ed.), Chap. 3, Dekker, New York, 1970, p. 131.
29. Bhatnagar, S. S., C. G. J. Speechly, and M. Singh, *J. Hyg.,* **38,** 663 (1938).
30. Bhattacharjee, S. S., and T. E. Timall, *Can. J. Chem.* **43,** 758 (1965).
31. Black, W. A. P., *J. Marine Biol. Assoc. U.K.* **29,** 379 (1950).
32. Blumberg, P. M., and J. L. Strominger, *J. Biol. Chem.* **247,** 8107 (1972).
33. Bouveng, H. O., I. Bremer, and B. Lindberg, *Acta Chem. Scand.* **19,** 1003 (1965).
34. Braun, V., and V. Bosch, *Proc. Natl. Acad. Sci. U.S.* **69,** 970 (1972).
35. Braun, V., and H. Wolff, *Eur. J. Biochem.* **14,** 387 (1970).

36. Brennan, P. J., D. P. Lehane, and D. W. Thomas, *Eur. J. Biochem.* **13,** 117 (1970).
37. Brown, M. R., J. H. Foster, and J. R. Clamp, *Biochem. J.* **112,** 521 (1969).
38. Brown, W. C., and F. E. Young, *Biochem. Biophys. Res. Commun.* **38,** 564 (1970).
39. Brunnemann, H., and H. Urbach, *Arch. Hyg. Bakteriol.* **148,** 128 (1964).
40. Bush, D. A., and M. Horisberger, *Carbohydr. Res.* **22,** 361 (1972).
41. Campbell, J. N., M. Leyh-Bouille, and J. M. Ghuysen, *Biochemistry* **8,** 193 (1969).
42. Carlson, D. M., and L. W. Matthews, *Biochemistry* **5,** 2817 (1966).
43. Carter, H. E., R. H. Gigg, T. Nakayama, and E. Weber, *J. Biol. Chem.* **233,** 1309 (1958).
44. Catley, B. J., J. F. Robyt, and W. J. Whelan, *Biochem. J.* **100,** 5 (1966).
45. Chen, F. W., A. D. Strosberg, and E. Haber, *J. Immunol.* **110,** 98 (1972).
46. Chihara, G., J. Hamuro, Y. Y. Maeda, Y. Arai, and F. Fukuoka, *Cancer Res.,* **30,** 2776 (1970).
47. Chihara, G., J. Hamuro, M. Yukiko, A. Yoshiko, and F. Fukuoka, *Nature* **225,** 943 (1970).
48. Choy, Y. M., G. G. S. Dutton, A. M. Stephen, and M. T. Yang, *Anal. Lett.* **5,** 675 (1972).
49. Cifonelli, J. A., P. A. Rebers, M. B. Perry, and J. K. N. Jones, *Biochemistry* **5,** 3066 (1966).
50. Clarke, A. E., and B. A. Stone, *Rev. Pure Appl. Chem.* **13,** 134 (1963).
51. Clutterbuck, P. W., W. N. Haworth, H. Raistrick, G. Smith, and M. Stacey, *Biochem. J.* **28,** 94 (1934).
52. Cohen, G. H., and D. B. Johnstone, *J. Bacteriol.* **88,** 329 (1964).
53. Coley, J., M. Duckworth, and J. Baddiley, *J. Gen. Microbiol.* **73,** 587 (1972).
54. Conrad, H. E., J. R. Bamburg, J. D. Epley, and T. J. Kindt, *Biochemistry* **5,** 2808 (1966).
55. Cummins, C. S., *Amer. Rev. Resp. Dis.* **92,** Suppl. 63 (1965).
56. Darby, G. K., A. S. Jones, J. F. Kennedy, and R. T. Walker, *J. Bacteriol.* **103,** 159 (1970).
57. Davidson, E. A., *Carbohydrate Chemistry*, Rinehart and Winston, New York, 1967, pp. 336–340.
58. Davies, D. A. L., *Advan. Carbohydr. Chem.* **15,** 271 (1960).
59. Dintzis, F. R., and R. Tobin, *Biopolymers* **7,** 581 (1969).
60. Distler, J. J., and S. Roseman, *J. Biol. Chem.* **235,** 2538 (1960).
61. Doyle, R. J., Ph.D. Thesis, University of Louisville, Louisville, Ky., 1967.
62. Doyle, R. J., E. E. Woodside, and C. W. Fishel, *Biochem. J.* **106,** 35 (1968).
63. Doyle, R. J., E. E. Woodside, and C. W. Fishel, *Carbohydr. Res.* **5,** 274 (1967).
64. Doyle, R. J., E. E. Woodside, and C. W. Fishel, *Life Sci.* **8,** Pt. 2, 955 (1969).
65. Doyle, R. J., and E. E. Woodside, *Starke* **23,** 48 (1971).
66. Dudman, W. F., and M. Heidelberger, *Science* **164,** 954 (1968).
67. Elbein, A. D., and M. Mitchell, *J. Bacteriol.* **113,** 863 (1973).
68. Erlander, S. R., *Symposium on Carbohydrate–Protein Interactions,* American Association of Cereal Chemists, Excelsior Springs, Mo., 1970, pp. 1–45.

69. Erlander, S. R., R. M. Purimas, and H. L. Griffin, *Cereal Chem.* **45,** 140 (1968).
70. Erlander, S. R., and R. Tobin, *Makromol. Chem.* **111,** 194 (1967).
71. Estrada-Parra, S. A. Zamova and L. F. Bajalil, *J. Bacteriol.* **90,** 571 (1965).
72. Falk, M., D. G. Smith, J. McLachlan, and A. G. McInnes, *Can. J. Chem.* **44,** 2269 (1966).
73. Frank, N., H. Friebolin, G. Kerlich, J. P. Merle, and E. Stefert, *Org. Magn. Resonance* **4,** 725 (1972).
74. Fuchs, A., *Nature* **178,** 921 (1956).
75. Galanos, C., O. Luderitz, and O. Westphal, *Eur. J. Biochem.* **9,** 245 (1969).
76. Geritz, R. J., R. W. Ferraresi, and S. Raffel, *J. Exp. Med.* **131,** 189 (1970).
77. Ghuysen, J. M., *Biochim. Biophys. Acta* **83,** 132 (1964).
78. Ghuysen, J. M., J. L. Strominger, and D. J. Tipper, in *Comprehensive Biochemistry*, Vol. 26A, M. Florkin and E. H. Stotz (eds.), Chap. II, Elsevier, New York, 1968, pp. 53–104.
79. Glaser, L., *J. Biol. Chem.* **222,** 627 (1958).
80. Goebel, W. F., *Proc. Natl. Acad. Sci. U.S.* **49,** 464 (1963).
81. Gold, M. H., D. L. Mitzel, and I. H. Segel, *J. Bacteriol.* **113,** 856 (1973).
82. Goldemberg, S. H., in *Biochemistry of the Glycosidic Linkage*, PAABS Symposium, Vol. 2, R. Piras and H. G. Pontis (eds.), Academic Press, New York, 1972, pp. 621–627.
83. Gorin, P. A. J., and A. S. Perlin, *Can. J. Chem.* **34,** 1796 (1956).
84. Gorin, P. A. J., and J. F. T. Spencer, *Advan. Appl. Microbiol.* **13,** 25 (1970).
85. Gorin, P. A. J., and J. F. T. Spencer, *Advan. Carbohydr. Chem.* **23,** 367 (1968).
86. Gorin, P. A. J., and J. F. T. Spencer, *Can. J. Chem.* **44,** 993 (1966).
87. Gotschlich, E. C., T.-Y. Liu, and M. S. Artenstein, *J. Exp. Med.* **129,** 1349 (1969).
88. Graham, P. H., and M. A. O'Brien, *Antonie van Leeuwenhoek J. Microbiol. Serol.* **34,** 326 (1968).
89. Graves, I. L., *Amer. J. Vet. Res.* **30,** 647 (1969).
90. Graves, I. L., *Biopolymers* **6,** 1573 (1968).
91. Gray, G. R., and C. E. Ballou, *J. Biol. Chem.* **247,** 8129 (1972).
92. Greenwood, C. T., in *The Carbohydrates, Chemistry and Biochemistry*, Vol. 2–B, W. Pigman and D. Horton (eds.), Chap. 38, Academic Press, New York, 1970, pp. 471–513.
93. Greenwood, C. T., *Starke* **12,** 169 (1960).
94. Hall, E. A., and K. W. Knox, *Biochem. J.* **96,** 310, (1965).
95. Hall, D. A., P. F. Lloyd, H. Saxl, and F. Happy, *Nature* **181,** 470 (1958).
96. Hamuro, J., Y. Maida, Y. Arai, F. Fukuoka, and G. Chihara, *Chem. Biol. Interactions* **3,** 69 (1971).
97. Hanessian, S., *Advan. Carbohydr. Chem.* **21,** 143 (1966).
98. Hanessian, S., and T. H. Haskell, in *The Carbohydrates, Chemistry and Biochemistry*, Vol. 2–A, 2nd ed., W. Pigman and D. Horton (eds.), Academic Press, New York, 1970, pp. 139–211.
99. Hassid, W. Z., M. A. Joslyn, and R. M. McGready, *J. Chem. Soc.*, p. 1944, 1941.
100. Haworth, N. and M. Stacey, *Ann. Rev. Biochem.* **17:** 97 (1948).
101. Heath, E. C., *Ann. Rev. Biochem.* **40,** 29 (1971).
102. Hehre, E. J., and D. M. Hamilton, *J. Bacteriol.* **55,** 197 (1948).

103. Heidelberger, M., *Arch. Biochem.*, Suppl. 1, p. 169, 1962.

104. Heidelberger, M., Immunochemistry of Bacterial Polysaccharides, in*Research in Immunochemistry and Immunobiology,* Vol. III, J. B. G. Kwapinski (ed.), University Park Press, Baltimore, Md. 1972.

105. Heidelberger, M., and W. Nimmich, *J. Immunol* **109,** 1337 (1972).

106. Heidelberer, M., N. Roy, and C. P. M. Glaudemans, *Biochemistry* **8,** 4822 (1969).

107. Heidelberger, M., and M. E. Slodki, *Carbohydr. Res.* **24,** 401 (1972).

108. Heidelberger, M., J. M. N. Willers, and M. F. Michel, *J. Immunol.* **102,** 1119 (1969).

109. Hellerquist, C. G., J. Hoffman, A. A. Lindberg, B. Lindberg, and S. Svensson, *Acta Chem. Scand.* **26,** 3282 (1972).

110. Hellerqvist, C. G., B. Lindberg, S. Svensson, T. Holme, and A. A. Lindberg, *Carbohydr. Res.* **9,** 237 (1969).

111. Hellerquist, C. G., B. Lindberg, S. Svensson, T. Holme, and A. A. Lindberg, *Carbohydr. Res.* **14,** 17 (1970).

112. Hewett, M. J., K. W. Knox, and A. J. Wicken, *J. Gen. Microbiol.* **60,** 315 (1970).

113. Herbert, D., P. J. Phipps, and R. E. Strange, in *Methods in Microbiology*, Vol. 5–B, J. R. Norris and D. W. Ribbons (eds.), Chap. 3, Academic Press, New York, 1971.

114. Heyns, K., and G. Kiessling, *Carbohydr. Res.* **3,** 340 (1967).

115. Hickman, J., and G. Ashwell, *J. Biol. Chem.* **241,** 1424 (1966).

116. Higginbotham, J. D., and M. Heidelberger, *Carbohydr. Res.* **23,** 165 (1972).

117. Holdsworth, E., *Biochim. Biophys. Acta* **8,** 10 (1952).

118. Horton, D., in *The Amino Sugars*, Vol. 1–A, R. W. Jeanloz (ed.), Academic Press, New York, 1969, p. 1.

119. Horton, D., and M. L. Wolfrom, in *Comprehensive Biochemistry*, Vol. 5, M. Florkin and E. H. Stotz (eds.), Chap. 7–B, Elsevier, New York, 1963, pp. 189–232.

120. How, M. J., M. T. Withnall, and C. N. D. Cruickshank, *Carbohydr. Res.* **25,** 341 (1972).

121. Hughes, A. H., I. C. Hancock, and J. Baddiley, *Biochem. J.* **132,** 83 (1973).

122. Hughes, R. C., *Biochem. J.* **96,** 700 (1965).

123. Hughes, R. C., and P. F. Thurman, *Biochem. J.* **117,** 441 (1970).

124. Jackson, G. D. F., and C. R. Jenkin, *Aust. J. Exp. Biol. Med. Sci.* **47,** 91 (1969).

125. Janczura, E., H. R. Perkins, and H. J. Rogers, *Biochem. J.* **80,** 82 (1961).

126. Jann, K., B. Jann, I., Orskov, and F. Orskov, *Biochem. Z.* **346,** 368 (1966).

127. Jeanes, A., *Encyl. Polym. Sci. Technol.* **8,** 693 (1968).

128. Jeanloz, R. W., in *Comprehensive Biochemistry*, Vol. 5, M. Florkin and E. H. Stotz (eds.), Chap. 8, Elsevier, New York, 1963.

129. Jeanloz, R. W., in *The Carbohydrates, Chemistry and Biochemistry*, Vol. 2–B, W. Pigman and D. Horton (eds.), Chap. 42, Academic Press, New York, 1970, pp. 589–625.

130. Johnston, J. H., R. J. Johnston, and D. A. R. Simmons, *Biochem. J.* **105,** 79 (1967).

131. Jolles, P., D. Samour, and E. Lederer, *Arch. Biochem. Biophys.* **98,** 283 (1962).

132. Jones, J. K. N., and M. B. Perry, *J. Amer. Chem. Soc.* **79,** 2787 (1957).

133. Juni, E., and G. A. Heym, *J. Bacteriol.* **87,** 461 (1964).

134. Kabat, E. A., H. Kaiser, and H. Sirkorski, *J. Exp. Med.* **80,** 299 (1944).

135. Kadis, S., G. Weinbaum, and S. J. Ajl (eds.), *Microbial Toxins*, Vol. 5, Academic Press, New York, 1971.

136. Kane, J. A. and W. W. Karakawa, *J. Immunol.* **106,** 900 (1971).

137. Kanetsuna, F. T. Imaeda, and G. Cunto, *Biochim. Biophys. Acta* **173,** 341 (1969).

138. Kashkin, A. P., *(USSR) Ref. Zh. Biol. Khim. Abstr.* No. 12F541, 1968.

139. Kauffmann, F., *The Bacteriology of Enterobacteriaceae*, Munksgaard, Copenhagen, 1966.

140. Kendall, J. F., M. Heidelberger, and M. H. Dawson, *J. Biol. Chem.* **118,** 61 (1937).

141. Kent, P. W., *J. Chem. Soc.*, p. 364 1951.

142. Khuller, G. K., and P. J. Brennan, *Biochem. J.* **127,** 369 (1972).

143. King, R. D., and E. A. Grula, *Can. J. Microbiol.* **18,** 519 (1972).

144. Knox, K. W., M-J. Hewett and J. Wicken, *Gen. Microbiol.* **60,** 303 (1970).

145. Komatsu, N., S. Okubo, S. Kikumoto, K. Kimura, G. Saito, and S. Sakai, *Gann* **60,** 137 (1969).

146. Kotani, S., T. Kato, T. Matsuda, K. Kato, and A., Misaki, *Biken's J.* **14,** 379 (1971).

147. Krause, R. M., *Bacteriol. Rev.* **27,** 369 (1963).

148. Krzywy, T., *Arch. Immun. Immunotherap. Exp.* **11,** 521 (1963).

149. Kwapinski, J. B. G., *Pathol. Microbiol.* **23,** 158 (1960).

150. Kwapinski, J. B. G., *Analytical Serology of Microorganisms*, Vols. 1 and 2, Interscience, New York, 1969.

151. Kwapinski, J. B. G., *J. Bacteriol.* **88,** 1211 (1964).

152. Kwapinski, J. B. G., A. Cheng, A. Alcasid, and J. Dowler, *Can. J. Microbiol.* **17,** 1537 (1971).

153. Kwapinski, J. B. G., and J. Dowler, *Can. J. Microbiol.* **18,** 305 (1972).

154. Kwapinski, J. B. G., and M. Merkel, *Bull. Acad. Pol. Sci.* **5,** 335 (1957).

155. Kwapinski, J. B. G., and C. S. Simmons, *Antonie van Leeuwenhoek J. Microbiol. Serol.* **33,** 100 (1967).

156. Larm. O., B. Lindberg, and S. Svensson, *Carbohydr. Res.* **22,** 391 (1972).

157. Laskov, R., and E. Margoliash, *Bull Res. Counc. Israel*, Sect. A–11, p. 391, 1963.

158. Law, J. H., *Ann. Rev. Biochem.* **29,** 131 (1960).

159. Lehmann, I. R., and E. A. Pratt, *J. Biol. Chem.* **235,** 3254, (1960).

160. Lehmann, V., G. Hammerling, M. Nurminera, I. Minner, E. Ruschmann, O. Luderitz, T. T. Kuo, and B. A. O. Stocker, *Eur. J. Biochem.* **32,** 268 (1973).

161. Leidy, G., E. Hahn, S. Zamenhof, and H. E. Alexander, *Ann. N. Y. Acad. Sci.* **88,** 1195 (1960).

162. Leive, L., V. K. Shovlin, and S. E. Mergenhagen, *J. Biol. Chem.* **243,** 6386 (1968).

163. Lindberg, B., J. Lonngren, and W. Nimmich, *Carbohydr. Res.* **24,** 47 (1972).

164. Lindberg, B., J. Loenngren, E. Romanowska, and U. Ruden, *Acta Chem. Scand.* **26,** 3808 (1972).

165. Lindberg, B., J. Lonngren, J. L. Thompson, and W. Nimmich, *Carbohydr. Res.* **25,** 49 (1972).

166. Linton, R., D. L. Shrivesteva, and S. C. Seal, *Indian J. Med. Res.* **25,** 509 (1938).

167. Liu, T. -Y., E. C. Gotshlich, E. K. Jonssen, and J. R. Wysocki, *J. Biol. Chem.* **246,** 2849 (1971).

168. Lofkvist, T., and J. Sjoquist, *Acta Pathol. Microbiol. Scand.* **56,** 295 (1962).

169. Love, J., W. Mackie, J. P. McKinnell and E. Percival, *J. Chem. Soc.*, p. 4177, 1963.

170. Luderitz, O., K. Jann, and R. Wheat, in *Comprehensive Biochemistry*, Vol. 26–A, M. Florkin and E. H. Stotz (eds.), Chap. 3, Elsevier, New York, 1968, pp. 105–228.

171. Lüderitz, O., O. Westphal, A. M. Staub, and H. Nikaido, in *Microbiol Toxins,* Vol. 4, G. Weinbaum, S. Kadis, and S. J. Ajl (eds.), Chap. 4, Academic Press, New York, 1971.

172. Mackie, J. M., and E. Percival, *J. Chem. Soc.*, p. 3141, 1960.

173. Maeda, Y. Y., J. Hamuro, and G. Chihara, *Int. J. Cancer* **8,** 41 (1971).

174. Malchow, D., O. Luderitz, B. Kickhofen, O. Westphal, and G. Gerisch, *Eur. J. Biochem.* **7,** 239 (1968).

175. Markowitz, H., and M. Heidelberger, *J. Amer. Chem. Soc.* **76,** 1313 (1954).

176. Martinez-Segovia, Z. M., G. Sokol, I. L. Graves, and W. W. Ackerman, *Biochim. Biophys, Acta* **95,** 329 (1965).

177. McCarty, M., and S. I. Morse, *Advan. Immunol.* **6,** 249 (1964).

178. McCluer, R. H., in *Biochemistry of Glycoproteins and Related Substances*, E. Rossi and E. Stoll (eds.), Karger, New York, 1968, p. 203.

179. McGuire, E. J., and S. B. Binkley, *Biochemistry* **3,** 247 (1964).

180. McKibbin, J. M., in *The Carbohydrates, Chemistry and Biochemistry*, Vol. 2–B, W. Pigman and D. Horton (eds.), Chap. 44, Academic Press, New York, 1970, pp. 711–738.

181. McNeely, W. H., in *Microbial Technology*, H. J. Peppler (ed.), Chap. 16, Reinhold, New York, 1967, pp. 381–401.

182. Meeuse, B. J. D., *Acta Bot. Neerl.* **12,** 315 (1963).

183. Migliore-Samour, D., and P. Jolles, *FEBS Lett.* **25,** 301 (1972).

184. Miller, G. A., and R. W. Jackson, *J. Immunol.* **110,** 148 (1972).

185. Mills, G. T., and E. E. B. Smith, *Bull. Soc. Chim. Biol.* **47,** 1751 (1965).

186. Milner, K. C., J. A. Rudbach, and E. Ribi, in *Microbial Toxins*, Vol. 4, G. Weinbaum, S. Kadis, and S. J. Ajl (eds.), Chap. 1, Academic Press, New York, 1971, pp. 1–65.

187. Mirelman, D., B. D. Beck, and D. R. D. Shaw, *Biochem. Biophys. Res. Commun.* **39,** 712 (1970).

188. Miwa, T., Y. Iriki, and T. Suzuki, *Colloq. Int. Centre Nat. Rech. Sci. (Paris)* **103,** 135 (1961).

189. Montgomery, R., in *The Carbohydrates, Chemistry and Biochemistry*, Vol. 2–B, W. Pigman and D. Horton (eds.), Chap. 43, Academic Press, New York, 1970, pp. 627–709.

190. Mordoh, J., C. R. Krisman, and L. F. Leloir, *Arch. Biochem. Biophys.* **113,** 265 (1966).

191. Mosser, J. L., and A. Tomasz, *J. Biol. Chem.* **245,** 287 (1970).

192. Murphy, D., C. T. Bishop, and G. A. Adams, *Can. J. Biochem. Physiol.* **34,** 1271 (1956).

193. Naumova, I. B., *Antibiotiki* (Russ.). **6,** (1961).

194. Nimmich, W., *Acta Biol. Med. Ger.* **22,** 191 (1969).

195. Nordin, J. H., and R. G. Hansen, *J. Biol. Chem.* **238,** 489 (1963).

196. Norris, R. F., M. Sipin, F. Zilliken, T. S. Harvey, and P. Gyorgy, *J. Bacteriol.* **67,** 159 (1954).

197. Nowotny, A., in *Microbial Toxins*, Vol. 4, G. Weinbaum, S. Kadis, and S. J. Ajl (eds.), Academic Press, New York, 1971, pp. 309–330.

198. Osborn, M. J., and L. I. Rothfield, in *Microbial Toxins*, Vol. IV, G. Weinbaum, S. Kadis, and S. J. Ajl (eds.), Chap. 8, Academic Press, New York, 1971.

199. Osborn, M. J., and I. M. Weiner, *J. Biol. Chem.* **243,** 2631 (1967).

200. Pardoe, G. I., G. W. G. Bird, and G. Uhlenbruck, *Z. Immunitaetsforsch.* **136,** 488 (1968).

201. Partridge, M. O., A. L. Davison, and J. Baddiley, *J. Gen. Microbiol.* **74,** 169 (1973).

202. Pazur, J. H., A. Cepure, J. A. Kane, and C. G. Hellerquist, *J. Biol. Chem.* **248,** 279 (1973).

203. Percival, E., in *The Carbohydrates, Chemistry and Biochemistry*, Vol. 2–B, W. Pigman and D. Horton (eds.), Chap. 40, Academic Press, New York, 1970, pp. 537–568.

204. Percival, E. G. V., and S. K. Chanda, *Nature* **166,** 587 (1950).

205. Percival, E., and R. H. McDowell, *Chemistry and Enzymology of Marine Algal Polysaccharides, Academic Press, New York, 1967.*

206. Perkins, H. R., *Biochem. J.* **86,** 675 (1963).

207. Plackett, P., and S. H. Butterly, *Nature* **812,** 1236 (1958).

208. Posternak, T., *J. Biol. Chem.* **188,** 317 (1950).

209. Pusztai, A., and W. B. Watt, *Eur. J. Biochem.* **10,** 523 (1969).

210. Rapin, A. M. C., and H. M. Kalckar, in *Microbial Toxins*, Vol. 4, G. Weinbaum, S. Kadis, and S. J. Ajl (eds.), Academic Press, New York, 1971, pp. 267–308.

211. Rebers, P. A., and M. Heidelberger, *J. Amer. Chem. Soc.* **83,** 3056 (1961).

212. Rees, D. A., *Advan. Carbohydr. Chem. Biochem.* **24,** 267 (1969).

213. Rees, D. A., *Rep. Prog. Chem.* **62,** 469 (1965).

214. Reeves, R. E., *J. Biol. Chem.* **154,** 49 (1944).

215. Reeves, R. E., and W. F. Goebel, *J. Biol. Chem.* **139,** 511 (1941).

216. Ribi, E., R. L. Anacker, K. Fukushi, W. T. Haskins, M. Landy, and K. C. Milner, in *Bacterial Endotoxins*, M. Landy and W. Braun (eds.), Rutgers University Press, New Brunswick, N. J., 1964, p. 16.

217. Robbins, P. W., and A. Wright, in *Microbial Toxins*, Vol. IV, G. Weinbaum, S. Kadis, and S. J. Ajl (eds.), Chap. 9, Academic Press, New York, 1971.

218. Roelofsen, P. A., *Biochim. Biophys. Acta* **10,** 477 (1953).

219. Romanowska, E., and T. M. Lochowicz, *FEBS Lett.* **8,** 293 (1970).

220. Rosenberg, E., and S. Zamenhof, *J. Biol. Chem.* **236,** 2845 (1961).

221. Rothfield, L., M. Takeshita, M. Pearlman, and R. W. Horne, *Fed. Proc.* **25,** 1495 (1966).

222. Roy, M., W. R. Carroll, and C. P. J. Glaudemans, *Carbohydr. Res.* **12,** 89 (1970).

223. Schmidt, G., B. Jann, and K. Jann, *Eur. J. Biochem.* **10,** 501 (1969).

224. Schramm, G., O. Westphal, and O. Luderitz, *Z. Naturforsch.* **7b,** 594 (1952).

225. Schultz, J. C., and H. Ankel, *J. Bacteriol.* **113,** 627 (1973).

226. Serrano, J. A., R. V. Tablante, A. A. Serrano, G. C. Samplas, and T. Imaeda, *J. Gen. Microbiol.* **70,** 339 (1972).

227. Shands, J. W., Jr., *Infect. Immunol.* **4,** 167 (1971).

228. Sharon, M., in *The Amino Sugars*, Vol. 2–A. E. A. Balazs and R. W. Jeanloz (eds.), Chap. 18, Academic Press, New York, 1965, pp. 1–45.

229. Sharpe, M. E., J. H. Brock, and K. W. Knox, *J. Gen. Microbiol.* **74,** 119 (1973).

230. Siegel, S. M., in *Comprehensive Biochemistry*, Vol. 26–A, M. Florkin and E. H. Stotz (eds.), Elsevier, New York, 1968, pp. 1–48.

231. Simmons, D. A. R., *Bacteriol. Rev.* **35,** 117 (1971).

232. Smith, H., and H. T. Zwartouw, *Biochem. J.* **63,** 447 (1956).

233. Spencer, J. F. T., and P. A. J. Gorin, *Biotechnol. Bioeng.* **15,** 1 (1973).

234. Stacey, M., and S. A. Barker, *Polysaccharides of Microorganisms*, Oxford University Press, London, England, 1960.

235. Stoddart, J. F., *Stereochemistry of Carbohydrates*, Wiley-Interscience, New York, 1971.

236. Strominger, J. L., Y. Higashi, H. Sanderman, K. J. Stone, and E. Willoughby, in *Biochemistry of the Glycosidic Linkage*, PAABS Symposium, Vol. 2, R. Piras and H. G. Pontis (eds.), Academic Press, New York, 1972, pp. 135–154.

237. Sutherland, I. W., *Biochem. J.* **115,** 935 (1969).

238. Sutherland, I. W., *J. Gen. Microbiol.* **70,** 331 (1972).

239. Sutherland, I. W., *Nature* **228,** 280 (1970).

240. Sutherland, I. W., and M. L. Smith, *J. Gen. Microbiol.* **74,** 259 (1973).

241. Sutherland, I. W., and J. F. Wilkinson, *Biochem. J.* **110,** 749 (1968).

242. Sutherland, I. W., and J. F. Wilkinson, in *Methods in Microbiology*, Vol. 5–B, J. R. Norris and D. W. Ribbons (eds.), Chap. 4, Academic Press, New York, 1971, pp. 359–368.

243. Suzuki, S., H. Hatsukaiwa, H. Sunayama, M. Uchiyama, F. Fukuoka, M. Nakanishi, and S. Akita, *Gann* **60,** 65 (1969).

244. Takeda, T., M. Funatsu, S. Shibata, and F. Fukuoka, *Chem. Pharm. Bull.* **20,** 2445 (1972).

245. Taylor, W. H., and E. Juni, *J. Bacteriol.* **81,** 688 (1961).

246. Trott, G. F., M.S. Thesis, University of Louisville, Louisville, Ky., 1968.

247. Trott, G. F., Ph.D. Thesis, University of Louisville, Lousiville, Ky., 1971.

248. Trott, G. F., and E. E. Woodside, *J. Colloid Interface Sce.* **36,** 40 (1971).

249. Trott, G. F., E. E. Woodside, K. G. Taylor, and J. C. Deck, *Carbohydr. Res.* **26,** 111 (1973).

250. Tsumita, T., and M. Ohashi, *J. Exp. Med.* **119,** 1017 (1964).

251. Vacheron, M. J., M. Guinand, G. Michel, and J. M. Ghuysen, *Eur. J. Biochem.* **29,** 156 (1972).

252. Vahlne, G., *Acta Pathol. Microbiol. Scand.*, Suppl., **62,** 1 (1945).

253. Vasko, P. D., J. Blackwell, and J. L. Koenig, *Carbohydr. Res.* **23,** 407–416 (1972).

254. Volk, W. A., *J. Bacteriol.* **91,** 39 (1966).

255. Volk, W. A., C. Galanos, and O. Luderitz, *FEBS. Lett.* **8,** 161 (1970).

256. Wada, S., M. J. Pallansch, and I. Liener, *J. Biol. Chem.* **233,** 395 (1958).

257. Wang, C. S., and P. Alupovic, *Biochemistry* **12,** 309 (1973).

258. Watson, G., and H. W. Scherp, *J. Immunol.* **81,** 331 (1958).

259. Watson, J. D., *Molecular Biology of the Gene*, 2nd ed., Benjamin, New York, 1970.

260. Weidel, W., and J. Primosigh, *J. Gen. Microbiol.* **18,** 513 (1958).

261. Westphal, O., F. Kaufmann, O. Luderitz, and H. Stierlin, *Zentralbl. Bakteriol.* **179,** 336 (1960).

262. Westphal, O., and K. Jann, in *Methods in Carbohydrate Chemistry*, Vol. 5, R. L. Whistler, J. N. BeMiller, and M. L. Wolfrom (eds.), Academic Press, New York, 1965, p. 83.

263. Whistler, R. L., and R. M. Rowell, in *Glucuronic Acid, Free and Combined*, G. J. Dutton, (ed.), Chap. 2, Academic Press, New York, 1966, pp. 137–182.

264. White, B., *The Biology of Pneumococcus*, Commonwealth Fund, New York, 1958.

265. Whyte, J. N. C., and G. A. Strasdine, *Carbohydr. Res.* **25,** 435 (1972).

266. Wicken, A. J., *Biochem. J.* **99,** 108 (1966).

267. Wicken, A. J., J. W. Gibbens, and K. W. Knox, *J. Bacteriol.* **113,** 365 (1973).

268. Wietzerbin-Falszpan, J., B. C. Das, C. Gros, J. F. Petitit, and E. Lederer, *Eur. J. Biochem.* **32,** 525 (1973).

269. Wilkinson, J. F., *Bacteriol. Rev.* **22,** 46 (1958).

270. Wilkinson, J. F., W. F. Dudman, and G. O. Aspinall, *Biochem. J.* **59,** 446 (1955).

271. Wilkinson, S. G., L. Galbraith, and G. A. Lightfoot, *Eur. J. Biochem.* **33,** 158 (1973).

272. Williamson, A. R., and S. Zamenhof, *J. Biol. Chem.* **238,** 2255 (1963).

273. Wolfrom, M. L., and H. El Kahdem, in *Starch, Chemistry and Technology*, Vol. 1, R. L. Whistler and E. F. Paschall (eds.), Academic Press, New York, 1965, pp. 251–278.

274. Woodside, E. E., *Can. J. Microbiol.* **11,** 243 (1965).

275. Woodside, E. E., in *Research in Immunochemistry and Immunobiology*, Vol. 2, J. B. G. Kwapinski (ed.), University Park Press, Baltimore, Md., 1972, pp. 123–207.

276. Woodside, E. E., R. J. Doyle, and G. F. Trott, *J. Colloid Interface Sci.* **36,** 24 (1971).

277. Woodside, E. E., C. A. Frick, and C. W. Fishel, *Can. J. Microbiol.* **13,** 1641 (1967).

278. Woodside, E. E., C. A. Frick, and W. F. Kocholaty, *Proc. Soc. Exp. Biol. Med.* **119,** 327 (1965).

279. Woodside, E. E., and W. F. Kocholaty, *J. Bacteriol.* **87,** 1140 (1964).

280. Woodside, E. E., G. F. Trott, R. J. Doyle, and C. W. Fishel, *Arch. Biochem. Biophys.* **117,** 125 (1966).

281. Woodside, E. E., G. F. Trott, R. J. Doyle, and C. W. Fishel, *Carbohydr. Res.* **6,** 449 (1968).

282. Yamamura, Y., K. Onoue, and I. Azuma, *Ann. N.Y. Acad. Sci.* **154,** 88 (1968).

283. Yoshida, A., S. Mudd, and N. A. Lenhart, *J. Immunol.* **91,** 777 (1963).

284. Zamenhof, S., and G. Leidy, *Fed. Proc.* **13,** 327 (1954).

CHAPTER

SIX

Microbial Lipids

W. M. O'LEARY

1. INTRODUCTION

A. History

Lipids have long been recognized as major and ubiquitous constituents of all types of microbial cells. Indeed, even many viruses contain lipids that appear essential to their nature and function. In spite of their prevalence, however, microbial lipids have proved to be most enigmatic with respect to both nature and function. The appreciable concentrations of lipids in microbial cells, their often unusual and energetically "expensive" structures, their essential and commonly stereospecific relations to microbial existence, and many other aspects argue in favor of these lipids having major and critical functions in the lives of microorganisms. Unhappily it has been (and still is) extremely difficult to determine just what these functions are.

Until relatively recently, studies of microbial lipids were concerned almost exclusively with characterizing the chemical nature of these substances and with elucidating their modes of biosynthesis. These studies, which occupied numerous investigators for the last 20 years, proved unexpectedly difficult and yielded many unanticipated results. We now can see that it is in the areas of lipid composition, metabolism, and presumably function that microorganisms, particularly bacteria, differ most from other forms of life. Also, it seems daily more and more likely that here the explanations for many as yet poorly understood but major aspects of microorganisms may lie, including taxonomic uniqueness and species differentiation, modes of pathogenesis, and antibiotic sensitivity/resistance, to cite only a few.

Today it is both exasperating and instructive to reflect on the fact that for a great many years it was believed (on the basis of no real evidence) that microbial lipids were trivial components differing in no important way from the lipids of higher forms with respect to chemical nature, biosynthesis, or physiological function. Perhaps curiously, the first real impetus to the study of microbial lipids was concern over anomalies in microbiological assays for biotin. Although this singular story has been related in detail elsewhere (39, 40), a brief recapitulation is relevant here. Some 25 years ago it was noted that "false positives" in microbiological assays of biotin occurred if the extracts being assayed contained unsaturated fatty acids. This suggested that such compounds and biotin might be interrelated in bacterial metabolism. Studies of this problem quickly bogged down, however, because of the dearth of reliable data on the nature of bacterial lipids. Years of investigation of the chemical aspects of microbial lipids then followed; these studies demonstrated the presence in bacteria not only of many familiar lipoidal substances but also of various curious, uncommon, and even unique compounds. These in turn implied the existence of curious and unique biosynthetic pathways, and the search for and characterization of these pathways consumed more years of research—and indeed continues today. A great deal of information in these areas has been accumulated, and the accretion continues steadily. This wealth of chemical and biosynthetic information, while greatly desirable in itself, is not an unmixed blessing because it tells much but explains little. We now find ourselves solemnly but uneasily contemplating a dazzling array of data that tells us at once how much we know about microbial lipids and how little we understand about microbial lipids. It is becoming more and more clear what they are and how they are made, but what are they for? What does it all mean in terms of cell function?

Functional studies are always difficult both technically and intellectually, and ideally they rest on a good knowledge of the substances whose functions are under study. Perhaps because of the difficulties and certainly because of the years of effort expended in the basic studies described above, productive work on the functions of microbial lipids has come late, and our knowledge is sketchy and is in flux. Nevertheless, function is the heart of the matter, not only biologically speaking but also in the context of the present chapter. Accordingly, the bulk of this discussion will deal with what we presently know (or suspect) regarding the roles of lipids in the structure and function of microbial cells.

Any discussion of "microbial" lipids is of necessity largely a treatment of *bacterial* lipids, since the great preponderance of work in this area has been done using bacteria. Bacteria are easy to grow, give essentially nondifferentiated cells, and possess interesting and distinctive lipids and lipid metabolic pathways. In contrast, fungi have been differentiated while their lipids have seemed "plantlike" and relatively uninteresting (although we are belatedly beginning to realize that things are more complicated here than once was thought). Viral lipids received relatively little attention until quite recently, mostly because of the great difficulty in amassing sufficient viral particles for analysis and in freeing the viruses from host cell lipids. The latter problem has been especially vexing, and for years it was believed that there were no viral lipids as such, but that viruses merely picked up host cell lipids "on the way out," so to speak. We now realize that this, too, is not so, at least in some viruses and perhaps in all viruses that contain so-called essential lipids. At any rate, the foregoing explains why this treatment of "microbial" lipids is so heavily dependent on studies of bacteria.

B. The Nature of Microbial Lipids

It was once virtually impossible to find detailed or even accurate information on microbial lipids in the literature. Happily (or unhappily) this is now true only of textbooks (one can only hope that authors of texts eventually discover that there is a large and important body of information in this area!). Today the interested reader can consult two extensive monographs (2, 40) and numerous reviews (16, 26, 27a, 33, 35, 35a, 41–43, 45, 50, 53, 54, 54b, to list only a few). Consequently, an extensive presentation of this subject here would be superfluous, redundant, and, indeed, impossible within the space limitations of the chapter. Accordingly, for details on such matters, these many other sources should be consulted.

However, for the more casual reader and for purposes of general orientation, a summry treatment of microbial lipids is appropriate here.

Bacterial Lipids. As mentioned, these are the microbial lipids most extensively studied to date, and the bulk of our knowledge concerns the constituent fatty acids (Tables 1 to 5). In most organisms, fatty acid chain lengths range up to 26 carbon atoms, with the majority containing 18 or fewer. Many familiar straight-chain saturated fatty acids, from caproic to stearic, are seen in almost all organisms. Unsaturated fatty acids are almost exclusively monoenoic compounds, primarily hexadecenoic and octadecenoic aids. Depending on the particular species, certain other fatty acids occur and even predominate. These include hydroxy, branched methyl (especially iso and anteiso forms), and, most singularly, cyclopropane ring-containing fatty acids. These ring-containing compounds are of varying chain lengths but largely consist of 17 or 19 carbon compounds. They seem to be as distinctive of bacteria as muramic acid, but are of unknown function.

The biosyntheses of all these fatty acid types have been investigated. The saturated straight-chain acids are made by the familiar coenzyme-A pathway. The unsaturated fatty acids are sometimes formed by the so-called aerobic dehydrogenation of saturated fatty acids, familiar in higher forms, but more commonly by the anaerobic pathway, which involves progressive lengthening of already unsaturated precursors. The cyclopropane ring-containing acids are formed by the construction of a methylene bridge across the site of a double bond in unsaturated acids, the added carbon being derived from the methyl group of methionine in the form of *S*-adenosylmethionine. Interestingly, this addition takes place only when the monoenoic precursor is incorporated into a phospholipid. Synthetic pathways for branched methyl and hydroxy fatty acids have also been studied and are described, as are the others above, in the references already cited.

Without getting into a detailed discussion of the matter, it should be mentioned that many aspects of both the chemical natures and the biosyntheses of bacterial fatty acids are characteristic of these organisms. This is true to such an extent that the fatty acid spectra of different species are beginning to be favored as taxonomic criteria of identity, and unusual synthetic pathways are being considered as promising sites of chemotherapeutic attack.

The quantitative and qualitative aspects of bacterial fatty acid composition are remarkably (and exasperatingly) responsive to the age of the culture and to many physical and chemical factors in the environment. Generally speaking, young cultures have high levels of unsaturated fatty acids and low levels of the derivative cyclopropane ring-containing fatty acids. With increasing age, this ratio progressively reverses, so that the age of the culture is critical in determining the concentrations of these compounds. In addition, bacterial fatty acid composition and lipid composition vary greatly, depending on the nonlipid composition of the medium, that is, the ratio of carbohydrate to nitrogenous compounds, and so on. Finally, bacteria show a disconcerting readiness to pick up and incorporate into their substance fatty acids from the medium. These exogenous fatty acids may either be used as such or be modified to varying degrees.

These properties can result in nearly kaleidoscopic variations in the fatty acid and lipid contents of bacteria, and the situation may seem like uninterpretable chaos. Actually, however, this is not so; like so much else in the natural world, the variations occasioned by a given environmental situation are biologically rational and predictable. The important thing is that the individual investigator be aware of this plasticity of response.

In the matter of the lipid composition of bacteria, we see various noteworthy characteristics. As a general statement, bacteria contain only modest, even negligible, amounts of free fatty acids and glycerides. (Note that earlier reports of high concentrations of these substances in certain bacteria have not been verified by more recent studies and were presumably artifactious observations due to inadvertent chemical or cell-derived-enzyme hydrolysis.) The major lipid category in bacteria consists of phospholipids.

When one examines the phospholipid categories found in bacteria (Table 6), many familiar types are seen: phosphatidic acid (PA), phosphatidylserine

Table 1. The Saturated Straight-Chain Fatty Acids of Microorganisms

Systematic Name	Common Name	Carbon Atoms	Structural Formula
	Formic	1	$HCOOH$
	Acetic	2	CH_3COOH
	Propionic	3	CH_3CH_2COOH
n-Butanoic	Butyric	4	$CH_3(CH_2)_2COOH$
n-Hexanoic	Caproic	6	$CH_3(CH_2)_4COOH$
n-Octanoic	Caprylic	8	$CH_3(CH_2)_6COOH$
n-Decanoic	Capric	10	$CH_3(CH_2)_8COOH$
n-Dodecanoic	Lauric	12	$CH_3(CH_2)_{10}COOH$
*n*Tetradecanoic	Myristic	14	$CH_3(CH_2)_{12}COOH$
n-Hexadecanoic	Palmitic	16	$CH_3(CH_2)_{14}COOH$
n-Octadecanoic	Stearic	18	$CH_3(CH_2)_{16}COOH$
n-Eicosanoic	Arachidic	20	$CH_3(CH_2)_{18}COOH$
n-Docosanoic	Behenic	22	$CH_3(CH_2)_{20}COOH$
n-Tetracosanoic	Lignoceric	24	$CH_3(CH_2)_{22}COOH$
n-Hexacosanoic	Cerotic	26	$CH_3(CH_2)_{24}COOH$
n-Octacosanoic	Montanic	28	$CH_3(CH_2)_{26}COOH$

Table 2. The Branched-Chain Fatty Acids of Microorganisms

Systematic Name	Common Name	Carbon Atoms	Structural Formula
3,5-Dihydroxy-3-methyl pentanoic	Mevalonic	6	$HO—CH_2—CH_2—\overset{OH}{\underset{CH_3}{C}}—CH_2—COOH$
2-Methyl-3-hydroxy-pentanoic		6	$CH_3—CH_2—\underset{OH}{CH}—\underset{CH_3}{CH}—COOH$
6-Methylheptanoic	Isooctanoic	8	$CH_3\underset{CH_3}{CH}—(CH_2)_4—COOH$
6-Methyloctanoic		9	$CH_3—CH_2—\underset{CH_3}{CH}—(CH_2)_4—COOH$
10-Methylhendecanoic	Isolauric	12	$CH_3—\underset{CH_3}{CH}—(CH_2)_8—COOH$
11-Methyldodecanoic		13	$CH_3—\underset{CH_3}{CH}—(CH_2)_9—COOH$
12-Methyltridecanoic	Isomyristic	14	$CH_3—\underset{CH_3}{CH}—(CH_2)_{10}—COOH$

12-Methyltetradecanoic		15	$CH_3{-}CH_2{-}\underset{CH_3}{\underset{\vert}{CH}}{-}(CH_2)_{10}{-}COOH$
13-Methyletradecanoic		15	$CH_3{-}\underset{CH_3}{\underset{\vert}{CH}}{-}(CH_2)_{11}{-}COOH$
14-Methylpentadecanoic	Isopalmitic	16	$CH_3{-}\underset{CH_3}{\underset{\vert}{CH}}{-}(CH_2)_{12}{-}COOH$
14-Methylhexadecanoic		17	$CH_3{-}CH_2{-}\underset{CH_3}{\underset{\vert}{CH}}{-}(CH_2)_{12}{-}COOH$
15-Methylhexadecanoic		17	$CH_3{-}\underset{CH_3}{\underset{\vert}{CH}}{-}(CH_2)_{13}{-}COOH$
16-Methylheptadecanoic	Isostearic	18	$CH_3{-}\underset{CH_3}{\underset{\vert}{CH}}{-}(CH_2)_{14}{-}COOH$
9-Methyloctadecanoic		19	$CH_3{-}(CH_2)_8{-}\underset{CH_3}{\underset{\vert}{CH}}{-}(CH_2)_7{-}COOH$
10-Methyloctadecanoic	Tuberculostearic	19	$CH_3{-}(CH_2)_7{-}\underset{CH_3}{\underset{\vert}{CH}}{-}(CH_2)_8{-}COOH$
	Diphtheric	35	$(C_{25}H_{51}){-}CH{=}\underset{CH_3}{\underset{\vert}{C}}{-}(CH_2)_6{-}COOH$

Table 2. (*Continued*)

Systematic Name	Common Name	Carbon Atoms	Structural Formula
	Corrinic	35	?(Empirical formula: $C_{35}H_{68}O_2$)
	Corynomycolic	32	$CH_3{-}CH_2)_{14}{-}\underset{OH}{CH}{-}\underset{C_{14}H_{29}}{CH}{-}COOH$
	Corynomycolenic	32	$CH_3{-}(CH_2)_5{-}CH{=}CH{-}(CH_2)_7{-}\underset{OH}{CH}{-}\underset{C_{14}H_{29}}{CH}{-}COOH$
	Corynolic	52	$CH_3{-}(CH_2)_{14}{-}\underset{CH_3}{CH}{-}(CH_2)_{17}CH{-}COOH$ $CH_3{-}\underset{OH}{CH}{-}(CH_2)_7{-}\underset{CH_3}{CH}{-}\underset{OH}{CH}{-}\underset{CH_3}{CH}{-}\underset{CH_3}{CH}$
	Phthioic	26	$CH_3{-}(CH_2)_3{-}\underset{CH_3}{CH}{-}(CH_2)_5{-}\underset{CH_3}{CH}{-}(CH_2)_9{-}\underset{CH_3}{CH}{-}CH_2{-}COOH$
	Phthienoic (several)	27, 29	$CH_3{-}(CH_2)_6{-}\underset{C_5H_{11}}{CH}{-}(CH_2)_8{-}\underset{CH_3}{CH}{-}CH{=}\underset{COOH}{C}{-}CH_3$ (C_{27} acid)

Mycolipenic (several)	25, 27	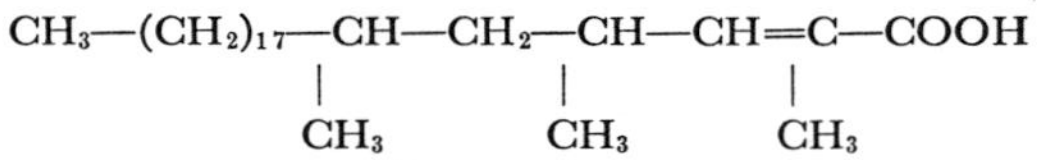 $CH_3-(CH_2)_{17}-CH(CH_3)-CH_2-CH(CH_3)-CH=C(CH_3)-COOH$ (C_{27} acid)
Mycocerosic	32	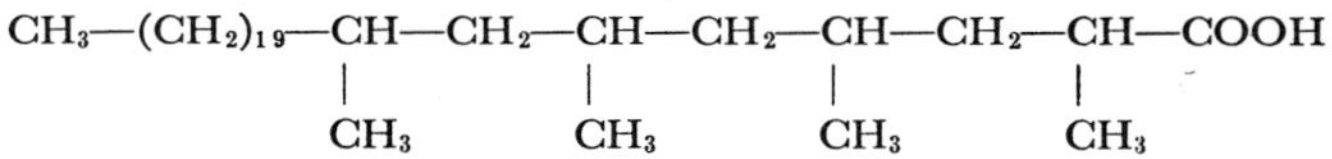 $CH_3-(CH_2)_{19}-CH(CH_3)-CH_2-CH(CH_3)-CH_2-CH(CH_3)-CH_2-CH(CH_3)-COOH$
Mycolic (many)	87, 88, etc.	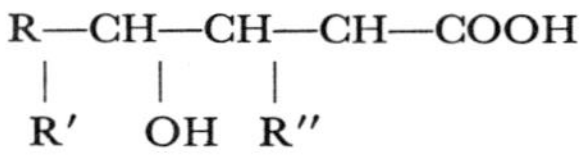 $R-CH(R')-CH(OH)-CH(R'')-COOH$ (where R, R′, and R″ may be of varying but appreciable length and may or may not contain OH groups, double bonds, etc.)

Table 3. The Hydroxy Fatty Acids of Microorganisms

Systematic Name	Common Name	Common Atoms	Structural Formula
3-Hydroxybutanoic	β-Hydroxybutric	4	$CH_3—CH(OH)—CH_2—COOH$
2-Methyl-3-hydroxypentanoic		6	$CH_3—CH_2—CH(OH)—CH(CH_3)—COOH$
3,5-Dihydroxy-3-methylpentanoic	Mevalonic	6	$HO—CH_2—CH_2—C(CH_3)—CH_2—COOH$
3-Hydroxyoctanoic		8	$CH_3—(CH_2)_4—CH(OH)—CH_2—COOH$
3-Hydroxydecanoic		10	$CH_3—(CH_2)_6—CH(OH)—CH_2—COOH$

3-Hydroxydodecanoic		12	$CH_3—(CH_2)_6—CH(OH)—CH_2—COOH$
3-Hydroxy-5-dodecenoic		12	$CH_3—(CH_2)_5—CH{=}CH—CH_2—CH(OH)—CH_2—COOH$
3-Hydroxytetradecanoic	*β*-Hydroxymyristic	14	$CH_3—(CH_2)_{10}—CH(OH)—CH_2—COOH$
3-Hydroxyhexadecanoic	*β*-Hydroxypalmitic	16	$CH_3—(CH_2)_{12}—CH(OH)—CH_2—COOH$
3-Hydroxyoctadecanoic	*β*-Hydroxystearic	18	$CH_3—(CH_2)_{14}—CH(OH)—CH_2—COOH$
Dihydroxyoctadecanoic	Dihydroxystearic	18	$C_{17}H_{33}(OH)_2COOH$
	Corynomycolic Corynomycolenic Corynolic Mycolic	See Table 2	

Table 4. The Unsaturated Fatty Acids of Microorganisms

Systematic Name	Common Name	Carbon Atoms	Structural Formula
cis-3-Decenoic		10	$CH_3—(CH_2)_5—CH{=}CH—CH_2—COOH$
cis-3-Dodecenoic		12	$CH_3—(CH_2)_7—CH{=}CH—CH_2—COOH$
cis-5-Dodecenoic		12	$CH_3—(CH_2)_5—CH{=}CH—(CH_2)_3—COOH$
5-Tetradecenoic		14	$CH_3—(CH_2)_7—CH{=}CH—(CH_2)_3—COOH$
cis-7-Tetradecenoic		14	$CH_3—(CH_2)_5—CH{=}CH—(CH_2)_5—COOH$
cis-9-Tetradecenoic	Myristoleic	14	$CH_3—(CH_2)_3—CH{=}CH—(CH_2)_7—COOH$
cis-7-Hexadecenoic		16	$CH_3—(CH_2)_7—CH{=}CH—(CH_2)_5—COOH$

cis-9-Hexadecenoic	Palmitoleic	16	$CH_3—(CH_2)_5—CH{=}CH—(CH_2)_7—COOH$
cis-10-Hexadecenoic		16	$CH_3—(CH_2)_4—CH{=}CH—(CH_2)_8—COOH$
cis-11-Hexadecenoic		16	$CH_3—(CH_2)_3—CH{=}CH—(CH_2)_9—COOH$
9-Heptadecenoic		17	$CH_3—(CH_2)_6—CH{=}CH—(CH_2)_7—COOH$
cis-9-Octadecenoic	Oleic	18	$CH_3—(CH_2)_7—CH{=}CH—(CH_2)_7—COOH$
cis-11-Octadecenoic	*cis*-Vaccenic	18	$CH_3—(CH_2)_5—CH{=}CH—(CH_2)_9—COOH$
cis-9,12-Octadecadienoic	Linoleic	18	$CH_3—(CH_2)_4—CH{=}CH—CH_2—CH{=}CH—(CH_2)_7—COOH$
cis-9,12,15-Octadecatrienoic	Linoleic	18	$CH_3—CH_2—CH{=}CH—CH_2—CH{=}CH—CH_2—CH{=}CH—(CH_2)_7—COOH$
cis-5,11-Octadecadienoic		18	$CH_3—(CH_2)_7—CH{=}CH—(CH_2)_2—CH{=}CH—(CH_2)_3—COOH$

Table 4. (*Continued*)

Systematic Name	Common Name	Carbon Atoms	Structural Formula
cis-5,9-Octadecadienoic		18	$CH_3—(CH_2)_5—\underset{H}{C}=\underset{H}{C}—(CH_2)_4—\underset{H}{C}=\underset{H}{C}—(CH_2)_3—COOH$
cis-5,8,11,14-Cicosatetraenoic	Arachidonic	20	$CH_3—(CH_2)_4—\underset{H}{C}=\underset{H}{C}—CH_2—\underset{H}{C}=\underset{H}{C}—CH_2—\underset{H}{C}=\underset{H}{C}—CH_2—\underset{H}{C}=\underset{H}{C}—(CH_2)_3—COOH$

Table 5. Cyclopropane Fatty Acids

Systematic Name	Common Name	Carbon Atoms	Structural Formula
Known to Occur in Bacteria			
cis-9,10-Methylenehexadecanoic		17	$CH_3-(CH_2)_5-\overbrace{\underset{H}{C}-\underset{H}{C}}^{\overset{H\quad H}{C}}-(CH_2)_7-COOH$
cis-11,12-Methyleneoctadecanoic	Lactobacillic	19	$CH_3-(CH_2)_5-\overbrace{\underset{H}{C}-\underset{H}{C}}^{\overset{H\quad H}{C}}-(CH_2)_9-COOH$
cis-9,10-Methyleneoctadecanoic	Dihydrosterculic	19	$CH_3-(CH_2)_7-\overbrace{\underset{H}{C}-\underset{H}{C}}^{\overset{H\quad H}{C}}-(CH_2)_7-COOH$

Table 5. *(Continued)*

Systematic Name	Common Name	Carbon Atoms	Structural Formula
	Others		
8,9-Methylene-8-heptadecenoic	Malvalic	18	$CH_3—(CH_2)_7—C=C—(CH_2)_6—COOH$ (C-8 and C-9 bridged by CH_2 (C with two H))
8,9-Methyleneheptadecanoic	Dihydromalvalic	18	$CH_3—(CH_2)_7—CH—CH—(CH_2)_6—COOH$ (C-8 and C-9 bridged by CH_2 (C with two H))
cis-9-10-Methylene-9-octadecenoic	Sterculic	19	$CH_3—(CH_2)_7—C=C—(CH_2)_7—COOH$ (C-9 and C-10 bridged by CH_2 (C with two H))

Table 6. Types of Phospholipid

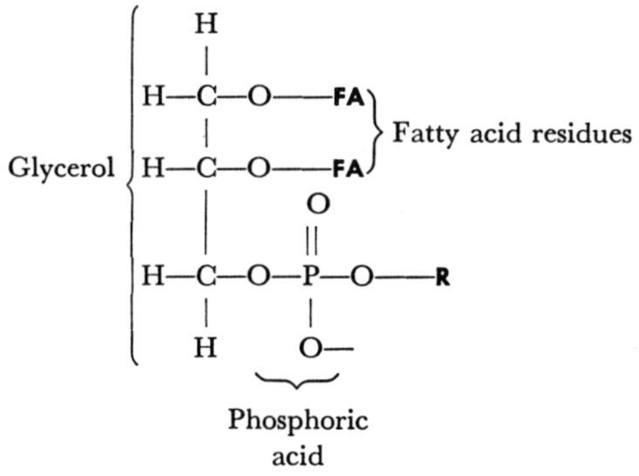

$$\text{Glycerol}\left\{\begin{array}{l} \text{H} \\ | \\ \text{H—C—O——FA} \\ | \\ \text{H—C—O——FA} \\ |\qquad\quad \text{O} \\ |\qquad\quad \| \\ \text{H—C—O—P—O——R} \\ |\qquad\quad | \\ \text{H}\qquad\ \text{O—} \end{array}\right.$$

(FA, FA: Fatty acid residues; P with O, O— : Phosphoric acid)

R	Compound
$—H$	Phosphatidic acid
$—CH_2—CH_2\overset{+}{N}—(CH_3)_3$	Phosphatidylcholine
$—CH_2—CH_2—NH_3^+$	Phosphatidylethanolamine
$—CH_2—CH(COOH)—NH_3$	Phosphatidylserine
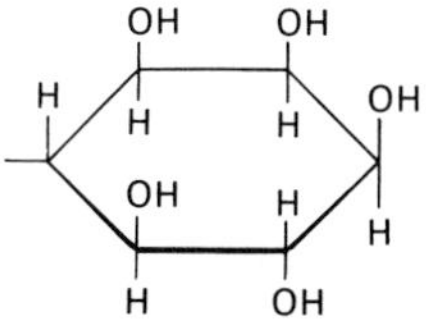 inositol ring (cyclohexane with H/OH substituents: OH, OH, OH, OH, OH; H, H, H, H, H, H)	Phosphatidylinositol
$—CH_2—CHOH—CH_2OH$	Phosphatidylglycerol
$—CH_2—CHOH—CH_2—O—\overset{O}{\overset{\|}{P}}(OH)—O—CH_2—\underset{}{C}H(O—FA)—CH_2(O—FA)$ [H—C—O—FA / H—C—O—FA / CH₂]	Diphosphatidylglycerol (cardiolipin)
$—CH_2—CHOH—CH_2—O—\overset{O}{\overset{\|}{C}}—\overset{H}{C}(NH_2)—R'$	3′-*O*-Aminoacylphosphatidylglycerol

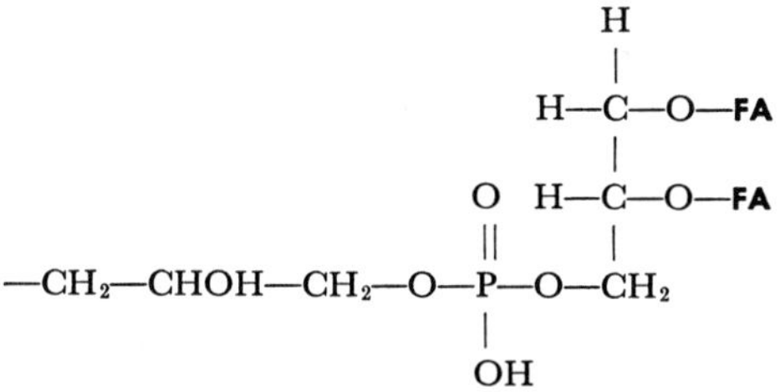

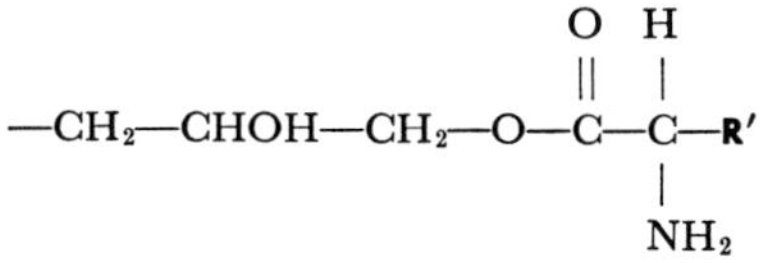

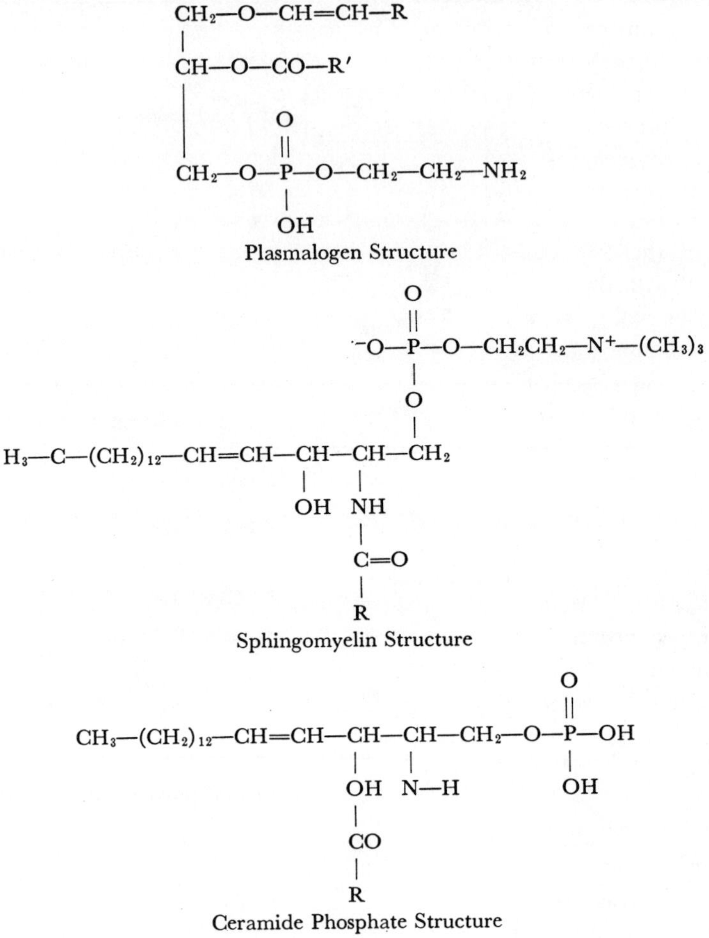

$$\begin{array}{l} CH_2\text{—}O\text{—}CH{=}CH\text{—}R \\ | \\ CH\text{—}O\text{—}CO\text{—}R' \\ | \qquad\quad O \\ | \qquad\quad \| \\ CH_2\text{—}O\text{—}P\text{—}O\text{—}CH_2\text{—}CH_2\text{—}NH_2 \\ \qquad\quad | \\ \qquad\quad OH \end{array}$$

Plasmalogen Structure

$$\begin{array}{l} H_3\text{—}C\text{—}(CH_2)_{12}\text{—}CH{=}CH\text{—}CH(OH)\text{—}CH(NH\text{—}C(=O)\text{—}R)\text{—}CH_2\text{—}O\text{—}P(=O)(\text{—}O^-)\text{—}O\text{—}CH_2CH_2\text{—}N^+\text{—}(CH_3)_3 \end{array}$$

Sphingomyelin Structure

$$CH_3\text{—}(CH_2)_{12}\text{—}CH{=}CH\text{—}CH(OH)\text{—}CH(N\text{—}H)\text{—}CH_2\text{—}O\text{—}P(=O)(OH)\text{—}OH$$
$$\text{(with OH—CO—R attached below the CH bearing OH, as drawn)}$$

Ceramide Phosphate Structure

Fig. 1. Three additional types of lipids found in bacteria.

(PS), phosphatidylethanolamine (PE), phosphatidylcholine (PC), phosphatidylglycerol (PG), diphosphatidylglycerol (cardiolipin), phosphatidyl inositol, and so on. To some extent investigators have noted new or uncommon phospholipids [e.g., aminoacylphosphatidylglycerol (= "lipoamino acids"), various lysophosphatides, and a number of glycolipids many of which are as yet not too well understood], but most of these compounds have been familiar in other forms of life. [It might be noted that even plasmalogens, sphingolipids, ceramides, fatty alcohols, and waxes are encountered in some bacteria (Fig. 1). There is indeed "something for everyone."].

Perhaps one generalization might be made regarding the phospholipid compounds: It is their relative abundance that characterizes at least many bacteria. Prominent in this respect is the prevalence of nonlecithin phospholipids. In higher forms, a dominant type of phospholipid is lecithin (phosphatidylcholine). However, PC is uncommonly encountered, particularly in appreciable concentrations, in bacteria. In these organisms, one more commonly sees the unmethylated or incompletely methylated precursors—PA, PS, PE, *N*-methyl-PE, and *N,N*-dimethyl-PE. It seems that bacteria are not well endowed with the ability to methylate PE, and perhaps this is an important step in the evolutionary progress to higher forms.

In discussing what fatty acids and lipids bacteria have, it is perhaps no less important to comment on what they lack. Two notable deficiencies merit mention here. First, bacteria are markedly bereft of polyunsaturated fatty acids, which are, of course, extremely important in higher forms of life. Also, it seems safe to say that the true bacteria do not normally contain and do not require sterols, sterol esters, or steroids. Note, however, that this is not to say that bacteria cannot utilize or, in the case of certain auxotrophic mutants, even require either polyunsaturated acids or cyclopentanoperhydrophenanthrene derivatives as nutrients. Many bacteria can utilize either or both, but they do not contain them as essential components. In pointing out that bacteria do not contain sterols, it might be mentioned that this raises some questions about membranes. In view of what we see as the important role of sterol in plasma membranes, what does this absence of sterols mean with respect to bacterial cell membranes? Are nonsterol membranes weaker, less efficient? Is the "step up" to sterol-containing membranes a major evolutionary event? This question will be discussed further below.

Finally, it should be mentioned in this brief treatment of bacterial lipids that the lipids of certain bacteria are "superunusual." I especially have in mind the mycobacteria, whose lipoidal constituents are immensely varied and complicated. In these organisms a great many different lipids with large and intricate molecules have been reported. I must here repeat what I have said elsewhere: I feel very strongly that in our studies of the mycobacteria we are missing something important. On the basis, admittedly, of no experimental evidence but only intellectual unease, I have come to wonder at the great variety of different and yet similar lipid substances reported to occur in mycobacteria. Do these organisms really make so many compounds? It seems that our present views of mycobacterial lipids violate the concept of biochemical economy. Perhaps we are seeing various portions of a lipid polymer of considerable complexity.

In any event, mycobacterial lipids certainly are different from other bacterial lipids. Interestingly, they are strikingly similar to the lipids found in species off the corynebacteria, which are sequestered in our taxonomy far from the mycobacteria on the basis of currently conventional criteria. On the basis of chemical composition (and, indeed, morphology) it would seem that they are badly misplaced.

Lipids of *Mycoplasma* and L Forms. These microorganisms are often simplistically thought of as "bacteria without cell walls." The fact is that there is nothing simple about these forms, as various compendia make abundantly clear (21, 54a). The lipoidal contents of these organisms differ strikingly from those of bacteria in general and, in the case of L forms, from the parent strains of bacteria as well. Prominent among the differences is a marked tendency to adsorb from the medium and retain cholesterol as such. These organisms, and their lipoidal properties as well as anomalies, will be of considerable interest to us in later considerations of microbial lipid function in that they are much used and considered in studies germane to this subject.

Lipids of the Fungi. The lipids of yeasts and filamentous fungi have been studied extensively for many years, and an appreciable literature on them exists (see ref. 23 for a summary of earlier work). After a quiescent period in this area we are now seeing a resurgence of interest and activity, utilizing the more advanced methods of analysis and separation. However, most of the newer work expands and quantitates what was already generally understood on the basis of earlier investigations. (see, e.g., ref. 5a, 8a, 23a, 36a, 56c).

In general what we observe in these organisms is a sharp departure from the lipid picture seen in the bacteria. The yeasts and filamentous fungi have lipids quite similar to those of plants in that they produce, store, utilize, and require polyunsaturated acids, phosphatidyl choline, sterols, etc.

Two areas in particular have begun to interest investigators of fungal lipids. One is the possible taxonomic use to which detailed data on lipid composition might be put in order to reduce the degree of reliance on subtle morphologic detail for identification. So far, this approach has not proved very promising in that there is a disconcerting similarity in lipid content from one fungus to another (e.g., 8a). Thus analytic taxonomy based on lipid content is likely to prove more useful in identifying the bacteria, whose lipids have many distinctive features, than the fungi, which are much less varied in this respect.

The other area of special interest at present is the relation of fungal lipids to pathogenesis and the response to antifungal therapeutic agents. The latter is particularly tantalizing in that some of the more effective antifungal agents either act on lipoidal structures (Nystatin, Amphotericin B) or are themselves lipoidal (undecylenic acid). Yet their modes of action are obscure and their effects often limited.

One factor that particularly complicates studies of fungal properties, including lipid composition and metabolism, is that fungi are at the threshold of structural differentiation. Therefore, in attempting to characterize the lipids of a fungus, the question that arises is, Are these actually the lipids, the lipid metabolism, of the cell, the mycelium, the spore, or what? In the dimorphic fungi the problem is even more acute. Understandably, therefore, most microbiologists and biochemists have preferred to work on the less complicated bacteria for the time being.

At any rate, in the light of our current knowledge and for present purposes, it suffices to say that yeast and fungal lipids are, overall, similar to plant lipids and quite dissimilar to bacterial lipids.

Lipids of the Viruses. If our knowledge of fungal lipids is less than optimal, our knowledge of viral lipids is pathetically minimal. Certainly research on these substances is in its infancy if not *in utero*. As indicated above, there have been two major problems here (aside from the fact that the nature of viral lipids has not been, shall we say, the principal concern of virologists). One of these major problems has been getting enough viral material with which to work. Happily, for some years the ability of virologists to make more and more virus has increased, and the ability of biochemists to perform reliable analyses on less and less material has also been improving. consequently, today we are just about at the point where these two curves of progress intersect.

The other problem has been the difficulty of freeing virus from the lipids of the host cell. Until recently, such contamination was so great and analytical techniques were so insensitive that it was believed by many, including me, that there were no such things as viral lipids—that is, lipids that were specifically characteristic of viruses and whose syntheses were directed by viruses, as in the case of viral proteins. In this view, such viral lipids as were known were merely host cell lipids that were coated on or perhaps incorporated into viral particles as they left the cells (e.g., 27).

However, with the much more sensitive analytical procedures non available, it seems that this may not always be true (or perhaps is never true). In some cases, it appears that viruses contain lipids differing quantitatively from those of the cells in which the viruses were grown; in other cases, it seems that there may even be qualitative differences (see, e.g., refs. 4a, 12b, 30).

For the present, however, our knowledge of viral lipids has to be regarded as extremely minimal and tentative. It should be noted that our indicision as to whether viral lipids are specific or merely gleanings of host cell lipids in no way mitigates the fact that they are essential to the nature and function of lipid-containing viruses, most notably their ability to infect cells under normal conditions.

C. Properties of Lipids

In assessing the roles of lipids in the molecular biology of microorganisms, some consideration should be given to the chemical, physical, and physiological properties of lipids that seem to be of special or peculiar significance. It is these properties that should help to explain lipoidal function or at least lead the way to an understanding of it.

One aspect that might be dealt with early on is the matter of "energy storage," particularly with respect to the triglycerides, which is so familiar and important in higher forms of life. It should be noted that this is not a major aspect of lipids in the bacteria. Although some instances are known in which lipids do serve as energy reserves, most notably in species that contain poly-β-hydroxybutyric acid (40), most bacteria store energy—insofar as they store it at all—in nonlipid forms such as glycogen. The reader will also recall that glycerides are not commonly prominent components of bacterial cellular lipids. Finally, it may be noted that in many species studied to date lipid contents actually increase with age and with various environmental stresses, and this is not what one would expect of energy reserves.

In the lipid-containing viruses (unfortunately more commonly, and more nebulously, known as the "ether-sensitive" viruses), the concept of energy storage is of course meaningless. These entities are concerned with host cell penetration and subsequent direction of cellular metabolism, utilizing cellular energy processes and reserves. Actually, however, at least the matter of penetration seems to require the presence of lipid in these viruses for some as yet unclear purpose, presumably the maintenance of some critical structural relationship.

The yeasts and filamentous fungi do contain glyceride deposits and do utilize them for energy storage in much the same fashion as higher plants and animals. As pointed out above, this is one more item of evidence that at this taxonomic level we are seeing the chiaroscuro transition from the distinctively microbial world to the more advanced forms of life.

At any rate, what the above says is that in the bacteria (and let us include here various related forms such a the rickettsia, mycoplasmas, and L forms), the major, significant, and essential properties of lipids are concerned with something other than a high energy content—that is to say, their value as structural components with metabolic utility, the latter being mentioned because the common and convenient term has an excessively nonkinetic connotation. Notable in this regard are the marked abilities of lipids, especially phospholipids, to orient, to associate, and to aggregate or pack.

Phospholipids and other lipids possess both polar and nonpolar "ends." When such compounds are placed in aqueous systems, they orient themselves in a more or less orderly fashion with the polar hydrophilic ends facing the water molecules and the nonpolar hydrophobic ends pointed away. Phospholids alone will do this and can be made to form a film or simple membrane on still water surfaces in such devices as the Langmuir film balance. Even these simple films can have surprising strength.

In addition to this orientation property, lipids also will readily associate with other molecular species, particularly proteins and, to a lesser extent, polysaccharides (9, 10, 28, 29, 56). The effect of the associated structures is to greatly increase the stability of films.

The third ability mentioned above is to aggregate or pack. When in sufficient concentration per unit surface area, lipids, lipoproteins, and lipopolysaccharides assume optimal interlocking arrangements, the architectural details of which are dependent on the stereochemical natures of the component molecules. Once again, this aggregation or packing, as it is commonly called, contributes to the integrity, durability, and even, it seems selective function of membranes.

Of course, what the foregoing says is that lipids seem to be most important in bacteria as components of the cell-limiting membranes. Certainly this is a devestatinglysimplistic statement, if only that we have so little definitive knowledge of exactly what the bacterial membranes are like and what they do. They are, as the cell biology cliche' states, the semipermeable membranes that regulate the passage of substances into and out of the cell (whatever that means), but they are far more than that. For example, in many species the membrane invaginates and convolutes into the cytoplasm to form mitochondria-like structures called mesosomes. Both the membrane and the mesosomes are sites of varied enzymatic activities (50). Also, the membrane seems to be the site of cell wall synthesis. Suffice it to say that these membranes are major and active cellular structures.

What is the nature of the membrane structure? This is no easier to answer in a bacterial context than it is when considering any plant or animal cell membrane. A casual examination of any of the prime references dealing with biological membranes (see above) will show, perhaps surprisingly that we are not at all sure just what membranes are like structurally. We do know what they look like in electron micrographs, and we know what they could be like on the basis of what we understand of the lipid and lipid aggregate properties described earlier.

Many years ago, Gorter and Grendel (17), on the basis of their studies of artificial membranes, suggested that membranes were composed of bimolecular layers of lipids in which the polar ends of the molecules faced out on each side, while the nonpolar ends faced in toward each other. Thus the membrane had polar surfaces and a nonpolar center. Such a structure could account for the existence of a membrane of some kind, could delimit the cell proper from the environment, and could provide the sort of electrical insulation characteristic of biomembranes. However, there was something wrong with this bilamellar concept; it soon was found that synthetic membranes following this model differed in many properties—chemical, physical, and biological—from natural membranes.

Several years later, Davson and Danielli (7) added the important concept of associated protein. In their model, membranes were composed of Gorter's and Grendel's bimolecular leaflet with a layer of protein on either side, thus giving a trilaminar structure of protein–lipid–protein. This concept was acclaimed as accommodating many factors that had caused unease with the old bimolecular lipid leaflet. It could be reconciled to a greater extent with analyses of the relatively crude membrane preparations of that day; it better suited the known physical properties of membranes (the Gortner–Grendel model seemed too weak and did not fit well with data on interfacial tension); and when electron micrographs of membrane cross sections became available, they looked trilaminar. Indeed, today this is still widely accepted concept of membrane structure. However, it should be noted that other concepts have been and still are being advanced, and the end has not yet been reached. Most of the more recent concepts entail a less static and far more complicated membrane and are prompted by our increasingly kinetic concept of membrane nature and function (17a, 28, 29, 56a). For example, it has been postulated that, as a concomitant of membrane transport, portions of the bimolecular lipid "middle layer" of the conventional trilaminar structure "bud off" to form discrete shperules of lipid in a proteinaceous matrix. Obviously, much remains to be done here.

A matter of considerable interest to the comparative biochemist is the apparent universality of structure—whatever its details may be—among biological membranes. Wherever one looks, among microorganisms, plants, or animals, one sees this basic protein–lipid–protein trilaminar boundary. Indeed, one gets the impression that the original evolutionary solution to the life–environment interface has not been improved on as other aspects of biology have progressed.

This section may well be concluded with some reference to Dr. Mitchell's recent and excellent consideration of this subject (36). He has pointed out that the membranes that define structurally and functionally the boundary between the cellular contents and the environment have four main attributes.

1. They represent regions of considerable mechanical stress (and so must be equal to it). There are marked differences of pressure across these membranes and corresponding tensions in their planes.
2. They represent regions of chemical anisotropy between compartments of differing chemical compositions. This anisotropy contributes to, if it does not totally determine, the "sidedness" of membranes.

3. They act as osmotic barriers, impeding or regulating the diffusion of various solutes in either direction.

4. They act as physicochemical links, catalyzing the translocation of specific substances into or out of the cell.

The properties of lipids already described, particularly in conjunction with both passive ("structural") and active (enzymatic) protein, are obviously peculiarly and specially suited to these roles.

2. LOCATION OF MICROBIAL LIPIDS IN THE CELL

As already indicated, the prime location of microbial lipids is the limiting structures where the microorganism meets its environment. I have already commented on the energy storage aspects. For the purposes of the present discussion, these will be regarded as, if not trivial, at least as of secondary importance and will not be considered further. This interface location of microbial lipids obtains whether one is considering viruses (27), fungi (3), or algae (32), and certainly prevails for bacteria, about which by far the most is known.

Within the Eubacteriales, a great variety of lipid compounds have been found and characterized both chemically and metabolically (6, 26, 35, 40); if one includes the Actinomycetales (2), an even greater profusion is seen. To fit this numbing array of diverse substances into the architecture of a limiting membrane for the moment far exceeds our ability, and even more do the functions of these substances elude us. Accordingly, for the present we must be satisfied with merely locating them cytologically.

In the Gram-positive bacteria this is relatively easy. In these organisms, wall and membrane are clearly separate and distinct structures, and membranes can relatively easily be obtained from "cell-wall-less" forms produced by using either lysozyme on normal cells to "stripoff" the wall, or penicillin in sublethal concentrations to produce protoplasts that can be burst to obtain membranes (47, 50). With such techniques, quite clean membrane preparations can be made. Although analyses vary with species, such membranes contain 20 to 40% lipid and 40 to 60% protein; within the 20 to 40% lipid is essentially 100% of the total cellular lipid (except for organisms containing intracytoplasmic granules of lipid).

Even "cleaner" membrane preparations can be obtained from mycoplasmas and L forms (42, 43) because these are totally lacking in cell wall material that could contaminate membrane preparations. Accordingly, such organisms are becoming more and more favored for membrane studies.

Unhappily, it is not easy—and, for the present, not possible—to obtain equally clean membranes from Gram-negative bacteria. Indeed, when considering these bacteria, the term "membrane" can be used only in a very loose and even theoretical sense; no one has ever obtained pure cytoplasmic membranes from Gram-negative bacteria, even though it seems almost certain that they exist. The problem in these organisms is that their external layers are so complex and numerous that it is hard to say just where the membrane is, and even harder to devise a way in which to isolate it. Accordingly, until such time as we perfect methods of getting cleaner preparations of Gram-negative membranes, various workers have encouraged the use of the term "envelope" to designate the complex outer layers that limit the cells of this group (see ref. 50 for an extensive discussion of this topic).

For some time, investigators attempted to produce wall and membrane preparations from Gram-negative bacteria by osmotic lysis, sonication, or mechanical disruption followed by differential centrifugation. Such preparations always proved, however, to be a melange of wall–membrane–envelope components that were ill suited to interpretive study. More recently better results have been obtained by the use of enzymes (especially lysozyme) and ethylenediamine tetraacetic acid, sodium dodecyl sulfate, and other reagents that can be employed to (in effect) strip off various layers in a progressive and relatively mild fashion (8, 47, 50). As Salton points out (50), considerable progress is being made on this troublesome problem, and it should not delay too much longer our resolution of the surface layers of Gram-negative bacteria.

However, as frustrating as this problem has been we do have at hand a considerable amount of information. The envelopes contain large amounts of protein, phospholipid, lipopolysaccharide, and peptidoglycan; both electron microscopy and chemical methods indicate that these components are assembled in several layers, although authors differ in regard to the number and nature of these layers. One of the more recent and highly regarded papers in this area (8) describes an outer layer composed of a mosaic of lipids, proteins, and polysaccharides (individually and in association), then a "rigid layer" of mucopeptide and protein, and finally a trilaminar "membrane." These main layers are then divided into numerous substrata—altogether a very complex matter, and one still subject to revision, to say the least. Of importance here with respect to our present consideration is that, while the site of lipids in the Gram-negative bacteria is the cell-limiting structure(s), they apparently are not limited strictly to the membrane as such, but are scattered about also on the wall–envelope surface in patches of lipoprotein and lipopolysaccharide.

Some comment on mesosomal structures is appropriate here in that these are generally agreed to be extensions of the cytoplasmic membrane and therefore also sites of lipid in the bacterial cell. There has been spirited debate over the nature and origin of these organelles, and not all of the debaters are yet reconciled to a single point of view. However, certainly most would agree that mesosomes (also called "chondrioids" and simply "intracytoplasmic membranes") with various differences in structural detail are found in many species of Gram-positive bacteria and also in Gram-negative bacteria (although here mesosomes are less extensive in size and number). In both types of organisms, mesosomes may eventually exhibit a stacked lamellar appearance or consist of an aggregation of packed vesicles (50). All these internal membranes exhibit the same trilaminar or "unit membrane" appearance seen in the cytoplasmic membrane.

The nature of membrane lipids has been extensively reviewed elsewhere (8, 16, 26, 31, 31a, 35b, 40, 56), and no attempt will be made to cover the same ground here. It is unfortunately true there have been wide differences in the reports of membrane lipid composition for a given organism, and one must assume that these are due to variations in the methods for culture of organisms, membrane preparation, and analysis, as practiced by different laboratories. In spite of such variations in the data appearing in the literature, it has been shown that under carefully controlled and duplicated conditions and techniques lipid analyses can be reliably duplicated on successive cultures of the same strain. An interesting fact that has come out of such analyses is that, in any given strain, one type of phospholipid always predominates (i.e., accounts for more than 50% of the membrane's content of phospholipid), although the nature of the predominating phospholipid can vary greatly from species to species and even from strain to strain within a species (50).

3. ASSOCIATIONS OF MICROBIAL LIPIDS

It has long been known that the bulk of the lipids in microorganisms is firmly bound to other components of the cell substance, while a much smaller proportion is free. This tight binding has been recognized from the first attempts to extract microbial lipids for analysis. If solvents such as ether and chloroform are used alone for phospholipid recovery from microbial cells, a large portion of phospholipid will be left with the cell residue after extraction. For more nearly complete recovery it is necessary to add a polar solvent such as methanol or ethanol. Even then an investigator in this field is always faced with The Dilemma: if he uses chemical and physical conditions strong enough to ensure complete lipid recovery, he is likely to degrade the material he is extracting; if he uses conditions sufficiently gentle for him to recover the cellular lipids in a state more nearly resembling their native one, he will certainly fail to recover a large portion of the lipids.

It seems that this "bound" state of cellular lipids is largely due to the polar end groups of the phospholipid molecules, which, being fit for ionic interaction, are well suited for associations with both proteins and polysaccharides. The resultant combinations often exhibit astonishing firmness; and, as Van Deenen (56) has discussed extensively, actually a variety of attractive forces have been shown to be operative, including hydrophobic bonding.

Although this firm attachment of lipids to other cellular components is an exasperation for would-be lipid analysts, it is, of course, the natural state of affairs in the functioning microorganism. Accordingly, it would be much to our advantage if we could gain some understanding of these macromolecular assemblies in which lipids play an important role. From varying sources, information is slowly accumulating on these aspects (18, 20, 22a, 38, 43, 47), but it is a difficult field, and specific facts accumulate slowly.

A. Lipid-Protein Associations

It appears that all bacterial membranes have associated structural proteins with high affinities for lipids. As already pointed out, these lipids are characterized by one predominating phospholipid with a great capacity for forming oriented bilayers, or micelles, and for undergoing phase changes. Whether the proteins assemble separately or on a preorganized lipid foundation is not presently clear (50). Physical and immunological data indicate that the protein moiety of a given species is quite specific, and the presence in a discrete structure of only certain proteins implies a mechanism for assembling them together with the membrane lipids in a highly organized fashion. Presently, a problem in resolving this further lies in the methods available for degrading the membrane structure, it being doubtful whether the "subunits" now being obtained are truly representative of the functional membrane. It is possible to reaggregate these fragments into triple-layered structures with the aid of divalent cations (5, 43), but how meaningful these experiments are is in question.

Studies with membrane components from *Bacillus subtilis* (34) have shown that up to a third of the protein is alpha-helical. On the basis of optical rotatory dispersion and circular dichroism, it was suggested that the alpha-helical portion of these proteins is located in special environments within the hydrophobic portions of the basic lipid bileaflet.

In this context, it should be noted that various workers have leaned toward a membrane concept that involves repeating arrays of lipoprotein subunits (see, e.g., ref. 19) instead of the more conventional random or homogeneous arrangement. This organization of functional proteins at specific sites is congruent both with the nature of products from ultrasonic disruption of bacteria (51) and with present concepts of transport and synthesis in the membrane. Whether or not specific lipids are associated with specific proteins is yet unanswered. It seems that little confidence can be placed in lipoprotein complexes remaining associated during membrane dissolution (47). Indeed, studies with *Mycoplasma* (11, 44) indicate that dissociation does occur.

Another aspect of lipid–protein association and their contributions of these compounds to the nature of membranes is the role of the apolar portions of the lipids. As already pointed out, a variety of lipids exist in the membranes, and the predominating one varies from strain to strain (as do the lesser components). Actually there is a twofold aspect to this variation in that (*a*) the lipid types vary, and (*b*) so also do the natures of the constituent fatty acids in the lipids. Thus it is possible for bacteria to contain both the "normal" and "lyso" forms of any of the lipids listed previously, in any ratio, and with varying fatty acid contents. This makes for a great variety of membrane lipid makeup, and this in turn for a great variety in membrane physical properties. The natures of the lipids present in any given form naturally express the differences in biosynthetic capabilities, and (perhaps even more important) they contribute to the distinctiveness of membrane nature and function. As discussed extensively by Van Deenan (56), variations in the apolar portions of membrane lipids can markedly affect both packing characteristics (which depend heavily on molecular structure and cross sections) and protein association.

Consequently, we have at least two major factors that can affect the structure and function of microbial membranes: the specific proteins incorporated into membranes by each strain, and the natures of the lipids. When one considers that variations in each of these can be qualitative, quantitative, and spatial, it can be seen that an almost infinite variety of membranes is possible. To what extent this diversity actually exists among microorganisms, however, remains to be seen. Indeed, we have now just enough evidence from the relatively few bacteria whose phospholipids, phospholipid fatty acids, and lipoproteins have been studied to advance the foregoing concept. However, the volume and variety of work being done in this area by microbial anatomists, biochemists, biophysicists, and geneticists should soon yield pertinent results.

When considering the impressively large number of variations in membrane structure made possible by these aspects of lipid–protein association, one cannot but think that these could be second only to genetic determinants as agents of differentiatioı . It is instructive to examine the essay of Fitz-James (12) regarding this aspect of bacterial membranes, wherein the manifold aspects of membrane function are considered as they relate to cell function. This prospect of considerable diversity in the makeup of microbial membranes also fits well with the large and increasing number of enzymatic activities that we are coming to associate with the cytoplasmic membrane and mesosomes (47, 50, 56).

B. Lipid-Polysaccharide Associations

Associations of microbial lipids with polysaccharides are seen primarily in the Gram-negative bacteria, where they are prominent in the outer layers of the envelope. These substances are often referred to as the "endotoxins" and as the "O-antigens." They are indeed highly toxic; in mammals and in minute amounts they collectively produce a severe syndrome known as endotoxin shock. The O-antigen designation derives from the fact that they are the immunologically distinctive somatic antigens of these bacteria.

The details of structure for these lipopolysaccharides are still in question. It is known that they are high-molecular-weight substances with at least four main components: the immunologically specific carbohydrate side chains (*ergo*, the O-antigen terminology); a core polysaccharide of glucose, galactose, *N*-acetylglucosamine, and heptose; and an end group of phosphate, ethanolamine, 2-keto-3-deoxyoctonoate (KDO), and, linked through the KDO, the lipid moiety (Fig. 2) (41, 41a, 46). There seem to be three discernible lipid fractions, at least on the basis of what can be extracted by different methods. The major component and the one assumed to be connected through the KDO is called lipid A. This covalently bound component is thought to contain a backbone of *N*-β-hydroxymyristylglucosamine phosphate. The available hydroxyl groups of the glucosamine residues and of the hydroxy acids seem to

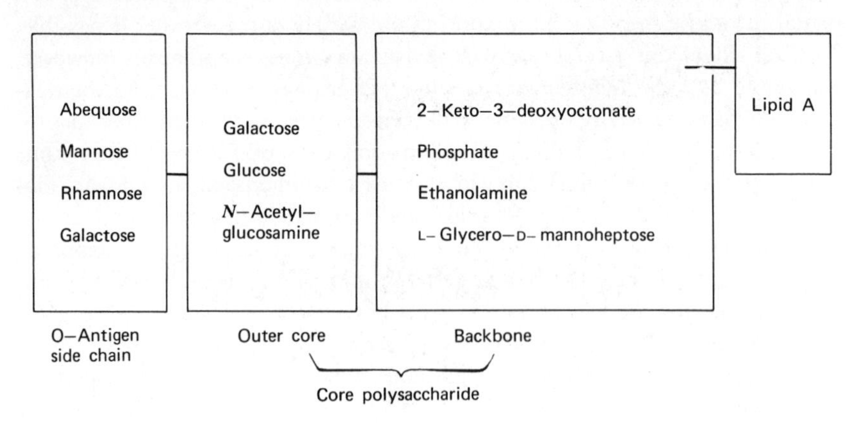

Fig. 2. Schematic structure of a lipopolysaccharide (from *Salmonella typhimurium*).

be esterified with fatty acids. Although increasingly sophisticated analyses may yet uncover some differences, presently available evidence indicates that lipid A is of similar composition in all lipopolysaccharides extracted from Gram-negative bacteria, regardless of species (15, 41a, 46). A lipid B that can be dissociated from lipopolysaccharide with formamide and ethanol, or with alkaline ethanol, has been reported. It seems to contain palmitic and oleic acids and about 1.5% phosphorus. A third and minor lipid component is an uncharacterized one that seems to be necessary for the assembly of monosaccharides into the O-antigen immunologically specific side chains (46).

A major amount of research effort has been expended on these lipopolysaccharides—originally because of their seemingly important role in pathogenesis and immunity where Gram-negative pathogens were involved, and later because of the discovery that they were involved in various biosynthetic activities (see below). Unfortunately, virtually all the work that has been done in this area has dealt with the carbohydrate portion of these lipopolysaccharides, and frustratingly littlewith the lipid end (38, 46).

When lipid A fractions are separated from the rest of the lipopolysaccharides, it is seen that the lipid A is primarily responsible for the toxic nature of these compounds, and that the carbohydrate portion is immunologically active. The toxic effects produced by these compounds include pyrexia, leucocytosis, cell death, vascular disturbances, and a shocklike prostration. Endotoxin-induced cell death is particularly marked in neoplastic tissue and has been considered as a palliative for otherwise nonresponsive malignancies, although the side effects are generally regarded as prohibitive.

The lipid A of *Escherichia coli* strain 0111 has been purified and characterized (6). It has a molecular weight of about 1700, has a reducing end group, and contains phosphate, glucosamine, and acetyl and long-chain acyl groups in the ratio of 1:2:3–4:5. A major amount of the fatty acid in this lipid proves to be β-hydroxymyristic acid, and hydroxy acids of varying chain lengths have been found in the lipid A fractions from other Gram-negative bacteria (15, 35). An interesting deficiency in these substances is the minimal concentration of cyclopropane ring-containing fatty acids. These acids are commonly present in appreciable amounts in the total fatty acids of Gram-negative bacteria, and it was expected that they would figure prominently in the other distinctive lipoidal substances in these organisms, the lipopolysaccharides. However, the fact is that cyclopropane ring-containing fatty acids have been found in only trivial amounts in lipopolysaccharides.

There was a time when these lipopolysaccharides (and their lipid A portions) were eagerly regarded as a likely area of study in order to understand Gram-negative infections at the molecular level. Certainly this area—the understanding of infectious disease in biochemical terms—is one in which we are grievously deficient. However, this approach proved to be less productive than originally thought, at least from the standpoint of understanding why there are different clinically recognizable diseases caused by Gram-negative pathogens. We see now that endotoxins from different Gram-negative pathogens produce essentially the same pharmacologic effects, and even that lipopolysaccharides from nonpathogenic Gram-negative bacteria produce effects scarcely discernible from those of similar preparations from virulent pathogens. This does not negate the fact that there very definitely is a severe disorder called endotoxin shock, nor do we mean to say that patients suffering from Gram-negative infections are not afflicted by this type of disorder. However, we find that endogoxin shock is a nonspecific unpleasantness, essentially the same from one Gram-negative pathogen to another, and that this syndrome is superimposed on other mechanisms, still unknown, that are responsible for the specific diseases associated with the different infectious agents in this group.

4. FUNCTIONAL ASPECTS OF MICROBIAL LIPIDS

Aye, there's the rub. Now that all the foregoing material has been assembled, categorized, and presented, what *are* the functions of microbial lipids? One is sorely tempted to answer. "Who knows?"—and not be far wrong at that. We have seen that in microorganisms, particularly in bacteria, energy storage is not especially important or distinctive in lipid function. In the viruses, although lipids seem to be important in facilitating cell entrance for certain types, again they appear to be of less that prime significance. In the bacteria, they seem to perform their roles, whatever they may be, at the cell surface, and usually in association with proteins or polysaccharides.

Accordingly, we would expect to see the physiological importance of microbial lipids to be manifested primarily at the membrane and/or envelope of the cell. Certainly there is a wealth of evidence showing that many enzymatic activities are associated with the membranes and mesosomes, including synthesis of cell wall material, initiation of cell division, electron transport, oxidative metabolism, and protein synthesis (12, 17a, 35b, 43, 50). Just what role lipids may have in these activities, aside from structural considerations, is not known. However, in the areas of transport and carbohydrate metabolism, some involvements of lipids have been noted or hypothesized.

Various investigators favor the idea that phospholipids may transport anions and cations across the membrane barrier. In model systems it has been shown that various inorganic species can be carried across a membrane from one aqueous system to another (56). Other studies have demonstrated actual or potential involvement of membrane lipids in amino acid transport and protein biosynthesis (13, 14, 17a, 25, 35b, 52).

It is in the areas of sugar transport and polysaccharide biosynthesis that mostof our present knowledge concerning lipid involvement in microbial metabolism has been obtained. In carbohydrate metabolism in *Salmonella*, lipopolysaccharide combines with phosphatidylethanolamine, followed by glycosylation involving either glucose or galactose nucleotides. It seems that an intermediate complex consisting of lipopolysaccharide, phosphatidylethanolamine, and glycosyl transferase is formed (34, 48, 49). Also in *Salmonella*, formation of the core polysaccharide of the lipopolysaccharide, (D-mannosyl-L-rhamnosyl-D-galactosyl)$_n$, involves di- and trisaccharide intermediates that are attached through phosphate to a lipid of unknown nature (57).

Nikaido (37) reported some years ago that he had noted in studies of β-galactoside permease that, during periods of maximal permease activity, incorporation of ^{32}P into phospholipids also increased, and suggested that a phospholipid might be involved in this permease activity. However, others have questioned this direct role of lipid (34).

Lipid-bound intermediates have been reported in the synthesis of Gram-positive cell walls (1). Both *N*-acetylmuramyl–pentapeptide–lipid and *N*-acetylglucosaminyl–*N*-acetylmuramyl–pentapeptide–lipid complexes have been observed. The latter complex is incorporated into cell wall glycopeptide, at which point the"carrier" lipid is released.

Indeed, numerous instances have been cited in the literature of the last few years wherein one lipid or another serves as a cofactor, carrier, or other type of accessorv in enzymatic processes; see ref. 35 for a summary of these reports. In this review, Lennarz suggests, "It could be that one of the most important functions of lipids in cell membranes is to serve as a 'solid state cofactor' providing a highly oriented, hydrophobic environment essential for certain enzymatic reactions."

One should also make mention here of lipids serving in the synthesis of other lipids, that is, such operations as the modification of intact lipids in order to complete the synthesis of subgroups. Notable in this regard is the methylation of monoenoic acids already incorporated into phospholipids in order to form cyclopropane-ring-containing acids (16, 33).

Another area in which microbial surface lipids may be involved is antibiotic sensitivity and resistance. Bacterial resistance to antibiotics is usually ascribed to enzymatic destruction of the antibiotic, to an alteration in an antibiotic-sensitive metabolic activity, or to "decreased permeability" of the cell to the antibiotic (e.g., 12a). The last suggestion, of course, would at least encourage one to wonder about the makeup of these cells with "decreased permeability."

Several reports have suggested that there may be a relationship between antibiotic resistance and the lipid content of bacterial cells (4, 24, 55). Recent studies in my own laboratory (10) have shown a pronounced correlation in certain Gram-negative bacteria between lipid composition and sensitivity or resistance to tetracyclines and polymyxin. How significant these correlations are in terms of cause and effect, and exactly what they may mean in terms of membrane composition, remain to be determined.

One last "function" of microbial lipids that could be included here is the rather anthropocentric one of being useful for taxonomic purposes. More and more interest is being shown in the analysis of microbial lipids as a means of identification and as a way of checking and improving on our present schemes of classification. Particularly well regarded at the present is gas–liquid chromatography of cellular fatty acids, supplemented by mass spectrometric examination of individual GLC fractions (56b). Computer evaluation of such data and supplementation by other instrumental methods such as infrared spectroscopy have given some encouraging results and may well prove to be the harbinger of the future in microbial identification and taxonomy. Needless to say, it is critical that culture conditions and extraction methods be carefully standardized (some thought has been given to GLC analysis of pyrolysis products, particularly in the case of viruses). Otherwise, variations in results will obscure the differences between species.

For the time being, this is a most pragmatic "function" for microbial lipids, but it is not as far from the main theme of this discussion as it may seem. Actually, this is in the nature of the ultimate taxonomy, that is, identification based on the actual substance of the organism being identified. By extension, we can reasonably expect compositionto be largely related to function, so in this case "one hand is washing the other," and it remains to be seen which approach will benefit the other more. Certainly our efforts to understand the functions of microbial lipids can only be furthered by the wealth of information of the nature of these lipids that should be a "spin-off" of the new taxonomy.

5. CONCLUSIONS

At present, we know very little about microbial lipids except with respect to their chemical compositions. We are achieving a respectable if not yet exhaustive knowledge of what these lipids are like, and undoubtedly in another 15 to 20 years or less we shall have a respectable idea of what they are for. As Bloch once commented, earlier generations of German chemists used to refer exasperatedly to lipid chemistry as *Schmierchemie*. That particular opprobrium is fading, but we are probably now in the stage of *Schmierphysiologie*.

We could make much use of information on just how, physically and chemically, the lipids whose natures we see fairly well are associated with proteins and polysaccharides; and we need a clearer and an accurate knowledge of molecular arrangements in the membranes and envelopes of microbial cells. Such information would go far toward the construction of more useful ideas of lipid function.

For the moment, we can see that most bacterial lipids are phospholipids, that most of these are in the membranes and/or envelopes, that these membranes are required for sequestering cellular mechanism from a fluctuating environment and for performing a variety of metabolic and transport functions, and that these phospholipids are essential to the integrity and durability of these membranes. We know that these membranes, especially the phospholipids, are not static, but participate actively in a variety of vital processes, no matter how obscure this participation is at present. We know that, at bottom, microbial lipid composition—and presumably physiological role—are biogenetically determined, but we also know that within wide limits numerous environmental factors also affect at least lipid composition. Beyond this, however, we are hard pressed to make confident statements.

In this notably interdisciplinary field it is clear that we can expect the accretion of pertinent information from many specialties—ultrastructural cytology, crystallographic studies, immunochemistry, biophysics, and the likewhere investigations will be directed specifically toward aspects of the question of lipid and lipid-associated function. In the interim some significant advances will probably be made made by relatively serendipitous routes, that is, by observing the dependence of isolated phenomena on one lipid fraction or another (e.g., specific transport relationships, or antibiotic resistance/lipid content correlations), long before we know in detail about the lipid–protein–whatever sheathing of the cell.

REFERENCES

1. Anderson, J. S., M. Matsuhasi, M. A. Haskin, and J. L. Strominger, *Proc. Natl. Acad. Sci. U.S.* **54,** 881 (1965).
2. Asselineau, J., *The Bacterial Lipids*. Holden-Day, San Francisco, 1966.
3. Bartnicki-Garcia, S., *Ann. Rev. Microbiol.* **22,** 87 (1968).
4. Benbough, J. E., and G. A. Morrison, *J. Gen. Microbiol.* **49,** 469 (1967).

4a. Blough, H. A., and D. E. Lawson, *Virology* **36,** 286 (1968).

5. Brown, J. W., *Biochim. Biophys. Acta* **94,** 97 (1965).

5a. Bunn, C. R., J. J. McNeill, and G. H. Elkan, *J. Bacteriol.* **102,** 24 (1970).

6. Burton, A. J., and H. E. Carter, *Biochemistry* **3,** 411 (1965).
7. Davson, H., and J. F. Danielli, *The Permeability of Natural Membranes*, 2nd ed., University Press, Cambridge, England, 1952.

8. De Petris, S., *J. Ultrastruct. Res.* **19,** 45 (1967).

8a. Domer, J. E., and J. G. Hamilton, *Biochim. Biophys. Acta* **231, 12.** Fitz-James, P. C., in *Function and Structure in Microorganisms*, M. R. Pollock, and M. H. Richmond (eds.), University Press, Cambridge, England, 1965, pp. 369 ff.

12a. Franklin, T. J., and G. A. Snow, *Biochemistry of Antimicrobial Action*, Academic Press, New York, 1971.

12b. Fraenkel-Conrat, H. L., *The Chemistry and Biology of Viruses*, Academic Press, New York, 1969.

13. Gaby, W. L., and E. H. McCurley, *Nature* **199,** 1187 (1963).

14. Gaby, W. L., R. N. Naughten, and C. Logan, *Arch. Biochem. Biophys.* **82,** 34 (1959).

15. Gallin, J. I., and W. M. O'Leary, *J. Bacteriol.* **96,** 660 (1968).

16. Goldfine, H., *Ann. Rev. Biochem.* **37,** 303 (1968).

17. Gorter, E., and F. Grendel, *J. Exp. Med.* **41,** 439 (1925).

17a. Green, D. E. (ed), Membrane Structure and Its Biological Applications, *Ann. N. Y. Acad. Sci.*, p. 195, 1972.

18. Green, D. E., and S. Fleischer, in *Metabolism and Physiological Significance of Lipids*, R. M. C. Dawson, and D. N. Rhodes (eds.), Wiley, New York, 1964.

19. Green, D. E., and J. F. Perdue, *Proc. Natl. Acad. Sci. U.S.* **55,** 1295 (166).

20. Gurd, F. R. N., in *Metabolism and Physiological Significance of Lipids*, R. M. C. Dawson and D. N. Rhodes (eds.), Wiley, New York, 1964.

21. Hayflick, L. (ed.), *The Mycoplasmatales and the* L-*Phase of Bacteria*, Appleton-Century-Crofts, New York, 1969.

22. Hendler, R. W., *Protein Biosynthesis and Membrane Biochemistry*, Wiley, New York, 1968.

22a. Henn, F. A., and T. E. Thompson, *Ann. Rev. Biochem.* **38,** 241 (1969).

23. Hilditch, T. P., and P. N. Williams, *The Chemical Constitution of Natural Fats*, Wiley, New York, 1964.

23a. Holtz, R. B., and L. C. Schisler, *Lipids* **6,** 176 (1970).

24. Hugo, W. B., and R. J. Stretton, *J. Gen. Microbiol.* **42,** 133 (1966).

25. Hunter, G. D., and R. A. Goodsall, *Biochem. J.* **78,** 564 (1961).

26. Kates, M., *Ann. Rev. Microbiol.* **20,** 13 (1966).

27. Kates, M., A. C. Allison, D. A. J. Tyrell, and A. T. James, *Cold Spring Harbor Symp. Quant. Biol.* **27,** 293 (1962).

27a. Kates, M., and M. K. Wassef, *Ann. Rev. Biochem.* **39,** 323 (1970).

28. Kavanau, J. L., *Structure and Function in Biological Membranes*, Holden-Day, San Francisco, 1965.

29. Kavanau, J. L., *Fed. Proc.* **25,** 1096 (1966).

30. Klenk, H. D., and P. W. Choppin, *Virology* **40,** 939 (1970).

31. Korn, E. D., *Theor. Exp. Biophys.* **2,** 49 (1969).

31a. Korn, E. D., *Ann. Rev. Biochem.* **38,** 263 (1969).

32. Lang, N. J., *Ann. Rev. Microbiol.* **22,** 15 (1968).

33. Law, J., in *The Specificity of Cell Surfaces*, B. D. Davis and L. Warren (eds.), Prentice-Hall, Englewood Cliffs, N. J., 1967.

34. Lenard, J., and S. J. Singer, *Proc. Natl. Acad. Sci. U.S.* **56,** 1828 (1966).

34a. Smith, P. F., *The Biology of Mycoplasmas*, Academic Press, New York, 1971.

35. Lennarz, W. J., *Advan. Lipid Res.* **4,** 175 (1966).

35a. Lennarz, W. J., *Ann. Rev. Biochem.* **39,** 359 (1970).

35b. Lennarz, W. J., in *Lipid Metabolism*, S. J. Wakil (ed.), Academic Press, New York, 1970.

36. Mitchell, P., in *Organization and Control in Prokapyotic and Eukaryotic Cells*, H. P. Charles, and B. C. J. G. Knight (eds.), University Press, Cambridge, England, 1970.

36a. Mumma, R. O., C. L. Fergus, and R. D. Sekura, *Lipids* **5,** 100 (1969).

37. Nikaido, H., *Biochem. Biophys. Res. Commun.* **9,** 486 (1962).

38. Nowotney, A. (ed.), *Ann. N.Y. Acad. Sci.*, p. 133, 1966.

39. O'Leary, W. M., *Bacteriol. Rev.* **26,** 421 (1962).

40. O'Leary, W. M., *The Chemistry and Metabolism of Microbial Lipids*, World, Cleveland, Ohio, 1967.

41. O'Leary, W. M., *Compr. Biochem.* **18,** 229 (1970).

41a. Osborn, M. J., *Ann. Rev. Biochem.* **38,** 501 (1969).

42. Panos, C., in *The Mycoplasmatales and the* L-*Phase of Bacteria*, L Hayflick (ed.), Appleton-Century-Crofts, New York, 1969, p. 503.

43. Razin, S., in *The Mycoplasmatales and the* L-*Phase of Bacteria*, L. Hayflick (ed.), Appleton-Century-Crofts, New York, 1969, p. 317.

44. Razin, S., H. J. Morowitz, and T. T. Terry, *Proc. Natl. Acad. Sci. U.S.* **54,** 219 (1965).

45. Reeves, H. C., R. Rabin, W. S. Wegener, and S. J. Ajl, *Ann. Rev. Microbiol.* **21,** 225 (1967).

46. Roantree, R. J., *Ann. Rev. Microbiol.* **21,** 443 (1967).

47. Rothfield, L., and A. Finkelstein, *Ann. Rev. Biochem.* **37,** 463 (1968).

48. Rothfield, L., and B. L. Horecker, *Proc. Natl. Acad. Sci. U.S.* **52,** 939 (1964).

49. Rothfield, L., and M. Pearlman, *J. Biol. Chem.* **241,** 1386 (1966).

50. Salton, M. R. J., *Ann. Rev. Microbiol.* **21,** 417 (1967).

51. Salton, M. R. J., and A. Netschey, *Biochim. Biophys. Acta* **107,** 539 (1965).

52. Silberman, R., and W. L. Gaby, *J. Lipid Res.* **2,** 172 (1961).

53. Smith, P., *Advan. Lipid Res.* **6,** 69 (1968).

54. Smith, P., in *The Mycoplasmatales and The* L-*Phase of Bacteria*, L. Hayflick (ed.), Appleton-Century-Crofts, New York, 1969, p. 469.

54a. Smith, P. F., *The Biology of Mycoplasmas*, Academic Press, New York, 1971.

54b. Stumpf, P. K. *Ann. Rev. Biochem.* **38,** 159 (1969).

55. Vaczi, L., *Postepy Mikrobiol.* **5,** 2 (1966).

56. Van Deenen L. L. M., *Prog. Chem. Fats Other Lipids* **8,** 1 (1966).

56a. Vandenheuvel, F. A., *Advan. Lipid Res.* **9,** 161 (1971).

56b. VanLear, G., and F. W. McLafferty, *Ann. Rev. Biochem.* **38,** 289 (1969).

56c. Weete, J. D., D. J. Weber, and J. L. Laseter, *J. Bacteriol,* **103,** 536 (1970).

57. Weiner, I. M., T. Higuchi, L. Rothfield, M. J. Osborn, and B. L. Horecker, *Proc. Natl. Acad. Sci. U.S.* **54,** 228 (1965).

CHAPTER

SEVEN

Molecular Virology

H. FRAENKEL-CONRAT

1. BASIC FEATURES OF VIRAL NUCLEIC ACIDS

A. The Properties of Viral Nucleic Acids in Solution and in Virions

All single-stranded polynucleotides, including RNA and DNA, show two main types of molecular interactions in aqueous solution. The bases on the polynucleotide chain interact intramolecularly with their neighbors, a phenomenon termed base stacking, which is nonspecific except for the low affinity of uridine residues to interact with one another. And there is also complementary base pairing, which favors conformations in which the greatest possible numbers of adenosine–uridine (or adenosine–thymidine) and guanosine–cytidine pairs are formed through hydrogen bonding between the 1, (2), 6 positions of the purines and the (2), 3, 4 positions of the pyrimidines (Fig. 1). This type of bonding occurs intra- and intermolecularly. Both of these phenomena decrease the ultraviolet absorbance of the polymer, both increase its stability, and both (but particularly the complementary base pairing) increase the negative optical rotation and lower the availability of the component bases toward ions, reagents, or enzymes.

Although we started this discussion with the words "single-stranded," it may have become evident to the reader that the double-stranded state is only one extreme of the range of possible interactions: when two completely complementary polynucleotides are available (the simplest example being a mixture of polyadenylic and polyuridylic acids, and the most complex the two strands of a mammalian chromosome), optimal base pairing, in conjunction with base stacking, will ensue under favorable conditions, and a double-stranded helical and rod-like molecule will form.

Various factors are unfavorable for nucleotide interactions in solution. Among these are high and low pH; the presence of certain dispersing solvents, such as dimethylsulfoxide or dimethylformamide; chemical agents that react with the amino groups on the bases, such as formaldehyde; and high temperature. Also, low ionic strength tends both to disrupt double-strandedness and to diminish base-stacking tendencies, because the negative charges on the phosphates are not dampened out sufficiently and they tend to stretch the molecule. Nucleic acid bases also can be prevented from interacting by being encapsidated in protein in a manner that does not permit either stacking or complementary base interactions. This is the case for the nucleic acids inside helical viruses such as tobacco mosaic virus (TMV) (149) (Fig. 2)

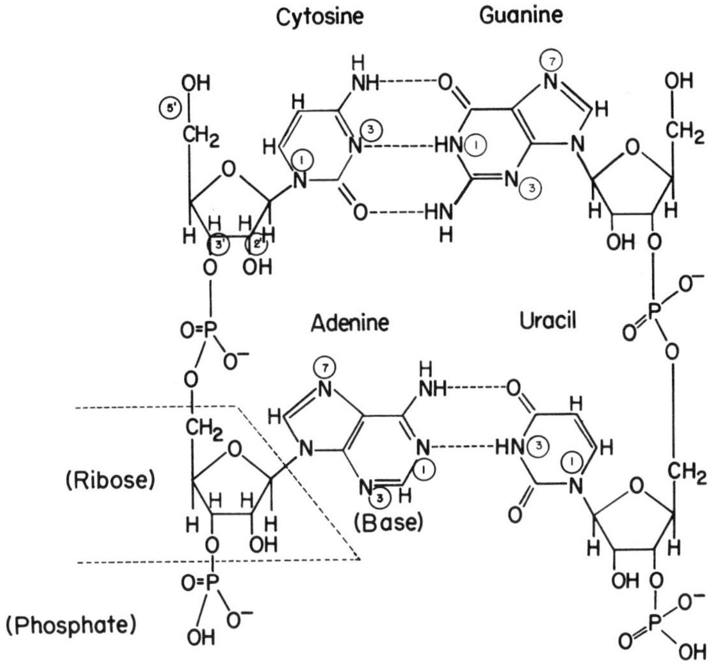

Fig. 1. Schematic presentation of complementary base pairing. Extension of the chain in the vertical direction would yield RNA. Removal of the 2′-OH groups and methylation of the uracil in position 5 (= thymine) would produce DNA.

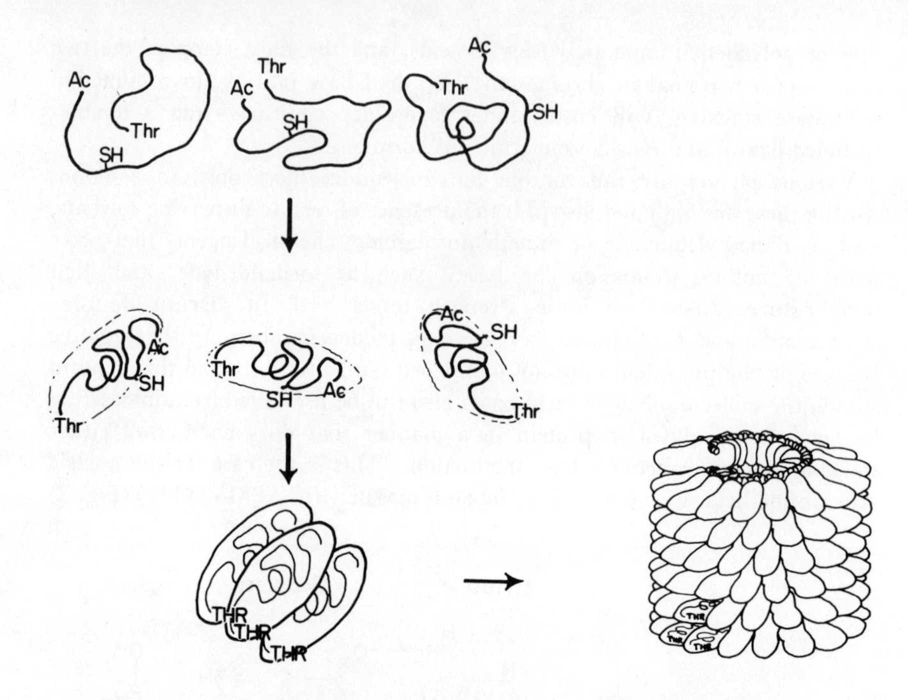

Fig. 2. Schematic presentation of tobacco mosaic virus protein aggregation. Top line: Denatured protein (N-terminal acetyl group, single cystein (—SH) residue, and C-terminal threonine are indicated. Middle line: Renatured TMV protein (arbitrarily assumed conformation). Bottom line: A protein (trimer) aggregating to TMV rod (49 protein molecules per 3 terms, pitch of helix 2.3 nm, maximal diameter of particle 18 nm, of nucleic acid helix 8 nm, of central hole 4 nm).

and probably also in good part in isometric viruses, for example, turnip yellow mosaic virus (TYMV) (43) (Fig. 3).

Thus, in considering various properties of viral nucleic acids, their physical states and conformations are at least as important as whether they contain uracil and ribose or thymine and deoxyribose (RNA vs. DNA). The states to consider for any nucleic acid are intraviral locations in a more or less tight protein environment, in free solutions of various ionic strengths and pHs, or in the presence of antistacking and/or antipairing agents. The biological and chemical properties of specific viral nucleic acids to be discussed below are all consequences of these molecular interactions.

B. Biological Activity of Nucleic Acids

Viruses carry certain biological properties of which, by definition, infectivity is the most important. It was suggested by the experiments of Hershey and

Chase (108) on T2 phage infection, and proved by the experiments with TMV RNA performed in the author's laboratory (72, 79) and by Gierer and Schramm (91), that the infectivity of viruses resides in their nucleic acids. This does not necessarily mean that any viral nucleic acid is able to initiate infection in any cell or even in its typical host cell. It appears that the single RNA molecule of most plant viruses (as well as the DNA of cauliflower mosaic virus) (230, 231) is able to cause the typical disease if introduced, as is usual for transmission of these viruses, through mechanical wounding of the host cell. Animal cells are also infectable by the RNA or DNA of many typical viruses. Failure to obtain infection may be attributable to one or more of three possible causes.

1. The nucleic acid complement of certain RNA viruses, with particular reference to influenza virus, Rous sarcoma virus, and the double-stranded RNA of the reoviruses, consists of multiple molecules (see p. 293), and upon disruption of the virus the probability of introducing into the host cell a complement of the entire viral genome is extremely low.

2. Nucleic acid, RNA or DNA, of more than 5×10^6) molecular weight, for example, the RNA of SV5 or Newcastle disease virus (7×10^6) or the DNA of the adenoviruses (23×10^6), may simply be too big to enter intact into a host cell, and therefore fails to initiate infection (46).

Fig. 3. Turnip yellow mosaic virus. Model of the virus with its 32 capsomeres (20 hexamers and 12 pentamers of the 20,000-molecular-weight protein chains = structural subunits); and cross section to illustrate the location of the RNA (black strand) (43).

3. Several groups of RNA viruses (e.g., the rhabdo-, myxo-, paramyxo-, oncorna-, and diplornaviruses) carry specific polymerases that are absent from RNA preparations and are required for replications (see p. 281)*

The nucleic acids of all coliphages are unable to infect *Escherichia coli*, regardless of their size and nature. The resistance of these Gram-negative bacteria to infection by nucleic acid is evidently due to the nature of their thick and multilayered cell wall. If this wall is partly degraded by means of lysozyme and EDTA, the resultant so-called spheroplasts appear to be susceptible to infection by all phage nucleic acids. Thus the single-stranded RNA of f2 and related phages (mol. wt. about 1.2×10^6), the DNA of the ØX174 and fd groups (1.7×10^6), and the double-stranded DNA of T1 (30×10^6), as well as the double- or single-stranded circular form of λ DNA, can all infect *Escherichia coli* spheroplasts (148, 299). Infection by the largest phage DNA, that of the T-even phages, has also recently been achieved, although only under very specific conditions (304, 352).

Since for all the coliphages here discussed there exists evidence of an infection organ, be it a single molecule of maturation factor in f2, a spike made up of 11 protein molecules on ØX174 or as complicated an organ as the tail of T2 (see p. 248), it appears reasonable to regard this organ as a requirement for infection of intact *E. coli*. Spheroplasts, in turn, lacking the attachment sites for this organ (the lock to fit the key), are not infectible by the intact phages, in contrast to their susceptibility to the free nucleic acids. One may reason, on this basis, that plant and animal viruses which are infective in the form of the naked nucleic acid lack specific and obligatory infection organs.

C. The Structure of Viral RNAs

The simplest viruses consist of one molecule of nucleic acid encased in a single type of protein. The nucleic acid is RNA in all plant viruses [except cauliflower mosaic and related viruses (230, 231)] and in many small bacterial and small and medium-sized animal viruses. The RNA is always linear (noncircular) and usually single-stranded in the virion, although a number of seemingly related animal, insect, and plant viruses (the diplornavirus group) contain double-stranded RNA. The single-stranded RNA of phages and plant viruses, as well as of the simplest animal viruses (the picorna- and togavirus groups, see p. 291), is able to act directly as messenger when inside the host cell and has been defined as the plus strand; the complementary strand made upon its replication is termed the minus strand. The existence of some encapsi-

* For definition and classification of viruses see pp. 290–300.

dated minus strands is probable (207). The more complex animal RNA virions contain predominantly, if not solely, the minus strand, unless they contain double-stranded RNA (e.g., reoviruses).

In size, viral RNA molecules range from 0.3×10^6 to 7×10^6 daltons (900–23,000 nucleotides), the lower limit for a complete infectious molecule being about 10^6 daltons (excepting viroids). Sedimentation rates vary greatly, depending on the composition of the medium, but even under standard conditions they do not closely reflect the molecular weights of different viral RNAs (e.g., the largest component of brome mosaic virus (BMV) RNA, 1.1×10^6 in molecular weight, is 27S, and of TMV RNA, 2×10^6 in molecular weight, 30S), a fact that is attributed to differences in the extent of their conformation. The particular sensitivity of certain sites of the RNA of some isometric viruses to enzymatic attack supports this concept (18). Furthermore, some of the more enzyme-resistant polynucleotide segments isolated from R17 may seemingly form many complementary base pairs upon hairpin folding (3a, 130).

The RNAs of the phages and of all typical plant viruses have molecular weights between 1 and 2×10^6. The RNAs of the small animal viruses, for example, poliomyelitis, encephalomyocarditis, and foot-and-mouth disease virus, have molecular weights of about 2.6×10^6, somewhat larger than formerly believed (35, 36, 188). Those of the more complex animal and related plant viruses range up to 7×10^6.

The compositions of viral RNAs vary within reasonably close limits. Only the four typical bases have been found, and the extremes in mole per cent seem to be 17 and 42, with values near 25 for all four bases being quite predominant. In the double-stranded RNAs the A–U pairs considerably exceed the G–C pairs.

Our knowledge of the nucleotide sequences of viral RNAs is still fragmentary. Thus the longest continuous nucleotide sequences presently established represent less than 20% of the length of the smallest complete viral nucleic acid. The RNAs of all plant and bacterial viruses that have been investigated by more than one method and in more than one laboratory carry near their right end (termed the 3′-and or the 5′-linked end) many pyrimidines and specifically terminate with-pCpCpCpA, terminally (3′) unphosphorylated, and sometimes lacking the last pA* (51, 94, 156, 173, 244, 255, 257, 282, 339).

* We are here using A, G, U, C, and T to symbolize the nucleosides adenosine, guanosine, uridine, cytidine, and thymidine, respectively. The letter p on the left represents a 5′-phosphate, on the right a 3′-phosphate. Thus the adenosine at the 3′-OH of the above fragment is not phosphorylated.

Frequently, when only the base sequence is important, the phosphate symbols (-p-) are omitted, in which case the letters A, G, U, C, and T represent nucleotides.

All known plant viral RNAs start (left end), seemingly, with Ap-(256, 335). A known exception to these generalizations is the RNA of the defective satellite tobacco necrosis virus, which starts in part with pppApGpU—and in part with ppApGpU—. (158, 284, 286, 349) and ends in part with 3′ phosphorylated Cp (349). The 5′ ends of the bacteriophage RNAs generally start with a triphosphate, followed by several G residues (pppGpGp . . .) (50, 93, 208, 301). There are various indications that the ends play no decisive functional role in terms of exact nucleotide sequence, since molecules occur and are presumably viable which lack one nucleotide or carry an extra one near the chain end (G on the left, C on the right) (50, 285). This is also illustrated by the finding that the 3′-terminal adenosine can be removed from the RNA phages by a stepwise degradation procedure (244) without loss of infectivity, and with regeneration of the original A-terminated structure in the progeny virus (140, 278a). However, earlier experiments in which that technique was applied to TMV RNA showed loss of infectivity upon removal of the terminal adenosine from the viral RNA (244). yet several plant viruses terminate with C (323, 392, 428), and at least in one of these cases the C terminus can be enzymatically adenylated, as follows (345).

In recent years it has become evident that the 3′ terminus of several plant virus RNAs can be enzymatically charged with specific amino acids (TYMV RNA with valine, TMV RNA with histidine; BMV RNA with tyrosine). Thus these termini are not only chemically but also functionally similar to tRNA (345, 345a, 371, 372, 392, 928).

Although there exist resemblances in the termini of the RNAs of typical plant and bacterial viruses, there seems to be more variability in this regard in animal viruses. This may in part be related to the fact that the former but not all of the latter are direct messenger RNAs. It now appears that messenger RNAs active in animal cells generally carry 3′-terminal poly A sequences, and the RNAs of the enteroviruses (e.g., polio RNA) which act as direct messengers are constructed in that manner, carrying apparently from about 16 to 200 terminal adenylic acid residues (377, 427). The RNA strands of the diplornaviruses, on the other hand, which are not messengers, show at the 3′ ends pyrimidines (304a, 370, 378). The oncornaviruses again appear to carry 3′-terminal poly A, suggesting that they also act as messengers, a theory which is supported by a recent report on their *in vitro* translation (410).

Regarding the 5′ ends, all animal virus RNAs seem to carry phosphates [pA in polio virus (424), di- or triphosphorylated G in diplornaviruses (159a), pppA in influenza virus (300)], but a recent report indicates that an oncornavirus terminates, like most plant viruses, in unphosphorylated A (411).

The methods developed in Sanger's laboratory (3, 3a, 130) and further developments contributed from Fiers' laboratory (178a, 318, 319, 379) have made the determination of long internal nucleotide sequences possible, and it is thanks to these that we now know more than one third of the nucleotide sequence of the group I RNA phages (f2, MS2, R17, etc.). This includes the entire coat-protein gene, the specific binding sites for ribosomes, and those of the coat protein and the replicase when acting as translation regulators, as well as various other terminal, intergenic, and intragenic sequences (see Fig. 4). Also, comparisons between equivalent sequences of various members of this group have become possible and have proved illustrative. Thus many more nucleotide differences were found between strains than the one or two amino acid exchanges in the coat proteins of these phages would have led one to expect. However, to the extent that these differences are located in areas coding for known amino acid sequences, it is clear that they usually involve the third base and are thus of no informational consequence (383a).

The principles of the techniques used for these studies are as follows. Controlled digestion with very specific nucleases (131) or with the T1 ribonuclease, which selectively splits at the more available G residues, was used to obtain large fragments, which were separated and purified. Good separations of oligonucleotides could be achieved only under conditions that abolished base-pairing interactions, such as high concentrations of urea (379). The resultant oligonucleotides were further degraded by complete digestion with specific enzymes and the fragments subjected to sequence analysis (131). Several of the resultant fragments, up to 74 nucleotides long, have been identified with various internal and terminal amino acid sequences of the coat proteins of the viruses under study, as well as with the apparently untranslated 5′-terminal regions (Fig. 4)

In another approach, particular attention was paid to the beginning of each gene by analyzing the three fragments which become more or less effectively protected against ribonuclease action by ribosomes attaching themselves to R17 RNA before its translation (245). A similar study with Qβ RNA yielded a single enzyme-resistant oligonucleotide sequence (110). A single ribosome-protected sequence at the coat-protein-initiating site was also found in f2 RNA which was very similar to the corresponding one of R17 (97), in accord with expectation, since these are very closely related viruses (265). It has now become clear that for conformational reasons phage RNAs combine with ribosomes initially and readily only at the coat-protein-initiation site and that partial degradation may have resulted in the earlier success of Steitz (245) in characterizing all three ribosome attachment sites. However, the other initiation sites can be made available on the intact RNA through agents that diminish the conformational protection of the molecule, such as formaldehyde (164).

Each of the six initiation fragments so far elucidated (three each in group I and group III phages) carried the protein-initiating triplet (AUG) approximately in the middle, and this was preceded, at slightly varying distances,

```
       10                20                30        40
pppGGGUGGGACCCCUUUCGGGGUCCUGCUCAACUUCCUGUCGAGCU
   50              60               70              80
AAUGCCAUUUUUAAUGUCUUUAGCGAGACGCUACCAUGGCUAUCG
90              100              110              120              130
CUGUAGGUAGCCGGAAUUCCAUUCCUAGGAGGUUUGA·CCU·AUG·
CGA·GCU·UUU·AGU·G·······CU·AAG·GCC·CAA·AUC·UCA·GCC·
      (A-prot.) F  -Met -Arg  -Ala  -Lys  -Ala  -Gln  -Ile  -Ser  -Ala  -
AUG·CAU·CGG·GGA·GUA·CAA·UCC·GUA·UGG·CCA·ACA·ACU·
Met  -His  -Arg  -Gly  -Val  -Gln  -Ser  -Val  -Trp  -Pro  -Thr  -Thr  -
GGC·GCG·UAC·GUA·AAG·UCU·CCU·UUC·UCG·AUG·GUC·CAU·
Gly  -Ala  -Tyr  -Val  -Lys  -Ser  -Pro  -Phe  -Ser  -Met  -Val  -His  -
ACC·UUA·GAU·GCG·UUA·GCA·UUA·AUC·AGG·CAA·CGG·CUC·
Thr  -Leu  -Asp  -Ala  -Leu  -Ala  -Leu  -Ile  -Arg  -Gln  -Arg  -Leu  -
Jer  -Arg
                                   G
UCU·AGA·UAG·AGCCCUCAACCGGAGUUUGA·AGC·AUG·GCU·
                                   (coat pr.) F  -Met  -Ala  -
  C**
UCU·AAC·UUU·ACU·CAG·UUC·GUU·CUG·GUC·GAC·AAU·GGC·
                                           10
Ser  -Asn  -Phe  -Thr  -Gln  -Phe  -Val  -Leu  -Val  -Asp  -Asn  -Gly  -
GGA·ACU·GGC·GAC·GUG·ACU·GUC·GCC·CCA·AGC·AAC·UUC·
                              20
Gly  -Thr  -Gly  -Asp  -Val  -Thr  -Val  -Ala  -Pro  -Ser  -Asn  -Phe  -
GCU·AAC·GGG·GUC·GCU·GAA·UGG·AUC·AGC·UCU·AAC·UCG·
Ala  -Asn  -Gly  -Val  -Ala  -Glu  -Trp  -Ile  -Ser  -Ser  -Asn  -Ser  -
  C***
CGU·UCA·CAG·GCU·UAC·AAA·GUA·ACC·UGU·AGC·GUU·CGU·
          40
Arg  -Ser  -Gln  -Ala  -Tyr  -Lys  -Val  -Thr  -Cys  -Ser  -Val  -Arg  -
                                                      U***
CAG·AGC·UCU·GCG·CAG·AAU·CGC·AAA·UAC·ACC·AUC·AAA·
50                                                  60
Gln  -Ser  -Ser  -Ala  -Gln  -Asn  -Arg  -Lys  -Tyr  -Thr  -Ile  -Lys  -
                    G***             U***
GUC·GAG·GUG·CCU·AAA·GUG·GCA·ACC·CAG·ACU·GUU·GGU·
                                           70
Val  -Gln  -Val  -Pro  -Lys  -Val  -Ala  -Thr  -Gln  -Thr  -Val  -Gly  -
GGU·GUA·GAG·CUU·CCU·GAU·GCC·GCA·UGG·CGU·UCG·UAC·
                              80
Gly  -Val  -Glu  -Leu  -Pro  -Val  -Ala  -Ala  -Trp  -Arg  -Ser  -Tyr  -
          C**       C**      U***
UUA·AAU·AUG·GAA·UUA·ACC·AUU·CCA·AUU·UUC·GGU·ACG·
          Leu**     90
Leu  -Asn  -Met  -Glu  -Leu  -Thr  -Ile  -Pro  -Ile  -Phe  -Ala  -Thr  -
  C***
AAU·UCC·GAC·UGC·GAG·CUU·AUU·GUU·AAG·GCA·AUG·CAA·
```

100
Asn -Ser -Asp -Cys -Glu -Leu -Ile -Val -Lys -Ala -Met -Gln -
GGU·CUC·CUA·AAA·GAU·GGA·AAC·CCG·AUU·CCC·UCA·GCA·
110 120
Gly -Leu -Leu -Lys -Asp -Gly -Asn -Pro -Ile -Pro -Ser -Ala -
U*** C**
AUC·GCA·GCA·AAC·UCC·GGC·AUG·UAC·UAA·UAG·AUG·CCG·
129
Ile -Ala -Ala -Asn -Ser -Gly -Ile -Tyr
GCC·AUU·CAA·ACA·UGA·GGA·UUA·CCC·AUG·UCG·AAG·ACA·
(replicase) F -Met -Ser -Lys -Thr -
ACA·AAG·AAG·UUC·AAC·UCU·UUA·UGU·AUU·GAU·CUU·CCU·
Thr -Lys -
CGC·GAU·CUU·UCU·CUC·GAA·AUU·UAC·CAA·UCA·AUU·GCU·
−100 −90
UCU·GUC·GCU·ACU·GG······GCUCCACCGAAAGGUGGGCGGGCU
−80 −70 −60 −50 −40
UCGGCCCAGGGACCCCUCCCUAAAGAGAGGACCCGGGAUUCUCCCGA
−30 −20 −10
UUUGGUAACUAGCUGCUUGGCUAGUUACCACCCA

Unknown locations

10 20 30 40
GCUCCUACCUGUAGGUAACAUGUUGCUCGGAGGCCUUACGGCCUCC
50 60 70 80
GGAUGCUCCUACAUGUCAGGAACAGUUG(UUACUG)
10 20 30 40
GCAUAUGAGAUGCUUACGAAGGUUCACCUUCAAGAGUUUCUUCC
50
UAUGAG

In polymerase gene

10 20 30 40
CACAGUGACUUUACAGCAAUUGCUVACUUAAGGGACGAAUUGCUCA
50 60 70
CAAAGCAUCCGACCUAAGGUUCUGGU

Fig. 4 Nucleotide sequences of the RNAs of phages of group I (MS2, R17, and f2). Where the nucleotide sequences correspond to known amino acid sequences, the amino acid sequences of A-protein, coat protein, and replicase are indicated under the corresponding triplets. The sequences given are those for MS2. Identical sequences were obtained to the extent investigated for R17, and for f2 at the 5′ and 3′ ends (1–129, -50 to -1). Differences detected within the group (R17, f2, MS2) are indicated by asterisks, as follows. Both R17 and f2 appear to have an A in the position marked as,* and concerning the subsequent sequence of about 10 nucleotides data are in part preliminary and agreement is not complete for MS2, R17, and f2. All other differences between MS2, on the one hand, and R17 or f2, on the other, occur at different nucleotides as indicated by ** for f2 and by *** for R17. A spontaneous mutation was observed in R17 in the course of its sequence studies, that is, G → A, which occurred 10 nucleotides upstream from the coat-protein initiation codon. The only amino acid difference between these phage coats—residue 88 being leucine in f2, as contrasted to methionine in MS2, R17, and M12—was borne out by the AUG → CUG change.

by either or both of the sequences UUUGA and AGGA. No better unifying feature was found to account for the recognition of these particular segments for ribosome binding and the initiation of protein synthesis.

The same principle of determining sequences of selectively protected nucleotides made use of the specific binding of the Qβ polymerase (421) and of the group I phage coat protein to the RNAs of these phages. The latter study indicated that the regulatory role of this process, first reported by Sugiyama and Nakada (259, 261), resided in a protein capsomere attaching itself to the beginning of the replicase gene (307).

The same methods that were so successfully applied to group I phages were also used with Qβ RNA. However, the greater part of our knowledge of Qβ RNA sequences comes from a different and ingenious method that derives sequence information from the study of the replication products of the RNA. This is done by stopping the replication of either the plus or the minus strand of Qβ RNA (using purified Qβ replicase and labeled triphosphates) after short intervals (5–30 sec), and observing the nature of the variously labeled polynucleotide products after enzymatic degradation by Sanger's methodology (23). Since the rate of RNA synthesis under these conditions was found to be six nucleotides per second, manageably short and progressively elongating pieces are obtained, the sequencing of which presents no problem. The only important requirement for this approach is synchronization of the enzyme's action, and this was achieved by initially allowing a quite limited extent of replication through the use of only two triphosphates, for example, GTP and ATP if the 3′-terminal sequence of the template was CCCCUGG. Thus replication stopped after the first five steps and was restarted in synchrony upon addition of the pyrimidine triphosphates. Later a method to synchronize the enzyme's action at internal sites was developed; this was based on arresting its action of incorporating unlabeled triphosphates by means of ribosome complexing of the RNA, and removing that arrest in the presence of the labeled triphosphates by dissociation of the ribosomes. By these techniques, as well as the purely degradative ones, about 15 to 20% of the nucleotide sequence of Qβ RNA has been established (see Fig. 5) (363, 423).

These studies revealed clearly, in support of what had earlier been suggested by terminal sequence studies (3a, 23, 94a), that both ends of viral RNAs carry no genetic message. Thus the first initiation codon, AUG, occurred only after 62 nucleotides from the 5′ end of Qβ RNA, and after 129 nucleotides in group I phage RNA.

These results also gave the first indication concerning the order of the three genes on Qβ RNA, since the protein initiated at this site was not the coat protein. Since then a variety of methods has shown that in all these phages the order of genes from the 5′ end is as follows: maturation protein, coat protein, replicase peptide chain (131, 151, 348). In Qβ a fourth protein was also ob-

10 20 30 40 50

pppGGG (G) ACCCCCCUUUAGGGG**G**U CAC (AC) (AC) (CUC) AGCAGUACUUCACUGAGUAUAAGAGGA

(A-prot.)

60 70 80 90 100

CAU· AUG· CCU· AAA· UUA· CCG· CGU· GGU· CUG· CGU· UUC· GGA· GCC· GAU· AAU· GAA· AUU·

F-Met-Pro-Lys-Leu-Pro

110 120 130 140 150 160

CUU· AAU· GAU· UUU· CAG· GAG· CUC· UGG· UUU· CCA· GAC· CU (UUC)U · AU**C** · GAA · UCU · UCC ·

170 180 190 200 210

GAC · ACG · CAU · CCG · UGG · UAC · ACA · CUG · AAG · GGU · CGU · GUG · UUG · AAC · GCC · ACC · CUU ·

220 230 240 250 260

GAU · GAU · CGU · CUA · CCU · AAU · GUA · GGC · GGU · CGC · CAG · GUA · AGG · CGC · ACA · CUC · CAU ·

270 280 290 300

CGC · GUC · ACC · GUU · CCG · AUU · GC [(UU) (CU) (C)] C · AGG · CCU · UCG · UCC · GGU · AAC · AAC ·

310 320 330

CGU · UCA · GUA · UGA · UCC · CGC · AG· · · · · · · AUCUUGAUACUACCUUUAGUUCGUUUAAACACGUU ·

CUUGAUAGUAUCUUUUUAUUAACCCAACGCGUAAAGCGUUGAAACUUUGGGUCAAUUUG · AUC · AUG ·

(coat prot.) F-Met

GCA · AAA · UUA · GAG · AC · · · · · · · · GCU · UAG · UAA · CUA · AGG · AUG · AAA · UGC · AUG · UCU · AGG ·

Ala-Lys-Leu-Glu-Thr- (Replicase) F-Met-Ser-Lys-

-160 -140 -120

ACA · GC · · · · · · · CCGUGUUCUGGCACCCUACGGGGUCUUCCAGGGCACGAAGGUUGCGUCUCUACAC

Thr-Ala

-100 -80 -60 -40

GAGGCGUAACCUGGGGGAGGGCGCCAAUAUGGCGCCUAAUUGUGAAUAAAUUAUCACAAUUACUCUUAC

-20 -10

GAGUGAGAGGGGGAUCUGCUUUGCCCUCUCUCCUCCC (A)

Fig. 5. Nucleotide sequences of the RNA of Qβ phage. Where the nucleotide sequences correspond to known amino acid sequences, the corresponding amino acid sequences of A-protein, coat protein, and replicase are indicated under the corresponding triplets.

served, but it has recently been demonstrated that this protein results from faulty termination of, or rather continuing translation beyond, the coat-protein gene (350, 382, 422). Thus the actual number and size of the three gene products of the RNA phages, of approximately 43,000, 14,000, and 65,000 dalton molecular weights, account for about all of the RNA in these viruses.

For several reasons plant viral RNAs have not as yet been successfully subjected to extensive sequence analysis. Their typical size, as exemplified by TMV RNA, and their presumed contents of genes and initiation sites are double those of the RNA phages. Furthermore TMV RNA lacks the conformational stabilization due to hairpin folding, which protects large segments of the phage nucleic acids against gentle enzyme action, at least when located in the TMV rod, where the RNA is unable to form such folded structures. Two groups of workers have studied the largest oligonucleotides resulting from complete T1 ribonuclease digestion of TMV RNA. Mandeles and Lloyd (163, 172) report the largest segment to have 69 nucleotides with a terminal G residue. The next largest group consists of 26 residues, which could be resolved by farther chromatography into two components of similar composition and sequence. Through controlled stripping of protein from the TMV rod, progressively uncoating the RNA from the 3′ end, Mandeles was able to establish the approximate location of the T1-resistant polynucleotides in the RNA chain. The conclusion was that the largest, ω, was about 3% from the 3′ end, and the other two, Ψ1 and ψ2, 6% and 44% from that end. It has been suggested by others that the gene for local lesion formation on *Nicotiana sylvestris*, and probably also that for the coat protein of this virus, are located in the left (5′) third of the molecule (137, 138).

Mundry and Priess (182, 383) now confirm the conclusions of Mandeles and Lloyd concerning the size of the largest T1 fragment of TMV RNA, and report minor differences in the composition of this fragment as isolated from three different strains of TMV, all three fragments showing a great predominance of A (about half) (182, 383).

D. The Structure of Viral DNA

The smallest DNA viruses, the spherical phages of the ØX174 and the threadlike phages of the fd group, contain single-stranded circular DNA with a molecular weight of 1.7 to 2.0 × 10^6 daltons, and the same is true of the DNA of the very small animal DNA viruses of the parvo group (45, 176, 205). Some inconsistencies regarding these, including the defective adeno-associated virus, AAV, were resolved when it was shown that individual particles contain either a single-stranded plus strand or its complementary minus strand, and that these nucleic acid molecules tend to become double stranded only as an artifact of isolation (21, 175, 210).

On the hand, a recent finding indicates that at least one type of parvoviruses, the AAV viruses, are terminally self-complementary and tend to close circles by means of a double-stranded segment (362). This appears also to be the case with adenovirus DNA, terminally single stranded (338). All larger DNA viruses and phages (particle weight $> 40 \times 10^6$ daltons) have double-stranded DNA, circular up to 5×10^6 daltons and linear within the virion for most larger DNA molecules (Fig. 6). The smallest of these, the DNA of polyoma virus (Fig. 7) and SV 40, contain about as many base pairs as ØX174 contains bases, even though their particle weights differ by a factor of 6.

In terms of composition, only a few of the bacteriophages contain appreciable amounts of untypical bases [5′-hydroxymethylcytosine, variously glu-

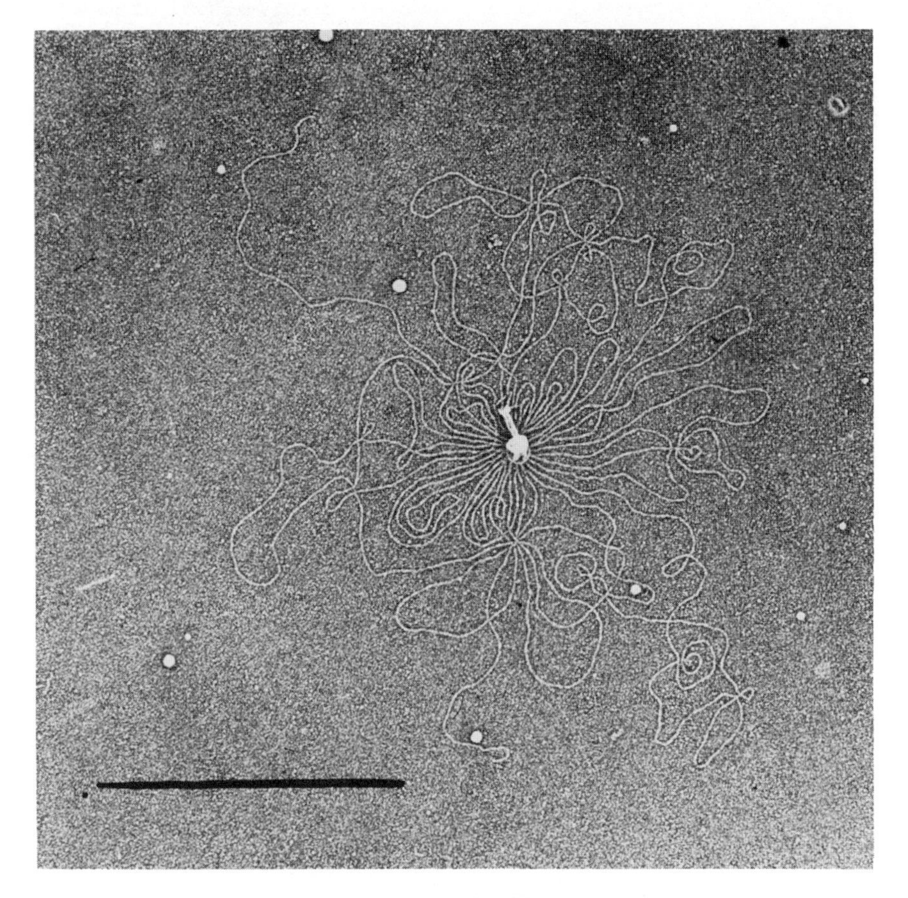

Fig. 6. T2 bacteriophage ghost with extruded DNA. The marker is 1 μm in length, the DNA 54 μm. Note the ends of the molecule at the top and bottom. (Courtesy of A. K. Kleinschmidt.)

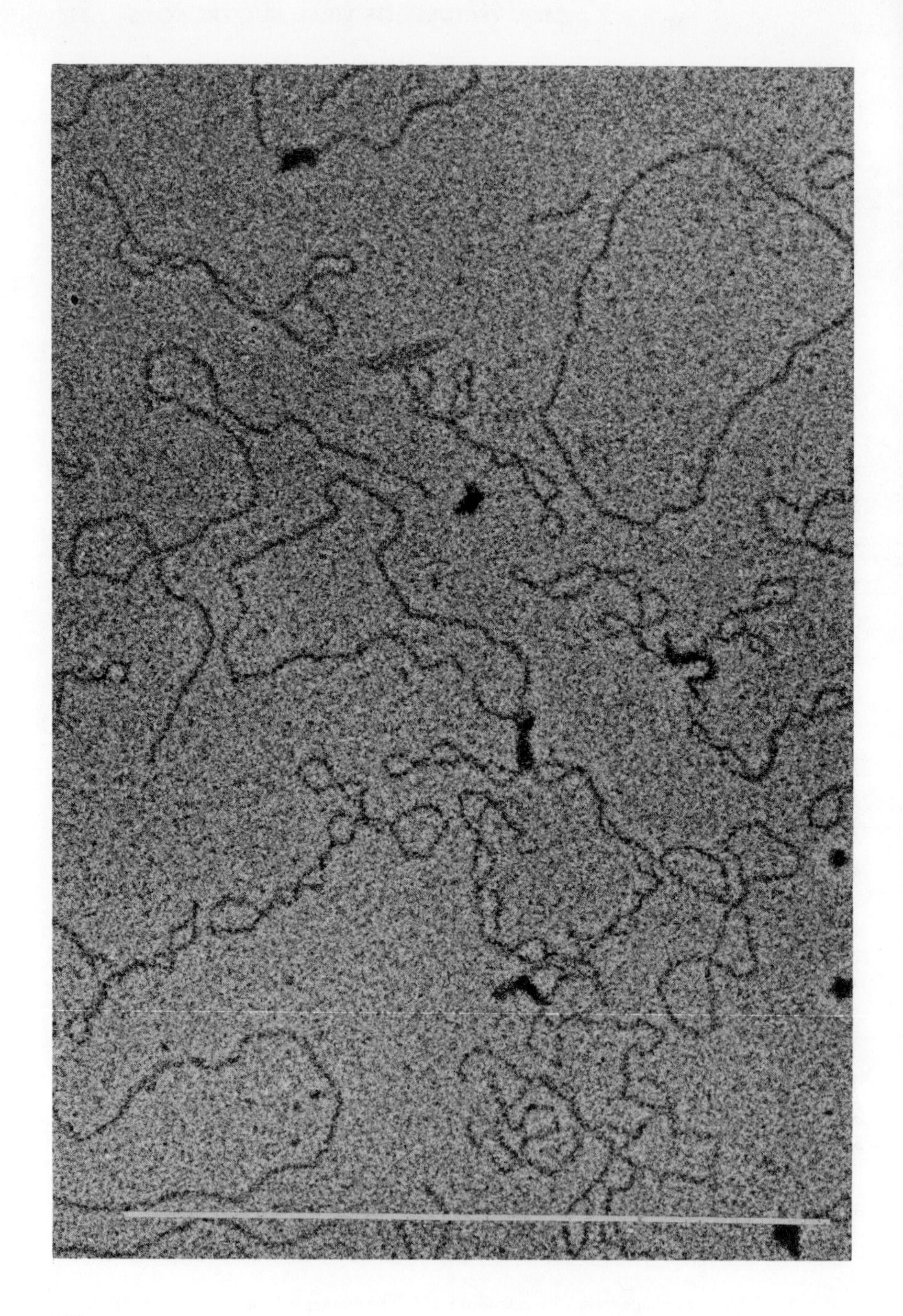

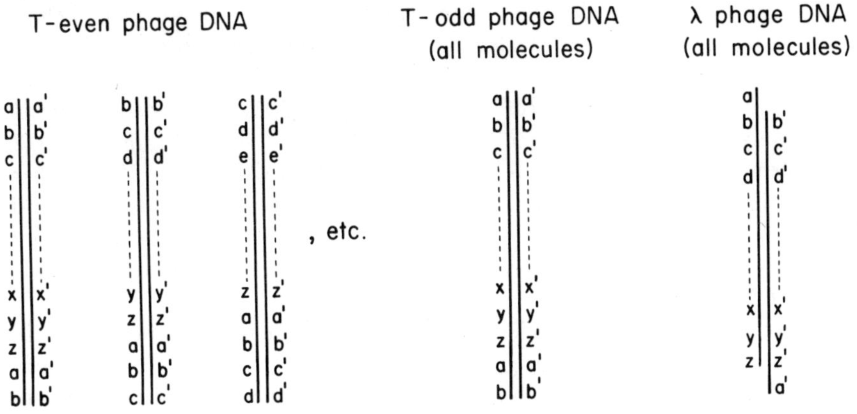

T-even phage DNA

T-odd phage DNA (all molecules)

λ phage DNA (all molecules)

, etc.

Fig. 8. Schematic presentation of circular permutation (in T-even phage DNA), terminal redundancy (in T-even and T-odd phages), and cohesive ends (in λ phage). The letters and primed letters represent genes and their complements in terms of nucleotide sequences.

cosylated, in the T-even phages, (157), 5′-hydroxymethyluracil, etc.], but minor amounts of methylated bases occur more frequently. In the double-stranded DNAs the A–T pairs represent from 36 mole % in the tipula iridescent virus to 65 and over 70 mole % in cowpox and T2 phage and herpesvirus.

The sedimentation behavior of many viral DNAs is complex, largely as a consequence of their circularity. Single-stranded circular DNA tends to be contaminated by nicked and thus linear molecules (e.g., ØX174: circular 12S, linear 10S). Double-stranded circular molecules (Fig. 7) are normally supercoiled and are transformed by a single nick in one strand into open circular forms, and by a nearby break in the other strand into linear forms (e.g., polyoma: 20.3S, 15.8S, and 14.4S, respectively).

The latter occurrence yields a linear double-stranded molecule with single-stranded but complementary terminal sections. This is the actual nature of the DNA of phage λ in the virion (107). The terminal segments are complementary and have been termed cohesive ends, since they tend to reform a circle in dilute solution, while at high DNA concentration intermolecular cohesion will lead to linear dimers and higher states of aggregation (Fig. 8). This is only one illustration of the fact that linearity and circularity of double-stranded

---

Fig. 7. The DNA of polyoma virus. Samples prepared by the protein film methods and shadowed with platinum. The micrograph shows supercoiled circular molecules, simple circular molecules, and linear molecules. The bar in the micrograph represents 1 μm. ×100,000. (Courtesy of E. A. C. Follett.)

DNA are closely related states; other aspects of this will be discussed below. The sequence of the deoxynucleotides on the "right-hand sticky end" of λ phage has now been reported as -GGGCGGCGACCT (295, 426), and the left hand is, by definition, complementary to it (295). A related lysogenic coli-phage, 186, shows a rather different "sticky end" with 19 rather than 12 complementary base pairs (-GGCGTGGCGGGGAAAGCAT) (388). Other chemical features first discovered with λ but later found for various other viral DNAs are the difference in GC content and thus the difference in buoyant density of the two strands, as well as of the two halves resulting from shearing of the DNA (107, 111, 112).

Methods for deoxynucleotide sequencing are only now being developed. In addition to the methods relying on the use of DNA polymerases, which have yielded the above data, as well as those concerning the terminal sequences in T7 DNA, others not depending on enzymes have recently yielded partial sequence data on ØX174 DNA (397, 431).

Other medium-sized phages (e.g., T1 and T7) carry DNA with a molecular weight of about 25 to 30 × 10^6 daltons, lacking single-stranded regions and cohesive ends, and end groups have been determined in several of these (281). Finally, the DNA of the T-even phages (120 × 10^6 in mol. wt.) is not, in chemical terms, homogeneous, since individual molecules start with different genes and thus different nucleotide sequences (see below). Both groups of phages show the phenomenon called "terminal redundancy." This means that genes and nucleotide sequences from one end recur at the other end. The two classes of viruses differ in that only the T-even phages generally show "circular permutation," the term denoting the heterogeneity mentioned above. This is illustrated schematically in Fig. 8.

The evidence for terminal redundancy and circular permutation was originally purely genetic, based on the finding that certain gene pairs could map both distant and close (252). That there existed complementary base sequences at both ends was later demonstrated in a physical manner by limited enzymatic degradation of each strand from the 3′ end by means of *E. coli* exonuclease III (167, 202). The resultant single-stranded oligonucleotide tails were found, as in the case of λ DNA, to show complementarity in their nucleotide sequences. In the cases of T1 and T7 all molecules had the same two complementary sequences, and both inter- and intramolecular annealing could be observed, depending on DNA concentration. In the permuted T-even phage DNA, each molecular tail was much more likely to find a complementary sequence on the other end of the same molecule than on another molecule, and thus circularization resulted under all conditions of annealing (Fig. 8).

The mechanism that very probably accounts for circular permutation is the intracellular synthesis and accumulation of enormously long, repeating viral

DNA molecules (a to z followed by a to z, etc.), which are then cut into phage-head-sized segments (81, 82). If the total genome is only 90% of such a segment, circular permutation with 10% terminal redundancy will be the consequence.

2. BASIC FEATURES OF VIRAL PROTEINS

A. Plant Virus Proteins

All viruses consist of, or contain, in addition to nucleic acid one or several proteins.* Usually one (but in some instances several) of the virus proteins is in close association with the nucleic acid. This combination represents the nucleocapsid. Simple viruses consist of only the nucleocapsid. Its protein then constitutes 60 to 96% of the mass of the virus. The coat protein of one plant virus family, the tobacco mosaic virus, has been exhaustively analyzed, and its amino acid sequence determined (6, 7, 86, 186, 271). We know that the TMV coat protein of 17,530 molecular weight consists of 158 residues, except for a few distantly related strains such as the Holmes ribgrass (HR) and various cucumber viruses, which show deletion and insertions so that they are said to range from 151 to 161 residues (87, 128, 288). All but one member of the subgroups of TMV have acetyl-seryl-tyrosyl at the N terminus and prolyl-alanyl-threonine or prolyl-alanyl-serine at the C terminus. In the strain that lacks the acetylated end, proline is N-terminal (201, 209) (Fig. 9). All other plant viral proteins except the tobacco necrosis virus (159) seem to carry blocked N termini, demonstrated to be like TMV protein acetylated in some instances. In the coat proteins of animal and bacterial viruses the N termini are also usually quite unreactive, although this is frequently due only to the conformation of these proteins. It appears that chemical or physical masking of end groups represents an advantage for viral coat proteins. The proline at the N terminus of the U2 strain of TMV and the prolines in third position from the C terminus in all TMV strains seem to play similar roles in representing residues inhibitory to degradation by many exoproteolytic enzymes.

In common with many viral coat proteins, TMV protein lacks methione and histidine, although the former occurs in other members of the family, and the HR strain contains also a histidine residue. The four completely and the two

* Some viruslike diseases (e.g., potato tuber spindle virus) have been found to be associated with naked nucleic acids of only about 10S and 50,000 daltons molecular weight (52, 53, 327, 407). Also, defective virus infections are known which are characterized by lack of production of functional virus coat proteins (234).

```
                                   (10)                                          (20)                                  (30)
                  **        *  **  *   *              **              **  **  **            *  **          **
Wild Type:   AcSer-Tyr-Ser-Ile-Thr-Thr-Pro-Ser-Gln-Phe-Val-Phe-Leu-Ser-Ser-Ala-Trp-Ala-Asp-Pro-Ile-Glu-Leu-Ile-Asn-Leu-Cys-Thr-Asn-Ala-Leu-Gly-
Dahlemense:    "   "   "   "   " Ser "   "   "   "   " Phe "   "   " Val "   "   "   "   "   "   " Leu " Val "   " Ser-Ser "   "
U2:           Pro  "  Thr  " Asn "   "   "   "   "   " Tyr "   "   " Ala-Tyr "   "   " Val "   " Ile " Leu "   " Asn-Ala "   "
HR:         AcSer  "  Asn  " Thr-Asn-Ser-Asn " Tyr-Gln " Phe-Ala-Ala-Val-Trp " Glu " Thr-Pro-Met-Leu " Gln " Val-Ser "   " Ser
-----------------------------------------------------------------------------------------------------------------------------------
                                      (40)                           (50)                             (60)
              **                                   **                         *           *  **     **      **
Wild Type:   Asn-Gln-Phe-Gln-Thr-Gln-Gln-Ala-Arg-Thr-Val-Val-Gln-Arg-Gln-Phe-Ser-Gln-Val-Trp-Lys-Pro-Ser-Pro-Gln-Val-Thr-Val-Arg-Phe-Pro-Asp
Dahlemense:   "   "   "   "   "   "   "   "   "   " Thr  "   " Gln  "   " Ser-Glu-Val "   "   " Phe " Gln-Ser "   "   "   "   " Gly
U2:           "   "   "   "   "   "   "   "   "   "   "   "   "   "   " Ala-Asp-Ala "   "   " Ser " Val-Met "   "   "   "   " Ala
HR:          Gln-Ser-Tyr "   "   " Ala-Gly " Asp "   " Arg "   "   "   " Asn-Leu-Leu-Ser-Thr-Ile-Val-Ala-Pro-Asn-Gln "   "   " Asp
-----------------------------------------------------------------------------------------------------------------------------------
                               (70)                               (80)                              (90)
                                                                                                                        *
CGMM-W:                                                                                                 Asn-Arg-Val-Ile-Glu-Val
CGMM-C:                                                                                                  "   "  Ala  "   "   "
              ** *                      **                         **
Wild Type:   Ser-Asp-Phe-Lys-Val-Tyr-Arg-Tyr-Asn-Ala-Val-Leu-Asp-Pro-Leu-Val-Thr-Ala-Leu-Leu-Gly-Ala-Phe-Asp-Thr-Arg "   " Ile  "   "   "
Dahlemense:  Asp-Val-Tyr-Lys "   "   "   "   " Ala-Val "   "   "   " Ile "   "   "   " Gly-Thr "   "   "   "   "   "   "   "   "
U2:          Ser-Asp-Phe-Tyr "   "   "   "   " Ser-Thr "   "   "   "   "   "   "   "   " Asn-Ser "   "   "   "   "   "   " Glx "
HR:          Thr-Gly-Phe-Arg "   " Val-Asn-Ser-Ala-Val-Ile-Lys "   " Tyr-Glu "   " Met-Lys-Ser "   "   "   "   "   "   " Gln-Thr-
-----------------------------------------------------------------------------------------------------------------------------------
                     (100)                              (110)                             (120)
CGMM-W:      Val-Asp-Pro-Ser-Asn-Pro-Thr-Thr-Ala-Glu-Ser-Leu-Asn-Ala-Val-Lys-Arg-Thr-Asp-Asp-Ala-Ser-Thr-Ala-Ala-Arg
CGMM-C:      Glx-Asx "   " Asp " Ser " Gly " Ala "   "   " Thr "   " Asp "   "   "   "   "   "   " His
              **      *                      **                                                   *            *   *
Wild Type:   Glu-Asn-Gln-Ala-Asn " Thr " Ala " Thr " Asp "   " Arg " Val "   "   " Thr-Val " Ile-Arg-Ser-Ala-Ile-Asn-Asn-Leu
Dahlemense:   "   "   " Gln-Ser "   "   "   "   "   "   "   "   "   "   "   "   "   "   "   "   "   "   "   "   "   "   "
U2:          Asx-Asx-Glx-Ala-Asx "   "   "   " Val(Thr,Pro,Asn,Ile)Glx-Glx "   "   "   "   "   "   "   "   " Ala-Ser "   "   "
HR:          Glu-Glu-Gln-Ser-Arg " Ser-Ala-Ser-Gln-Val-Ala-Asn-Ala-Thr-Gln "   "   "   "   "   "   "   "   " Ser-Gln " Gln-Leu "
-----------------------------------------------------------------------------------------------------------------------------------
                (130)                          (140)                             (150)
              **                **      *    **  *  *                            *   *                 **          **
Wild Type:   Ile-Val-Glu-Leu-Ile-Arg-Gly-Thr-Gly-Ser-Tyr-Asn-Arg-Ser-Ser-Phe-Glu-Ser-Ser-Ser-Gly-Leu-Val-Trp-Thr-Ser-Gly-Pro-Ala-Thr-
Dahlemense:  Val-Asn "   " Val "   "   "   " Leu-Tyr " Gln-Asn-Thr "   " Ser-Met "   "   "   "   "   "   " Ala "   " Ser
U2:          Ala  "   "   "   "   "   "   "   " Met-Phe " Gln-Ala-Gly "   " Thr-Ala "   "   "   "   "   " Thr-Thr "   " Thr
             Leu  "   "   " Ser-Asn-His-Gly " Tyr-Met " Arg " Glu "   " — Ala-Ile — " Pro "   "   " Ala "   "   "
-----------------------------------------------------------------------------------------------------------------------------------
```

Fig. 9. The amino acid sequences of the coat proteins of TMV strains. The wild-type, Dahlemense, U2, and HR strains are representatives of classes A, B, C, and D. CGMM viruses, partially sequenced, are cucumber green mottle mosaic virus strains. Asterisks indicate amino acids that have been found to be replaced by others once (*) or more frequently (**) in chemically produced mutants (usually only one, two, or quite rarely three replacements in one mutant).

partially sequenced TMV strains (Fig. 9) share the locations of most of the prolines and of the larger aromatic amino acid residues (positions 2, 10, 12, 17, 35, 48, 67, 70, 87, 139, 144, and 152), as well as of characteristic oligopeptide segments at 36–38, 61–63, 87–95, 113–122, etc., which presumably play critical roles in the folding of these proteins and enable them to function as coats for the RNA.

Indications concerning the conformation of TMV protein in the virus rod have been derived also from other data (Fig. 2). Various facts point to the availability, presumably at the surface of the rod, of the C-terminal segment of the chain. First among these indications is the great susceptibility of the C-terminal threonine to carboxypeptidase (101). The fact that only this one amino acid is released (1 per subunit, or 2130 per virus rod) represented the key observation that initiated structural studies on viruses. As stated above, the block in digestibility of the protein after release of this one threonine is probably due to the location of proline in third position from that end. The enzymatically dethreoninated virus is still fully viable and has normal stability. Mutants in which the proline in position 156 is replaced by leucine (see p. 242), however, lose three amino acids, and much of the stability, under carboxypeptidase attack (269). Another indication of the external location of the C-terminal region is that only the tyrosine in position 139 can be iodinated in the intact virus (74). Also, a lysine introduced in position 140 is quite reactive (390). Furthermore, the C-terminal hexapeptide plays a determinant role in the antigenic action of intact TMV, as does another apparently superficial segment comprising residues 60 to 65 (5). Other protein groups, in turn, are masked and/or unreactive in the intact virus; among these are the lysine in position 63 (73) and, particularly, the only cysteine (residue 27), which shows quite unusual properties (70). The main antigenic determinant in the isolated protein, as contrasted to the intact virus, is presented by residues 98 to 105 (297, 298).

The endgroups and some partial sequences have been elucidated in several other plant virus proteins (see ref. 68 and Table 1) (365). The complete amino acid sequence has been established for the turnip yellow mosaic virus (100, 391, 393). Most plant virus proteins studied seem to consist of single chains containing zero to four cysteine residues but no disulfide bonds; most are between 15×10^3 and 25×10^3 in molecular weight. However, at least two groups of plant viruses, exemplified by the tomato bushy stunt and cowpea mosaic viruses, have two coat proteins; in the first case, a minor component is arranged in a manner resembling the ØX174 structure (12 spikes per particle) (312), and in the other the two proteins appear to be present in proportional amounts (171, 425), as is the situation prevailing for the animal picornaviruses, exemplified by the poliomyelitis virus.

Table 1. Peptide Chains of RNA Viruses and Simple DNA Viruses[a]

Virus Class	Virus	Nucleocapsid (or Internal Protein) Mol. Wt. $\times 10^{-3}$ (and End Groups)[b]	Envelope (or External Protein) Mol. Wt. $\times 10^{-3}$	References[c]
1	Tobacco mosaic	17.5 (for sequence see Fig. 9) (2130)	—	271, 6
	Tobacco rattle (strain SP5)	24 (-Pro)	—	224
	Alfalfa mosaic (strain 425)	25 (Ac-Ser-····-Arg-His)	—	(122) 365
	Potato virus X	27 (Ac-····-Pro)	—	(178) 419
	Potato virus Y (tobacco etch)	22	—	(49) 322a
2	Turnip yellow mosaic	20.4 (Ac-Met-····-Ser-Thr)[d]	—	100
	Tobacco necrosis	33.3 (Ala-····-Met)	—	159
	Satellite of tobacco necrosis	~23 (Ala-Lys-····-Ala)	—	200, 272, 360
	Cowpea mosaic	42 (60)	—	425
		22 (60)		
	Tomato bushy stunt	38 (180)	—	(177) 312
		28 (12)		
	Mouse Elberfeld (and other enteroviruses, e.g. polio, mengo, EMC, foot-and-mouth disease)	35, 29, 25, 7 (60 each)	—	416 (262, 169, 37, 213, 214, 187)
	MS2, f2, fr phage	14 (for sequence see Fig. 10) (180)	—	279, 289
		43 (A-protein, 1)	—	203
	Qβ	14 (for sequence see Fig. 10) (180)	—	277, 423
		44 (A-protein, 1)	—	403
3	Sindbis	30	53, 53	249, 251
	Kunjin (encephalitis B)	18	65	282a
	Dengue (type 2)	13.5	8, 59	248
	Semliki Forest	32 (20%)	51 (80%)	1
4	Vesicular stomatitis	55 (N)[e]	69 (G)[e]	(41, 141, 183)
			32 (M)[e]	420

Class	Virus	Protein	Protein	References
5	Newcastle disease	56 (N)	74 (hemaggl.)	(103, 22)
		41 (M)	56	(56) 382
	SV5 (para-influenza)	61 (N)	67 (hemaggl. + neuram.)	(39, 103)
			56	
6	Influenza	60 (20%, M?)	60 (neuram. 7%)	406, 359
		26 (46%)	53 (hemaggl. 5%)	(117, 280)
		88 (2%)		417, 315, 316
7	Rous sarcoma	14 (9%, 35)	90–105[f]	(61, 56) 309, 324
		27 (36%, M?)	(76, 45, 35, *26*, 22, 12, 10)	
		35 (10%)		
	Mammalian oncornaviruses	Same general pattern		384
8	Reovirus type 3	155 (16%, 113)	80 (2%, 23)	166, 242
		140 (10%, 80)	72 (36%, 550)	
			42 (1%, 31)	
		38 (7%, 202)	34 (28%, 890)	
9	Isometric DNA phage (φX174)	48 (60)		33
		19 (12 × 5)		
		5 (12 × 5)		
		36 (12)		
	Filamentous DNA phage (fd)	5.2		
		70 (A-protein, 1?)		29, 105, 196, 212

[a] Typical examples of viruses of classes 1 to 9 (see pp. 000–000).
[b] A number in parentheses represents the number of molecules per virus particle or the percentage. All values are rough approximations, except for TMV and TYMV.
[c] Obsolete references in parentheses.
[d] Sequence known.
[e] These letters have been suggested to indicate nucleocapsid (N), glyco (G), and membrane (M) proteins of animal viruses. However, the envelope proteins of other virus groups are also probably all glycoproteins.
[f] Strain-specific differences, larger for transforming than for nontransforming strains.

```
                              (10)                          (20)                         (30)
Qβ:  Ala-Lys-Leu-Glu-Thr-Val-Thr-Leu-Gly-Asn-Ile-Gly-Lys-Asp-Gly-Lys-Gln-Thr-Leu-Val-Leu-Asp-Pro-Arg-Gly-Val-Asn-Pro-Thr-Asn-Gly-Val-
                                   (10)                           (20)
f2:        Ala-Ser-Asn-Phe  "  Gln-Phe-Val-Leu-Val-Asn  "  "  Gly-Thr-Gly-Asn  "  Thr-Val-Ala-Pro—— Ser  "  Phe-Ala  "  "  "
fr:         "  "  "  "  Glu-Glu  "  "  "  "  "  "  "  "  "  "  Asp  "  Lys  "  "  "      "  "  "  "  "  "  "

                  (40)                                  (50)                                    (60)
Qβ:  Ala-Ser-Leu-Ser-Gln-Ala-Gly-Ala-Val-Pro-Ala-Leu-Glu-Lys-Arg-Val-Thr-Val-Ser-Val-Ser-Gln-Pro-Arg———————Asn-Arg-Lys———Asn-Thr-
     (30)                                (40)                      (50)                                 (60)
f2:   "  Gln-Trp-Ile-Ser-Ser-Asn-Ser————Arg-Ser-Gln-Ala-Tyr-Lys  "  "  Cys  "  "  -Arg  "  Ser-Ser-Ala-Gln  "  "  "  Tyr-Thr-Ile-
fr:   "  "  "  "  "  "  "  "      "  "  "  "  "  "  "  "  "  "  "  "  "  "  "  "  Asn  "  "  "  "  "  Val-

                                                                                                        (90)
Qβ:  Lys-Val-Gln-Val———Lys-Ile-Gln-Asn-Pro-Thr-Ala-Cys-Thr-Ala-Asn-Gly-Ser-Cys-Asp-Pro-Ser-Val-Thr-Arg-Gln-Ala-Tyr-Ala-Asp-Val-Thr-
                              (70)                                          (80)
f2:   "  "  Glu  "  Pro  "  Val-Ala-Thr-Gln  "  Val-Gly——Gly-Val-Gln-Leu—————  "  Val-Ala-Ala-Trp-Arg-Ser  "  Leu-Asn-Leu-Glu-
fr:   "  "  "  "  "  "  "  "  "  "  Val-Gln  "      "  "  "  "            "  "  "  "  "  "  "  "  Met  "  Met  "

Q                   (100)                                                                          (120)
Qβ:  Phe-Ser-Phe-Thr-Gln-Tyr-Ser-Thr-Asp-Glu-Glu-Arg-Ala-Phe———Val-Arg-Thr-Glu-Leu-Ala-Ala-Leu-Leu-Ala-Ser-Pro-Leu-Leu-Ile-Asp-Ala-
     (90)                              (100)                          (110)                              (120)
f2:  Leu-Thr-Ile-Pro-Ile-Phe-Ala  "  Asn-Ser-Asp-Cys-Gln-Leu-Ile  "  Lys-Ala-Met-Gln-Gly-Leu  "  Lys-Asp-Gly-Asn-Pro-Ile-Pro-Ser  "
fr:   "  "  "  "  Val  "  "  "  Asx-Asp  "  "  Ala  "  "  "  "  "  Leu  "  "  Thr-Phe  "  Thr  "  Ile-Ala-Pro-Asn-Thr  "

                 (130)
Qβ:  Ile-Asp-Gln-Leu-Asn-Pro-Ala-Tyr
f2:   "  Ala-Ala-Asn-Ser-Gly-Ile  "
fr:   "  "  "  "  "  "  "  "
```

Fig. 10. The amino acid sequences of the coat proteins of RNA phages (Qβ, f2, fr). These are arranged in a manner to show the similarities between Qβ (group III) and the two related phages of group I.

B. Bacteriophage Proteins

The simplest bacterial viruses, the RNA phages, also consist of one predominant nucleocapsid protein of about 15 × 10^3 molecular weight. The sequences of two members of the group I phages (f2, fr) have been established (129 residues) (Fig. 10) (279, 289, 290). Similarly to variations among TMV strains, these two differ by 23 amino acid replacements, but nevertheless show great similarities (location of all prolines and most of the large residues, e.g., Tyr, Trp, Phe, Met, Leu, Lys, Arg). Also, as with TMV, several natural isolates have poved to be even closer relatives, differing by one to three amino acid exchanges in the coat protein (R17, MS2, M12, f2). Serologically not related to this group is Qβ (group III) (222). Nevertheless its amino acid sequence, recently reported, shows some similarities to the sequences of the above group (150a), although the difference between Qβ and f2 is considerably greater than that between fr and f2. The protein of Qβ consists of 131 amino acids and lacks methionine, histidine, and tryptophane (Fig. 10).

The RNA phages present a more complex protein structure than the typical plant viruses, in that they contain one molecule (per particle) of a second protein, termed the A-protein or maturation factor. This protein in group I phages contains five histidine residues and has a molecular weight of about 42 × 10^3 (246, 247). The corresponding protein in Qβ is of 44 × 10^3 molecular weight (89). A second, similarly sized protein occurring in Qβ is an enlarged coat protein (38 × 10^3 mol. wt.), due to an error in termination (see p. 246).

The simplest group of DNA phages (f1, fd, M13), though grossly different in architecture, shows some similarity to the RNA phages. These fibrous viruses (5.5 × 800 nm), containing circular single-stranded DNA, consist largely of a very small coat protein lacking cysteine, histidine, and arginine (49 residues in fd, molecular weight 5169, Fig. 11), which has been sequenced (10, 30). The coat protein of ZJ2 is one residue longer and differs in only two amino acids from that of fd (412a). In addition, evidence has been reported for a second protein having a molecular weight of 60,000 to 70,000 and containing histidine and arginine, located at the end of the fiber and probably

Ala-Glu-Gly-Asp-Asp-Pro-Ala-Lys-Ala-Ala-Phe-Asp-Ser-Leu-Gln-Ala-

Ser-Ala-Thr-Glu-Tyr-Ile-Gly-Tyr-Ala-Trp-Met-Val-Val-Val-Ile-Val-

Gly-Ala-Thr-Ile-Gly-Ile-Lys-Leu-Phe-Lys-Lys-Phe-Thr-Ser-Lys-Ala-Ser

Fig. 11. The amino acid sequence of fd virus coat protein.

fulfilling in the fibrous DNA phages functions similar to those of the maturation factor in the infection with isometric RNA phages (29, 105, 196, 212).

The biologically similar but physically quite different smallest isometric DNA virus of the ØX174 group shows much greater protein complexity, the virion consisting of three major and one minor component (193, 194). End group data have been reported as Ser-Asp- and Met-Ser- (3:1). The molecular weight of the main capsid protein, of which there are 60 per particle, is about 48×10^3. The 12 spikes evident on the surface of the particle, which may represent attachment organs, appear to be composed of five molecules each of a 19×10^3 (with N-terminal methione) and a 5×10^3 dalton species, and one molecule of 36×10^3 daltons (33). A typical medium-sized phage, T7, is made up of about ten proteins (254, 311, 314, 417). Phage λ consists of three major proteins, assigned to head, tail, and internal roles, respectively (38–45, 31, and 12×10^3 daltons; 60, 19, and 19%). Mutation to petit capsids caused the absence of the smallest component (352a).

The large T-even phages obviously show the greatest protein complexity. The main head protein, representing 85%, consists predominantly of peptide chains of 42×10^3 molecular weight with N-terminal alanine and C-terminal glycine (19, 145, 152). It appears to be identical in T2 and T4. Its assembly into heads proceeds in conjunction with the cleavage of a precursor protein, and certain mutants, characterized by elongated so-called polyheads, contain a larger protein (55×10^3 daltons) which was not properly cleaved to size. Three minor and smaller proteins, two of 18×10^3 and one of 11×10^3 daltons (the "internal protein"), and polypeptides differ among strains, and their roles have not yet been established. One of the 18×10^3 dalton proteins has N-terminal methionine; the others have alanine. The contractile tail sheath consists largely of 144 molecules of a protein of 50 to 80×10^6 daltons containing one histidine residue, C-terminal glycine and serine, and no detectable terminal amino group, possibly because these groups carry carbohydrate residues (63, 211). Contraction seems to result from a conformational change of this protein, probably leading from a helix of 24 turns containing 12 molecules to one of 12 turns containing 24 protein molecules per turn. The roles of the bound Ca^{2+} and of the ATP released during this process are not yet clear. Also, the roles of the dihydrofolate reductase and the pteroylhexaglutamate associated with the base plate have not been clearly defined. The tail plates (77S) consist of three proteins (53, 31, and 17×10^3 daltons); the tail tubes, of one predominant protein of about 20×10^3 daltons (322).

The characteristically bent tail fibers (13S) consist of two main structural proteins, making up two 70-nm-long pieces (155×10^3 and 120×10^3 daltons), as well as minor components. The lysozymes associated with T2 and T4 (164 residues long) have been completely sequenced (125, 270) and found to differ in only three aminoacid replacements from one another, in contrast to their very marked difference from egg white lysozyme, which nevertheless shows very similar enzymatic properties. Thus the phage lysozymes have two cysteine residues and three tryptophanes, compared to four disulfide bonds and six tryptophanes in egg white lysozyme. The specific functional role of phage lysozyme remains to be elucidated.

C. Animal Virus Protein

The small animal RNA viruses (picornaviruses), exemplified by the mouse-Elberfeld (213, 214) virus, which are isometric viruses similar in dimensions to many plant viruses and the RNA phages, show greater complexity than these in protein structure, since they all typically consist of 60 units of four components of molecular weights near 35, 30, 24, and 7×10^3 (169, 262, 331, 400, 416), always adding up to 96×10^3 daltons. They differ in composition but have not been studied in great detail. The poliomyelitis virus, encephalomyocarditis virus (37), and rhinoviruses are all probably built up in this manner. It now appears probable that the foot-and-mouth disease virus too is coated by four proteins (154, 283).

The now extensively used technique of polyacrylamide gel electrophoresis of proteins heated and electrophoresed in the presence of sodium dodecylsulfate and mercaptoethanol has provided a tool for the resolution and characterization (in terms of peptide chain molecular weight and composition) of minute amounts of proteins, detectable only by radioactivity. This method has supplied the evidence for the number and molecular weight(s) of many viral coat proteins and other intracellular virus-specific products; a representative selection of recent data obtained for the proteins of the simpler viruses, largely by this technique, is summarized in Table 1. Unfortunately, agreement between different laboratories or at different times in the same laboratory is far from perfect, and the data listed are not necessarily beyond question. It must be stressed that the method as usually employed dissociates most protein aggregates, an essential feature when dealing with such powerful aggregating systems as virus proteins. However, the molecular weights obtained in this manner are not those of functional units, such as enzymes, but only those of their building blocks, the individual peptide chains.

It has generally been assumed that each viral protein represents the product of (at least) one gene. Yet the number and size of animal virus coat proteins and of virus-specific proteins made in infected animal cells, as detected on polyacrylamide gels, have in many instances added up to considerably more genetic information than could be carried by the viral RNA. For example, the upper limit in the case of poliomyelitis virus with an RNA of 2.6×10^6 daltons would be about 250×10^3 daltons of protein. Strong evidence has been adduced that a single protein approaching this size is the primary gene product (126, 127, 169). This protein is believed to be subsequently and gradually fragmented, presumably by specific proteolytic action, yielding proteins of intermediate size and finally the small structural proteins of the virion. This is illustrated by the finding that an intermediate in the maturation of the poliomyelitis virus, the procapsid, a particle that still lacks RNA, contains one larger protein in lieu of two of the smaller ones found in the mature virus particle (126, 127) (α, γ, and ϵ, of approximate mol. wts. 35, 24, and 37×10^3, respectively, instead of α, β, γ, and δ of 35, 30, 24, and 7×10^3). Indications have been reported that this process of posttranslational cleavage is operative also in the togaviruses (38, 250, 404). It appears possible that animal cells and their viruses differ from bacterial systems in lacking mechanisms for the translation of polygenic messengers into multiple protein products.

In contrast to the picorna- and togaviruses, the virions of the more complex classes of RNA viruses do not contain encapsidated messenger RNAs, but rather their complements, so-called negative strands, and these virions also carry specific transcriptases for their RNAs. Thus it appears probable that these RNAs are transcribed as monogenic messengers. Nothing definite is known concerning plant cells in these regards, except that the complex plant viruses resembling animal rhabdo- and diplornaviruses also carry enzymes and are probably not direct messenger strands (337, 361).

The main surprise afforded by recent studies concerning animal virion proteins is that even the more complex RNA viruses, the nucleocapsids of which are enclosed in an envelope, consist of a small and definite number of virus-specific proteins and carry few if any host proteins (117). Most of the viruses in the toga-, myxo-, paramyxo-, and rhabdovirus groups mature by budding through the cell wall and membrane, and these viruses were formerly believed to incorporate many proteins from the host cell and particularly from its membrane. Yet analyses of these viruses by gel electrophoresis did not reveal the presence of appreciable amounts of host proteins, quite in contrast to their contents of carbohydrates and particularly lipids, which are largely host derived. Instead typical togaviruses (arboviruses A and B), exemplified by Sindbis (249, 251) and Semliki Forest viruses (1), as well as Kunjin and Dengue viruses (248), consist of only one nucleocapsid and probably two envelope proteins (403, 404). In the Sindbis virus, the internal protein (about 30×10^3 in molecular weight) constitutes about one fifth of the total, and the rest consists of the lipoglycoproteins forming the envelope (about 53×10^3 mol. wt.). These viruses, like the myxoviruses, have characteristic type-specific antigens, and the glycoproteins of the envelope could be identified with the antigenic type specificity of the virus, differences in electrophoretic properties mirroring those in antigenic specificities.

Influenza virus, the typical myxovirus, contains about six proteins, but agreement about their sizes and functions has not been reached. It has been suggested that the main glycoproteins of the envelope are the neuraminidase, consisting of peptide chains of 54 to 63 × 10^3 daltons, and the slightly smaller hemagglutinin, 50 to 56 × 10^3 daltons (155a, 220), the latter possibly associated with a chain of half that size. Four chains disulfide-cross-linked in pairs may make up the neuraminidase enzyme (about 10S), and two disulfide-linked chains the hemagglutinin (8S) (280, 316, 369a, 406). Each of these and additional larger envelope proteins represents between 6 and 10%. However, other molecular weights, percentages, and assignments of functions and native associations of these polypeptide chains are also suggested (359). All glycoproteins can be removed by proteases without a loss of lipids from the virion (359) (Table 1).

The bulk of the not glucosylated proteins is associated with the nucleocapsid (26 × 10^3 daltons), representing 46% of the total, and possibly with the membrane (53 × 10^3 daltons, 20%), not to mention minor components.

The proteins of the paramyxoviruses (e.g., Newcastle disease virus, Sendai virus, and SV5) appear similar to and somewhat simpler than those of the myxoviruses, having only four main components (39, 44). The largest, a glycoprotein of 67 to 74 × 10^3 daltons, appears to carry both hemagglutinin and neuraminidase activity (103, 402a), although some authors believe the latter to be a property of the second envelope glycoprotein of 53 to 56 × 10^3 daltons. These proteins are of about 9S and 6S. The capsid protein, the main component, is of 56 to 61 × 10^3 daltons, and there is a possible membrane protein of 38 to 41 × 10^6 daltons (22, 382).The three glycoproteins of both the myxo- and paramyxoviruses contain glucosamine, sucrose, galactose, and probably other sugars (317) (Table 1).

The situation is not much clearer in regard to the rhabdoviruses, exemplified by the vesicular stomatitis virus (41, 141). Agreement has been reached concerning the existence of a large glycoprotein (G, about 69 × 10^3 daltons), which appears to build up the spikes and determines the serological specificity of the virion (358); a membrane glycoprotein (M, 29–34 × 10^3 daltons); and a capsid protein (50–60 × 10^3); the largest (L, 190 × 10^3 daltons) appears to be the transcriptase. However, there are also minor components that may well be of major functional importance; one of these is termed NS (420).

In the RNA tumor viruses (leukoviruses) the presence in the envelope of two main glycoproteins of about 100 and 32 × 10^3 daltons has been demonstrated (56, 61, 309, 324). These are distinctly different in electrophoretic behavior

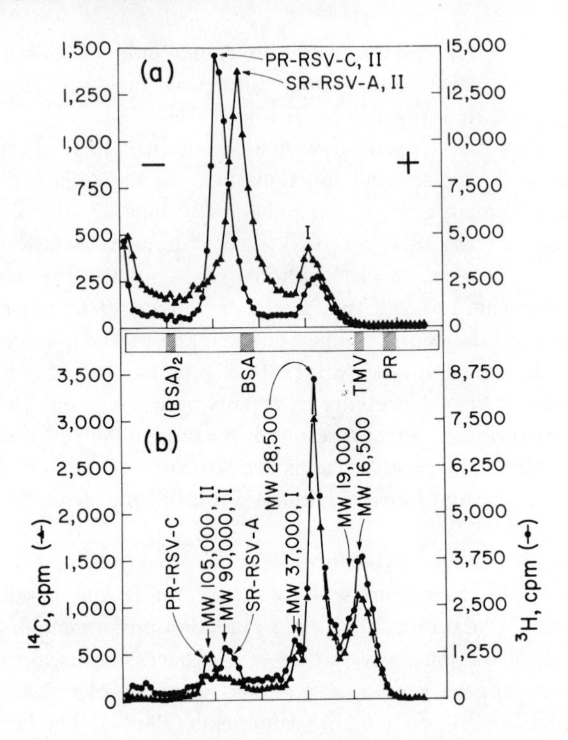

Fig. 12. Proteins of Rous sarcoma virus. (*a*) Glycoproteins of two strains (^{3}H–glucosamine labeled). (*b*) Molecular weights of all proteins of the two strains (labeled with ^{14}C and ^{3}H amino acids, respectively). A strain difference is evident only for the larger of the glycoproteins (56).

and composition from the four internal proteins (one of 28×10^3 daltons greatly predominating) and from host cell membrane proteins (Fig. 12). In contrast, the group-specific antigens, identical for all members of each of the three groups of RNA tumor viruses (avian leukosis and sarcoma, murine leukemia and sarcoma, and mouse mammary tumor viruses), were identified with the internal carbohydrate-free proteins. Differences in the sizes of the glycoproteins (and the glycopeptides derived from them) of transforming and nontransforming strains of these viruses were also recently noted (367).

Very complex protein patterns were obtained with the reoviruses, which consist of at least seven proteins ranging from 155 to 34×10^3 in molecular weight, from 36 to 1% in amount, and from 890 to 23 in number of molecules per virion (166, 242). The outer shell consists of two major proteins of 72 and 34×10^3 daltons (36 and 28%); the core, of proteins of 155, 140, and 38×10^3 daltons (16, 10, 7%) (Table 1).

The simplest animal DNA virus proteins that have been analyzed are those of the polyoma virus and SV40. The main polyoma coat protein is of about 50×10^3 daltons (66, 266). For SV40, polyacrylamide gel electrophoresis indicates one predominant protein of molecular weight 45×10^3 (75%), minor amounts of two smaller proteins, and traces of three yet smaller components (92). These results are in disagreement, however, with earlier data obtained by other methods, which indicated that the main component was of 17×10^3 molecular weight and could be separated according to charge, yielding three polypeptide chains of different amino acid compositions (216).

The best characterized of the more complex animal DNA viruses are the adenoviruses, which contain about seven components in their inner cores and three main components in their exterior shells (120, 170, 171, 192, 351). The latter occupy specific geometric positions, illustrated in Fig. 8, with 240 hexons (six neighbors), 12 pentons at the apices (five neighbors), and the fibers attached to these (see p. 300). The molecular weights of the polypeptides making up these structures are 120, 70, and 62×10^3 daltons.

The herpes virions appear to consist of about 24 proteins, ranging in molecular weights from 275 to 25×10^3 daltons, 9 of which are glycoproteins (413). The fact that herpesviruses are largely assembled in the nucleus has made the origin of these proteins of particular interest (190, 232). It appears that one membrane of the herpes virion is synthesized intranuclearly and another at the nuclear membrane (20). It seems of interest that the host affects the glycoprotein pattern of these viruses (147) without this influence being attributable to host protein contamination, a phenomenon also observed with simpler animal RNA viruses. The vaccinia viruses are yet more complex, having an internal substructure and containing at least 30 proteins of 200 to 8×10^3 daltons.

The virus-specific proteins that we have discussed can be classified in a general manner as follows.

1. The nucleocapsid proteins, as well as the outer-shell proteins of double-shelled viruses such as the adenoviruses and reoviruses, are predominantly structural.

2. The attachment proteins, ranging from the maturation protein to the envelope and spike proteins and the phage tails, combine structural and functional features. Some of these have enzymatic activities. They may at times carry components determined by the host cell and not only by the viral genome, but they contain no appreciable amounts of host proteins.

3. Virion-carried enzymes involved not in attachment but rather in transcription and occasionally other early intracellular functions may be present in all RNA viruses, except the picornaviruses of plants, animals, and

bacteria and the togaviruses of animals; such enzymes also occur in the more complex DNA virions.

4. Enzymes, repressors, and other factors that are not part of the mature virion but are coded by the viral genome and function during the intracellular vegetative phase of viral development are also present.

The total number of virus-specific proteins ranges from 3 for the RNA phages (lacking a representative of class 3) to probably about 70 for the T-even phages, while the virion-associated proteins range from 1 to about 20 in number.

3. BASIC ARCHITECTURE OF VIRUS PARTICLES

Viruses seem to be constructed according to one or the other of two main architectural principles, both resulting in highly efficient types of "biological containers," the helical tube and the icosahedral shell (42, 149). Helical viruses consist of a single protein, the dimensions of which determine the diameter of the helical rod, other parameters being the pitch of the helix and the number of protein molecules per turn (Fig. 2). The length of the stable rod is a consequence of the length of the RNA molecule. In addition to rod-shaped and fibrous plant viruses and bacteriophages, the nucleocapsids of the rhabdoviruses (183, 243) and paramyxoviruses fall into this category.

The assembly of identical protein molecules into isometric shells, the geometric basis of the "spherical" viruses, can be tetrahedral, octahedral, or icosahedral, accommodating 12, 24, or 60 subunits, respectively. These are insufficient numbers for most virus particles, which are composed of 180 (or higher multiples of 60) protein molecules. That icosahedral shells of such magnitudes, including double-shelled, asymmetric, and even elongated particles, can be constructed on the basis of this geometry is explained by the theory of quasi-equivalent bonding (42, 43). This means a regular pattern of pentameric and hexameric protein aggregates, occupying ideally the verticles and facets of an icosadeltahedron (see Fig. 3). The triangulation number, *T*, represents the multiple of the minimal unit of 60 residues characterizing a particular type of particle. It is 3 (i.e., 180 protein residues) for many of the simple isometric plant animal and bacterial viruses.

The outer shell of the adenoviruses consists of the two different proteins making up the pentamers and hexamers (penton and hexon), the former, on the apices, also carrying typical fibrous appendices (see p. 300). Other proteins constitute the internal nucleocapsid, also probably in icosahedral array. In the reoviruses only the cores may be of icosahedral symmetry (374). The envelopes of the myxo-, paramyxo-, and poxviruses probably lack such symmetry. Also the toga- and leukoviruses seem to lack all symmetry, and attempts to isolate a nucleocapsid from the latter group have been unsuccessful, although nucleoprotein cores of 130S, containing 20% RNA and five proteins with a 14×10^3 species greatly predominating (10% of the total RSV proteins), have been isolated (324).

4. DEGRADATION AND RECONSTITUTION OF VIRUS PARTICLES

"Degradation" and "reconstitution" are terms designating laboratory practices, but they find their equivalents in intracellular processes connected with infection of a cell and assembly and maturation of the progeny virus. Considerably more is known about the *in vitro* processes than about the intracellular ones, but the same principles surely hold for both.

A. Degradation of Viruses

Since virus particles represent orderly aggregates of protein subunits around nucleic acid strands, degradation consists in favoring protein disaggregation, and reconstitution is the reverse. Viruses differ greatly in their stability, a consequence of the specific conformation of their proteins and of the nature, number, and strength of the bonds holding the subunits together. A group of plant viruses characterized by bromegrass mosaic virus is most stable at pH 5 and swells and then degrades at pH values of 7 and higher. These viruses also are degraded by molar salt solutions at low temperatures (296). This group, as well as others, is susceptible to pancreatic ribonuclease at neutrality (26, 83). Some animal viruses are unstable at all pHs below neutrality. However, most typical viruses are degraded only at low or high pH (e.g., < 3 or > 9), at high temperature, or in the presence of typical protein dispersants, such as anionic detergents (1% SDS), urea ($>5M$), guanidine halides ($>4M$), or two-phase protein solvents (e.g., phenol).

As indicated, virus degradation involves at least partial protein denaturation, but most virus proteins tend to renature more or less readily if the denaturant is removed, as shown by TMV protein that can be recovered in functional form after treatment with urea or guanidine halides (4, 69). Nevertheless, the least severe method of virus degradation that will serve the purpose at hand is usually selected. If virus nucleic acids are to be isolated, high and low pHs should be avoided, and phenol extraction, often coupled with detergent treatment, is the method of choice (91). For the isolation of renaturable protein, acetic or formic acid or weak alkali is preferred, and de-

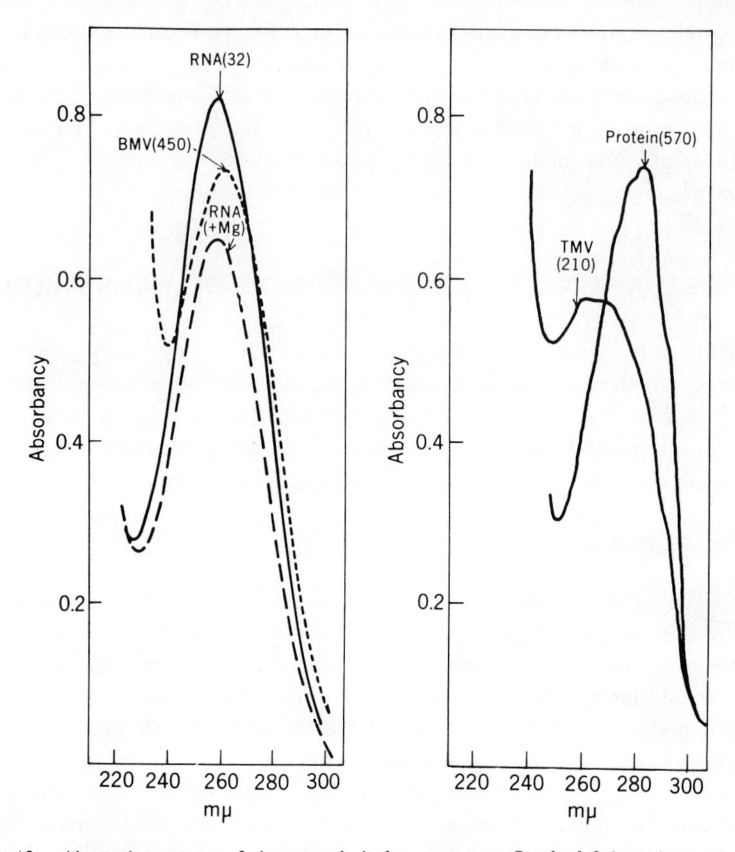

Fig. 13. Absorption spectra of viruses and viral components. On the left is an isometric virus containing 21% RNA (BMV) and viral RNA (TMV) in hyperchromed state in H_2O, and the same sample on addition of $MgCl_2$ to 10^{-3} *M*. On the right is TMV (whole virus, 5% RNA) and TMV protein. The numbers refer to the concentrations (in μg/ml) of the respective materials used. Note that the maximal absorbance per milligram is about 1:2:20 for TMV protein:TMV:TMV RNA.

tergents are shunned (71). The degradation of a virus is often visually apparent in the disappearance of the typical opalescence of virus solutions, which is due to the light-scattering effect of the virus particles. Usually proteins and nucleic acids must be separated after the particles have been degraded. This can be done most easily after phenol treatment, since the protein is found in the phenol and interphases, while the nucleic acid remains in the aqueous phase. In 67% acetic acid most proteins remain dissolved, whereas the nucleic acid precipitates. In other systems, varying salt concentrations (e.g., ammonium sulfate) can be used to precipitate one or, successively, both components.

It is of the greatest importance to protect the integrity of viral RNA against nuclease attack. Bentonite appears to be the most useful of the various agents advocated for this purpose (78, 237).

The purity of both the viral proteins and the nucleic acids is checked most easily by means of a recording spectrophotometer. Not only the E_{max} (characteristically somewhere between 275 and 282 nm for proteins, about 260 nm for nucleic acids), but also the location and depth of the minimum (E_{min}, about 248 nm for proteins, 228 nm for nucleic acids) and details of the spectrum (e.g., maximum/minimum ratio, slight shoulder at 290 nm) are critical factors in identifying viral components and evaluating their purity (Fig. 13).

Certain virions, for instance, those of the turnip yellow mosaic virus, represent remarkably stable particles which can be freed of their nucleic acid without degrading the protein shell (142). Frequently protein shells occur naturally, together with the complete virus (top components), and in most of these instances it has not yet been established whether these shells are intermediates or by-products of virus maturation. The degradation of certain other viruses is markedly affected by the use of mercury derivatives having high affinity for —SH groups (54, 142).

B. Reconstitution of Viruses

The tobacco mosaic virus remains the only virus that can be degraded to pure protein and nucleic acid and can be reconstituted from these components in good yield (about 50%) to fully infective virus indistinguishable from the *in vivo* assembled material (75–77, 80). Prerequirements are (*a*) intact nucleic acid of about 30S, and (*b*) native protein, clearly soluble in H_20 at pH 7.5. Favorable conditions are 0.1 *M* pyrophosphate or phosphate of pH 7.2 to 7.0. The reaction requires about 4 to 6 hr to go to completion at 30° and does not proceed below 15°.

Recent years have witnessed an intensive reinvestigation of the reconstitution reaction, with some important new conclusions and some stubbornly upheld erroneous hypotheses. Agreement appears complete on the fact that TMV assembly *in vitro*, and presumably also *in vivo*, starts with an interaction of the 5′ end of the RNA with a coat-protein aggregate, probably usually a double disc of 34 molecules. Subsequent elongation of the rods is caused not by discs but by the protein trimer, the so-called A-protein, or yet smaller units (313, 332, 386, 387, 396, 398, 418).

This new understanding of the process enables the researcher to better control it experimentally so as to achieve very rapid or slow reconstitution, to isolate intermediates, and so on. The nature of the specific affinity of the 5′ end of the RNA for the protein complex has led to an interest in the nucleotide

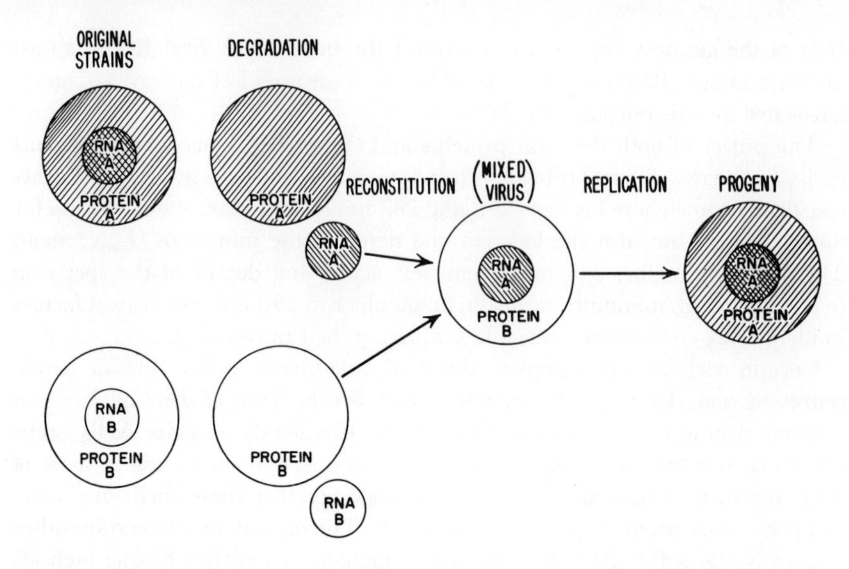

Fig. 14. Mixed reconstitution. Mixed virus is reconstituted from RNA of strain A and protein of strain B. The progeny of such mixed virus has all the properties, including the amino acid sequence, of strain A.

sequence at that end, not apparently dampened by the sequence heterogeneity reported for the 5' end (335).

The protein of TMV forms rods also with quite unrelated nucleic acid (from MS2, TYMV, etc.), as well as with synthetic polynucleotides containing appreciable amounts of purines. These pseudoviruses do not always form under the optimal conditions for TMV reconstitution, nor is their formation as consistently reproducible (67, 75). In the absence of all polyribonucleotides, TMV protein aggregates to viruslike rods only below pH 6, and these are stable only over the range of pH 4 to 6. These protein rods may at times show a stacked-disc rather than the typical helical structure.

The protein of TMV is able to reconstitute in a more genuine manner with related nucleic acids. Mixed virus can be prepared by the use of protein and nucleic acid from different strains of TMV, the yield and stability of the reaction product being correlated with the proximity in relationship (and thus in protein composition) of the two strains (76, 118). Such mixed virus has been of great historical importance in the demonstration that RNA represents the only and the fully competent carrier of genetic information in RNA viruses (Fig. 14). Also, the use of mixed viruses has facilitated the study of host specificity (136, 303), which is no longer regarded as being primarily dependent on the protein coat of a virus.

Quite different and yet related to these studies of *in vitro* assembly of mixed virus particles are investigations of *in vivo* formation of mixed viruses. Thus it has become apparent that the genome of certain defective or temperature-sensitive strains can be rescued by being encapsidated with the protein of a different strain of the same virus. This type of phenotypic mixing has been achieved with different TMV strains (303, 357). A related phenomenon may be the mixed viruses obtained upon simultaneous infection of plants by BMV and the closely related cowpea chlorotic mottle virus (305). The appearance of pseudovirions of polyoma virus as well as TMV, particles containing host DNA (266) or RNA (409), respectively, instead of the viral genome may also be regarded as related phenomena, as may the mixed viruses obtained from vesicular stomatitis with oncornaviruses (430) or with paramyxoviruses (315), as well as the typical packaging of Rous sarcoma RNA in the proteins of RAV. Other helical viruses have not lent themselves as readily to reconstitution as TMV, possibly the best results being obtained with tabacco rattle virus.

The *in vitro* assembly of two groups of isometric viruses has been studied intensely in recent years. These are the BMV group of plant viruses previously mentioned because of their unusual instability and the RNA phages. The BMV group of viruses is reformed *in vitro* from RNA and protein in the form of particles physically almost indistinguishable from the original viruses (16, 17, 109, 278). Viruses of this class are said to be no more infective than their isolated nucleic acids, and thus degradation and reconstitution do not cause, as with TMV, a dramatic loss and regain in infectivity. The only biological evidence for successful reconstitution is the loss and regain in resistance of the infectivity to the action of snake venom phosphodiesterase. In that regard true reconstitution of virions has been achieved. The nonspecificity of the assembly of these viruses is somewhat surprising. Thus their proteins, even though chemically and physically greatly different, can interact in heterologous fashion with any of their or of other viral or nonviral nucleic acids, be they RNA or DNA. The proteins can even form mixed-coat particles.

The reconstitution of the crystallographically very similar RNA phages shows an important point of difference: the presence of one molecule or a very few molecules of a special protein(s), the A-protein (maturation factor) of the bacteriophages. Electronmicroscopically viruslike particles form readily from the RNA and coat protein of the RNA phages (MS2, R17, f2, and the quite unrelated phage Qβ); however, these show sedimentation constants of 70S rather than the 80S of typical virus, and they are noninfective (113–115, 123, 258). If the maturation protein is present, a small amount of infective virus sedimenting with 80S is formed, but the yield of this truly reconstituted virus is maximally only about $10^{-3}\%$ (123, 203). It is believed that the A-protein serves either as an attachment organ for the phage or as a stopper needed to contain the RNA within the particle, or fulfills both functions. In the absence of this protein the RNA seems to become shortened, be it inside or protruding from the particle, and the particles resemble the defective mutants described by Heisenberg (104). It has been reported that, with techniques which allow the isolation of nuclease-free and thus stable RNA and protein, phage particles containing undegraded protruding RNA are formed, but the absence of the A-protein renders them biologically ineffective and susceptible to host exonucleases (113, 115).

Studies on mixed particles similar to those performed with TMV and the BMV group have again shown that heterologous particles can be formed, even with unrelated phage components (e.g., mixing protein and nucleic acid from Qβ and MS 2) (161). The question of whether nucleic acid is essential for particle formation seems to have found the same answer for both types of isometric viruses. Polyanions greatly favor the *in vitro* formation of typical particles from protein subunits. Even polyvinyl sulfate can serve this purpose. Polyuridylic acid, short polynucleotides, and tRNA catalyze particle formation, but these particles seem to diffuse out of the reconstituted virion. Nucleic acid longer than the homologous one causes the formation of various monster particles. Under special conditions particles can also form without any nucleic acid (106), but these are unlike the typical empty shells in density or sedimentation constant and are believed to form around a primary protein aggregate of 60 subunits serving as the nucleating agent instead of a polynucleotide (114). However, typical empty R17 shells have been reported to be reformed from carefully and not completely dissociated viral protein.

The problems of reconstituting the two-protein-component phages are naturally magnified when dealing with the four-component animal picornaviruses. Nevertheless some success in dissociating and reassembling poliomyelitis virus particles *in vitro* has been reported (328).

C. Maturation and Autoassembly of Viruses

The ease of spontaneous assembly of representatives of the two main geometrical types of viruses, TMV and BMV, the helical and the isometric, demonstrates the principle of autoassembly of biological particles. The reconstitution of ribosomes from mixtures of the component proteins represents an even more remarkable illustration of this principle (181). Yet many viruses of even limited complexity in terms of number of protein components do not form at all readily *in vitro*, and it is thus not surprising that the 20 to 30 components of the T-even phages do not spontaneously reassemble (294). This is, however, not only a consequence of the complexity of the system but is also due to the fact that several if not many of the steps in virus assembly are under the control of specific enzymes. One illustration of this principle is the procapsid of the polio virus, a protein shell in which a precursor protein has not yet been split into the two characteristic proteins of the complete virion (127). Other examples of specific proteolytic steps required for virion maturation have been encountered in the T-even phages (146). Possibly only very few of the steps in this process take place spontaneously *in vitro* through the interaction of purified components. One of these spontaneous events is the attachment of heads to complete tails (294). The formation of the tail fibers alone requires 5 genes, 2 of which apparently account for the specific enzymatic modification of precursors, and about 45 genes are needed for the complete assembly of this phage.

5. MULTICOMPONENT VIRUSES

As stated, the typical minimal plant virus consists of one molecule of nucleic acid encased in a shell composed of a definite number of identical protein molecules. The varying complexity of the coat proteins in animal and bacterial viruses has already been discussed. Thus in the animal viruses this shell usually consists of at least four components. In the phages there exist at least one molecule and in most instances greater numbers of several minor components, which serve important functions during infection and achieve remarkable complexity in the case of the T-even phages, which are built up of 20 to 30 proteins.

In regard to the nucleic acid, the entire viral genome is in most instances a single molecule, single-stranded RNA or single- or double-stranded DNA, as the case may be. This is true for the simplest as well as the most complex plant, animal, and bacterial viruses with molecular weights ranging from 10^6 to 200×10^6.

Yet, for many plant and animal viruses of intermediate complexity, this is not the case. Two types of variants of this principle have been observed, both presently only with RNA viruses. Several nonidentical molecules of nucleic acid may occur in all particles of a given virus, or they may occur singly in different particles, encased by the same coat protein. These two types of viruses may be termed multi-nucleic-acid viruses and complementary multicomponent viruses or coviruses, respectively.

A. Multinucleic Acid Viruses

Two groups of viruses long recognized as showing unusual biological activities have in recent years been found to carry multiple strands of RNA. These are

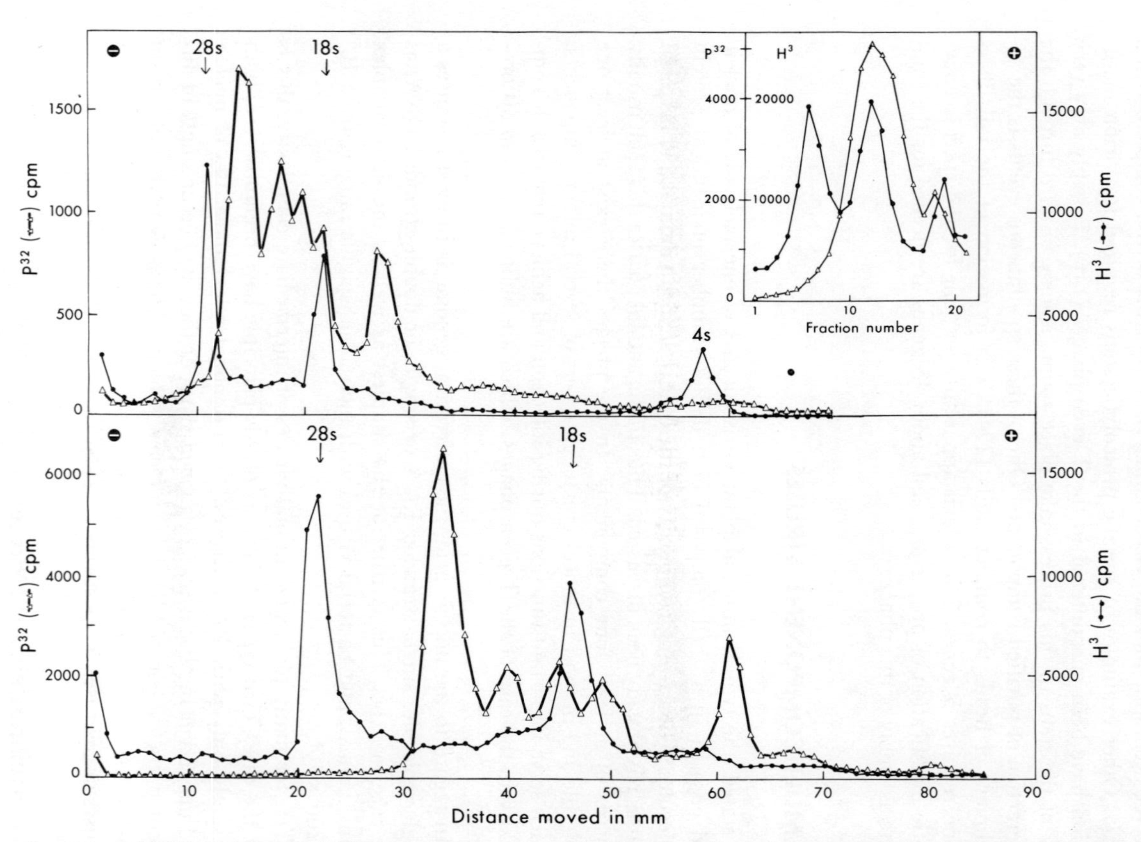

Fig. 15. Separation of RNA components of influenza virus (on polyacrylamide gels or, in the inset, on a sucrose gradient). Solid circles represent ribosomal and tRNA markers (28S, 18S, 4S); open triangles, ^{32}P-labeled RNA isolated from the virion. The upper graph was obtained with a shorter time of electrophoresis; the lower, with longer electrophoresis to improve resolution (58).

the myxoviruses, particularly exemplified by influenza virus, and the RNA tumor viruses, such as the Rous sarcoma virus. The RNA of each influenza virus particle can be isolated as a single unit of 38S in the presence of divalent metals, but this large-molecular-weight material has been definitely recognized as an artifact. In the presence of EDTA, at least five to six characteristic RNA species ranging from 18S to 6S have been identified (Fig. 15) (58, 195), and each of these is specifically associated with a probably helical nucleocapsid component of corresponding length and separable on sucrose gradients or by gel electrophoresis after the virion has been dissociated by removing its envelope (57). Surprisingly, the RNA is not protected against nuclease in these nucleocapsids [or in the nucleocapsids of Semliki Forest virus (2)], while the protein is quite resistant against pronase. Influenza virus differs from other RNA viruses in showing high levels of genetic recombination (up to 50%) for certain markers (24, 155, 168). It appears that the multiple nature of its RNA and of its nucleocapsid would supply a possible explanation for this biological pecularity and related ones of influenza virus (see p. 293). The genuine multiplicity of the RNA strands in influenza virus is supported by the recent finding that there exists about one pppAp-chain end per 800 nucleotides (300).

Rous sarcoma virus and all other oncornaviruses (avian and murine leukemia and tumor viruses, including Bittner's mammary tumor factor) seem to contain a single molecule of 60S to 70S RNA corresponding to the nucleic acid content of the virus particle. However, if treated with heat or dispersing agents which break base-pairing bonds, these nucleic acids sediment with an average of 30S to 35S, and are of 2 to 3 rather than 10×10^6 daltons in molecular weight (59). This is definitely their true molecular character and is indicative of multiple strands held together probably by base-paired regions. The multiplicity of the oncornavirus nucleic acids is not an artifact, since typical virus nucleic acids (e.g., TMV RNA and the RNA of paramyxovirus, such as Newcastle disease virus) show unchanged 30S and 56S sedimentation properties, respectively, under the same conditions of dispersing treatment that dissociate the tumor virus nucleic acids. Although the mechanism of transformation and oncogenesis caused by these viruses is not yet fully understood, it appears probable that their multiple nature represents an evolutionary advantage in this regard. There is an additional aspect to the multiple nucleic acid nature of the oncornaviruses, and that is the existence of a somewhat larger species of about 35S in the transforming strains, the *a* component, as contrasted to the *b* component (about 30S) present in nontransforming strains (330). Isolates seemingly containing *a* and *b* represent strain mixtures. The *a* and *b* components show considerable similarity in their patterns of large oligonucleotides after T1 digestion, with a few characteristic additional oligonucleotides in component *a* (368).

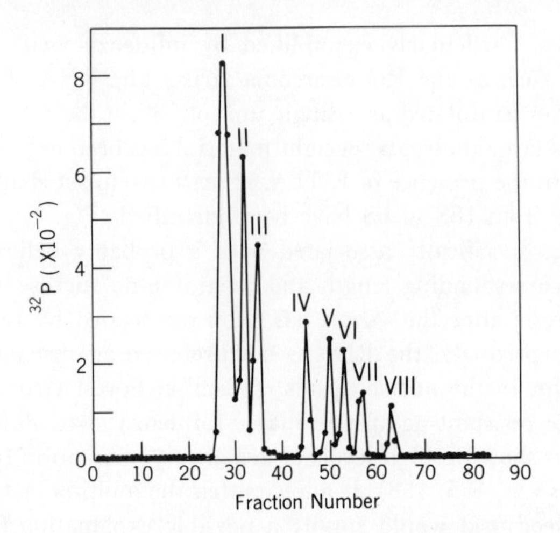

Fig. 16. Electrophoretic separation of ^{32}P-labeled cytoplasmic polyhedrosis virus RNA subunits. Migration from left to right in 2.25% polyacrylamide–1% agarose gels of virus treated with 1% SDS. The molecular weights ranged from 2.4 to 0.48×10^6 daltons (160).

Another group of viruses that contain multiple RNA components are the double-stranded RNA viruses of animals (reoviruses) (228), plants (wound tumor and rice dwarf viruses) (139, 180, 293), and insects (cytoplasmic polyhedrosis virus) (139, 160, 179). All these viruses seem to contain 10 to 15 double-stranded RNA components that can be separated on polyacrylamide gels into distinct molecular weight fractions (Fig. 16). The fact that these are separate genomic entities and not products of fragmentation was established by determining the number of 3′ end groups (either unphosphorylated U and C, or only C) of the viral RNA in the virion and in the infected cell (370, 378). The 5′ end groups have also been identified (pppG-) (159a) (see p. 230).

B. Complementary Multicomponent Viruses or Covirus Systems

An ever-increasing number of plant viruses have been observed to occur in multiple forms in infected tissue and in purified preparations. Among these are isomatric viruses like cowpea mosaic virus and elongated viruses like alfalfa mosaic (276, 308) and tobacco rattle virus (215). When methods for the separation of virus particles of slightly varying dimensions had been sufficiently refined, it became evident that the various components of each of these viruses showed great chemical similarity but were biologically clearly different and, in varying manner, interdependent. It is now clear that the genome of such a virus, its RNA, consists of two, three, or even four molecules which are separately encapsidated by the same coat protein. The larger and smaller nucleoprotein components of the cowpea mosaic virus (32, 273, 274) (115S and 95S) contain 32 and 23% of RNA and 34S and 26S, 2.58 and 1.45 $\times 10^6$ molecular weights, and different base compositions. Neither of the two types of particles, when completely free of the other component, is able to initiate an infection. Upon remixing of the two, however, infectivity is restored. Beanpod mottle virus was found to be similarly dependent on the synergism of two particles of different sizes (380).

It should be noted that the cowpea mosaic and beanpod mottle viruses contain two proteins, whereas all other typical covirus systems, to be discussed below, have a single coat protein. The two proteins of the cowpea virus group occur, however, equally in both the 95S and 115S components and are not the respective coats of two different viruses.

Many viruses related to these, as well as to tobacco ringspot and tobacco streak virus, also show nucleoprotein particles of two sizes, one of which, usually the larger, is reported as sufficient for infectivity. In most of these cases, however, addition of the second component greatly increases infectivity. Since it is difficult to achieve quantitative separation of similar particles differing by possibly only 10% in sedimentation value, the residual activity of a single one of these components is not easily interpreted. It appears likely that most of these will prove to be genuine covirus systems with the genome distributed over two types of molecules.

Another group of plant viruses, exemplified by BMV, has only recently been recognized as also representing coviruses. This is so because the various particles are so very similar that they have not yet been quantitatively separated from one another. These viruses were long known (296) to contain RNA of three sizes, now recognized as actually representing four molecular species, rather than fragments. Their molecular weights are about 1.05, 0.95, 0.7, and 0.3 $\times 10^6$, the last two occurring in the same particle. Infectivity results only upon inoculation of the three biggest components. Although the role of the smallest one is somewhat enigmatic, it was recently shown to be a monogenic, very efficient messenger for the coat protein (408a), which is, however, also encoded in the 0.7 $\times 10^6$ molecule (408). The latter codes more efficiently for another protein, of molecular weight 35,000, which happens to be the molecular weight of a protein found to be associated with RNA replicase activity (346). If this attribution of genes is correct, the question is, What essential functions are coded for by the two 10^6-dalton components?

The tubular covirus system, the alfalfa mosaic virus (AMV), actually resembles the isometric BMV group most closely in also requiring three components for infectivity. The typical virus preparation consists of five components, ranging from nearly spherical particles to 70-nm-long rods (122). The cooperative functions of the heaviest and the lightest RNA fractions obtained from AMV ("bottom *a*" and "top *a*") have been studied. A more recent reinvestigation of this problem, based at first on the separation of all five classes of virions in the presence of EDTA, has indicated that three of them (bottom, middle, and top *b* components of 95S, 82S, and 73S) must be combined to obtain infectivity (276).

However, when the corresponding RNA fractions were tested, all four (24S, 20S, 17S, and 12S, the RNA of the top *a* component) were necessary for infectivity. This paradox was resolved when it was shown that the function of the top *a* RNA component could be performed by the viral coat protein alone (308). It now appears that this protein plays an important, possibly a regulatory or enzymatic, role early in the infection process, and that only the RNA of the top *a* component can serve as a rapid messenger for the necessary production of this protein (275), notwithstanding the fact that the same gene is also carried in the 17S top *b* RNA. Thus, also in this regard, the AMV system resembles the BMV system, both having a monogenic piece of RNA which is not strictly essential and the coat-protein information of which is carried also by a larger genome component.

The situation is slightly different, and yet more interesting, for tobacco rattle virus (TRV), which always occurs in the form of helical particles of two lengths (162, 215). The large component of this rod-shaped virus (usually about 200 nm) is by itself infective, but produces an unstable progeny that proved to consist of nucleic acid alone. Upon remixing of the two classes of particles, the long and the short, stable protein-coated virus was produced. Thus the complete genome for RNA replication was carried in the large particle, and the coat-protein gene in the smaller particle. The small particle, however, differs greatly in length (45–115 nm) among various strains of TRV, and the RNA of even the smallest is considerably in excess of the amount calculated as necessary for the information pertaining to coat protein. Thus genes of unknown but seemingly nonessential function and of variable number or size appear to occur in this particle. Alternatively, such particles, as well as many other plant viruses, may contain considerable amounts of genetically nonfunctioning or redundant RNA.

The evolutionary advantage of distributing the genomes of all these viruses over several particles is not obvious. However, a major disadvantage is evident. Whereas a single particle of a typical virus is able to initiate an infection, at least two and in some instances three particles of the viruses under discussion must enter the same cell for complete virus replication. Such a requirement renders virus infection considerably less efficient, and almost improbable at low virus concentration. This was borne out by the infectivity dilution curves obtained with these viruses. Both alfalfa and cowpea mosaic virus showed typical steep multiple-hit curves (Fig. 17); with TRV the total infectivity followed, like that of typical viruses, single-hit kinetics, but the frequency of lesions containing complete and thus stable virus showed multiple-hit characteristics (215). The properties and relationships of the various multicomponent and defective viruses are summarized in Table 2.

Table 2 also includes a special type of cooperative viral system not yet mentioned and distinctly different from those discussed above. This consists of the satellites of tobacco necrosis, very small and defective viruses that are completely dependent on an unrelated group of virus strains, the tobacco necrosis virus. This parasitism differs distinctly from coviruses in that the coat proteins of the satellite and its infective helper are of different molecular weights (about 20 vs. 32×10^6 daltons) and also show no serological relationship (200, 272). The preferred working hypothesis is that the satellite codes only for its coat protein and that it requires the specific RNA polymerase induced in the various common hosts (tobacco, beans, etc.) by the tobacco necrosis virus. No other plant viruses activate this satellite virus, and even the strains of either virus are not functionally freely interchangeable.

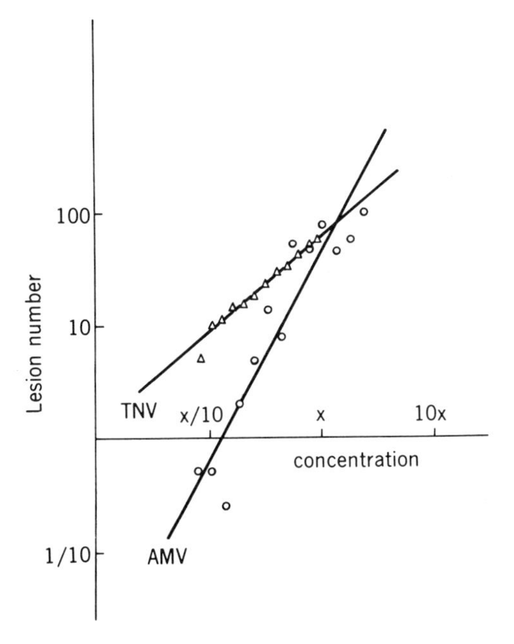

Fig. 17. Relation of virus concentration to lesion number. Response to inoculation of various concentrations of a covirus system (AMV), as compared to a typical single-component virus (TNV) (see Table 2).

Table 2. Modification Reaction[a]

Amino Acid Exchange	Frequency of observed Coat Protein Mutants				Codon Exchange	
	HNO_2	Br_2	CH_3–	Others		
Gly	4	3	1		A → G	GGPu
↗						↗
Arg						AGPu
↘						↘
Lys	1!	1			G → A	AAPu
Ala	(4)!				A → C	GCPy
↗						↗
Asp						GAPy
↘						↘
Gly	(2)				A → G	GGPy
$AspNH_2$ → Ser	4 (2)	3			A → G	AAC → AGC
Gly	1 (1)				A → G	GGPu
↗						↗
Glu						GAPu
↘						↘
Val	(2)!				A → U	GUPu
Arg	1				A → G	CGPu
↗						↗
$GluNH_2$						CAPu
↘						↘
His	1!			2!	Pu → Py	CAPy
Met	(2)				A → G	AUG
↗						A ↗
Ileu → Thr		1		1	U → C	AU → ACPy
↘						Py ↘
Val	3 (4)				A → G	GUPy
Leu → Phe	(1)				C → U	CUPy → UUPy
Leu	3 (4)	6	4	(1)	C → U	CUX
↗						↗
Pro						CCX
↘						↘
Ser	(4)				C → U	UCX
Gly		1		(1)	A → G	AGC → GGC
↗						
Ser → Leu	(2)				C → U	UUG
↘						↗
						UCX
						↘
Phe	4 (4)	2	2		C → U	UUPy

Table 2. Continued

Amino Acid Exchange	Frequency of observed Coat Protein Mutants				Codon Exchange	
	HNO_2	Br_2	CH_3^-	Others		
Ala	2			(1)	A → G	GCX
↗						↗
Thr → Ileu	(9)				C → U	ACX → AUPy
↘						↘
Met	(3)				C → U	AUG
Try → Cys	(1)				A → C	UAPy → UGPy
Ala				(1)	U → C	GCX
↗						↗
Val						GUG
↘						↘
Met	1!				G → A	AUG

[a] Reprinted, with additional data from Ref. 88. The data in parentheses are from Wittmann and coworkers (Tübingen), the others were obtained in Berkeley. Exclamation marks indicate unexpected exchanges.

6. VIRUS INFECTION AND REPLICATION

A. Entry and Uncoating

Although we have generally stressed the marked similarities in the natures of plant, animal, and bacterial viruses, distinct differences do become evident in considering the process of infection. Plant viruses usually seem to lack specific mechanisms of entry into host cells and rely on the presence of wounds in the cell wall. Many plant viruses have become adapted to various vectors (arthropods, nematodes, fungi, etc.) as a means of transmittal, and for others the probability of mechanical transmission to a new host organism is maximized by their very high intracellular concentrations. These factors probably account for the successful transmission of the multicomponent viruses discussed in Section 5.

Many animal viruses are also arthropod-borne and are transmitted by their vectors. Most animal viruses, however, have particular affinities to features on the potential host cell surface and enter through these more or less specific sites (116). The neuraminidase and/or hemagglutinin spikes or peplomers on the surfaces of many virus groups (see Fig. 18) very probably represent such in-

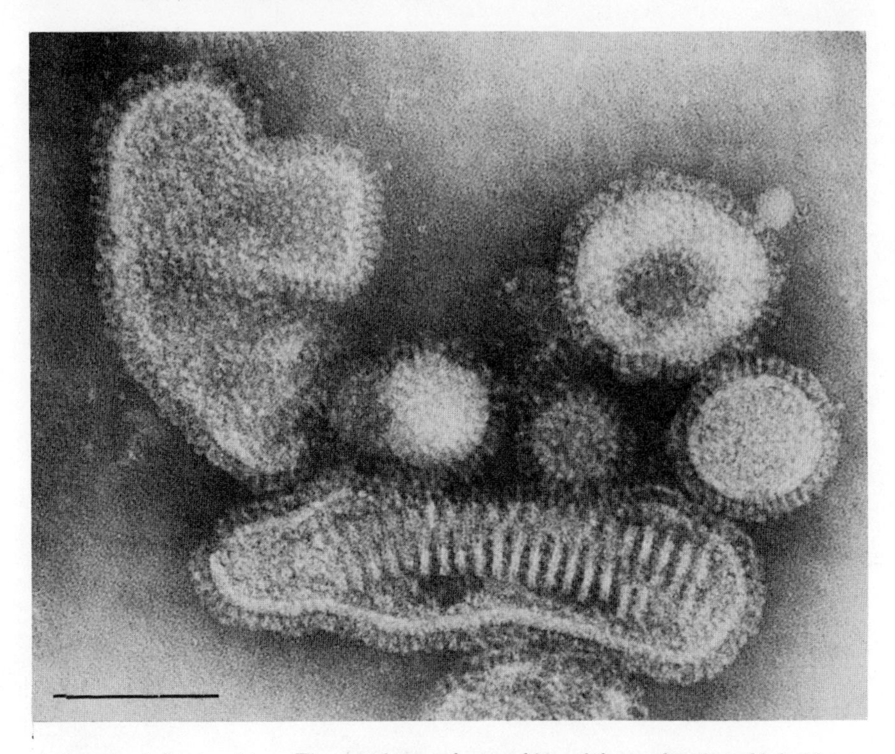

Fig. 18. Influenza virions. The particles are pleomorphic, and the envelope is studded with projections, spikes, or peplomers. Penetration of the envelope of one particle by phosphotungstate reveals the coiled internal nucleocapsid. The bar is 100 nm. (Courtesy of J. D. Almeida and F. Fenner.)

fection organs. Animal cells react to other viruses on their surface (e.g., pox- and herpesviruses) by engulfing them (47–49).

Gram-positive bacteria seem to be susceptible to entry by large molecules and small particles, although receptor sites on their wall play a positive role in such entry. In contrast, Gram-negative bacteria have very impermeable cell walls, and their viruses have had to develop specific organs to enable them to pass through these thick, multilayered walls. Most highly developed are the tails of the T-even phages, which consist of hinged fibers for long-distance attachment, a base plate with pins for close-range attack, and a tail core inside the contractile sheath for piercing the wall, so that the contents of the phage head, largely DNA, can enter the bacterial cell (Figs. 19 and 20). Many bacteriophage tails are noncontractile, others quite short, and both types are thus of lesser complexity; in the case of the small DNA phages, they have become reduced to the 12 geometrically located spikes, which consist of pentamers of two proteins forming the base and one molecule of another protein the apex (33). Still simpler is the infection organ of the RNA phages, presumably the one molecule of maturation factor. The fibrous DNA phages also have a special protein (of 10S), located probably at only one end of the 800-nm-long particles (347). It was recently shown that this protein accompanies the DNA to its site of intracellular attachment and may play an important functional role there (353). The viruses of this group may be the only ones to enter the *E. coli* cell intact, rather than injecting only their nucleic acid (268), although this is still uncertain (347). They are certainly singular in leaving the *E. coli* cell upon maturation without causing lysis. Both the DNA- and RNA-containing small phages, lacking the powerful infecting organ of the T-even phages, have found a comparatively soft site of attack in the F-pili (or other pili) on which to attach themselves, the isometric RNA phages sticking to their sides and the fibrous phages to their tips (40) (Fig. 21). The subsequent steps in this infection process are still under active study (191).

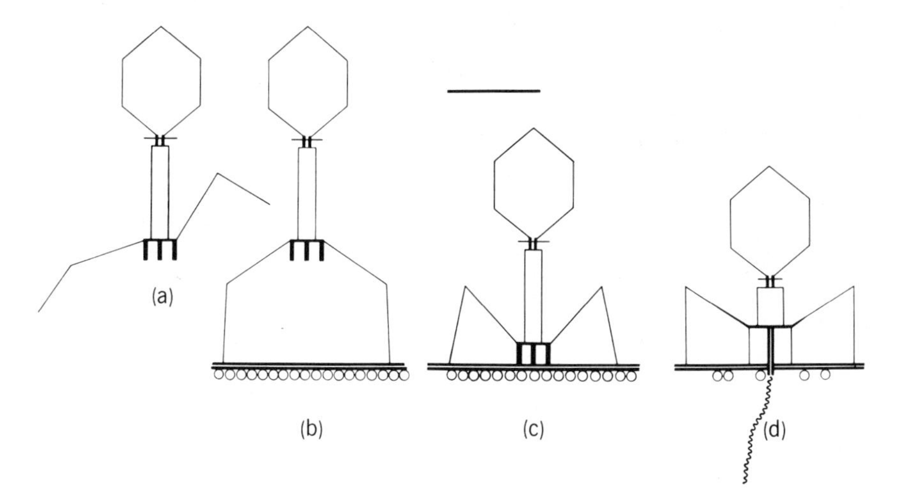

Fig. 19. Schematic illustration of the steps of attachment, penetration, and injection of T4 and T2 bacteriophages. (*a*) An unattached phage. (*b*) The long, thin tail fibers have attached to the host's cell wall. They are kinked slightly near their middle. The base plate of the phage is over 100 nm from the cell wall. (*c*) The phage has moved closer to the cell wall, and the pins extending from its base plate are in contact with the wall. The long tail fibers are shown kinked and bent sharply at their hinge on the base plate. (*d*) The tail sheath has contracted, and the sheath has retracted up the tail tube, carrying with it the freed base plate. The phage is still linked to the host cell by short and long tail fibers. The inner tail tube has penetrated the outer three layers of the host cell wall.A decrease in the density of the rigid portion of the cell wall is also shown. (Courtesy of L. D. Simon and T. F. Anderson.)

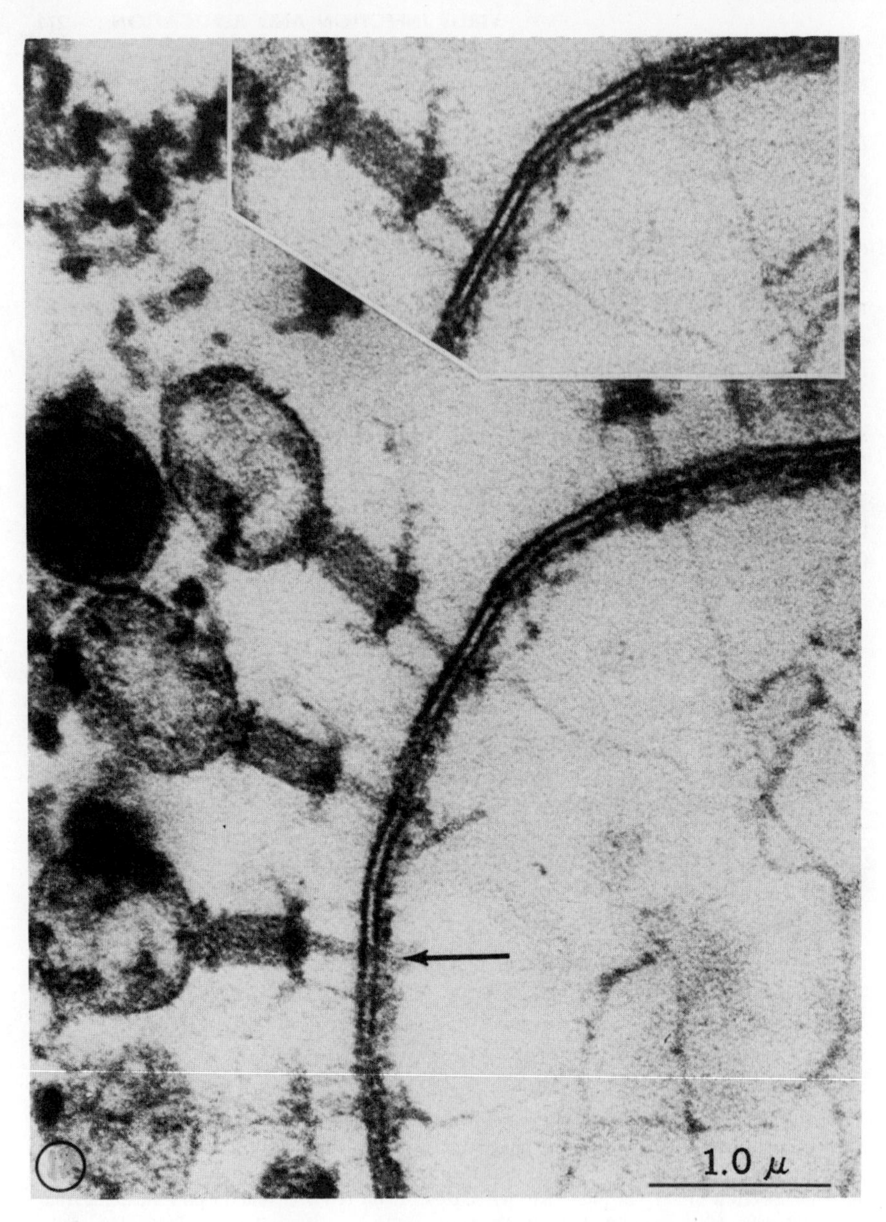

Fig. 20. Thin-section electronmicrograph of T4 phages, adsorbed on *E. coli* B cell wall. The phages are bound to the bacterial surface by short fibers extending from the base plate. The base plates have a "boat-type" profile in which the tips of the tail plate point away from the cell surface. The arrow indicates the tail tube of one of the phages, which has penetrated through the cell wall.The thin fibrils visible (3-nm diameter) are probably the phage DNA. (Courtesy of L. D. Simon and T. F. Anderson.)

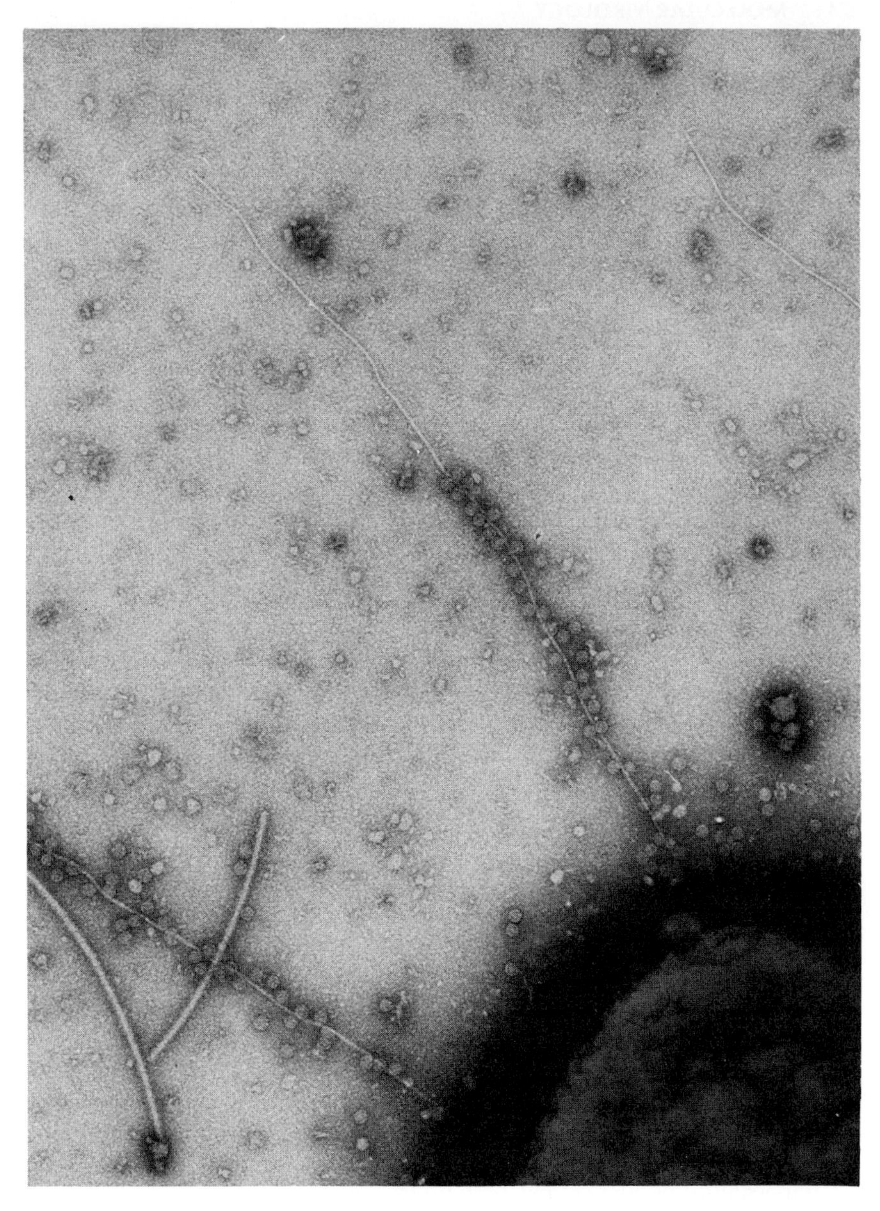

Fig. 21. Infection process with small bacteriophages. Two pili are shown emerging from an *E. coli* C600 cell, both carrying many attached MS2 (RNA) phages. A rod-shaped f1 (DNA) phage is also seen at the tip of one pilus. ×41,895. (Courtesy of L. G. Caro).

Remarkably little is known about the mechanisms whereby the simple plant and animal viruses are uncoated. In the case of TMV this process takes several hours, although some of the coat protein is released within minutes (229). The complex DNA viruses of animals seem to require special uncoating enzymes, which are probably synthesized early upon infection when part of the viral DNA has become released and is being transcribed and translated (134, 135). The first stage of uncoating of several animal viruses is accomplished through host lysosomal enzymes in cytoplasmic vacuoles. This usually removes only the external envelope or protein shell, since most viral nucleocapsids are remarkably resistent to enzymes. Small viruses lacking envelopes, or double shells, such as the poliomyelitis virus, do not become enclosed in vacuoles when entering the cell (47). These viruses are altered in some unknown manner by the cellular receptors so that they release their RNA into the cytoplasm. There, apparently, the chance for the RNA to encounter ribosomes and, later, RNA polymerases (i.e., to become translated and replicated; see below) is greater than the likelihood of encountering nucleases (i.e., to become degradated), since infection frequency is not forbiddingly low, namely, one focus per 10 to 100 virus particles.

B. Direct Translation of the RNAs of the Picornaviruses of Plants, Bacteria, and Animals and the Togaviruses of Animals

In the cases of the simple viruses, the invading viral single-stranded RNA is regarded by the host cell metabolic machinery as fully competent messenger RNA, and it is readily translated into proteins by the host's ribosomes, tRNAs, amino-acid-activating enzymes, and factors. There are strong indications that animal cells translate certain of these viral RNA molecules (if not all) without interruptions (126, 169). It appears possible that animal cells do not normally deal with polygenic messengers. Only secondarily are such monster proteins cut by specific proteolytic action into several structural and enzymatic proteins, the origin and the specificity of these proteases remaining unexplored. The order of the genes on all enterovirus RNAs appears to be the same: A, the precursor of the coat proteins, nearest the single initiation site, followed by F and C (313a, 394a).

Study of the nature of the plant viral translation products is hampered by the impossibility of suppressing host protein synthesis to a sufficient extent. Thus success in detecting protein synthesis resulting from viral infection has been achieved only by double-labeling procedures, in which, for example, ^{3}H–leucine-labeled uninfected leaf extracts or fractions thereof were coelectrophoresed with identically prepared ^{14}C–leucine-labeled infected leaf tissue, the gels were sliced and counted, and the ratio of ^{3}H/^{14}C counts was plotted.

Unfortunately this method does not differentiate between virus-coded proteins and those plant proteins the synthesis of which is stimulated or inhibited as the result of viral infection, and it has been shown that such effects are not uncommon. When this technique was applied to intact infected leaves, only the viral coat protein formed a detectable peak upon infection of tobacco with wild-type TMV (412); presumably other gene products were produced below the threshold of detectability, that is, about 2% of the coat protein. When protein synthesis was studied in shredded leaf tissue, isolated cells, or protoplasts, or upon inoculation with various TMV mutants, several or many larger protein products, respectively, were detected, *in toto* far in excess of the genome's coding capacity, which is about 200×10^3 daltons of protein (401, 429). These response patterns are regarded as probably due largely to host responses, particularly since several of them are characterized by valleys rather than peaks and thus can be caused only by suppression of a normal synthetic process (446).

The definite finding of the 17,500-dalton TMV coat protein without equivalent amounts of other proteins under a great variety of different conditions of infection and fractionation is a strong argument against the hypothesis that plants respond in a manner similar to animal cells to picornavirus infection. Great amounts of the coat protein cannot well be made by posttranslational cleavage without large precursors or other split products becoming evident.

In contrast to the RNAs of the picornaviruses of plants and animals and the togaviruses of animals, those of the more complex plant and animal viruses (rhabdo-, diplorna-, myxo-, paramyxoviruses) appear to represent the negative strands, which are first transcribed to messenger RNAs. The enzymatic mechanisms for this transcription will be discussed in the next section. It appears that the subsequent translation of the messages is largely or entirely performed by the host's machinery.

Bacteria, finally, translate viral RNA in the form of several specific proteins apparently relying on one or several initiation (AUG, UUG) and termination triplets (UAG, UAA, UGA) in tandem. At least five protein factors have been isolated from *E. coli* that play specific roles at the various stages (initiation, elongation, termination, and release) of translation of messenger RNAs by the ribosomal system.

In the case of the small RNA phages three virus-specific proteins of characteristic molecular weights and identified functions have been recognized intracellularly at all stages of infection (277) (Fig. 22). These account for most of the genome. Two of these three proteins are the frequently mentioned coat and maturation proteins of the virion, and the third is the specific RNA replicase (or synthetase) which is needed for this most essential step of virus replication

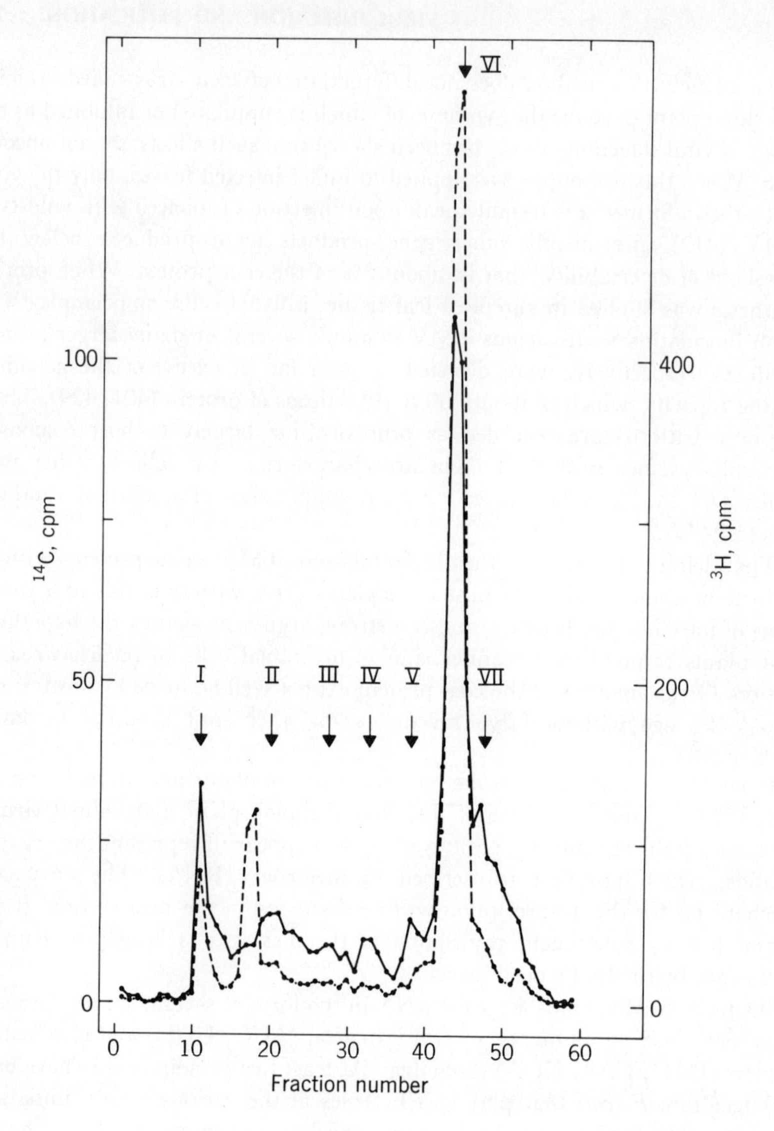

Fig. 22. Comparison of *in vivo* and *in vitro* products of translation of MS2 RNA. The dashed line indicates the three proteins made intracellularly (as separated on polyacrylamide gels). The solid line shows the more heterogeneous mixture of proteins resulting from *in vitro* translation of MS2 RNA, only the first (the replicase) and the third (the coat protein) being clearly present (260). Products II, III, IV, V, and VII are unidentified.

and will be discussed on p. 000. Thus for the simple RNA viruses it is clearly established that translation must precede replication.

Although the order of these three genes has now been established by several methods (131, 151, 348) as maturation factor—coat protein—synthetase (from the 5′ to the 3′ terminus of the RNA), the amount of coat protein made always predominates greatly, *in vivo* and *in vitro*, and no detectable amounts of maturation factor are produced *in vitro* by *E. coli* ribosomes (164) (see Fig. 22). In contrast, ribosomes from *Bacillus stearothermophilus* synthesize only that protein *in vitro*, a finding that was of great value in the elucidation of the structure and gene order of the RNA of those phages (165).

The question of the control mechanisms that lead to preferential translation of one or the other gene is obviously very important and intriguing. One means of control of the rate of translation of the various genes has been found to be exercised by the coat protein. (259–261). About six protein molecules combine with the RNA in a specific manner and at a specific site (307), even at intracellular concentrations, and thereby make the RNA functionally monogenic, in that essentially only coat protein is being translated from this RNA–coat protein complex in the *in vitro* system. Thus it would appear that the required great excess of coat protein over the other gene products can be attained through this type of feedback control in RNA-phage-infected cells.

There is evidence, however, that additional factors related to the conformation of the RNA affect the relative availability of the three genes to the ribosomes. Thus the maturation factor gene is believed to be of too compact a conformation to be translated, except when the RNA is being replicated and is thus in the double-stranded form; the replicase gene initiation site, in turn, is bound to the coat-protein gene and becomes available only upon translation of the latter (164, 379). This factor is believed to account for the so-called polarity effect: that arrest of the translation of the coat protein at residue 6 through mutation to a termination codon prevents the replicase protein from becoming translated, but that similar arrests further along in the coat-protein sequence have no such effect (165). It has also been suggested that the coat protein may form a specific complex with the replicative or double-stranded form of the viral RNA (see Section 6.D) and thus exert a regulatory role during replication. (204).

C. In Vitro Homologous and Heterologous Translation of Viral RNA

Although most of the conclusions summarized in the preceding section were derived from *in vivo* studies of the translation of viral RNAs in their host, most of the principles could be verified *in vitro* by observing the products formed when the messenger RNA was added to the solubilized active extracts plus ribosomes, prepared according to Nierenberg, Capecchi, and others. The cell-free system from *E. coli* was able to synthesize the three proteins coded for by the RNA phage genome, although normally only two genes were translated to appreciable extent, because the conformation of the RNA made the maturation protein gene inaccessible to ribosomal attachment. Thus some regulatory mechanisms are not easily controllable *in vitro*, but others, such as the inhibition of reading of the replicase gene by the coat protein, were first demonstrated *in vitro*.

Although the multiplicity of the virion proteins of animal viruses renders their identification much more difficult, it appears probable that cell-free systems derived from animal cells are able to translate the RNA of animal picornaviruses in proper manner (332a, 358a, 412). Only quite recently has a cell-free host-derived plant virus system been shown to translate a plant viral RNA in efficient, complete, and accurate manner; this was a wheat germ system and the messenger RNA was the small component of BMV RNA, having a molecular weight of 0.28×10^6 daltons and apparently not coding for any other protein (408a).

In contrast to these homologous systems, heterologous *in vitro* translation has been attempted ever since the nature of the polypeptides that result from TMV RNA being added to the Nierenberg *E. coli* system was studied. Here the question was whether plant and, later, animal virus RNAs would be correctly translated by the *E. coli* protein-synthesizing system. The first, apparently positive indications were soon corrected (221), and there remains no doubt that most, if not all, of the protein products synthesized under the direction of TMV RNA cannot be identified with TMV coat protein. However, for three particularly small, incomplete plant genomes—the top *a* component of alfalfa mosaic (275), the small component of BMV (253a), and the satellite tobacco necrosis virus (STNV)—all 0.3 to 0.4×10^6 daltons and at most bigenic—good indications were obtained that their protein products, when digested with trypsin, gave peptide mixtures resembling those of the respective coat proteins (275, 360, 375).

This question has recently been critically reinvestigated, and the conclusion is that there are sufficient similarities in the peptides from proteins synthesized in infected tobacco leaves and *in vitro* with *E. coli* components to establish that the codons on the messenger STNV RNA are read in proper manner (395). However, although a small amount of the product, of the order of 5%, appeared to represent genuine though probably formylmethionylated STNV protein, it appears that the bulk of the product was variably shorter than the complete STNV protein. It was suggested, therefore, that either initiation or termination was incorrect, with our laboratory favoring the working hypothesis that most peptide chains were initiated at internal (presumably AUG) codons and thus yielded shorter peptides.

Surprisingly, a plant incorporation system derived from wheat embryos also yielded products with STNV RNA that were predominantly shorter than the STNV coat protein (300, 375). It may be that this is a consequence of the fact that TNV and STNV do not replicate in wheat, and the recent positive data obtained with a wheat virus (BMV) in that homologous system (see above) seem to support this explanation.

Data obtained with poliomyelitisvirus RNA in the *E. coli* system also gave ambigious results (394), but a recent study using an oncornavirus RNA was interpreted as indicating correct translation of that RNA by the *E. coli* system (410).

D. Replication of the RNA of Simple Viruses

The process of replication of the RNA of simple phages has been studied in considerable detail in recent years. It appears that the viral RNA, called the plus (+) strand if it acts as messenger RNA, first serves as a template for the formation of the complementary or minus (−) strand. The complex of these two, the replicative form (RF), then gradually releases the plus strand from its 5′ end while initiating simultaneous formation of up to six additional positive strands. This frayed molecular complex with much unpaired polynucleotide material is termed the replicative intermediate (RI). These products are schematically illustrated in Fig. 23.

Several protein factors are required for the control of RNA replication. Yet enzyme preparations have been isolated from Qβ-infected *E. coli* Q13 by Haruna and Spiegelman (102), as well as by others, which must contain all necessary components since they are able to achieve *in vitro* the synthesis of considerable amounts of infective (and thus plus) Qβ RNA from the four nucleoside triphosphates and a small amount of parental Qβ plus-strand RNA. The host protein factors are needed only for the first stage, the synthesis of the minus strand (11, 12, 84, 85, 226). The latter, while itself noninfectious (for it is not functional as messenger RNA), serves as a template for the formation of infective viral plus strands without lag when the purified polymerase complex is added *in vitro* (64). This complex is now recognized as consisting of one virus-coded peptide chain repeatedly mentioned (between 62 and 67 $\times$ 10^3 daltons for various RNA phages), and three host components (of about 72, 45, and 35 $\times$ 10^3 daltons) (356, 364).

As casually mentioned in the preceding sections, each step in the replication and assembly of simple RNA phages can be made to occur *in vitro*. The macromolecular components required for the replication of viral RNA are a

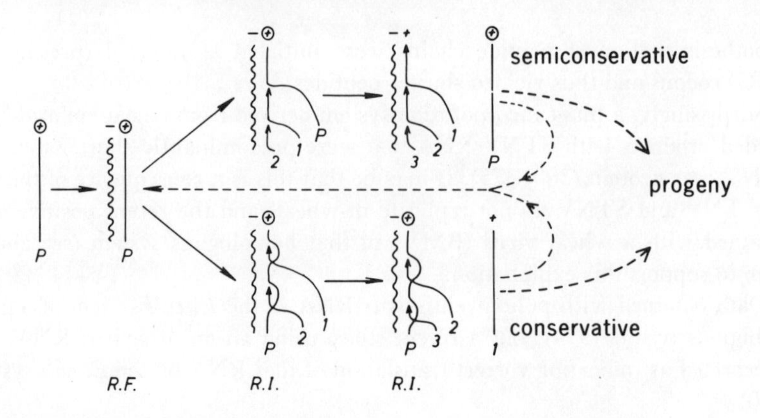

Fig. 23. Mechanism of replication of viral RNA. The natures of RF and RI, the replicative form and the replicative intermediate, one fully and the other partially double stranded, are illustrated. The latter allows the simultaneous synthesis of several (probably about six) plus strands on one minus strand. The principle of conservative and semiconservative replication is also illustrated.

purified though complex enzyme and two protein factors. For the synthesis of viral proteins, ribosomes, amino-acid-activating enzymes, t-RNAs and more protein factors are needed. These proteins acquire their native and functional conformations spontaneously as a consequence of their amino acid sequences. They are then able to form the simple virus particles automatically through coassembly of protein and nucleic acid components. Thus it appears that the need for a living cell should be dropped from the definition of virus.

The replication of the RNA of small animal and plant viruses is also being actively studied, largely by means of determining the amount of RI, RF, and single-stranded RNA at various time intervals and in the presence of various inhibitors (15, 185). It is of interest that these replicative forms of animal and plant virus RNA are infective, in contrast to the double-stranded forms of the RNA phages, which cannot infect spheroplasts. In the case of plant viruses, only double-stranded RF seems to accumulate (28, 199).

RNA polymerases have also been isolated from several virus-infected plants (174, 223, 389). In most instances they also are largely bound to particles or membranes and require no template. However, in the case of barley infected with BMV, an RNA polymerase was solubilized by Triton X-100 treatment; this then required a template and was highly active only when the RNA of this virus or the closely related RNA of the cowpea chlorotic mottle virus was added (343). This enzyme had an approximate molecular weight of 150×10^3 daltons, similar to the weights of the corresponding enzymes found in Chinese cabbage and in cucumber cotyledons infected with cucumber mosaic or tobacco ringspot virus (302, 389). The BMV-induced enzyme seemed to consist of, or to contain, a 34.5×10^3 dalton peptide chain. It appears noteworthy that a protein of this molecular weight was also detected as the main *in vitro* synthesized product of BMV-RNA component 3 (0.7×10^6 daltons) (408a).

The animal picornavirus-induced polymerases, with particular reference to the polymerase of poliomyelitis virus, were largely membrane-bound and of about 6S when solubilized (339). Such fractions contained five polypeptide chains, four of which were identical in molecular weight to the four components of active Qβ replicase, that is, of 75, 65, 57, 45, and 35×10^3 daltons. This solubilized enzyme was also, like the corresponding phage enzymes, active with poly C as template.

The sizes of the protein components of the poliovirus replicase are such that they, together with the coat proteins, cannot all have been coded for by the enterovirus genome of 2.6×10^6. It thus appears probable that in this case, as in the *E. coli* RNA phages, a virus-coded polypeptide synergizes with host components to make the active enzyme. Although no evidence is this regard exists for plant viruses, it appears unlikely that these, some of which have genomes of only 1×10^6, carry the complete information for their polymerases, which appear to have native molecular weights of about 150×10^3. However, it is possible that the active enzyme is actually composed of multiple, identical virus-coded peptide chains, such as those of 34.5×10^3, detected in BMV-infected plants.

E. Transcription and Replication of Complex RNA Viruses

The larger and more complex RNA viruses of animals and plants carry their transcriptase in the virions. Thus the messenger RNA is formed at the onset of the intracellular infection process, and probably in monogenic form, rather than as a complete copy of the virion RNA, which in the case of the paramyxoviruses is over 20,000 nucleotides long. The RNAs of the paramyxoviruses, as well as those of the rhabdoviruses, are thus believed to be transcribed by their own enzymes without any participation of cellular DNA or enzymes in the process (13).

Surprisingly, also, the virions of the diplornaviruses that occur in mammals, arthropods, and plants carry transcriptases which replicate one of the strands to serve as messenger (27, 160, 227). Thus each of the 10 to 13 molecules of double-stranded RNA becomes associated with the transcriptase also present in the core of these viruses, which quickly produces single-stranded copies of one strand of these many RNA species when supplied with the four triphosphates in a favorable medium. This messenger RNA synthesis proceeds even *in vitro* at a high rate of at least 425 bases/min (compared to phage RNA *in vitro* replication of 2100 bases/min) and for as long as 48 hr (241).

The myxoviruses exemplified by influenza virus, in contrast to all the other viruses discussed above, require DNA synthesis during the early stages of their transcription or replication. This explains the fact that actinomycin D inhibits influenza virus RNA replication during the first 2 to 3 hr of infection (24, 60). It appears possible that the relatively small genome of such a virus (3 to 4×10^6 daltons) carries only part of the information for its transcriptase and that other parts are derived from the host's genome and must be transcribed and translated to further support the initial infection. The dependence of myxovirus replication on DNA function may also be related to the finding that the nucleocapsids are formed in the nucleus, although most evidence favors the cytoplasm as the site of RNA replication. The myxovirus RNA strands, as they occur both in the virion and intracellularly, negative or positive in terms of messenger activity, are variably short and probably monogenic (217).

The viruses most actively studied at present belong to a group for the naming of which no consensus has been reached: the RNA tumor viruses, leukoviruses, or oncornaviruses. The nomenclature problem is due to the fact that not all members of this group cause solid tumors nor do they all cause leukemias. They all have one thing in common which is pertinent to the present discussion, however: they contain reverse transcriptase, an enzyme the activity of which resides in a protein of 70×10^3 daltons (341, 342) and which makes DNA on an RNA template (14, 263a).

Since all viruses that contain this enzyme activity are now retroactively being acknowledged as honorary members of this group of viruses, and no typical nondefective member lacks this activity, the name reverse transcriptase or revtra-viruses appears logical for this group.

The need for such an enzyme activity in the replication of the oncornaviruses was clearly recognised (263) as soon as it became evident that these viruses, alone among the RNA viruses, require DNA synthesis to occur throughout their replication, as indicated by the sensitivity of this process to actinomycin D and other tests. It appears that DNA is being synthesized on the infecting RNA template; the short pieces of RNA which are always associated with these virions seem to act as primers, and the DNA is then freed from the RNA template by the ribonuclease H (= hybrid) activity of the same reverse transcriptase, replicated, and in some manner incorporated into the transformed host cell's genome. This series of steps appears to represent the special attribute of this group of viruses of transforming susceptible cells and causing malignancy. Indications that revtra-activity is associated with human malignancy are increasing (306, 402).

F. Transcription and Replication of Viral DNA

The single-stranded DNA of ØX174 can be replicated *in vitro* by means of Kornberg's DNA polymerase and the ligase of uninfected *E. coli* (96, 189). However, among the eight complementation groups (genes) of this phage and related ones, at least three are concerned with the replication of the phage DNA after the first cycle and the first steps of transcription (129). It appears most probable that all DNA viruses carry information for their specific DNA polymerases. Several of these have been isolated, particularly from various T phages, and compared with the host enzymes (9). These and other DNA phages (342) also require several new enzymes to synthesize the unusual nucleoside triphosphates needed for the synthesis of their DNA (e.g., hydroxymethycytidine triphosphate). Thus there must first occur at least partial transcription of viral DNA to messenger RNA, and translation of the latter into enzyme proteins, before viral replication begins.

It appears likely that most DNA bacteriophages and many animal viruses utilize host DNA polymerases and/or transcriptases at the earliest stages of their reproduction. In the particular case of *E. coli* the nature of this transcriptase, the DNA-dependent RNA polymerase, has been the subject of intensive studies in recent years (34). It appears that the enzyme, composed of five different subunits (total molecular weight 400×10^3), acquires specificity in recognizing the initiation sites of specific genes only upon combination with another protein, the σ factor of 95×10^3 molecular weight. Apparently variants of σ, as well as of at least one and probably all of the other component polypeptide chains, are coded by the phage genome and arise in the course of phage infection (34). Thus phage-specific components of the polymerase complex direct the transcription of later phage genes, and supply a means of positive control of transcription. These studies were performed with both T-even phages and lysogenic phages. Particularly the intracellular fate of the latter, as exemplified by λ, has been the subject of intensive study in recent years. The mechanism of integration of the λ genome into the *E. coli* DNA has been largely elucidated, as well as the nature of a repressor, which specifically binds to the DNA in a manner to prevent such integration (198). This complex field has been frequently reviewed (62, 235) and will not be considered further in this chapter.

A different solution to the problem of specific and early transcription is used by the large DNA viruses, particularly those replicating in the cytoplasm rather than the nucleus. Thus a virus-specific DNA-dependent RNA polymerase is carried by the vaccinia virion, which can initiate transcription and translation before complete uncoating of the vaccinia DNA (134, 143, 144, 182). The even more intriguing recent discoveries of an RNA-dependent

DNA polymerase in RNA tumor virus particles, and of the RNA polymerase in vesicular stomatitis virus, were discussed earlier. Among other enzymes coded for, but not carried by the vaccinia virus, are a thymidine kinase, a specific DNA polymerase, an ATPase, and several nucleases. Although viruses of the herpes group also carry information for such enzymes, no RNA polymerase has been demonstrated in the virion, and this enzyme may not be necessary since the herpesviruses, like most DNA viruses, are replicated in the nucleus, where the host's transcriptase and messenger RNA machinery are available. Most of the viral messenger RNAs seem to become translated in thee cytoplasm, but the coat proteins may reenter the nucleus, since the nucleocapsids of the herpersviruses, adenoviruses, and possibly other viruses are formed in the nucleus. The herpes viruses mature, in terms of acquiring their envelopes, at the nuclear membrane, in contrast to the enveloped RNA viruses, which mature at the cell membrane.

7. MUTATION OF VIRUSES

A. Naturally Occurring Strains and Mutants

Most viruses occur as more or less closely interrelated groups. Members of such groups of viral strains rarely differ in physical dimensions and appearance on electron micrographs, although one such case has been mentioned, the tobacco rattle virus (215, 224). Strains differ more frequently in various biological aspects related to their infectivity, such as symtomology, host range, vectors, or temperature sensitivity. Another criterion for distinguishing between virus strains utilizes differences in their antigenic specificity. Finally, and most decisively, all virus strains differ in their chemical fine structure. These differences may, in a given case, be as small as one nucleotide replacement and thus almost undetectable by present methods, but they are nevertheless the basis of for the existence of various strains and thus obligatory. Examples of families of virus strains characterized by from 1 to more than 80 amino acid exchanges in the coat protein are known for TMV, as well as for the RNA phages, and such coat-protein differences and similarities are illustrated in Fig. 9 and 10.

We now discuss the nature of mutation from this basic viewpoint, treating the terms "strain," "mutant," and "variant" as interchangeable, although "mutant" is preferentially used for laboratory products, and "strain" for genetically related "field" isolates. On the basis of our present understanding of the transcription and translation of genetic material (see p. 274, 281), any replacement of one nucleotide base by another is a potential mutagenic event. Because of the redundancy of the code, many such mutations particularly in the third place of the triplet) are ineffectual. Some of these, however, can secondarily become effective and thus may be regarded as latent mutations. This is exemplified as follows:

$$\text{CUA (Leu)} \xrightarrow{(a)} \text{CUG (Leu)} \xrightarrow{(b)} \text{AUG (Met); CUA (Leu)} \xrightarrow{(b)} \text{AUA (Ile).}$$

The base replacement termed *a* genotypically not mutagenic but is a latent mutation, since a second event, C → A (*b*), gives a different result after event *a* than without it.

All these changes in a single codon are termed point mutations, if they lead to a replacement of an amino acid in a protein. It is now apparent that not all nucleotides in viral RNA are translated into protein structure. This is the case for the first 140 and 62 nucleotides in group I and III phage RNAs, as well as for shorter pieces between genes (see Figs. 4 and 5). The function of these noncoding segments remains unknown. If some of them play a role in frameshift determination, their replacement may have serious phenotypic consequences.

All viral genomes code for several proteins which through point mutations may become altered in their primary structures. Depending on the role of the protein that is affected, the nature of the amino acid replacement, and its location in the peptide chain, a point mutation may be phenotypically undetectable, it may cause one or several of the biological effects discussed above, or it may finally be lethal so that no viable virus can be isolated under standard conditions. However, host strains may contain suppressor mutations that enable them to mistranslate inactivating (e.g., terminating) codons in a more or less harmless or potentially functional manner, so that the mutated virus becomes viable in this permissive host. Conditional lethal mutations of this type are usually termed *amber* and *ochre* mutations, although host-sensitive (*hs*) or suppressor-sensitive (*sus*) are more descriptive terms for this group. The other class, the temperature-sensitive (*ts*) mutants, often are the result of the altered protein, be it an enzyme or a viral coat protein, being functional at one but not the other temperature. This is frequently due to the mutated protein being more susceptible to denaturation and thus not functional at the higher temperature. An example is the TMV mutant *Ni*118, obtained by nitrous acid (see the next section), in which proline residue 20 has been replaced by leucine (CCX → UCX). This virus is viable only below 30°C, and the isolated coat protein was found to be susceptible to heat denaturation at temperatures not harmful to wild-type TMV. (132, 133).

B. Chemically Induced Mutations

As discussed above, a replacement of a single base by another (and thus a potential or actual point mutation) is produced when nitrous acid reacts with a cytosine residue and transforms it to uracil (90, 218, 219) Hydroxylamine can replace the amino group of cytosine by the hydroxyamino group (220). This compound,

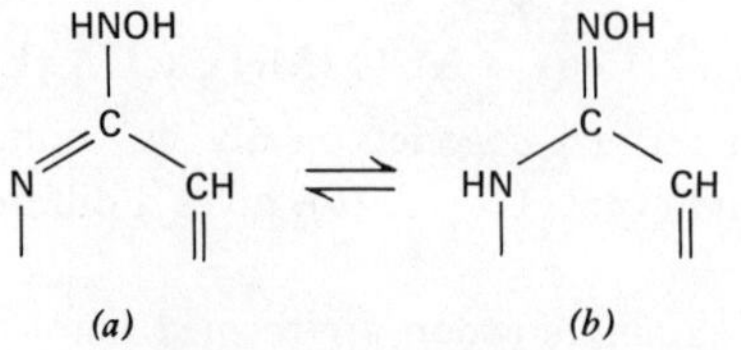

HNOH — C — N, CH ⇌ NOH — C — HN, CH

(*a*) (*b*)

N-4-hydroxycytosine (*a*), occurs preferentially in the tautomeric form resembling uracil (*b*) (31), and the mutagenicity of hydroxylamine and its *O*-methyl derivative has been attributed to this reaction (236, 336). The action of nitrous acid on adenine, transforming it to hypoxanthine, is believed to be mutagenic on similar grounds, since hypoxanthine, like guanine, exists predominantly in the opposite tautomeric state to adenine. It appears doubtful that tautomeric shifts can account for all the mutagenic effects of alkylating agents, including nitrosoguanidine, which are able to introduce alkyl groups into guanine at the 7 position, into cytosine at the 3 position, and into adenine at the corresponding 1 position and, to a lesser extent (except in double-stranded nucleic acids, which are unreactive at the 1 position), at the 7 and 3 positions (238, 240) (see Fig. 1 for identification of these positions). Recent evidence favors the theory that the methylation of cytosine is the most mutagenic of the various methylation reactions (239) and that this modification does not give cytosine a uracil-like character but rather renders it completely nonselective. The mechanism of mutagenesis, if any, of the recently recognized *O*-alkylation of guanosine is similarly unclear (373). A newly discovered specific and mutagenic reaction of cytosine is the addition of bisulfite to the double bond, which secondarily favors deamination (225).

The chemical agents discussed above, as well as bromination but not ultraviolet irradiation, lead to biologically detectable mutants of TMV (238). These mutations have also been demonstrated by *in vitro* studies, in that the same reagents, when acting on polynucleotides, affect their template activities in directing polynucleotide synthesis with RNA polymerase in the expected manner: nitrous acid and hydroxylamine treatment of poly C causes it to act as a template not only for GTP but also for ATP incorporation, while partially methylated poly C incorporates not only GTP but also ATP, CTP, and UTP (239).

The most straightforward mutagenic reactions appear to be adequately explained in this direct manner. However, most mutations are produced not *in vitro* but by allowing the mutagen to act in the living and metabolizing cell, and it appears that many other mechanisms are involved in the course of such *in vivo* mutagenesis. Thus reagents that are not highly mutagenic when acting on the virus or its RNA *in vitro* may be very effective mutagens in the living cell. One such group of mutagens consists of the acridines, which tend to squeeze between (intercalate) the base pairs of double-stranded DNA. These reagents lead predominantly to deletion or insertion of bases or groups of bases and have proved to be very useful tools in molecular genetics. Another group of *in vitro* mutagens are the base analogs, such as 2-aminopurine and 5-bromouracil (for DNA) or 5-fluorouracil (for RNA). Their action is also not necessarily or exclusively one of replacement of a base in the messenger molecule, since they are of much higher or different activity when present in the living cell than when occurring in polynucleotide linkage. Thus 40% of the uracil of TMV RNA can be replaced by 5-fluorouracil by growing the virus in leaves infused with this base (95), without significant numbers of mutants occurring in such virus isolates. In similar manner, much 5-bromouracil can be enzymatically incorporated in place of thymine into polydeoxyribonucleotides without gross effect. Yet the mutation frequency of T-even phages is greatly increased by this modification; also, RNA phages growing in *E. coli* (233) in the presence of fluorouracil, as well as intracellular poliomyelitisvirus (44), show characteristic mutants, and for vesicular stomatitis virus fluorouracil is a better mutagen than the alkylating agent, EMS (197).

The phenotypic results of chemically induced mutations have been studied in some detail with two RNA virus systems, TMV and the f2 class of phages, and with two DNA virus systems, the T-even phages and OX174. Many of the hundreds of TMV mutants isolated, but not all, showed aminoacid replacements in the coat protein, usually affecting only one residue and not adjacent ones if two or three were affected (69, 88, 269, 291). The natures of most but not all of the exchanges (Table 3) are in accord with the expected results of the mutagenic treatment. Thus deamination generally leads to replacement of aminoacids coded by C-containing triplets by others carrying more Us in their condons. This results in frequent replacements of proline by leucine (CCX → CUX), serine by phenylalanine (UCX → UUX), and so on. Yet some other exchanges expected on this basis have not been detected (e.g., alanine → valine, GCX → GUX). Obviously the site and the nature of exchanges that will not destroy a crucial functional aspect of the viral coat protein are limited. However, on protein-chemical grounds one would regard proline replacements as harmful and would also not expect that serines could readily be replaced by phenylalanine. It appears significant, therefore, that these frequent occurrences almost always involve the same few proline and serine residues near the ends of the protein molecule (see Fig. 9).

Strains of low viability, in which the protein is not properly functional have also been studied. Surprisingly, the aminoacid replacements in some of these

Table 3. Coviruses, Satellite, and Multinucleic Acid Viruses

Coviruses	
Cowpea mosaic virus group	Two dissimilar isometric particles or their two different RNAs. Both needed for infection (information for essential gene, possibly replicase, distributed over both RNAs).
Tobacco rattle virus group	Two dissimilar rod-shaped particles, two different RNAs. The longer is alone infective but yields unstable virus (sterile lesions) (coat-protein information in shorter RNA, longer RNA carrying all genes essential for infection and lesion formation).
Alfalfa mosaic virus group	Four dissimilar rod-shaped to isometric particles, four different RNAs. The three largest particles or all four RNAs needed for infection (coat-protein information in the two shortest RNAs).
Brome mosaic virus group	Three very similar isometric particles, four different RNAs. The three particles or the three larger RNAs needed for infectivity (coat-protein information in the two shortest RNAs).
Satellite and tobacco necrosis virus	Two unrelated particles and RNAs. Necrosis infectivity carried by large particle or RNA alone; satellite infectivity only by combination of both (the satellite believed to carry only its own protein information and to use the replicase and any other components needed for infection from the helper virus, tobacco necrosis).
Multinucleic acid viruses	
Influenza virus, Oncornaviruses, Diplornaviruses	One particle containing several or many (3–13) RNA molecules (single- or double-stranded), which carry different bits of information.
Typical RNA viruses	One particle containing one molecule of RNA.

proteins were not quantitatively different from the many others observed in functional coat proteins. However, one exchange in PM2, the first such strain to be studied in this regard, that is, Glu → Asp at position 95, is in an area of the molecule where six natural strains and all chemical mutants that were studied are the same (Fig. 9). Thus this exchange may be the key to the not properly functional protein; a second exchange in this mutant, Thr → Ile at position 28, is not singular in regard to its location (287). The introduction of a second cysteine residue in lieu of arginine at position 112 in PM5 is also a distinctive alteration (98). Another defective TMV strain was found to be viable when grown at 23°C but not at 35°C; it appeared that aminoacid replacements at positions 81 (Thr → Ala) and 143 (Ser → Phe) rendered the coat protein more salt-sensitive and must in some way account for the temperature sensitivity of this virus (99). In contrast, *N*i118 and the natural isolate, flavum, which had Pro → Leu and Asp → Ala replacements at positions 20 and 19, respectively, show lower denaturation points for their isolated coat proteins. Furthermore, the finding by Jockusch (132) that temperature-sensitive TMV mutants exist which have quite heat-resistant coat proteins (e.g., *N*i2519 with an Ile → Thr exchange in position 129) indicates that many mutants studied are likely to carry multiple mutations, and that no conclusions can be drawn solely on the basis of coat-protein analyses (Fig. 9).

The RNA phages were mutagenized by treatment with nitrous acid or by growing them in the presence of fluorouracil (119, 267). In contrast to strains of viruses of plants and animals, which are usually differentiated by the host's response, phage mutants are more frequently detected by their nonviability under restrictive (nonpermissive) conditions. Thus a change of CAPu (Gln) to UAPu leads to chain termination and abortive infection, except in suppressor strains which carry tRNAs that read UAPu as Ser (UCPu), Tyr (UAPy), or Gln (CAPu) (253). It is largely by these methods that phage point mutations have been characterized and the chemical nature of the mutagenic event has been identified. Thus analysis of *sus* mutants of T4 lysozyme has shown that this protein is functional (50% of the wild-type enzyme) after all three of its tryptophan residues have been replaced in the triple recombinant by tyrosine residues [UUG (Trp) → UAG (terminating), read as UAPy (tyrosine)] (124).

With the T-even phages, acridines were also used as mutagens; and when aminoacid differences were detected in the lysozymes produced by these mutated phages, these corresponded to the results expected for the deletion of one or two bases, corrected by insertion of the same number of bases in the vicinity. Aminoacid sequence analysis has demonstrated the validity of this genetically predicted mechanism of mutagenesis in a number of instances, as illustrated by one mutant of T4 lysozyme, where the deletion of an A is cor-

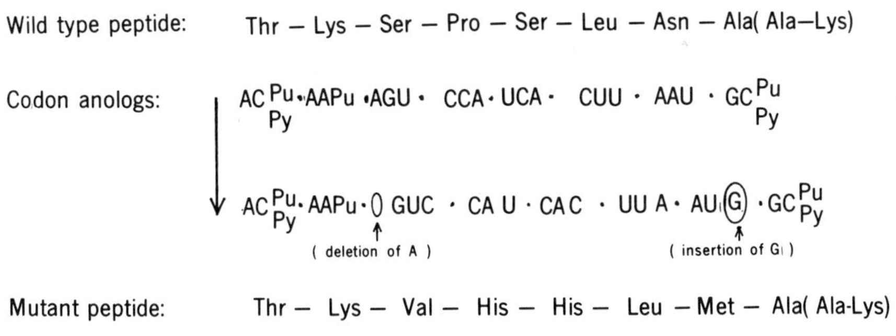

Wild type peptide: Thr — Lys — Ser — Pro — Ser — Leu — Asn — Ala(Ala—Lys)

Codon anologs: AC^{Pu}_{Py} • AAPu • AGU • CCA • UCA • CUU • AAU • GC^{Pu}_{Py}

↓ AC^{Pu}_{Py} • AAPu • () GUC • CA U • CAC • UU A • AU (G) • GC^{Pu}_{Py}

(deletion of A) (insertion of G)

Mutant peptide: Thr — Lys — Val — His — His — Leu — Met — Ala(Ala-Lys)

Fig. 24. Corresponding mutant peptides and mRNA analogs for the wild type and a frameshift mutant T-even lysozyme. The amino acid composition of the peptides can be accounted for on the basis of the codon dictionary only by the assigned triplets, with one deletion and one insertion (264).

rected by the insertion of a G in position 15, leading to an altered peptide sequence (264) (Fig. 24).

8. CLASSIFICATION AND BRIEF CHARACTERIZATION OF VIRUSES

Various criteria have at times been used for the classification of viruses. Nothing is known about the evolutionary relationship of viruses, and it is even possible that some viruses represent primitive forms of half-life, while others may be derived from cellular organelles of higher organisms. In any case no classification can at present be based on the evolutionary proximity of different types of viruses. Classification by host organism appears artificial, since physically very similar viruses, such as the wound tumor, reo, and cytoplasmic polyhedrosis viruses, occur and replicate in plants, animals, and insects; and another group of similar viruses, the small RNA viruses, occur in these, as well as in bacteria. Classification based on means of transmission (arboviruses = arthropod-borne viruses) or on components of the envelope (myxoviruses to include all viruses having carbohydrates or neuraminidase in their envelopes) appears equally limited and arbitrary.

The classification based on physical and chemical principles used in recent books appears least subject to question. Thus we have come to classify a virus first according to its nucleic acid (RNA, DNA, each either single- or double-stranded) and then according to its particle dimensions and other unifying physicochemical features. We have also incorporated the nomenclature suggestions of the International Committee on Nomenclature of Viruses (8). Viruses of classes 1 to 7 contain single-stranded RNA; class 8, double-stranded RNA; classes 9 and 10, single-stranded DNA. and classes 11 to 18, double-stranded DNA. These classes are listed according to increasing size and complexity. Examples of the various groups will be given, as well as information on their dimensions, architecture, and component parts.

Class 1 contains *rod-shaped plant viruses*, the virion representing a helical nucleocapsid ranging from long, flexible rods (potato viruses X and Y, 11 × 540 and 800 nm) to stiff rods (TMV, 18 × 300 nm) and stubby rods (tobacco rattle virus, 26 × 75–190 nm). These virus particles, up to 50×10^6 daltons (~200S) in weight, contain 4 to 6% of RNA as a single molecule of 2 to 2.4×10^6 molecular weight and one type of protein of 17 to 30×10^6 molecular weight (Fig. 25).

Class 2 represents the large group of *picornaviruses*, isometric small (pico) RNA viruses (~100S) consisting of a coat of 180 protein molecules (icosahedral symmetry, $T = 3$, and 1 molecule of RNA of 1 to 3×10^6 molecular

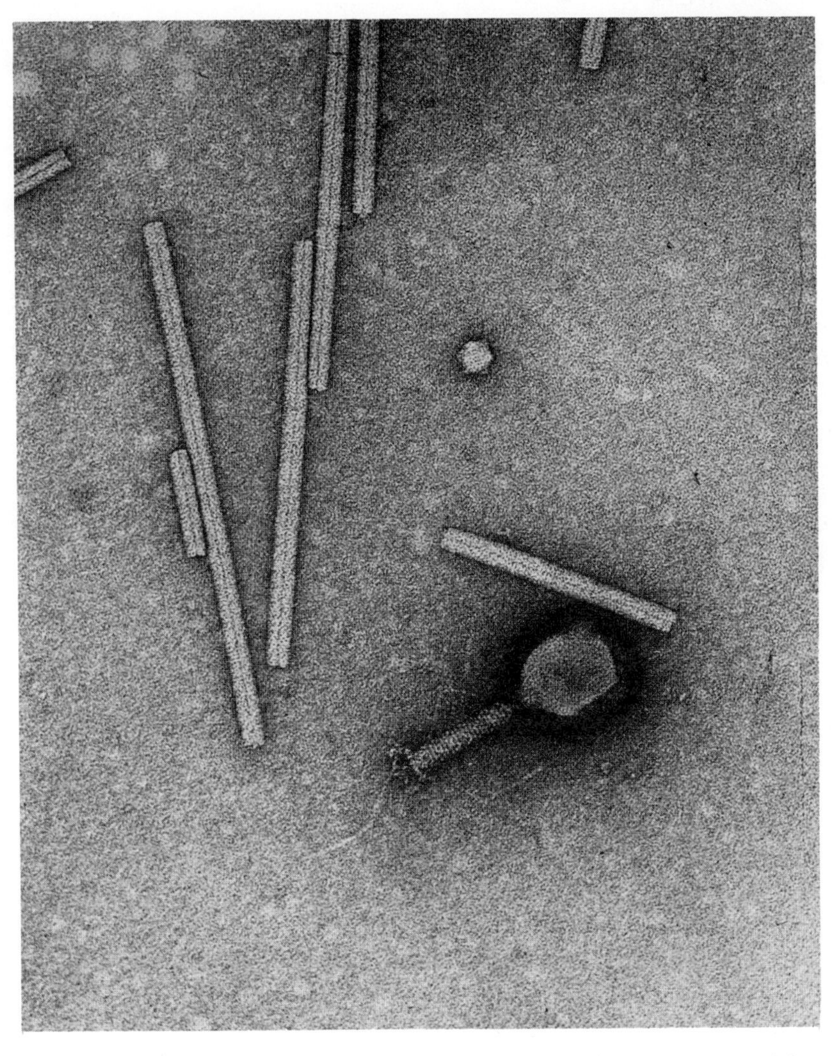

Fig. 25. Electron micrograph showing a small isometric virus (ØX174 phage), a rod-shaped virus (TMV, 30 nm long), and a complex virus (T4 phage).

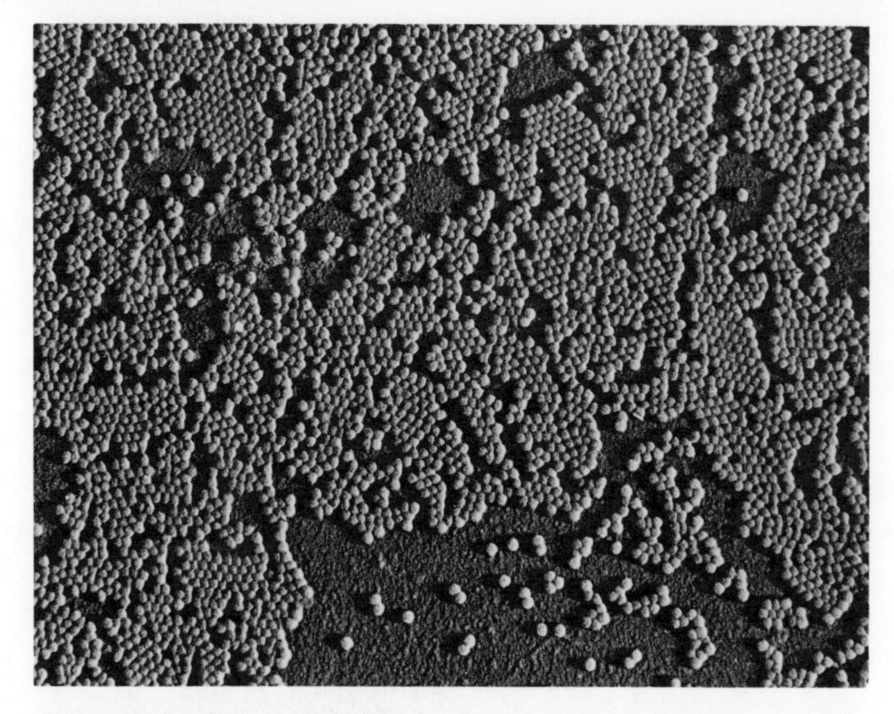

Fig. 26. Poliomyelitis virus. A crystalline alignment of particles can be observed.

weight. The diameters range from 20 to 30 nm, and the particle weights from 4 to 9 × 10^6 daltons. Many plant, animal, and bacterial viruses fall in this group (e.g., TYMV and BMV; poliomyelitis, rhino, and foot-and-mouth desease viruses; f2 and Qβ phages). The exceptionally small (50S) defective satellite virus of tobacco necrosis may be regarded as a subgroup of the picorna viruses. Its diameter is 17 nm, and its particle and RNA weights are 2.0 and 0.4 × 10^6. Most of the plant viruses in this class contain a single species of protein, and the bacterial viruses a single structural coat protein and a monomolecular tail (A-protein), but all of the animal viruses seem to consist of four coat proteins (Fig. 26).

Class 3, for which Fenner proposed the name encephaloviruses, is now termed *togaviruses* (*toga* = cloak) because they are covered by a membrane. This class contains many of the arboviruses (e.g., equine encephalitis, Sindbis virus, and yellow fever virus) occurring in animals and arthropods. The particle weights are about 50 to 10^6, and the dimensions are 35 to 80 nm

(~300S). The single molecule of RNA of these viruses is about 4×10^6 in molecular weight. These viruses consist largely of one nucleocapsid protein and one or two glycolipoproteins in their membranes, which act as hemagglutinin under certain conditions. The togaviruses are ether-sensitive.

Class 4, the *rhabdoviruses* (*rhabdos* = rod), represents a group of rather large bullet-or stubby-rod-shaped viruses. The vesicular stomatitis and rabies viruses are examples. Rhabdoviruses infect also many plants and insects. In addition to the typical particles (e.g., 70×200 nm, 600S), shorter ones are often seen, the significance of which is not yet known. The RNA is a single molecule of 4 to 5×10^6 daltons. The helical nucleocapsid, consisting of one or two main proteins, is covered by a membrane and a glycolipoprotein-containing envelope. The virions contain a transcriptase (Fig. 27).

Class 5 contains the *myxoviruses,* exemplified by the main representative, the influenza virus. These viruses, as well as those in the next two classes, are irregular in shape (pleomorphic) and are covered by a lipid-containing membrane with spikes consisting of two proteins, the hemagglutinin and the neuraminidase. The particle weights and dimensions of influenza virus are about 300×10^6 daltons and 80×100 nm, respectively. The RNA, amounting to about 3 to 5×10^6 daltons, exists in the virion in at least five or six seemingly helical nucleocapsid threads, consisting of a single protein and enclosed in a membrane formed by another protein. These viruses also contain a transcriptase and are ether-sensitive (Fig. 18).

Class 6 contains the *paramyxoviruses,* which are somewhat larger than the myxoviruses (200 nm in diameter, 500×10^6 in particle weight) and have similar envelopes. However, their nucleocapsid is a typical helix resembling that of TMV except for its greater flexibility. For example, Newcastle disease virus, 700×10^6 daltons in weight, contains a single molecule of RNA of about 7×10^6 daltons, coated by a single capsid protein. It appeears that in all viruses with helical nucleocapsids these consist of a single protein (classes 1, 4, 5, and 6). Several of these viruses, exemplified by the Sendai virus, lead to cell fusion and have thus proved to be very useful tools in experimental biology.

Class 7 contains the viruses termed the *leukoviruses, oncornaviruses*, the *RNA tumor viruses,* or the revtra (for reverse transcriptase) *viruses*. Examples are the Rous sarcoma, mammary tumor, and murine leukemia viruses. These viruses are somewhat pleomorphic and are enveloped in a similar manner to the two preceding groups, all of which occur only in animals. However, the RNA of the revtraviruses, seemingly of about 10×10^6 daltons, can be dissociated to shorter pieces ($2–3 \times 10^6$ daltons). The viruses of this class are somewhat smaller than the paramyxoviruses (100–200 nm, about 500×10^6 daltons). They contain about seven proteins, with type-specific glycoprotein

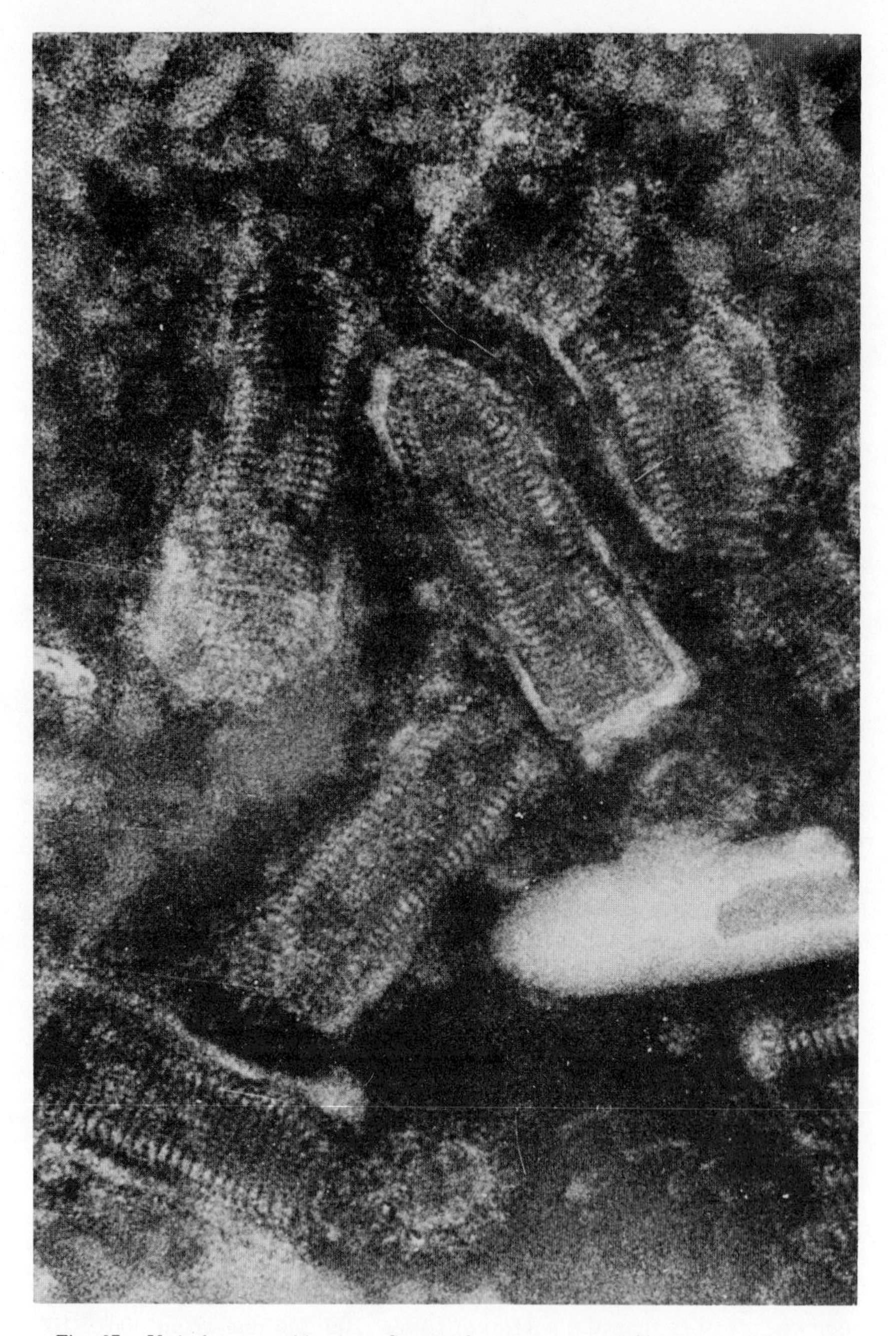

Fig. 27. Vesicular stomatitis virus. Structural components and filamentous disintegration products of VSV are shown. One intact particle has been penetrated by the phosphotungstic acid only in the area of the axial hole. (Courtesy of A. J. Hackett.)

antigens in the envelope. They are ether-sensitive and contain reverse transcriptase.

Class 8 contains the *double-stranded RNA viruses* (diplornaviruses) of mammals, plants, and insects, isometric with an icosahedral core, 80 and 60 nm in diameter, and 100 and 70 × 10^6 daltons in particle weight, respectively. The virions are of about 600S. The RNA complement of these viruses (14–20%) occurs in 10 to 13 characteristically different pieces. The double shell of these viruses consists of about five different proteins. Examples are the reovirus, wound tumor virus, and cytoplasmic polyhedrosis viruses (Fig. 28). They contain transcriptase.

Class 9, the *small DNA bacteriophages*, contains two subclasses, the isometric particles exemplified by ØX174, and the flexible rod particles by fd (Figs. 21 and 25). The dimensions are 30 nm and 5 × 800 nm, respectively; the particle weights are 6.2 and about 20 × 10^6. Each subclass contains a single molecule of circular single-stranded DNA, of 1.7 to 2.0 × 10^6 daltons. The ØX174 type is composed of one main structural protein and three others that make up its 12-apical attachment organ. The fd viruses are built of a very small coat protein (mol. wt. 5200) and probably one molecule of a maturation protein (70 × 10^3 daltons). The geometry is dodecahedral and alpha-helical, respectively.

Little is known of class 10, the *parvoviruses* (i.e., small viruses), including the *adeno-associated virus* (AAV), beyond their approximate dimensions (about 23 nm diameters, 7.5 × 10^6 particle weight). They are icosahedral and seemingly all contain single-stranded DNA of approximately 1.7 × 10^6 molecular weight. The DNA in individual particles may be either of the complementary strands, so that upon isolation a double-stranded molecule tends to be formed. Members of the parvoviruses (minute murine, etc.) may be oncogenic, while AAV occurs only in conjunction with adenoviruses and appears to be in some manner defective.

Class 11 comprises the *papovaviruses*, icosahedral double-stranded DNA viruses, including the polyoma and SV40 (40 × 10^6 dalton particles, 45 nm in diameter) and the slightly larger papilloma viruses. These viruses contain circular double-stranded DNA of 3.4 and 5.0 × 10^6 molecular weight. They are oncogenic in many of their mammalian hosts (Fig. 29).

Classes 12 and 13 contain the *medium-sized* and *large bacteriophages*. Their head dimensions are about 50 × 50 and 65 × 95 nm, and their tails are up to 150 nm long. Particle weights range from 50 to 220 × 10^6. Those of class 12 have very short to long noncontractile tails; those of class 13 usually have contractile tails. Representative of class 12 are λ (Fig. 30) and T1; of class 13, the T-even phages (Fig. 20). The molecular weights of their double-stranded DNAs are about 30 and 120 × 10^6 daltons. The DNA in the particle is linear,

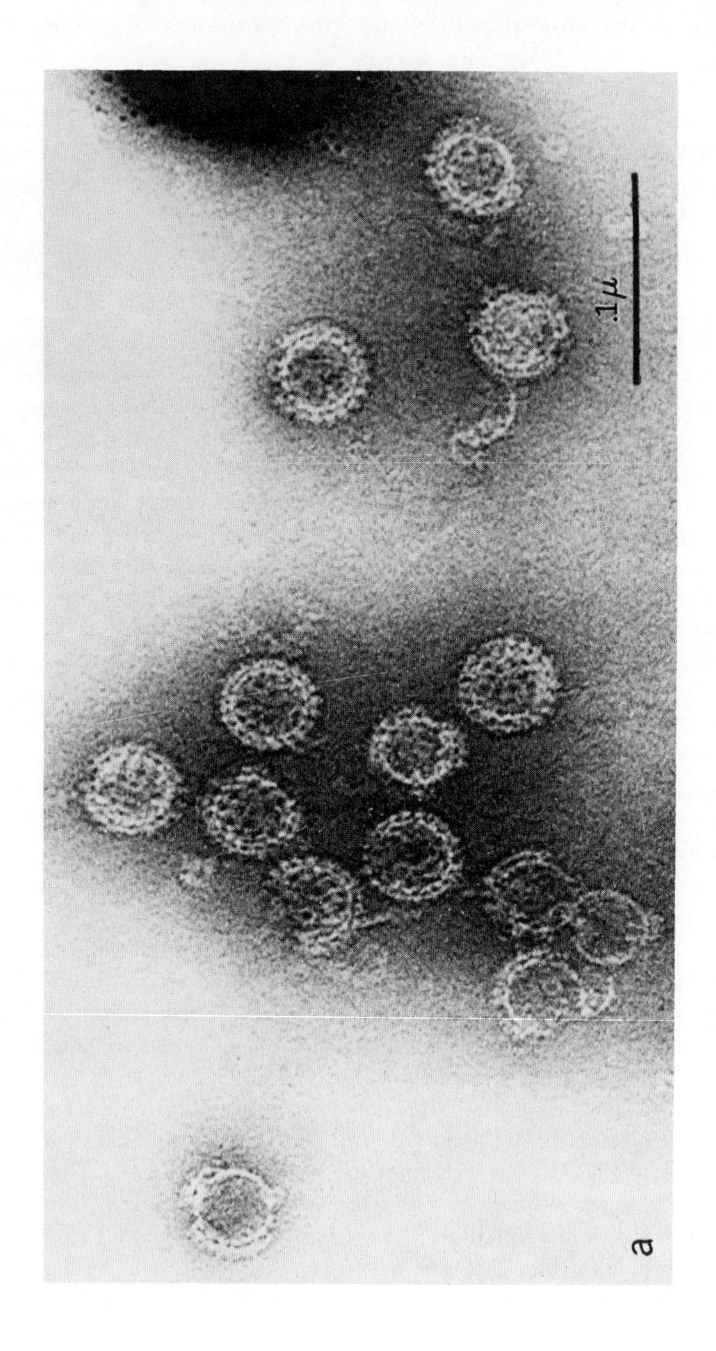
1μ
a

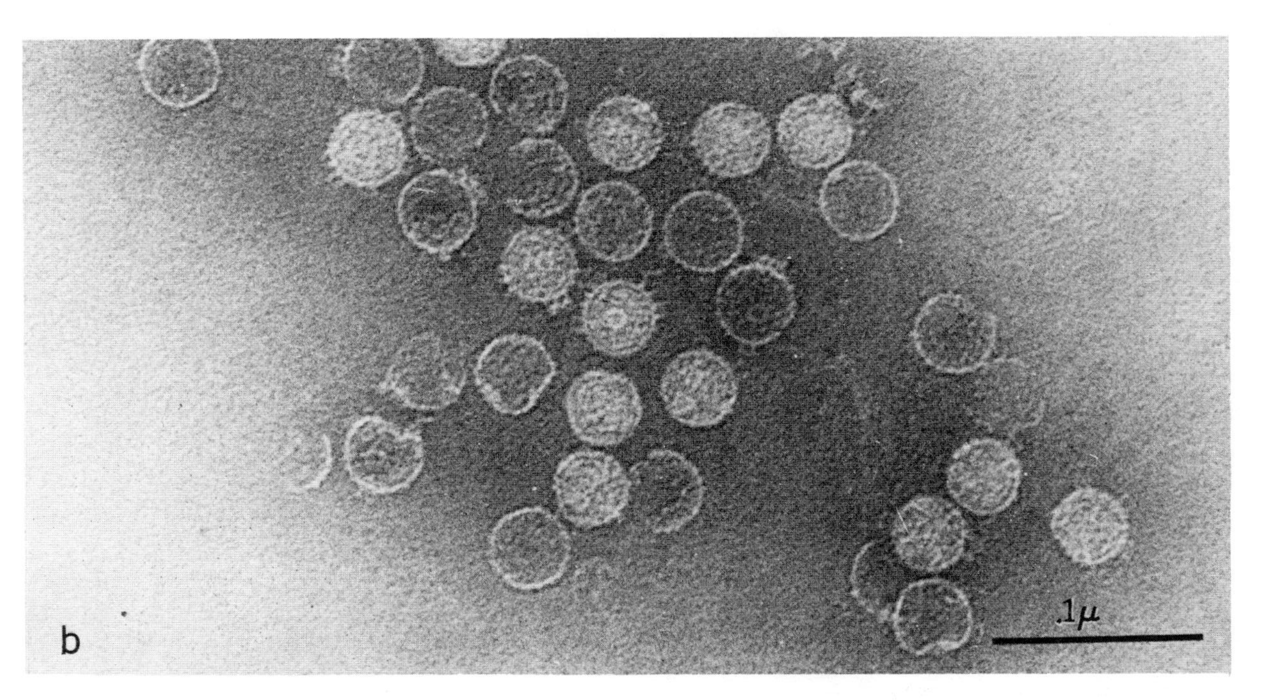

Fig. 28. Reovirus. (*a*) Particles stained to enhance double capsid construction. ×120,000. (*b*) Inner capsids of reovirus artificially produced by treatment with trypsin. The particles are 45 nm in diameter, and there is some evidence of subunit structure. The preparation is negatively stained with 2% ammonium molybdate. ×120,000 (Courtesy of H. D. Mayor.)

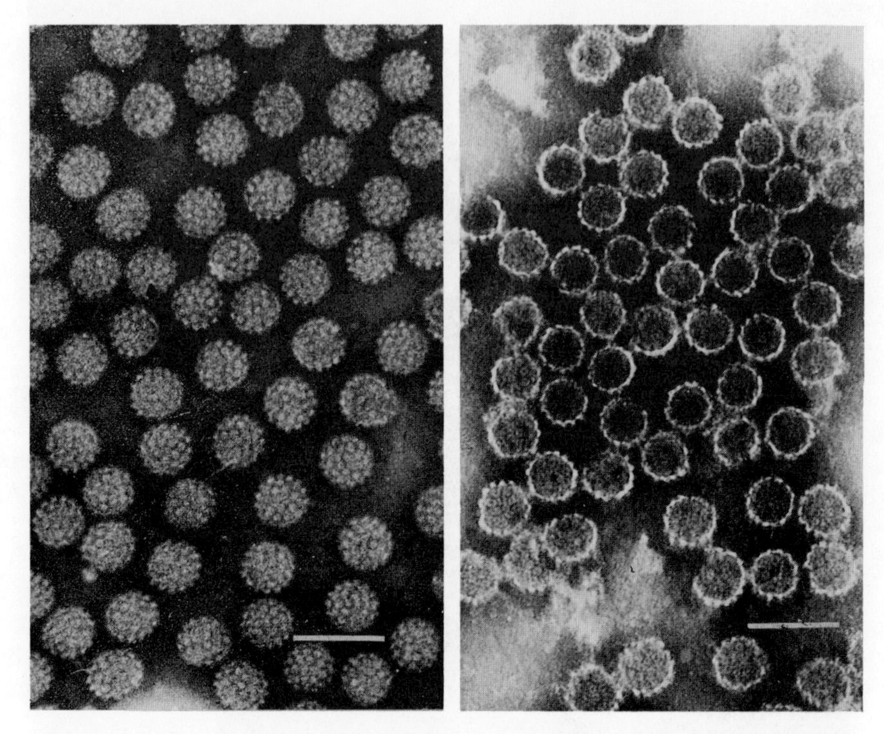

Fig. 29. Human papilloma virus particles. The micrograph on the left shows "full" particles; that on the right, "empty," nucleic-acid-free particles. The two preparations are negatively stained with potassium phosphotungstate. The bar in each micrograph represents 100 nm. (Courtesy of E. A. C. Follett.)

but that of the medium-sized phages tends to circularize *in vivo* and, through cohesive ends, frequently also *in vitro*. The DNAs of both classes show terminal redundancy, and those of the T-even phages also circular permutation (see p. 240). Among the medium-sized phages are many that are lysogenic. These phages do not generally replicate and lyse the host cell but become integrated in and replicated with the *E. coli* genome. They can be induced to virulence after many generations by UV light and other agents.

Class 14, the *adenoviruses* of mammals and birds, consists of double-shelled icosahedral particles of beautiful symmetry, 80 to 90 nm in diameter and having a particle weight about 175×10^6 and 800S (Fig. 31). Their DNA is linear and has a molecular weight of about 23×10^6. Many of these viruses are oncogenic.

Class 15, the *iridoviruses* of insects, is composed of icosahedral particles of 82-nm diameter (about 800×10^6 daltons), containing double-stranded DNA of 126×10^6 daltons (Fig. 32).

Class 16 consists of viruses for which the name *baculoviruses* has been proposed. These are the nuclear polyhedrosis and granulosis viruses of insects, which are bacilliform (about 50×300 nm), enveloped, and containing double-stranded DNA of 80 to 100×10^6 molecular weight. They are usually occluded in protein crystals.

Class 17, the *herpesviruses*, are isometric, have icosahedral symmetry, and are 100 to 150 nm in diameter and 500 to 1000×10^6 daltons in particle weight, containing double-stranded DNA of 54 to 92×10^6 molecular weight. The herpesviruses may or may not contain an envelope derived from the nuclear membrane. Their hosts are mammals.

Class 18, the *vaccinia viruses*, are brick-shaped, of 200×300 nm dimensions ($\sim 2500 \times 10^6$ daltons in particle weight) and thus visible in the microscope. They are of very complex structure and contain many enzymes, including transcriptase and DNA replicase, and double-stranded linear DNA of 160 to 200×10^6 daltons.

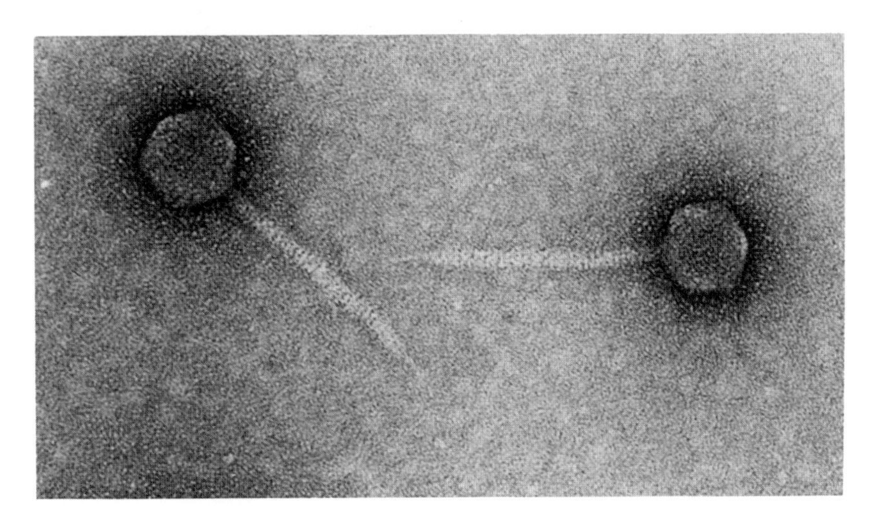

Fig. 30. Particles of λ phage. Diameter of head: 50 nm. (Courtesy of A. D. Kaiser.)

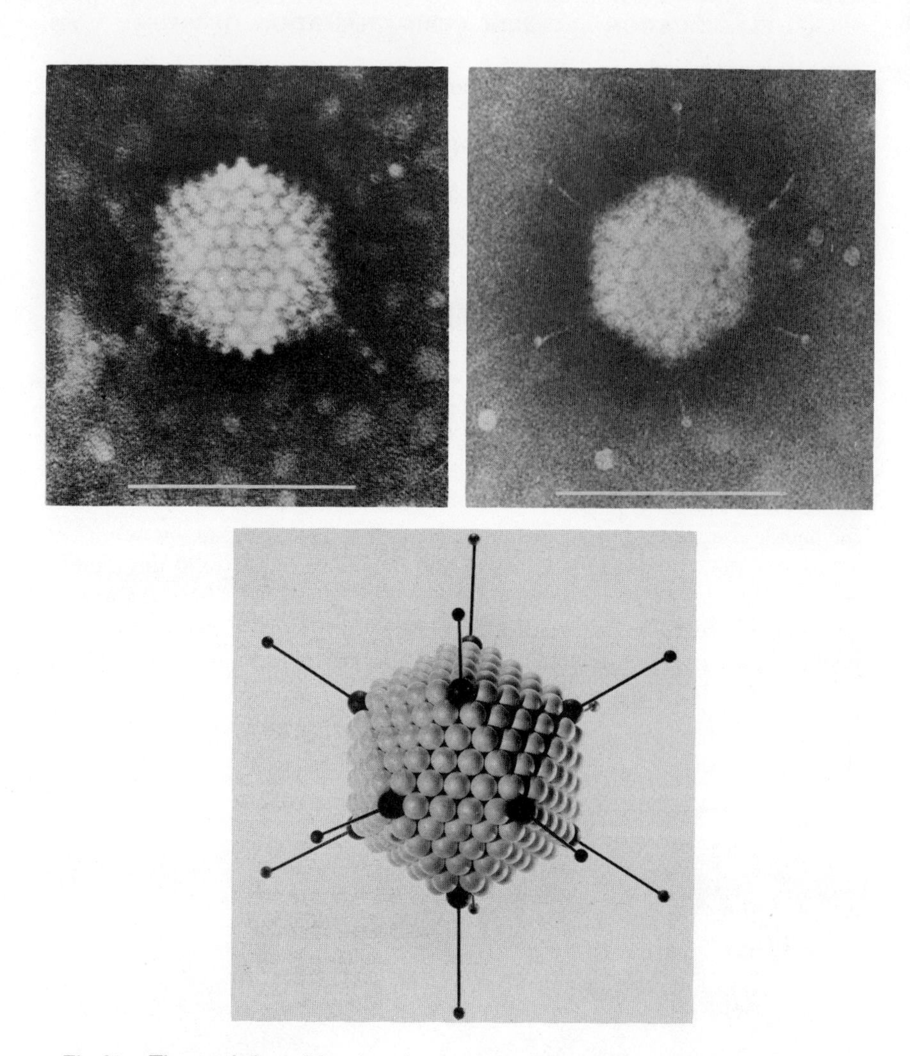

Fig. 31. The morphology of the adenovireus particles. The two upper pictures are electron micrographs of adenovirus type 5 particles in preparations negatively stained with sodium silicotungstate. The micrograph on the left shows the capsomeres and their icosahedral arrangement; the one on the right, the fibers projecting radially from the pentons. The bar in each micrograph represents 100 nm. The photograph below is a model of the adenovirus particle based on the electron microscopic observations. (Courtesy of R. C. Valentine.)

Fig. 32. Tipula iridescent virus and model icosahedron. Both the virus and the model are shadowed from two angles to illustrate the relationship of vertices to edges. The diameter of the shadowed particle is about 170 nm. (Courtesy of R. C. Williams and J. Toby.)

REFERENCES

1. Acheson, N. H., and I. Tamm, *Virology* **41,** 321 (1970).
2. Acheson, N. H., and I. Tamm, *J. Virol.* **5,** 714 (1970).
3. Adams, J. M., P. G. N. Jeppesen, F. Sanger, and B. G. Barrell, *Nature* **223,** 1009 (1969).

3a. Adams, J. M., and S. Cory, *Nature* **227,** 570 (1970).

4. Anderer, F. A., *Z. Naturforsch.* **14b,** 642 (1959).
5. Anderer, F. A., and H. d. Schlumberger, *Biochim. Biophys. Acta* **115,** 222 (1966).
6. Anderer, F. A., H. Uhlig, E. Weber, and G. Schramm, *Nature* **186,** 922 (1960).
7. Anderer, F. A., B. Wittmann-Liebold, and H. G. Wittmann, *Z. Naturforsch.* **20b,** 1203 (1965).
8. Andrewes, C. H., *Virology* **40,** 1070 (1970).
9. Aposhian, H. V., and A. Kornberg, *J. Biol. Chem.* **237,** 519 (1962).
10. Asbeck, F., K. Beyreuther, H. Kohler, G. von Wettstein, and G. Braunitzer, *Z. Physiol. Chem.* **350,** 1047 (1969).
11. August, J. T., *Nature* **222,** 121 (1969).
12. August, J. T., L. Eoyang, M. T. Franze de Fernandez, S. Hasegawa, C. H. Kuo, U. Rensing, and L. Shapiro, *Fed. Proc.* **29,** 1170 (1970).
13. Baltimore, D., A. S. Huang, and M. Stampfer, *Proc. Natl. Acad. Sci. U.S.* **66,** 572 (1970).
14. Baltimore, D., *Nature* **226,** 1209 (1970).
15. Baltimore, D., *J. Mol. Biol.* **32,** 359 (1968).
16. Bancroft, J. B., and E. Hiebert, *Virology* **32,** 354 (1967).
17. Bancroft, J. B., G. W. Wagner, and C. E. Bracker, *Virology* **36,** 146 (1968).
18. Bassel, B. A., Jr., *Proc. Natl. Acad. Sci. U.S.* **60,** 321 (1968).
19. Baylor, M. B., and P. F. Roslansky, *Virology* **40,** 251 (1970).
20. Ben-Porat, T., H. Shimono, and A. S. Kaplan, *Virology* **41,** 256 (1970).
21. Berns, K. I., and J. A. Rose, *J. Virol.* **5,** 693 (1970).
22. Bikel, I., and P. H. Duesberg, *J. Virol.* **4,** 388 (1969).
23. Billeter, M. A., J. E. Dahlberg, M. Goodman, J. Hindley, and C. Weissmann, *Nature* **224,** 1083, 1969.
24. Blair, C. D., and P. H. Duesberg, *Ann. Rev. Microbiol.*, 1970.
25. Blair, C. D., and W. S. Robinson, *J. Virol.* **5,** 639 (1970).
26. Bol, J. F., and H. Veldstra, *Virology* **37,** 74 (1969).
27. Borsa, J., and A. F. Graham, *Biochem. Biophys. Res. Commun.* **33,** 895 (1968).
28. Bové, J. M., C. Bové, and B. Mocquot, *Biochem. Biophys. Res. Commun.* **32,** 480 (1968)
29. Braunitzer, G., S. BRaig, F. Krug, and G. Hoborn, FEBS Lett., **7,** 83 (1970).
30. Braunitzer, G., F. Asbeck, K. Beyreuther, H. Köhler, and G. von Wettstein, *Z. Physiol. Chem.* **348,** 725 (1967).
31. Brown, D. M., M. J. E. Hewlins, and P. Schell, *J. Chem. Soc.* (C), p. 1925, 1968.
32. Bruening, G., and H. O. Agrawal, *Virology* **32,** 306 (1967).
33. Burgess, A. B., *Proc. Natl. Acad. Sci.* **64,** 613 (1969).
34. Burgess, R. R., and A. A. Travers, *Fed. Proc.* **29,** 1164 (1970).

35. Burness, A. T. H., *J. Gen. Virol.* **6,** 373 (1970).
36. Burness, A. T. H., and F. W. Clothier, *J. Gen. Virol.* **6,** 381 (1970).
37. Burness, A. T. H., and D. S. Walter, *Nature* **215,** 1350 (1967).
38. Burrell, C. J., E. M. Martin, and P. D. Cooper, *J. Gen. Virol.* **6,** 319 (1970).
39. Caliguiri, L. A., H.-D. Klenk, and P. W. Choppin, *Virology* **39,** 460 (1969).
40. Caro, L. G., and M. Schnos, *Proc. Natl. Acad. Sci. U.S.* **56,** 126 (1966).
41. Cartwright, B., P. Talbot, and F. Brown, *J. Gen. Virol.* **7,** 267 (1970).
42. Caspar, D. L. D., *Advan. Protein Chem.* **18,** 37 (1963).
43. Caspar, D. L. D., and A. Klug, *Cold Spring Harbor Symps. Quant. Biol.* **27,** 1 (1962).
44. Content, J., and P. H. Duesberg, *J. Virol.* **6,** 707 (1970).
45. Crawford, L. V., E. A. C. Follett, M. G. Burdon, and D. J. McGeoch, *J. Gen. Virol.* **4,** 37 (1969).
46. Compans, R. W., and P. W. Choppin, *Virology* **35,** 289 (1968).
47. Crowell, R. L., *J. Bacteriol.* **91,** 198 (1966).
48. Dales, S., *Prog. Med. Virol.* **7,** 1 (1965).
49. Dales S., and H. Silverberg *Virology* **37,** 475 (1969).
50. de Wachter, R., and W. Fiers, *nature* **221,** 233 (1969).
51. de Wachter, R., and W. Fiers, *J. Mol. Biol.* **30,** 507 (1967).
52. Diener, T. O., and W. B. Raymer, *Virology* **37,** 351 (1969).
53. Diener, T. O. and W. B. Raymer, *Science* **158,** 378 (1967).
54. Dorne, B., and L. Hirth, *C. R. Acad. Sci. (Paris)* **267,** 127 (1968).
55. Duesberg, P. H., and R. D. Cardiff, *Virology* **36,** 696 (1968).
56. Duesberg, P. H., G. S. Martin, and P. K. Vogt, *Virology* **41,** 631 (1970).
57. Duesberg, P. H., *J. Mol. Biol.* **42,** 485 (1969).
58. Duesberg, P. H., *Proc. Natl. Acad. Sci. U.S.* **59,** 930 (1968a).
59. Duesberg, P. H., *Proc. Natl. Acad. Sci. U.S.* **60,** 1511 (1968b).
60. Duesberg, P. H., and W. S. Robinson, *J. Mol. Biol.* **25,** 383 (1967).
61. Duesberg, P. H., H. L. Robinson, W. S. Robinson, R. J. Huebner, and H. C. Turner, *Virology* **36,** 73, 1968.
62. Echols, H., and A. Joyner, The Temperate Bacteriophage, in *Molecular Basis of Virology*, H. Fraenkel-conrat (ed.), Reinhold, New York, 1968, p. 526.
63. Farid, S. A. A., and L. M. Kozloff, *J. Virol.* **2,** 308 (1968).
64. Feix, G., R., Polett, and C. Weissmann, *Proc. Natl. Acad. Sci. U.S.* **59,** 145 (1968).
65. Finch, J. T., and A. J. Gibbs, *J. Gen. Virol.* **6,** 141 (1970).
66. Fine, R., and W. T. Murakami, *J. Mol. Biol.* **36,** 167 (1968).
67. Fraenkel-Conrat, H., *Ann. Rev. Microbiol.*, 1970.
68. Fraenkel-Conrat, H., in *The Chemistry and Biology of Proteins*, Academic Press, New York, 1970.
69. Fraenkel-Conrat, H., Chemical properties of the RNA of Small Viruses, in *Molecular Basis of Virology*, H. Fraenkel-Conrat (ed.), Reinhold, New York, 1968, p. 134.
70. Fraenkel-Conrat, H., The Masked —SH Group in Tobacco Mosaic Virus Protein, in *Symposium on Sulfur in Proteins*, R. Benesch (ed.), Academic Press, New York, 1959, p. 339.

71. Fraenkel-Conrat, H., *Virology* **4,** 1 (1957).

72. Fraenkel-Conrat, H., *J. Amer. Chem. Soc.* **78,** 882 (1956).

73. Fraenkel-Conrat, H., and M. Colloms, *Biochemistry* **6,** 2740 (1967).

74. Fraenkel-Conrat, H., and M. Sherwood, *Arch. Biochem. Biophys.* **120,** 471 (1967).

75. Fraenkel-Conrat, H., and B. Singer, *Virology* **23,** 354 (1964).

76. Fraenkel-Conrat, H., and B. Singer, *Biochim. Biophys. Acta* **33,** 359 (1959).

77. Fraenkel-Conrat, H., and B. Singer, *Biochim. Biophys. Acta* **24,** 540 (1957).

78. Fraenkel-Conrat, H., B. Singer, and A. Tsugita, *Virology* **14,** 54 (1961).

79. Fraenkel-Conrat, H., B. Singer, and R. C. Williams, *Biochim. Biophys. Acta* **25,** 87 (1957).

80. Fraenkel-Conrat, H., and R. C. Williams, *Proc. Natl. Acad. Sci. U.S.* **41,** 690 (1955).

81. Frankel, F. R., *J. Mol. Biol.* **18,** 109 (1966).

82. Frankel, F. R., *J. Mol. Biol.* **18,** 127 (1966).

83. Franski, R. I. B., *Virology* **34,** 694 (1968).

84. Franze de Fernandez, M. T., L. Eoyang, and J. T. August, *Nature* **219,** 588 (1968).

85. Franze de Fernandez, M. T., L. Eoyang, and J. T. August, *Nature* **219,** 675 (1968).

86. Funatsu, G., and H. Fraenkel-Conrat, *Biochemistry* **3,** 1356 (1964).

87. Funatsu, G., and M. Funatsu, *Phytopathol. Soc. Japan*, p. 1, 1968.

88. Funatsu, G., A. Tsugita, and H. Fraenkel-Conrat, *Arch. Biochem. Biophys.* **105,** 25 (1964).

89. Garwes, D., A. Sillero, and S. Ochoa, *Biochim. Biophys. Acta* **196,** 166 (1969).

90. Gierer, A., and K. W. Mundry, *Nature* **182,** 1457 (1958).

91. Gierer, A., and G. Schramm, *Z. Naturforsch.* **11b,** 138 (1956).

92. Girard, M., L. Marty, and F. Saurez, *Biochem. Biophys. Res. Commun.* **40,** 97 (1970).

93. Glitz, D. G., *Biochemistry* **7,** 927 (1968).

94. Glitz, D. G., A. Bradley, and H. Fraenkel-Conrat, *Biochim. Biophys. Acta* **161,** 1 (1968).

94a. Goodman, H. M., M. A. Billeter, J. Hindley, and C. Weissmann, *Proc. Natl. Acad. Sci. U.S.* **67,** 921 (1970).

95. Gordon, M. P., and M. Staehelin, *Biochim. Biophys. Acta* **36,** 351 (1959).

96. Goulian, M., A. Kornberg, and R. L., Sinsheimer, *Proc. Natl. Acad. Sci. U.S.* **58,** 2321 (1967).

97. Gupta, S. L., J. Chen, L. Schaefer, P. Lengyel, and S. M. Weissman, *Biochem. Biophys. Res. Commun.* **39,** 883 (1970).

98. Hariharasubramanian, V., and A. Siegel, *Virology* **37,** 203 (1969).

99. Hariharasubramanian, V., M. Zaitlin, and A. Siegel, *Virology* **40,** 579 (1970).

100. Harris, J. I., and J. Hindley, *J. Mol. Biol.* **13,** 894 (1965).

101, Harris, J. I., and C. A. Knight, *Nature* **170,** 613 (1952).

102. Haruna, I., and S. Spiegelman, *Science* **150,** 884 (1965).

103. Haslam, E. A., I. M. Cheyne, and D. O. White, *Virology* **39,** 118 (1969).

104. Heisenberg, M., *J. Mol. Biol.* **17,** 136 (1966).

105. Henry, T., and D. Pratt, *Proc. Natl. Acad. Sci. U.S.* **62,** 800 (1969).

106. Herrmann, R., D. Schubert, and U. Rudolph, *Biochem. Biophys. Res. Commun.* **30,** 576 (1968).

107. Hershey, A. D., E. Burgi, and L. Ingraham, *Proc. Natl. Acad. Sci. U.S.* **49,** 748 (1963).
108. Hershey, A. D., and M. Chase, *J. Gen. Physiol.* **36,** 39 (1952).
109. Hiebert, E., J. B. Bancroft, and C. E. Bracker, *Virology* **34,** 492 (1968).
110. Hindley, J., and D. H. Staples, *Nature* **224,** 964 (1969).
111. Hogness, D. S., and J. R. Simmons, *J. Mol. Biol.* **9,** 411 (1964).
112. Hogness, D. S., W. Doerfler, J. B. Egan, and L. W. Black, *Cold Spring Harbor Symp. Quant. Biol.* **31,** 129 (1966).
113. Hohn, T., *Eur. J. Biochem.* **8,** 552 (1969).
114. Hohn, T., *J. Mol. Biol.* **43,** 191 (1969).
115. Hohn, T., *Eur. J. Biochem.* **2,** 152 (1967).
116. Holland, J. J., *Virology* **15,** 312 (1961).
117. Holland, J. J., and E. D. Kiehn, *Science* **167,** 202 (1970).
118. Holoubek, V., *Virology* **18,** 401 (1962).
119. Horiuchi, K., H. F. Lodish, and N. D. Zinder, *Virology* **28,** 438 (1966).
120. Horwitz, M. S., M. D. Scharff, and J. V. Maizel, Jr., *Virology* **39,** 682 (1969).
121. Hull, R., *Virology* **40,** 34 (1970).
122. Hull, R., G. J. Hills, and R. Markham, *Virology* **37,** 416 (1969).
123. Hung, P. P., and L. R. Overby, *Biochemistry* **8,** 820 (1969).
124. Inouye, M., E. Akaboshi, M. Kuroda, and A. Tsugita, *J. Mol. Biol.* **50,** 71 (1970).
125. Inouye, M., and A. Tsugita, *J. Mol. Biol.* **37,** 213 (1968).
126, Jacobson, M. F., J. Asso, and D. Baltimore, *J. Mol. Biol.* **49,** 657 (1970).
127. Jacobson, M. F., and D. Baltimore, *Proc. Natl. Acad. Sci. U.S.* **61,** 77 (1968).
128. Jauregui-Adele, J., I. Hindenbach, and H. G. Witmann, *Z. Naturforsch.* **24b,** 870 (1969).
129. Jeng, Y. C., and M. Hayashi, *Virology* **40,** 406 (1970).
130. Jeppesen, P. G. N., J. L., Nichols, F. Sanger, and B. G. BArrell, *Cold Spring Harbor Symp. Quant. Biol.* **35,** 13 (1970).
131. Jeppesen, P. G. N., J. A. Argetsinger-Steitz, R. F. Gesteland, and P. F. Spahr, *Nature* **226,** 230 (1970).
132. Jockusch, H., *Virology* **35,** 94 (1968).
133. Jockusch, H., *Z. Vererbungsl.* **98,** 344 (1966).
134. Joklik, W. K., The Large DNA Animal Viruses: The Poxvirus and Herpesvirus Group, in *Molecular Basis of Virology*, H. Fraenkel-Conrat (ed.), Reinhold, New York, 1968, p. 576.
135. Joklik, W. K., *J. Mol. Biol.* **8,** 277 (1964).
136. Kado, C. I., and C. A. Knight, *Virology* **40,** 997 (1970).
137. Kado, C. I., and C. A. Knight, *J. Mol. Biol.* **36,** 15 (1968).
138, Kado, C. I., and C. A. Knight, *Proc. Natl. Acad. Sci. U.S.* **55,** 1276 (1966).
139. Kalmakoff, J., L. Lewandowski, and D. R. Black, *J. Virol.* **4,** 851 (1969).
140. Kamen, R., *Nature* **221,** 321 (1969).
141. Kang, C. Y., and L. Prevec, *J. Virol.* **3,** 404 (1969).
142. Kaper, J. M., and F. G. Jenifer, *Biochemistry* **6,** 440 (1967).
143. Kates, J., and J. Beeson, *J. Mol. Biol.* **50,** 1 (1970).
144. Kates, J., and B. R. McAuslan, *Proc. Natl. Acad. Sci. U.S.* **58,** 134 (1967).

145. Kellenberger, E., *Virology* **34,** 549 (1968).

146. Kellenberger, E., and C. Kellenberger-van-der Kamp, *FEBS Lett.* **8,** 140 (1970).

147. Keller, J. M., P. G. Spear, and B. Roizman, *Proc. Natl. Acad. Sci. U. S.* **65,** 865 (1970).

148. Kiger, J. A., E. T. Young, and R. L. Sinsheimer, *J. Mol. Biol.* **28,** 157 (1967).

149. Klug, A., and D. L. D. Caspar, *Advan. Virus Res.* **7,** 225 (1960).

150. Klug, A., W. Longley, and R. Leberman, *J. Mol. Biol.* **15,** 315 (1966).

150a. Konigsberg, W., T. Maitra, J. Katze, and K. Weber, *Nature* **227,** 271 (1970).

151. Konings, R. N. H., R. Ward, B. Francke, and P. H. Hofschneider, *Nature* **226,** 604 (1970).

152. Kozloff, L. M., Biochemistry of the T-even Bacteriophages of *Escherichia coli*, in *Molecular Basis of Virology*, H. Fraenkel-Conrat (ed.), Reinhold, New York, 1968, p. 535.

153. Kozloff, L. M., C. Verses, M. Lute, and L. K. Crosby, *J. Virol.* **5,** 740 (1970).

154. LaPorte, J., *J. Gen. Virol.* **4,** 631 (1969).

155. Laver, W. G., and E. D. Kilbourne, *Virology* **30,** 493 (1966).

155a. Laver, W. G., and R. C. Valentine, *Virology* **38,** 105 (1969).

156. Lee, J. C., H. L. Weith, and P. T. Gilham, *Biochemistry* **9,** 113 (1970).

157. Lehman, I. R., and E. A. Pratt, *J. Biol. Chem.* **235,** 3254 (1960).

158. Lesnaw, I. A., and M. E. Reichmann, *Proc. Natl. Acad. Sci. U.S.* **66,** 140 (1970).

159. Lesnaw, I. A., and M. E. Reichmann, *Virology* **39,** 729 (1969).

159a. Levin, D. H., G. Acs, and S. C. Silverstein, *Nature* **227,** 603 (1970).

160. Lewandowski, L. J., J. Kalmakoff, and Y. Tanada, *J. Virol.* **4,** 857 (1969).

161. Ling, C. M., P. P. Hung, and L. R. Overby, *Biochemistry* **8,** 4464 (1969).

162. Lister, R. M., *Virology* **28,** 350 (1966).

163. Lloyd, D. A., and S. Mandeles, *Biochemistry* **9,** 932 (1970).

164. Lodish, H. F., *J. Mol. Biol.* **50,** 689 (1970).

165. Lodish, H. F., and N. D. Zinder, *J. Mol. Biol.* **19,** 333 (1966).

166. Loh, P. C., and A. J. Shatkin, *J. Virol.* **2,** 1353 (1968).

167. MacHattie, L. A., D. A. Ritchie, and C. A. Thomas, Jr., *J. Mol. Biol.* **23,** 355 (1967).

168. Mackenzie, J. S., *J. Gen. Virol.* **6,** 63 (1970).

169. Maizel, J. V., Jr., and D. F. Summers, *Virology* **36,** 48 (1968).

170. Maizel, J. V., Jr., D. O. White, and M. D. Scharff, *Virology* **36,** 115 (1968).

171. Maizel, J. V., Jr., D. O. White, and M. D. Scharff, *Virology* **36,** 126 (1968).

172. Mandeles, S., *J. Biol. Chem.* **243,** 3671 (1968).

173. Mandeles, S., *J. Biol. Chem.* **242,** 3103 (1967).

174. May, J. T., J. M. Gilliland, and R. H. Symons, *Virology* **39,** 54 (1969).

175. Mayor, H. D., L. Jordan, and M. Ito, *J. Virol.* **4,** 191 (1969).

176. McGeoch, D. J., L. V. Crawford, and E. A. C. Follett, *J. Gen. Virol.* **6,** 33 (1970).

177. Michelin-Lausarot, F. P., C. Ambrosino, R. L. Steere, and M. E. Reichmann, *Virology* **41,** 160 (1970).

178. Miki, T., and C. A. Knight, *Virology* **36,** 168 (1968).

178a. Min Jou, W., and W. Fiers, *J. Mol. Biol.* **40,** 187 (1969).

179. Miura, K. I. Fuijii, T. Sakaki, M. Fuke, and S. Kawase, *J. Virol* **2,** 1211 (1968).

180. Miura, K., I. Kimura, and N. Suzuki, *Virology* **28,** 571 (1966).

181. Mizushima, S., and M. Nomura, *Nature* **226,** 1214 (1970).

182. Mundry, K. W., *Mol. Gen. Genet.* **105,** 361 (1969).

182a. Munyon, W., E. Paoletti, and J. T. Grace, Jr., *Proc. Natl. Acad. Sci. U.S.* **58,** 2280 (1967).

183. Nakai, T., and A. F. Howatson, *Virology* **35,** 268 (1968).

184. Nichols, J. L., *Nature* **225,** 147 (1970).

185. Noble, J., and L. Levintow, *Virology* **40,** 634 (1970).

186. Nozu, Y., and Y. Okada, *Virology* **40,** 1066 (1970).

187. O'Callaghan, D. G., T. W. Mak, and J. S. Colter, *Virology* **40,** 572 (1970).

188. Ohlbaum, A., F. Figueroa, C. Grado, and G. Contreras, *J. Gen. Virol.* **6,** 429 (1970).

189. Olivera, B. M., and I. R. Lehman, *Proc. Natl. Acad. Sci. U.S.* **57,** 1426 (1967).

190. Olshevsky, U., and Y. Becker, *Virology* **40,** 948 (1970).

191. Paranchych, W., P. M. Krahn, and R. D. Bradley, *Virology* **41,** 465 (1970).

192. Pettersson, U., and S. Hoglund, *Virology* **39,** 90 (1969).

193. Poljak, R. J., *Virology* **35,** 185 (1968).

194. Poljak, R. J., and A. J. Suruda, *Virology* **39,** 145 (1969).

195. Pons, M. W., and G. K. Hirst, *Virology* **34,** 385 (1968).

196. Pratt, D., H. Tzagoloff, and J. Beaudoin, *Virology* **39,** 42 (1969).

197. Pringle, C. R., *J. Virol.* **5,** 559 (1970).

198. Ptashne, M., *Proc. Natl. Acad. Sci. U.S.* **57,** 306 (1967).

199. Ralph, R. K., and S. Wojcik, *Virology* **37,** 276 (1969).

200. Rees, M. W., M. N. Short, and B. Kassanis, *Virology* **40,** 448 (1970).

201. Rentschler, L., *Mol. Gen. Genet.* **100,** 84, 96 (1967).

202. Ritchie, D. A., C. A. Thomas, Jr., L. A. MacHattie, and P. C. Wensink, *J. Mol. Biol.* **23,** 365 (1967).

203. Roberts, J. W., and J. E. Argetsinger Steitz, *Proc. Natl. Acad. Sci. U.S.* **58,** 1416 (1967).

204. Robertson, H., R. E. Webster, and N. D. Zinder, *Nature* **218,** 533 (1968).

205. Robinson, D. M., and F. M. Hetrick, *J. Gen. Virol.* **4,** 269 (1969).

206. Robinson, W. E., R. H. Frist, and P. Kaesberg, *Science* **166,** 1291 (1969).

207. Robinson, W. S., *Nature* **225,** 944 (1970).

208. Roblin, R., *J. Mol. Biol.* **31,** 51 (1968).

209. Rombauts, W. A., *Biochem. Biophys. Res. Commun.* **23,** 549 (1966).

210. Rose, J. A., K. I. Berns, M. D. Hoggan, and F. J. Koczot, *Proc. Natl. Acad. Sci. U.S.* **64,** 863 (1969).

211. Roslansky, P. F., and M. B. Baylor, *Virology* **40,** 260 (1970).

212. Rossomando, E. F., and N. D. Zinder, *J. Mol. Biol.* **36,** 387 (1968).

213. Rueckert, R. R., *Virology* **26,** 345 (1965).

214. Rueckert, R. R., and P. H. Duesberg, *J. Mol. Biol.* **17,** 490 (1966).

215. Sanger, H. L., *Mol. Gen. Genet.* **101,** 346 (1968).

216. Schlumberger, H. D., F. A. Anderer, and M. A. Koch, *Virology* **36,** 42 (1968).

217. Scholtissek, C., and R. Rott, *Virology* **40,** 989 (1970).

218. Schuster, H., and G. Schramm, *Z. Naturforsch.* **13b,** 697 (1958).

219. Schuster, H., and R. C. Wilhelm, *Biochim. Biophys. Acta* **68,** 554 (1963).

220. Schuster, H., and H. G. Witmann, *Virology* **19,** 421 (1963).

221. Schwartz, J. H., *J. Mol. Biol.* **30,** 309 (1967).

222. Scott, D. W., *Virology* **26,** 85 (1965).

223. Semal, J., and J. Kummert, *J. Gen. Virol.* **7,** 173 (1970).

224. Semancik, J. S., *Virology* **40,** 618 (1970).

225. Shapiro, R., R. E. Servis, and M. Welcher, *J. Amer. Chem. Soc.* **92,** 422 (1970).

226. Shapiro, L., M. T. Franze de Fernandez, and J. T. August, *Nature* **220,** 478 (1968).

227. Shatkin, A. J., and J. D. Sipe, *Proc. Natl. Acad. Sci. U.S.* **61** 1462 (1968).

228. Shatkin, A. J., J. D. Sipe, and P. C. Loh, *J. Virol.* **2,** 986 (1968).

229. Shaw, J. G., *Virology* **31,** 665 (1967).

230. Shepherd, R. J., G. E. Bruening, and R. J. Wakeman, *Virology* **41,** 339 (1970).

231. Shepherd, R. J., R. J. Wakeman, and R. R. Romanko, *Virology* **36,** 150 (1968).

232. Shimono, H., T. Ben-Porat, and A. S. Kaplan, *Virology* **37,** 49 (1969).

233. Shimura, Y., R. E. Moses, and D. Nathans, *J. Mol. Biol.* **28,** 95 (1967).

234. Siegel, A., M. Zaitlin, and O. P. Sehgal, *Proc. Natl. Acad. Sci. U.S.* **48,** 1845 (1962).

235. Signer, E., *Ann. Rev. Microbiol.* **22,** 451 (1968).

236. Singer, B., and H. Fraenkel-Conrat, *Prog. Nucleic Acid Res. Mol. Biol.* **9,** 1 (1969).

237. Singer, B., and H. Fraenkel-Conrat, *Virology* **14,** 59 (1961).

238. Singer, B., and H. Fraenkel-Conrat, *Virology* **39,** 395 (1969).

239. Singer, B., and H. Fraenkel-Conrat, *Biochemistry* **9,** 3694 (1970).

240. Singer, B., H. Fraenkel-Conrat, J. Greenberg, and A. M. Michelson, *Science* **160,** 1235 (1968).

241. Skehel, J. J., and W. K. Joklik, *Virology* **39,** 822 (1969).

242. Smith, R. E., H. J. Zweerink, and W. K. Joklik, *Virology* **39,** 791 (1969).

243. Sokol, F., H. D. Schlumberger, T. J. Wiktor, and H. Koprowski, *Virology* **38,** 651 (1969).

244. Steinschneider, A., and H. Fraenkel-Conrat, *Biochemistry* **5,** 2735 (1966).

245. Steitz, J. A., *Nature* **224,** 957 (1969).

246. Steitz, J. A., *J. Mol. Biol.* **33,** 923 (1968).

247. Steitz, J. A., *J. Mol. Biol.* **33,** 937 (1968).

248. Stollar, V., *Virology* **39,** 426 (1969).

249. Strauss, J. H., Jr., B. W. Burge, and J. E. Darnell, *J. Mol. Biol.* **47,** 437 (1970).

250. Strauss, J. H., Jr., B. W. Burge, and J. E. Darnell, *Virology* **37,** 367 (1969).

251. Strauss, J. H. Jr., B. W. Burge, E. R. Pfefferkorn, and J. E. Darnell, *Proc. Natl. Acad. Sci. U.S.* **59,** 533 (1968).

252. Streisinger, G., R. S. Edgar, and G. H. Denhardt, *Proc. Natl. Acad. Sci. U.S.* **51,** 775 (1964).

253. Stretton, A. O. W., S. Kaplan, and S. Brenner, *Cold Spring Harbor Symp. Quant. Biol.* **31,** 173 (1966).

253a. Stubbs, J. D., and P. Kaesberg, *Virology* **33,** 385 (1967).

254. Studier, F. W., and J. V. Maizel, *Virology* **39,** 575 (1969).

255. Sugiyama, T., *J. Mol. Biol.* **11,** 856 (1965).

256. Sugiyama, T., and H. Fraenkel-Conrat, *Biochemistry* **2,** 332 (1963).

257. Sugiyama, T., and H. Fraenkel-Conrat, *Proc. Natl. Acad. Sci. U.S.* **47,** 1393 (1961).

258. Sugiyama, T., R. R. Herbert, and K. A. Hartman, *J. Mol. Biol.* **25,** 455 (1967).

259. Sugiyama, T., and D. Nakada, *J. Mol. Biol.* **48,** 349 (1970).

260. Sugiyama, T., and D. Nakada, *J. Mol. Biol.* **31,** 431 (1968).

261. Sugiyama, T., and D. Nakada, *Proc. Natl. Acad. Sci. U.S.* **57,** 1744 (1967).

262. Summers, D. F., J. V. Maizel, Jr., and J. E. Darnell, Jr., *Proc.Natl. Acad. Sci. U.S.* **54,** 505 (1965).

263. Temin, H. M., *Virology* **23,** 486 (1964).

263a. Temin, H. M., and S. Mitzutani, *Nature* **226,** 1211 (1970).

264. Terzaghi, E., Y. Okada, G. Streisinger, J. Emrich, M. Inouyi, and A. Tsugita, *Proc. Natl. Acad. Sci. U.S.* **56,** 500 (1966).

265. Thirion, J. P., and P. Kaesberg, *J. Mol. Biol.* **47,** 193 (1970).

266. Thorne, H. V., E. Joyce, and D. Warden, *Nature* **219,** 728 (1968).

267. Tooze, J., and K. Weber, *J. Mol. Biol.* **28,** 311 (1967).

268. Trenkner, E., F. Bonhoeffer, and A. Gierer, *Biochem. Biophys. Res. Commun.* **28,** 932 (1967).

269. Tsugita, A., and H. Fraenkel-Conrat, *J. Mol. Biol.* **4,** 73 (1962).

270. Tsugita, A., and M. Inouye, *J. Mol. Biol.* **37,** 201 (1968).

271. Tsugita, A., D. T. Gish, J. Young, H. Fraenkel-Conrat, C. A. Knight, and W. M. Stanley, *Proc. Natl. Acad. Sci. U.S.* **46.** 1463 (1960).

272. Uyemoto, J. K., Grogan, R. G., *Virology* **39,** 79 (1969).

273. Van Kammen, A., *Virology* **34,** 312 (1968).

274. Van Kammen, A., and J. L. D. van Griensven, *virology* **41,** 274 (1970).

275. Van Ravenswaay Claasen, J. C., A. B. J. van Leeuwen, G. A. H. Duijts, and L. Bosch, *J. Mol. Biol.* **23,** 535 (1967).

276. van Vloten-Doting, L., A. Dingjan-Versteegh, and E. M. J. Jaspars, *Virology* **40,** 419 (1970).

277. Vinuela, E., I. C. Algranati, and S. Ochoa, *Eur. J. Biochem.* **1,** 3 (1967).

278. Wagner, G. W., and J. B. Bancroft, *Virology* **34,** 748 (1968).

278a. Weber, H., and C. Weissmann, *J. Mol. Biol.* **51,** 215 (1970).

279. Weber, K., and W. Konigsberg, *J. Biol. Chem.* **242,** 3563 (1967).

280. Webster, R. G., *Virology* **40,** 643 (1970).

281. Weiss, B., and C. C. Richardson, *J. Mol. Biol.* **23,** 405 (1967).

282. Weith, H. L., G. T. Asteriadis, and P. S. Gilham, *Science* **160,** 1459 (1968).

282a. Westaway, E. G. and B. M. Reedman, *J. Virol.,* **4,** 688 (1969).

283. Wild, T. F., J. N. Burroughs, and F. Brown, *J. Gen. Virol.* **4,** 313 (1969).

284. Wimmer, E., A. Y. Chang, J. M. Clark, Jr., and M. E. Reichmann, *J. Mol. Biol.* **38,** 59 (1968).

285. Wimmer, E., and M. E. Reichmann, *Nature* **221,** 1122 (1969).

286. Wimmer, E., and M. E. Reichmann, *Science* **160,** 1452 (1968).

287. Wittmann, H. G., *Z. Vererbungsl* **97,** 297 (1965).

288. Wittmann, H. G., I. Hindenbach, and B. Wittmann-Liebold, *Z. Naturforsch.* **24b,** 877 (1969).

289. Wittmann-Liebold, B., *Z. Naturforsch.* **21b,** 1249 (1966).

290. Wittmann-Liebold, B., and H. G. Wittmann, *Mol. Gen. Gent.* **100,** 358 (1967).

291. Wittmann-Liebold, B., and H. G. Wittmann, *Z. Vererbungsl.* **97,** 305 (1965).

292. Wittmann-Liebold, B., and H. G. Wittmann, *Z. Vererbungsl.* **94,** 427 (1963).

293. Wood, H. A., and G. Streissle, *Virology* **40,** 329 (1970).

294. Wood, W. B., R. S. Edgar, J. King, I. Lielausis, and M. Henninger, *Fed. Proc.* **27,** 1160 (1968).

295. Wu, R., *J. Mol. Biol.* **51,** 501 (1970).

296. Yamazaki, H., and P. Kaesberg, *J. Mol. Biol.* **7,** 759 (1963).

297. Young, E. T., and R. L. Sinsheimer, *J. Mol. Biol.* **30,** 147 (1967).

298. Young, J. D., E. Benjamini, and C. Y. Leung, *Biochemistry* **7,** 3113 (1968).

299. Young, J. D., E. Benjamini, M. Shimizu, C. Y. Leung, and B. F. Feingold, *Nature* **199,** 831 (1963).

300. Young, R., and J. Content, *Nature New Biol.* **230,** 140 (1971).

301. Young, R., and H. Fraenkel-Conrat, *Biochim. Biophys. Acta* (in press).

ADDITIONAL REFERENCES (1970–73) ADDED AT TIME OF FINAL UPDATING.

302. Astier-Manifacier, S., and P. Cornuet, *Biochim. Biophys. Acta* **232,** 484 (1971).

303. Atabekov, J. G., V. K. Novikov, V. K. Vishnichenko, and V. G. Javakhia, *Virology* **41,** 108 (1970).

304. Baltz, R. H., *J. Mol. Biol.* **62,** 425 (1971).

304a. Banerjee, A. K., R. Ward, and A. J. Shatkin, *Nature New Biol.* **232,** 114 (1970).

305. Bancroft, J. B., *J. Gen. Virol* **14,** 223 (1972).

306. Baxt, W., R. Hehlmann, and S. Spiegelman, *Nature New Biol.* **240,** 72 (1972).

307. Bernardi, A., and P.-F. Spahr, *Proc. Natl. Acad. Sci. U.S.* **69,** 3033 (1972).

308. Bol, J. E., L. van Vloten-Doting, and E. M. Jaspars, *Virology* **46,** 73 (1971).

309. Bolognesi, D. P., H. Bauer, H. Gelderblom, and G. Hüper, *Virology* **47,** 551 567 (1972).

310. Brezinski, D. P., and J. C. Wang, *Biochem. Biophys. Res. Commun.* **50,** 398 (1973).

311. Buchwald, M., H. Murialdo, and L. Siminovitch, *Virology* **42,** 390 (1970).

312. Butler, P. J. G., *J. Mol. Biol.* **52,** 589 (1970).

313. Butler, P. J. G., and A. Klug, *Nature New Biol.* **229,** 47 (1971).

313a. Butterworth, B. E., and R. R. Rueckert, *J. Virol.* **9,** 823 (1972).

314. Casjens, S., T. Hohn, and A. D. Kaiser, *Virology* **42,** 496 (1970).

315. Choppin, P. W., and R. W. Compans, *J. Virol.* **5,** 609 (1970).

316. Compans, R. W., H.-D. Klenk, L. A. Caliguiri, and P. W. Choppin, *Virology* **42,** 880 (1970).

317. Content, J., and P. H. Duesberg, *J. Virol.* **6,** 707 (1970).

318. Contreras, R., M. Ysebaert, W. Min Jou, and W. Fiers, *Nature New Biol.* **241,** 99 (1973).

319. Contreras, R., A. Vandenberghe, G. Volckaert, W. Min Jou, and W. Fiers, *FEBS Lett.* **24,** 339 (1972).

320. Cummings, D. J., A. R. Kusy, V. A. Chapman, S. S. DeLong, and K. R. Stone, *J. Virol.* **6,** 534 (1970).

321. Cummings, D. J., V. A. Chapman, S. S. DeLong, A. R. Kusy, and K. R. Stone, *J. Virol.* **6,** 545 (1970).

322. Cummings, D. J., *J. Virol.* **9,** 547 (1972).

322a. Damirdagh, T. S., and R. J. Shepherd, *Virology* **40,** 84 (1970).

323. Darby, G., and A. C. Minson, *J. Gen. Virol.* **14,** 199 (1972).

324. Davis, N., and R. R. Rueckert, *J. Virol.* **10,** 1010 (1972).

325. De Wachter, R., and W. Fiers, *Biochemistry* **49,** 184 (1972).

326. Dickson, R. C., S. L. Barnes, and F. A. Eiserling, *J. Mol. Biol.* **53,** 461 (1970).

327. Diener, T. O., and D. R. Smith, *Virology* **46,** 498 (1970).

328. Drzeniek, R., and P. Bilello, *Nature New Biol.* **240,** 118 (1972).

329. Duesberg, P. H., *J. Mol. Biol.* **42,** 485 (1969).

330. Duesberg, P. H., and P. K. Vogt, *Proc. Natl. Acad. Sci. U.S.* **67,** 1673 (1970).

331. Dunker, A. K., and R. R. Rueckert, *J. Mol. Biol.* **58,** 217 (1970).

332. Durham, A. C. H., and A. Klug, *Nature New Biol.* **229,** 42 (1971).

332a. Eggen, V. L., and A. J. Shatkin, *J. Virol.* **9,** 636 (1972).

333. Englund, P. T., *J. Mol. Biol.* **66,** 209 (1972).

334. Forrest, G. L., and D. J. Cummings, *J. Virol.* **8,** 41 (1971).

335. Fraenkel-Conrat, H., and E. Fowlks, *Biochemistry* **11,** 1733 (1972).

336. Fraenkel-Conrat, H., and B. Singer, *Biochim. Biophys. Acta* **262,** 264 (1972).

337. Francki, R. I. B., and J. W. Randles, *Virology* **47,** 220 (1972).

338. Garon, C. F., K. W. Berry, and J. A. Rose, *Proc. Natl. Acad. Sci. U.S.* **69,** 2391 (1972).

339. Glitz, D. G., and D. Eichler, *Biochim. Biophys. Acta* **238,** 224 (1971).

340. Grandgenett, D. P., G. F. Gerard, and M. Green, *J. Virol.* **10,** 1136 (1972).

341. Grandgenett, D. P., G. F. Gerard, and M. Green, *Proc. Natl. Acad. Sci. U.S.* **70,** 230 (1973).

342. Grippo, P., and C. C. Richardson, *J. Biol. Chem.* **246,** 6867 (1971).

343. Hadidi, A., and H. Fraenkel-Conrat, *Virology* (in press).

344. Hadidi, A., V. Hariharasubramanian, and H. Fraenkel-Conrat, *Intervirology* (in press).

345. Haenni, A. L., A. Prochiantz, O. Bernard, and F. Chapeville, *Nature New Biol.* **241,** 167 (1973).

345a. Hall, T. C., D. S. Shih, and P. Kaesberg, *Biochem. J.* **129,** 969 (1972).

346. Hariharasubramanian, H., A. Hadidi, B. Singer, and H. Fraenkel-Conrat, *Virology* (in press).

347. Henry, T. J., and C. C. Brinton, Jr., *Virology* **46,** 754 (1971).

348. Hindley, J., D. H. Staples, M. A. Billeter, and C. Weissmann, *Proc. Natl. Acad. Sci. U.S.* **67,** 1180 (1970).

349. Horst, J., H. Fraenkel-Conrat, and S. Mandeles, *Biochemistry* **10,** 4748 (1971).

350. Horiuchi, K., R. E. Webster, and S. Matsahashi, *Virology* **45,** 429 (1971).

351. Horwitz, M. S., J. V. Maizel, Jr., and M. D. Scharff, *J. Virol.* **6,** 569 (1970).

352. Hua, S., R. P. Mackal, B. Werninghaus, and E. A. Evans, Jr., *Virology* **46,** 192 (1971).

353. Huntley, G. H., and C. L. Kemp, *Virology* **46,** 298 (1971).

354. Jazwinski, S. M., R. Marco, and A. Kornberg, *Proc. Natl. Acad. Sci. U.S.* **70,** 205 (1973).

355. Jeppesen, P. G. N., B. G. Barrell, F. Sanger, and A. R. Coulson, *Biochem. J.* **128,** 993 (1972).

356. Kamen, R., *Nature* **228,** 527 (1970).

357. Kassanis, B., and M. Conti, *J. Gen. Virol.* **13,** 361 (1971).

358. Kelley, J. M., S. V. Emerson, and R. R. Wagner, *J. Virol.* **10,** 1231 (1972).

359. Klenk, H. D., R. Rott, and H. Becht, *Virology* **47,** 579 (1972).

360. Klein, W. H., C. Nolan, J. M. Lazar, and J. M. Clark, Jr., *Biochemistry* **11,** 2009 (1972).

361. Knudson, D. L., and R. MacLeod, *Virology* **47,** 285 (1972).

362. Koczot, F. J., B. J. Carter, C. F. Garon, and J. A. Rose, *Proc. Natl. Acad. Sci.* **70,** 215 (1973).

363. Kolakofski, D., M. A. Billeter, H. Weber, and C. Weissmann, article. *J. Mol. Biol.* **76,** 271 (1973).

364. Kondo, M., R. Gallerani, and C. Weissmann, *Nature* **228,** 525 (1970).

365. Kraal, B., J. M. de Graaf, T. A. Bakker, M. A. van Beynum, M. Goedhart, and L. Bosch, *Euro. J. Biochem.* **28,** 20 (1972).

366. Laemmli, U. K., *Nature* **227,** 680 (1970).

367. Lai, M. M. C., and P. Duesberg, *Virology* **50,** 359 (1972).

368. Lai, M. M. C., P. Duesberg, J. Horst and P. K. Vogt, *Proc. Natl. Acad. Sci.* **70,** 2266 (1973).

369. Lane, L. C., and P. Kaesberg, *Nature New Biol.* **232,** 40 (1971).

369a. Laver, W. G., *Virology* **45,** 275 (1971).

370. Lewandowski, L., and S. H. Leppla, *Virology* **10,** 965 (1972).

371. Lewandowski, L., and B. L. Traynor, *Virology* **10,** 1053 (1972).

371a. Litvak, S., D. S. Carre, and F. Chapeville, *FEBS Lett.* **11,** 316 (1970).

372. Litvak, S., A. Tarragó, L. Tarragó-Litvak, J. E. Allende, *Nature New Biol.* **241,** 88 (1972).

373. Lovelass, A., *Nature* **223,** 206 (1969).

374. Luftig, R. B., S. S. Kilham, A. J. Hay, H. J. Zwerrinck, and W. U. Joklik, *Virology* **48,** 170 (1972).

375. Lundquist, R. E., J. M. Laren, W. H. Klein, and J. M. Clark, *J. Biochem.* **11,** 2014 (1972).

376. Martin, G., and P. Duesberg, *Virology* **47,** 494 (1972).

377. Miller, R. L., and P. G. W. Plagemann, *J. Gen. Virol.* **17,** 349 (1972).

378. Millward, S., and A. F. Graham, *Proc. Natl. Acad. Sci. U.S.* **65,** 422 (1970).

379. Min Jou, W., G. Haegeman, M. Ysebaert, and W. Fiers, *Nature* **237,** 82 (1972).

380. Moore, B. J., and H. A. Scott, *Phytopathology* **61,** 831 (1971).

381. Moore, C. H., F. Farron, D. Bohnart, and C. Weissmann, *Nature New Biol.* **234,** 204 (1971).

382. Mountcastle, W. E., R. W. Compans, and P. H. Choppin, *J. Virol.* **7,** 47 (1971).

383. Mundry, K. W., and H. Priess, *Virology* **46,** 86 (1971).

383a. Nichols, J. L., and H. D. Robertson, *Biochim. Biophys. Acta* **228,** 676 (1971).

384. Nowinski, R. C., E. Fleissner, N. H. Sarkar, and T. Aoki, *J. Virol.* **9,** 359 (1972).

385. Ohno, T., Y. Nozu, and Y. Okada, *Virology* **44,** 510 (1971).

386. Ohno, T., H. Inoue, and Y. Okada, *Proc. Natl. Acad. Sci.* **69,** 3680 (1972).

387. Okada, Y., and T. Ohno, *Mol. Gen. Genet.* **114,** 205 (1972).

388. Padmanabhan, R., and R. Wu, *J. Mol. Biol.* **65,** 447 (1972).

389. Peden, K. W. C., J. T. May, and R. H. Symons, *Virology* **47,** 498 (1972).

390. Perham, R. N., *Biochem. J.* **131,** 119 (1973).

391. Peter, R., D. Stehelin, J. Reinbolt, D. Collot, and H. M. Duranton, *Virology* **49,** 615 (1972).

392. Pinck, M., P. Yot, F. Chapeville, and H. M. Duranton, *Nature* **226,** 954 (1970).

393. Reinbolt, J., R. Peter, D. Stehelen, D. Collot, and H. Duranton, *Biochim. Biophys. Acta* **207,** 532 (1970).

394. Rekosh, D. M., H. F. Lodish, and D. Baltimore, *J. Mol. Biol.* **54,** 327 (1970).

394a. Rekosh, D. J., *J. Virol.* **9,** 479 (1972).

395. Rice, R., and H. Fraenkel-Conrat, *Biochemistry* **12,** 181 (1973).

396. Richards, K. E., and R. C. Williams, *Virology* (in press).

397. Robertson, H. D., B. G. Barrell, H. L. Waith, and J. E. Donelson, *Nature New Biol.* **241,** 38 (1973).

398. Rodionova, N. P., N. E. Vesenina, O. B. Kichatova, and J. G. Atabekov, *Virology* **46,** 183 (1971).

399. Rosenberg, H., B. Diskin, L. Aron, and A. Traub, *Proc. Natl. Acad. Sci.* **69,** 3815 (1971).

400. Rueckert, R. R., in *Comparative Virology*, K. Maramorosch and E. Kurstark (eds.), Academic Press, New York, 1971.

401. Sakai, F., and T. Takebe, *Mol. Gen. Genet.* **118,** 93 (1972).

402. Sarngadharan, M. G., P. S. Sarin, M. S. Reitz, and R. C. Gallo, *Nature New Biol.* **240,** 67 (1972).

402a. Scheid, A., and P. W. Choppin, *J. Virol.* **11,** 263 (1973).

403. Schlesinger, M. J., S. Schlesinger, and B. W. Burge, *Virology* **47,** 539 (1972).

404. Schlesinger, S. J., and M. J. Schlesinger, *J. Virol.* **10,** 925 (1972).

405. Schubert, D., and H. Frank, *Virology* **43,** (1971).

406. Schulze, I. T., *Virology* **42,** 890 (1970).

407. Semancik, J. S., and L. G. Weathers, *Virology* **47,** 456 (1972); **49,** 622 (1972).

408. Shih, D.-S., L. C. Lane, and P. Kaesberg, *J. Mol. Biol.* **64,** 353 (1972).

408a. Shih, D.-S., and P. Kaesberg, *Proc. Natl. Acad. U.S.* (in press).

409. Seigel, A., *Virology* **46,** 50 (1971).

410. Seigert, W., R. N. K. Konigs, H. Baker, and P. H. Hofschneider, *Proc. Natl. Acad. Sci. U.S.* **69,** 888 (1972).

411. Silber, R., V. G. Malathi, L. H. Schulman, J. Hurwitz, and P. Duesberg, *Biochem Biophys. Res. Commun.* **50,** 467 (1973).

411a. Singer, B., *Virology* **46,** 247 (1971).

412. Smith, A. E., *Eur. J. Biochem.* **33,** 301 (1973).

412a. Snell, D. T., and R. E. Offord, *Biochem. J.* **127,** 167 (1972).

413. Spear, P. G., and B. Roizman, *J. Virol.* **9,** 143 (1971).

414. Staples, D. H., J. Hindley, M. A. Billeter, and C. Weissmann, *Nature New Biol.* **234,** 202 (1971).

415. Stavis, R. L., and J. T. August, *Ann. Rev. Biochem.* **39,** 527 (1970).

416. Stoltzfus, C. M., and R. Rueckert, *J. Virol* **10,** 347 (1972).

417. Studier, F. H., *Science* **176,** 367 (1972).

418. Thouvenel, J-C., H. Guilley, C. Stussi, and L. Hirth, *FEBS Lett.* **16,** 204 (1971).

419. Tung, J-S., and C. A. Knight, *Virology* **49,** 214 (1972).

420. Wagner, R. R., L. Prevec, F. Brown, D. K. Summers, F. Sokol, and R. MacLeod, *J. Virol.* **10,** 1228 (1972).

421. Weber, H., M. A. Billeter, S. Kahane, C. Weissmann, J. Hindley, and A. Porter, *Nature New Biol.* **237,** 166 (1972).

422. Weiner, A. M., and K. Weber, *Nature New Biol.* **234,** 206 (1971).

423. Weissmann, C., M. A. Billeter, H. M. Goodman, J. Hindley, and H. Weber, *Ann. Rev. Biochem.* **42,** 00 (1973).

424. Wimmer, E., *J. Mol. Biol.* **68,** 537 (1972).

426. Wu, R., and E. Taylor, *J. Mol. Biol.* **57,** 491 (1971).

427. Yogo, Y., and E. Wimmer, *Proc. Natl. Acad. Sci. U.S.* **69,** 1877 (1972).

428. Yot, P., M. Pinck, A.-L. Haenni, H. M. Duranton, and F. Chapeville, *Proc. Natl. Acad. Sci. U.S.* **67,** 1345 (1970).

429. Zaitlin, M., and V. Hariharasubramanium, *Virology* **47,** 296 (1972).

430. Zavada, J., *Nature New Biol.* **240,** 122 (1972).

431. Ziff, E. B., J. W. Sedat, and F. Galibert, *Nature New Biol.* **241,** 34 (1973).

CHAPTER

EIGHT

Molecular Aspects of Oncogenic Viruses

E. R. BROWN, R. A. ALBACH, and M. J. GERBER

1. INTRODUCTION

The infectious nature of cancer was first suggested in the seventeenth century, but it was not until the end of the nineteenth century, with the rise of modern bacteriology as developed by Pasteur and Koch, that methods became available whereby the infective theory of cancer could be empirically tested. Bacteria, protozoa, yeast, and spirochetes were alternatively proposed as etiological agents of neoplasia. It was in 1903 that Borrel (20), largely on the basis of his work on contagious avian epitheliosis, showed the presence of inclusions in neoplastic cells and thus claimed to have demonstrated the existence of the *virus cancéreux*. Borrel prophetically argued that the transplanted tumor (which was the common way to study tumor growth at that time) represented only a secondary stage of cancer, namely, malignant growth, whereas the actual solution to the cancer problem would be found by analysis of the viral processes transforming cells from the normal to the malignant state (21, 22). We now know this to be the core of the problem.

Three groups of viruses are involved in oncology: (*a*) the DNA naked viruses of various origins (murine, simian, human) oncogenic in nonnatural hosts, especially the hamster; (*b*) the herpes-type viruses associated, in animals and man, with different types of lymphoma and certain carcinomas; and (*c*) the animal viruses oncogenic in their natural hosts, commonly known as oncogenic RNA viruses or oncornaviruses.

A. The DNA Naked Viruses

These widespread viruses are generally latent but are fully expressed in their natural hosts, e.g., polyoma in mice, SV40 in monkeys, adenovirus in human beings. However, when experimentally inoculated into nonnatural hosts such as the hamster, they induce a tumor from which no infectious particle can be isolated under normal conditions. This phenomenon was first demonstrated by Stewart et al. (128) for the polyoma virus in 1957, by Girardi et al. for SV40 in 1962 (66), and by Trentin, Yake, and Taylor for adeno 12 virus in 1962 (139). Studies using this type of virus led virologists to establish an analogy between viral oncogenesis and lysogeny. The data supporting this current view will be reviewed later in the chapter.

B. The Herpesviruses

Lucke (99) suggested that the herpesvirus might be a latent virus potentially capable of producing neoplasia in frogs. He showed that the renal carcinoma of the frog was transmitted by a herpes-like virus. As time progressed, other animal tumors were also shown to be transmited by viruses resembling the herpesvirus. Among these diseases are chicken neurolymphomatosis (Marek's disease) (31), rabbit lymphoma (69), and simian lymphoma (82, 104). In diseases related to man, a herpes-like virus (EB virus) has been found to be associated with Burkitt's lymphoma (50, 51), nasophoryngeal carcinoma (36), Hodgkin's disease (52), and Kaposi's sarcoma (66). Recently there have been reports associating herpes simplex type II with carcinoma of the cervix (2).

C. The Oncornaviruses

Ellerman and Bang (47) provided evidence for the viral etiology of a neoplastic disease, erythromyeloblastosis in chickens. Unfortunately, they lost the virus strain and were unable to repeat their experiments. In 1911, Peyton Rous (115) was more fortunate in transmitting the first solid tumor, a chicken sarcoma, and his work with the virus was reproduced successfully in other laboratories. These results, restricted to the avian species, were followed by a long interval of time during which no other neoplasia could be detected that was transmitted by tumor-cell-free filtrate. Therefore the viral etiology at first appeared to be an isolated phenomenon, not at all applicable to mammalian and human disease. In the 1950s Joseph Beard (14) reported on the avian leukosis problem and noted that the disease was caused by a complex of viruses. This established the basis for further studies. These viruses appeared to be of a unique structure and morphology (RNA nucleic acid, presence of an envelope, size of 90 to 110 nm). They were characterized by their antigenic specificity, which was related to the envelope, and they could be grouped into four types, A, B, C, and D, according to the genetic strains of chicken they were able to infect. This was the first evidence of the relation between the envelope of the oncogenic viruses and the susceptible cell. The Rous sarcoma virus was shown by several workers to induce nonproductive transformation in mammalian cells, unless a chicken leukemia virus was present and coded for the envelope. This was the first notion of defective and helper virus, and also the first presumption that a virus could not be fully expressed as an entity even though some of its genetic information was still present in the host cell.

Subsequently, a second series of discoveries attracted even more attention because mammalian tumors were involved. In 1932 Shope (125) succeeded in isolating a cell-free extract with biological activity from a rabbit tumor. Bittner (19) reported that the mouse mammary carcinoma agent was filterable and could be transmitted through the milk of nursing mice that were highly tumor prone. This discovery was very important from the standpoint of viral etiology. As a matter of fact, it happened just at the time when inbred strains of mice with high and low incidences of neoplasia were developed, indicating the importance of genetic susceptibility. Bittner (19) demonstrated that only females transmitted the carcinoma. This type of transmission could be explained as a sex-linked factor, through the cytoplasm of the ovum or by contamination during intrauterine life. Subsequently, foster-nursing experiments fully demonstrated that the mammary carcinoma agent (Bittner agent) was in the milk. The C3H (high-incidence strain) animals removed from the mother by Caesarean section and nursed by C57 BL mice (low-incidence strain) did not present any tumor through five generations. The reverse experiments produced high incidence in the C57 BL.

The exciting reports in the 1950s concerning the discovery of the murine leukemia viruses threw a new and brighter light on the viruses implicated in cancer and provided strong arguments to support the viral oncogenesis theory. The first of the murine leukemia viruses was isolated in 1951 by Gross (72) from the Ak inbred strain of mice, which has a high incidence of leukemia. He demonstrated that the agent could infect CH3 newborn mice, an inbred strain with a low incidence of leukemia. The first passage of the agent did not demonstrate more than 30% of incidence even after a long period of incubation (8 to 10 months); Gross worked out his "passage A" in such a manner that he produced a more potent strain of virus, capable of producing a high (95%) incidence, in a shorter incubation time, 2 to 3 months.

After Gross's report, many other murine leukemia viruses were discovered, S63 and GC (23), as well as others, all from a variety of neoplastic tissues, by Graffi et al. (71), Schoolman et al. (120), Friend (58), Lieberman and Kaplan (97), and Moloney (105). A complex of leukemia and sarcoma viruses, comparable to the avian complex, has now been demonstrated. All of these viruses have been extensively studied, both individually and in relation to other murine and avian viruses. Later, feline, canine, and even bovine viruses (83, 98, 42, 53) were reported which produced leukemias and lymphosarcomas in their respective natural hosts. Progressively common characteristics were defined, permitting the constitution of a homogeneous group, the oncornaviruses. They all appear to have the same morphology (type A or C) and budding particles, as described by de Harven (35). All the mammalian viruses shared an antigenic specificity (interspecies antigen gs3 of the murine viruses) (62, 116).

The RNA viruses, for the reasons cited, deserve most of our interest, particularly in view of their peculiar enzymatic machinery. These RNA viruses will be reviewed in Section 3. The molecular biology of such viruses will be explained and discussed. Indeed, we believe that, even if viruses are not "the" cause of cancer but represent only one such cause, an understanding of their effects at the cellular and subcellular levels may shed some light on the general problem of oncogenesis.

2. DNA TUMOR VIRUSES

A. Interaction Between Molecules of Polyoma Virus and Host Cells

Interaction between molecules of polyoma virus and host cells may be depicted as "a breakdown of growth regulation of host cells caused by polyoma virus" (41). Two effects may be induced by polyoma virus in host cells, both *in vivo* and *in vitro*: lethal effect and/or transforming effect. The effect (lethal or transforming) on a particular host cell is closely related to, if not dependent on, the mammalian origin of the cell. In most cells, the virus induces a lytic or lethal effect; in hamster cells it causes a transforming effect.

The DNA of polyoma virus is a small molecule measuring 3×10^6 daltons (144, 145). It is circular and contains about 5000 nucleotide pairs. Because of its small size, it codes for a relatively few number of proteins. Therefore the genetic functions of the virus in infected cells are few. This limited number of genes offers the advantage of relative simplicity in identifying their functions in relation to growth regulations of the cell.

In Swiss mouse embryo cell culture infected by polyoma virus, small, dark pyknotic cells are visible from 4 to 7 days after virus inoculation on the surface of cells. The cells progressively die and fall off the glass in 2 or 3 weeks (46). Infectious virus is regularly produced and released into the medium in the lytic infection.

The virus is first absorbed to the cell surface and enters the cytoplasm, where it is probably uncoated. Six hours later viral DNA is detectable within the nucleus. By use of radioisotope labeling of RNA and DNA early in infection, small amounts of newly synthesized RNA are observed in the cell before DNA synthesis starts. A sharp increase in the amount of viral RNA synthesized appears during the later phase of virus development. Maturation of polyoma virus occurs in the cell nucleus (54). Antigen and enzymes necessary for DNA synthesis are produced early in the infection. Viral capsid protein synthesis appears in the late period of infection and is dependent on the occurrence of DNA synthesis, as shown by the prevention of capsid protein synthesis by 5-fluorodeoxyuridine, an inhibitor of DNA synthesis.

The transforming effect is studied mainly in hamster embryo cultures, since they readily reveal this phenomenon. The first obvious aberration in transformed cells is a loss of contact inhibition. Contact inhibition is a well-known physiologic phenomenon of normal cells, resulting in the arrest of growth when the population density reaches the stage of formation of complete monolayers. At that point, essentially all cellular DNA and protein synthesis stops.

By contrast, transformed cells continue to grow actively under crowded conditions. They pile up in several layers and have less adhesion to the substrate (loss of anchorage dependance) as a result of surface alteration. In addition, transformed cells show antigenic alterations that will be discussed later and other membrane alterations, such as increased agglutinability by vegetal lectins (concanavalin A, e.g.) and a higher rate of hexose uptake. These characteristics are inherited by the progeny cells, and successive indefinite transfers of the cells are possible as a consequence of this transformed character. Transformed cells do not contain and do not produce infectious virus, contrary to what occurs in lytic infection. It appears that the production of infectious polyoma virus is not compatible with the viability of the cell. In cells that do not allow the production of infectious virus, a specific event takes place, resulting in loss of control of cellular growth regulation.

The degrees of penetration of the virus into the cell in early stages appear comparable, whether the polyoma virus is infecting mouse cells and will induce a cytocidal effect, or is infecting hamster cells and will induce transforming effects.

The stimulation of a group of enzymes related to DNA synthesis, referred to as the "DNA complex" (41)—thymidine kinase, deoxcytidine deaminase, deoxycytidylic deaminase, deoxyadenylic deaminase, dihydrofolate reductase, thymidine synthetase, and DNA polymerase—is detected early in the stages of transformtion (56, 91, 92, 123, 124, 145).

Monolayer cultures of quiescent hamster cells show a low rate of DNA synthesis, as a result of contact inhibition. Upon addition of virus that will induce transformation, an active incorporation of labeled precursors into DNA begins after about 10 hr and leads to synthesis of DNA, which is about two thirds cellular and one third viral. Therefore viral infection abolishes growth inhibition caused by cell-to-cell contact and in some way inhibits the original regulatory mechanism. By using specific inhibitors of protein or RNA synthesis, some data whereby to approach an understanding of this induction have been obtained.

Puromycin reversibly blocks protein synthesis. As a consequence, DNA synthesis progressively declines, reaching a plateau within 3 to 4 hr. This two-hit curve suggests two processes that contribute to DNA synthesis in virus-infected cells: (*a*) initiation of the "DNA complex" in previously quiescent molecules, this step being inhibited by puromycin; and (*b*) completion of DNA, this step being resistant to puromycin.

Other data are collected from studies of the recovery of protein synthesis after the removal of this type of inhibition. If puromycin is removed when all enzymes of the DNA complex are considerably reduced, recovery of these enzymes is delayed for about 6 hrs. A similar lag exists for the resumption of the synthesis of DNA. By contrast, the recovery of total protein synthesis, as detected by incorporation of radioactive leucine into acid-insoluble proteins, is much faster; it occurs within 1 hr after the removal of puromycin, in contrast to an approximate 6 hr delay in the recovery of the enzymes of the DNA complex. Other inhibitors of protein synthesis reveal the same type of data; thus the delay in recovery of the DNA complex must be a consequence of the cessation of protein synthesis. Protein synthesis must resume and again reach a certain level to initiate the synthesis of the enzyme proteins of the DNA complex. The implication of such data is that some other protein, such as a regulatory protein, is required for the function of the genes of the DNA complex. Sufficient time is necessary for this protein to reach a value adequate to induce functioning of the genes of the DNA complex.

This hypothetical regulatory protein may be important in the further understanding of transformation. Is it specified by a viral or a cellular gene? If it is a virus-specified gene, the induction of the DNA complex depends directly on viral information. If it is a cellular gene, the regulatory protein should be found in normal, actively dividing, noninfected cells. To clarify the last point, the effects of puromycin on thymidine kinase (one of the enzymes of the DNA complex) activity were studied in uninfected cultures. As described previously for the DNA complex, there is a lag of about 6 hr before the resumption of thymidine kinase synthesis. Therefore regulatory protein does appear to exist in normal cells, and the virus may induce the DNA complex either by producing its own regulatory protein or by influencing the cellular regulatory gene.

The foregoing are the molecular events known to occur when the virus enters hamster cells and initiates the transforming effect. It is well known that an infective virus is not produced in these cells, and that the cells keep this transformed character and transmit it to their progeny. These data lead to some degree of understanding of neoplastic transformation. The function of virus molecules in this phenomenon will now be discussed and also will be extended to transformation by other DNA viruses in order to present a more extensive review.

B. Function of Virus Molecules in Neoplastic Transformation

In transformation by polyoma virus, two facts support the analogy with lysogenic conversion: (1) the inheritance of transformation by progeny cells, and (2) the disappearance of virus as an infectious particle.

In keeping with the comparison of transformation to lysogeny, the persistence and the integration of the viral genome should be demonstrable in the cell genome by some specific properties; inducing agents should be able to activate the viral genome to complete expression; and, finally, the phenomenon of transduction of cellular genetic material by activated virions should be detectable. The first experiment of induction performed on polyoma-virus-transformed cells was unable to activate any viral infectivity (142). For a short time this virus was called "hit and run" virus, since it did not leave any specific imprint on the genome of the cell. Sjögren et al. (126), however, demonstrated the presence of a new antigen on the surface of the transformed cell, called TSTA (tumor-specific transplantation antigen), which is specific for the viral strain inducing the transformation. The demonstration of this antigen is evidence for the presence of the viral genome. It is recognized that TSTAs are present in all transformed cells. Furthermore, such TSTAs (i.e., those induced by viruses) are virus-specific and are therefore related to the inducing viral genome.

Another antigen, called T-antigen (tumor antigen), is induced by the DNA oncogenic viruses. The T-antigen is in the same way specific for the transforming virus, but, unlike TSTA, it is not directly related to the transforming capacity of the virus. The T-antigen is demonstrated as an early product of the syntheses of the same viruses (polyoma, SV40, adenovirus) in lytic infection. Its presence in the nucleus of transformed cells is further evidence of the persistence of at least some viral genes.

More data have been provided by subsequent experiments. Production of infectious virus from SV40 tumor cells is induced by mixing hamster tumor cells with sensitive cells, such as African green monkey kidney (64). Inactivated Sendai virus is used to produce heterokaryons between hamster transformed cells and sensitive cells. This experiment demonstrates the presence of the complete genome of SV40 in the tumor cells. The mechanism of this induction is not yet well understood. Recent observations demonstrate that the synthesis of the infectious virions starts in the nucleus of the transformed cells (148). The infection, therefore, does not result from the passage of the genetic material from the transformed cell to the sensitive cell, but may be produced by activation through a factor diffused from the sensitive cell.

So far, it has not been demonstrated that adenovirus- and polyoma-transformed cells can produce infectious particles after experimental manipulation. Therefore the question may be posed, Is the viral genome fully represented in the transformed cell? Adenoviral RNA messenger can be isolated in polysomes of hamster adenovirus-induced tumor cells as complementary evidence of the presence of the viral genome (60), but hybridization of viral DNA with transformed cell DNA is not conclusive.

Data independently reported by Benjamin (16) and by Basilico and Di Mayorca (13) tend to demonstrate that in polyoma virus transformation only part of the viral genome is involved. By using radiation or chemical agents, these investigators evaluated the target size of the virus responsible for the transformation of the hamster cell and for viral multiplication in mouse embryo cells. The target size of transformation is about 60% that of virus multiplication. The data suggest that only half of the polyoma genes are active in the transformation process, and the authors assume that these were the only genes persisting in the transformed cells.

However, there is no evidence in this case that the other viral genes are absent. The only direct conclusion is that they are not functioning. Additional observations recently derived from polyoma virus mutants shed some light on this problem. One class of polyoma mutants consists of the host-range mutants (17). Although generally polyoma virus induces lytic infection in mouse cells, it has also been possible to transform mouse cells by polyoma virus. Polyoma virus mutants have been isolated that lost the capability to produce lytic infection on normal mouse cells but recovered this characteristic on polyoma-transformed mouse cells. It can be assumed that the mutant, lacking the genetic information for cytolysis, was "complemented" by the viral genome present in transformed cells. Consequently, albeit not expressed, the viral genes responsible for virion production ought to persist in nonproductive transformed mouse cells. Other types of mutants, temperature-sensitive ones (44, 37, 57), have been isolated and classified as early and late mutants according to the period of viral replication in which the temperature-sensitive function is involved; *tsa* is an early mutant unable to initiate transformation at nonpermissive temperature. When *tsa*-transformed cells are shifted to nonpermissive temperature and then fused with permissive cells, full recovery of the virus is obtained (132). Therefore, when the gene that is initiating transformation is not functioning, all viral genes can be expressed. The implications of this observation are twofold: (1) in transformed cells, the complete viral genome is present since it can be completely recovered; (2) transformation and repression of lytic viral genes are correlated, since the abolishment of transformation permits recovery of the full virion. When the gene function required for initiation of transformation is inactivated, full expression of the virus may occur.

Other data reported by Rapp and his coworkers (112, 113) about the biologic properties of the PARA particle (particle aiding and aided by the replication of adenovirus) demonstrate that only one part of the genome is involved in the oncogenic properties of SV40, although the presence of complete viral genomes is demonstrated by rescuing experiments with permissive cells. This PARA particle is composed of an adenovirus capsid and the SV40 genome, which is defective since it is unable to code for the provision of its own capsid. However, this particle is able to induce tumors in hamsters and to transform cells in tissue culture; both types of tumor cells exhibit SV40 T-antigens. Moreover, the PARA particle is able to induce the rejection of SV40-transformed cells in preimmunized hamsters. Thus in this case only one part of the SV40 genome is sufficient to induce transformation and the presence of new tumor antigens in the cells.

The DNA of the PARA particle E46 (88) was subjected to reannealing procedure with a nonhybrid adenovirus 7 to form a heteroduplex molecule in which one strand was provided by adeno 7 DNA and the other by $E46^+$ DNA. This technique showed that $E46^+$ DNA was 16% shorter than adenovirus 7 DNA. The 16% lacking was made up of the SV40 genome, estimated at 75% of the full SV40 genome. Since $E46^+$ shows the oncogenic properties of SV40, 75% of the SV40 genome appears to be sufficient to code for transformation.

Very recently another refined technique was applied to study the transcription of SV40 (147); separated strands were preparatively obtained, and the strand orientation as well as the extent of transcription, was demonstrated in acutely infected permissive monkey cells, in abortively infected nonpermissive mouse cells, and in virus-free transformed cells of various mammalian origins. In permissive and nonpermissive cells (89) the pattern of transcription is comparable; only 35 to 40% of the T minus strand is transcribed early in infection. This transcription persists at the late stage of infection associated with a transcription of 60 to 64% of the plus strand of SV40 DNA. In transformed cells, however, the pattern is quite different, since only the minus strand is transcribed, sometimes to a larger extent than in the permissive or nonpermissive cells (90). Thus there is strong evidence to indicate that all the viral genome is present in DNA virus-transformed cells but that only part of it is involved in transformation.

To return to the lysogeny analogy, it may be assumed that the part of the viral genome which is expressed and persists in transformed cells is integrated in cellular DNA. The DNA of the papova group is normally circular, but it seems unlikely that adenoviral DNA circularizes, in view of its large size, in order to be inserted in the cellular genome. Defendi et al. (34), Weiss, Ephrussi, and Scaletta (146), and other workers, using hybrid cells (constituted of normal mouse cells and SV40-transformed human cells), indicate that there is an integration of the viral genome within the DNA of the host cell. Their studies showed that, when the human chromosome is lost within the clone of hybrid cells, the T-antigen is no longer expressed. This strongly suggests that there is a correlation between the expression of the viral antigen and such a T-antigen loss. It is as if the viral genome became integrated in the human chromosome. One should note that it is not possible to exclude the fact that the viral DNA is integrated in the mitochondial DNA. As Weiss et al. (146) point out, the loss of the T-antigen may well be correlated to the loss of the human gene involved in the maintenance of human mitochondria.

That abortive transformation takes place in a high proportion of cells (130) and that cellular DNA synthesis is an important factor in the efficiency of transformation by SV40 (138) or by adenovirus (30) suggest that transformation could be a two-stage process. In the first step, the viral genome, although not integrated, could express its genetic information and would be integrated only in a second step with the help of cellular division. The low efficiency of cell transformation is consistent with this hypothesis. An argument for the integration of the viral genome is the loss of sensitivity of the T-antigen to interferon in transformed cells (108). Green advanced another very strong argument for the integration of viral DNA in adenovirus-transformed cells. Using the techniques of DNA–RNA hybridization, Green and his collaborators (140) demonstrated that the RNA transcribed from the DNA of transformed cells contains covalently linked viral and cellular sequences. The high efficiency of hybridization of this RNA with DNA from normal cells strongly suggests that highly reiterated cellular DNA sequences are adjacent to the viral DNA, and therefore that the viral DNA is integrated in the host–cell genome.

Recent experiments based on autoradiographic detection RNA–DNA hybrids in sections of tumors by Shope papilloma virus, a papova virus, strongly suggest a "provirus state for the viral genome" (107).

Dulbecco (41) suggests that the viral genome would act in two directions; (*a*) a permanent "DNA complex," which is necessary for DNA replication irrespective of regulatory influences, would be induced; and (*b*) another gene, or another function of the same gene, would cause transformation on the membrane of the transformed cells. It may be assumed from this that the cell surface may indeed control a possible regulatory gene. The insertion of virus-coded components can result in a modification of the cell membrane and, consequently, of the regulatory gene.

The study of two polyoma virus temperature-sensitive mutants provides experimental support of Dulbecco's prediction; they demonstrate that one viral gene function is required for early viral DNA synthesis and initiation of transformation [*tsa* mutant (43)]. Another gene function directs continuous cellular and viral synthesis in addition to membrane properties (45, 40). Cells transformed by *ts*3 mutant at permissive temperature, when transferred at nonpermissive temperature, demonstrate the loss of their membrane properties (absence of contact inhibition and anchorage dependence, increased agglutinability by vegetal lectins, and increased uptake of hexoses). Therefore the temperature-sensitive function of *ts*3 mutant appears to be required for maintenance of transformation and is related to membrane properties.

Thus solid experimental data strongly support the concept concerning the integration of the genome of a DNA oncogenic virus within the transformed cell. In complete parallel with this concept of lysogeny, a repressor protein should be available, as described in 1967 by Ptashne (111). The presence of a repressor substance has been demonstrated in hamster cell extracts from SV40-

induced tumor (28, 29). This substance is able to inhibit the multiplication of SV40 in permissive cells of African green monkeys. It was demonstrated that this repressor is also produced in permissive cells; however, another factor, which blocks the repressor, is also synthesized. Such data on SV40 support rather strongly the analogy with lysogeny, since the complete genome present in the tumor cells may be activated and since a repressor substance could be isolated.

Although the existence of this repressor has been contested (84), recent work by Suarez et al. (131) circumvents these criticisms. The authors hypothesize that the cells contain negative (repressor) and positive (anti repressor) factors regulating viral production. The negative factor is usually dominant; if not, virions can be produced either after infection or after transformation by a non-defective viral genome. This hypothesis is substantiated by previous data: (1) the repressor is detectable in SV40-transformed monkey and human cells; (2) SV40-transformed cells containing a nondefective virus genome may produce infectious virus upon treatment with extracts of normal monkeys or human cells but not with extracts of SV40-transformed monkey or human cells.

In polyoma transformation also, the existence of repressor is contested because hybrids of hamsters and mouse cells shown to contain full complements of hamster chromosomes were not transformed (12). Thus it appears that a repressor is not part of the hamster gene products. But the notion of antirepressor can explain this observation, since enough mouse information was present to oppose the possible repressor synthesized by hamster cell genome. The capability of mouse cells to undergo transformation induced by some polyoma viruses suggests, nonetheless, that a repressor dominant over antirepressor is not always necessary for transformation, unless some change occurred in the cells that negated the antirepressor effect.

In conclusion, some analogy exists between transformation by DNA virus and lysogeny. Ample evidence is at hand to support the assumption that malignant transformation subsequent to viral infection involves the function of some viral genes. Several studies support the hypothesis that these changes may take place by an indirect effect of viral function in control of cellular genetic expression. Extensive research activities in progress in several laboratories seek the answers to these newer aspects of the interaction between the virus and the host cell.

3. RNA TUMOR VIRUSES

In this section we restrict our discussion to a description of viral nucleic acid and its synthesis. We emphasize the recent findings on RNA-directed DNA polymerase. This discovery is described in terms of its implications for the transformation of the neoplastic cell on a theoretical basis, and from the viewpoint of its potential importance in tumor diagnosis. We review the actual theories of viral oncology before presenting our own study on nucleic acid metabolic systems within cells infected by a murine leukemia virus.

A. Structure of Oncogenic RNA Viruses

Oncogenic viral RNA is larger than the RNA extracted from other animal RNA viruses. Its molecular weight is 10^7 daltons, and its sedimentation constant is between 60S and 70S. Electron microscope studies disclosed that the RNA from Rauscher murine leukemia virus contains linear filaments with curled regions, suggesting that these filaments were originally coiled within the virus particle (85).

Exposures of Rous sarcoma virus (RSV) RNA to various denaturing treatments (dimethyl sulfoxide, urea, heating) change the viral RNA to a new structural form (38, 106). These units show sedimentation constants from 32S to 36S. A molecular weight of 2.5×10^6 daltons can be derived from a 34S value; therefore four such subunits could join in the formation of a whole RSV RNA molecule.

There has been disagreement regarding the significance of these subunits. They may result from fracture of the large viral RNA molecule at equidistant weak points in the nucleic acid, or from hydrogen bond dissociation, changing the viral RNA from a specific compact form to a more random, less compact state (3, 4). Workers are currently in agreement that these subunits probably correspond to replication units reassembled before virus maturation (102). It is likely, therefore, that these units code for different viral components and functions transcribing different polycistronic messenger RNAs. Two types of data support this hypothesis: (*a*) the determination that the target size of the transforming function of RSV is comparable to the target size of tobacco mosaic viral RNA infectivity, which in turn is the size of one subunit, suggesting that the transforming function of RSV virus could be coded by an RNA molecule about the size of one subunit (93, 94); and (*b*) the preparation of two deficient mutants by irradiation of the SR–RSV with an equal dose of gamma rays. Some mutants are no longer infective but are still oncogenic; the others are able to replicate and produce progeny, but are not transforming (67). One may assume that the infective capacity may also be carried by a single subunit.

The physiological function of such subunits is supported by two types of experiments. The first is the elucidation of the spontaneous origin of the defective particle. The technique of polyacrylamide gel electrophoresis made possible the distinction of two classes of subunits, *a* and *b*, in the RNA of avian tumor viruses (39). The transforming agents of this group of viruses, namely, the sarcoma viruses, possess both subunits, *a* and *b*, whereas the leukosis virus possesses only *b* subunit. The spontaneous loss of *a* subunit has been demonstrated by the repeated passage of the Schmitt–Ruppin strain of subgroup A on chick embryo fibroblast free of inducible endogenous avian leukosis virus (102). This finding suggests strongly that the RNA of oncornaviruses exists in subgenomic fragments. Second, there are indications that recombination may indeed occur among RNA viruses (87). This could be the explanation for the alpha form of Bryan high-titer RSV, found in media in which cells have undergone recombinational repair of genetic defects, and for the activation of endogenous leukosis virus when chicken cells are infected with replicating leukosis virus (75, 76). It was also found that the transformation marker of a helper-independent sarcoma virus and the host-range marker of a leukosis virus could be reassorted in a stable form (143).

B. RNA Synthesis

For a long while, little was known regarding the replication of oncogenic virus RNA. The use of specific inhibitors on virus production yielded the following evidence concerning the requirements for viral RNA replication:

1. Pretreatment of cells with mitomycin C, or small doses of ultraviolet light, or actinomycin D shortly after infection irreversibly inhibits viral growth, suggesting an absolute requirement for early cellular DNA transcription.
2. Actinomycin D and UV light inhibit virus production at later periods after infection, when RNA replication has already started. This irreversible effect suggests that another cell-dependent event is constantly required for the production of virions.
3. Specific inhibitors of DNA synthesis (iododeoxyuridine, bromodeoxyuridine, arabinoside—C) prevent virus production when used at the time of infection, suggesting that cellular DNA synthesis is required at an early stage.

The second theory of viral RNA replication considers fully the absolute requirement for DNA synthesis. The early requirement for DNA synthesis in the replication of oncogenic viral RNA led Temin (135) to postulate that a DNA intermediate was active in RNA replication and was transmitted as a "provirus" in multiplying transformed cells. Although a higher rate of hybridization of RSV RNA with DNA from transformed cells than with DNA from normal cells was observed, the difference was small, and there are reports of a degree of homology between RSV RNA and uninfected chick embryo (77). Two other studies confirm Temin's observation. One reports a higher rate of hybridization of the Rauscher virus with infected cells than with noninfected cells (49); another (8) describes the effect of 5-bromodeoxyuridine (BUDR) and light on the transforming ability of RSV. The selective incorporation of BUDR into DNA provirus during the resting phase of cells makes it susceptible only to the action of light when cell DNA synthesis is resumed. This selective susceptibility reduces by 50 to 90% the focus formation. These data, therefore, indirectly support the concept that a DNA provirus is formed and is dependent on cell DNA synthesis in order to perform cellular transformation. The experimental conditions may be criticized, however, because stimulation of cell DNA synthesis by RSV infection has been reported (86, 95, 100). Therefore some cellular DNA may incorporate BUDR and be destroyed by the action of the light, which obviously would prevent transformation.

Unequivocable evidence in support of this hypothesis has been provided by independent investigators. Temin (136), using Rous sarcoma virus, demonstrated an RNA-dependent DNA polymerase in a cell-free DNA-synthesizing substrate containing all four deoxyribonucleotide triphosphates, one of which was labeled with tritium (thymidine triphosphate). Activity was recovered in acid-insoluble material that was DNase-sensitive but RNase-resistant. To observe full activity of the enzyme, the virions must be disrupted, indicating that the activity is probably present in the nucleoid of the virions. Activity was susceptible to RNase but not to DNase, indicating that intact RNA was necessary for incorporation of thymidine triphosphate into the polymer. The polymerase was not capable of polymerization of RNA, as shown by the incubation of the virions with the four ribonucleotides. Concentrated pellets of supernatants of uninfected cells did not show such polymerase activity. Baltimore (9) independently, and concurrently, presented results with the Rauscher murine leukemia virus which further substantiated Temin's original hypothesis that RNA is the template for polymerase activity; using sucrose density gradients, he also demonstrated that the enzyme is localized in the virions. Spiegelman et al. (127), using seven RNA tumor viruses, obtained the same results. Furthermore, they demonstrated the complementary nature of the DNA product to the virion RNA and DNA–RNA hybrid structure of an early intermediate.

These three independent studies are important in that they firmly establish the existence of RNA-directed DNA polymerase in the RNA oncogenic viruses. The existence of this enzyme strongly supports Temin's original hypothesis that the synthesis of a unique species of DNA is an early event in the replication cycle of the RNA tumor viruses and that this "provirus" is the DNA template for further viral synthesis. This explains the hitherto unexplained absolute requirement (for viral RNA synthesis) of DNA synthesis.

The finding of RNA-directed DNA synthesis has been interpreted by some investigators as a heresy against the central dogma of molecular biology (32). There really is no conflict, however, inasmuch as the rather rare event of

RNA-directed synthesis can easily be included in the Watson and Crick theory. In part, the rejection of the hypothesis was due to a lack of information concerning the mechanism of RNA oncogenic virus replication and possible interpretations of it. As new discoveries on the subject are rapidly being reported in the literature, means for a hypothetical explanation are available. Four additional enzymes or, more probably, four different functions of the same enzyme have been experimentally demonstrated in the virion: (*a*) DNA-directed DNA polymerase, (*b*) an endonuclease, (*c*) a ligase, and (*d*) a ribonuclease. The first permits the conversion of single-stranded DNA to the double-stranded type and amplifies the DNA duplex once it is formed. The endonuclease and ligase might be only part of the necessary equipment for DNA synthesis, but could also be involved in the integration of the "provirus" within the host cell genome. The endonuclease trims away a gap in the cell chromosome into which the viral DNA is inserted and then is sealed into the host DNA by the ligase. The ribonuclease digests the RNA strand of the hybrid.

Spiegelman et al. (127) have shown that oncogenic viruses contain DNA polymerase, whose activity is directed by single-stranded RNA, double-stranded DNA, and DNA–RNA hybrids. They demonstrated that the synthetic hybrids are superior to the natural template in stimulating polymerization. Similar observations were recently reported by Bader (6). Using inhibitor of protein synthesis, he established that the virion-associated DNA polymerase is responsible for the transcription of the viral genomes. He noted that no early proteins are required for the synthesis of the viral DNA. This intravirion enzymatic activity was sufficient to establish at least one strand of DNA complementary to the viral RNA. This newly synthesized DNA thus serves as a template for additional viral DNA, as shown by the logarithmic rise in resistance to iododeoxyuridine with increasing time. The DNA synthesis is completed within 6 to 7 hr after virus infection. Regulatory processes controlling cellular DNA synthesis and also the availability of DNA precursors may stop viral DNA synthesis.

Such observations led Bader to propose a model for the mechanism of virus reproduction (6). When the first DNA copy of the viral RNA has been established by the RNA-directed DNA polymerase, other DNA strands are synthesized, using the first DNA as template and the DNA-directed DNA polymerase; the latter is probably of viral origin, and even is possibly another functioning site of the previous enzyme, although a cell DNA polymerase cannot be excluded at the present time. It seems unlikely (and there is no reason to believe) that all the new DNAs become integrated into the host cell chromosomes. Therefore most of these DNAs ought to be extrachromosomal and involved in viral replication. They are probably transcribed directly into viral RNA and viral proteins, with progeny virions resulting as end products. Extrachromosomal DNA could be discarded from the cell or diluted and lost by segregation during division. This phenomenon affords an explanation for the revertants described by MacPherson (101). On the other hand, some of the viral DNAs become integrated (supporting data for this assumption are described in Section 3. C). This DNA is now regulated by conditions governing chromosomal transcription. Integration could occur when the first DNAs are synthesized or after one or several cell divisions have occurred, the extrachromosomal viral DNA still being available. This model is compatible with the available experimental results reported on reproduction and transformation by RNA tumor viruses.

However, certain criticisms have been raised concerning "reverse transcription" being the unique answer to the question concerning the replication of RNA oncogenic viruses. Among them is the fact that the requirement for cell mitosis is not fulfilled by the existence of the reverse transcriptase template. This requirement still appears to be absolute for the release of infectious virus, in Temin's view (81), although recently Bader clearly stated that "cellular division is unnecessary for successful completion of the reproductive cycle of RSV" (5). Minute amounts of DNA have been isolated within the Rous sarcoma virus virion (96). Although such DNA may be the product of reverse transcriptase, it could also be a template for the viral DNA polymerase which is no longer RNA dependent (15). The large amount of RNase needed to specifically inactivate the polymerase reaction by denaturation of the viral RNA template has also been questioned (15). Particularly, criticism is leveled in terms of the quantity of impurity (DNase) that could be brought into the reaction. The DNA product of the reverse transcriptase activity does not necessarily contain all the viral genome, and Baluda and Markham (11) take exception to this. In addition, studies on the primer requirement of the DNA polymerase show that this requirement is similar to that of other DNA polymerases which are DNA directed (10). So far, these enzymes are known as repair enzymes, that is, they will replicate any single-stranded template provided that the appropriate deoxyribonucleotides are available and a 3′—OH is in position to initiate synthesis.

This raises the question of the physiological functions of viral RNA-directed DNA polymerase. If this enzyme is designed to copy the viral RNA, how is the synthesis initiated? Perhaps the small amount of DNA isolated in the virion serves this purpose (96). However, no substantiation of this hypothesis is at hand. Alternatively, the RNA itself could initiate synthesis, resulting in the formation of a covalently linked RNA–DNA molecule. Recent observations (27) suggest that this is the case. The DNA products of the viral RNA-directed DNA polymerase, synthesized on a 60S to 70S viral RNA template, appeared to be covalently linked to one RNA molecule much smaller than the viral template RNA. In the same report, it is suggested that < 10S RNA subunits of the 60S to 70S viral RNA can act as natural primers for the DNA synthesis in the host cell.

A recent observation lends support to the physiological function of RNA-directed DNA polymerase in viral replication (74). Strong evidence was presented that noninfectious particles of RSV alpha are devoid of the enzyme, whereas in all infectious particles that have so far been examined the enzyme is present.

In some instances the use of the potent synthetic hybrid duplexes has led to some confusion concerning interpretation of the results. Reverse transcriptase activity has been found in a Visna virus (121) and a primate syncytial virus, neither of which has been shown to be oncogenic in normal mouse cells or normal human diploid cells (though the Visna virus was reported to transform mouse cells *in vitro*). The discovery of a reverse transcriptase in normal human cells (122) cast serious doubts on the diagnostic significance of the previously described enzyme in human leukemic cells (61). Goodman and Spiegelman (70), among other scientists, claim that similarity is not identity. Only RNA oncogenic viruses contain a polymerase using natural single-stranded RNA as template. Scolnick et al. (121) have demonstrated (chromatographically and serologically) that DNase polymerase activities are present in normal, transform, and cancer cells, with synthetic DNA–RNA hybrid templates. They feel that these represent cellular enzymes unrelated to the reverse transcriptase of the animal RNA oncogenic viruses. This would indicate the potential importance of the use of *viral* RNA-directed DNA polymerase as an important diagnostic tool.

Evidence in this regard is found in the study by Schlom, Spiegelman, and Moore (118), working with Parsi women in India. They found that milk obtained from these women, who present a high incidence of breast cancer, shows a high content of virus particles when examined under the electron microscope. To avoid nonspecific reaction, they demonstrated RNase-sensitive endogenic DNA polymerase activity. These investigators treated the purified virion with a nonionic detergent to liberate the nucleoid. This reaction of the reverse transcriptase with purified 60S to 70S template and its activity with oligohomopolymer templates different from those used by the normal cells provide reasonable criteria for the specificity of its detection (114).

An even more refined technique was later presented by Schlom, Spiegelman, and Moore (119). This assay method simultaneously detects reverse transcriptase and 60S to 70S RNA and thus increases the accuracy with which positive results can be interpreted. Under these conditions, the type B particle from human milk, similar to the Bittner agent, showed endogenous reverse transcriptase activity. This discovery of the B particle in human milk indicated that there might well be a virus etiology in human breast cancer, a phenomenon long suspected since Bittner's early work on the mouse mammary tumor virus. Recently, an RNase-sensitive endogenous DNA polymerase system has been shown in leukemic cells (117), reemphasizing the importance of this diagnostic test.

Moreover, several type C particles have been demonstrated in tissue cultures of human tumors. The first one (110) was contested on an immunological basis as being a possible murine contaminant. Another one was produced in rhabdomyosarcoma cells prenatally inoculated in kittens. Immunological studies eliminated contamination by feline virus (103). The third one was induced in a human sarcoma cell line after treatment by 5-iododeoxyuridine (129). Lastly, viral type C particles were detected in three papillary cancers of the human renal pelvis (48). The particles were observed in tissue culture derived from the tumor. These viruses are not typical of oncornaviruses in that (1) they appear to produce a cytopathic effect when inoculated in human fibroblasts; and (2) budding is not observed at maturation. A reverse transcriptase, however, was shown to be active on endogenous template. Further studies are necessary before these new strains can be recognized as human oncogenic viruses.

Eventually, when the function of viral transcriptase has been clearly defined, it may be possible to prepare inhibitors against it. These inhibitors could provide a valuable therapeutic tool. No doubt such treatment would be prophylactic, that is, used before the insertion of the viral DNA into the host cell genome. Certain drugs, such as streptovaricin or rifampicin, are currently being studied, but because of toxicity the dosage is limited.

It should be noted that in the period of 1970–1973 more knowledge was uncovered in the field of viral oncology then in the previous half century. This surge of scientific discovery was essentially due to the capability of researchers to work within the cell at the molecular and genetic levels.

C. Mechanisms of Transformation

An anology can be drawn between RNA tumor viruses and lysogeny. This is based on the discovery of the DNA copy of the RNA genome. It is now possible to envision the insertion of this DNA copy into the genome of the host cell, which is then transformed into a malignant cell. Several investigations based on transformed nonpermissive cells support this view. Transformed nonpermissive cells are, as in the case of RSV, mammalian cells transformed by the virus. Under normal conditions these cells do not yield infectious viral particles. It should be noted that chicken cells can be transformed by RSV; such cells can produce viral progeny. These transformed nonpermissive cells have been shown to contain the group-specific antigen common to all avian oncogenic viruses (79, 133). Such cells are susceptible to virus induction when cultivated with permissive cells (chicken cells) with (134) or without (141) inactivated Sendai virus. Such observations strongly imply that the viral genome persists in all transformed cells and is constantly required for maintaining transformation.

Recent studies concerning the biology of temperature-sensitive mutants have shed further light on the mechanism of transformation by RNA tumor viruses. Both Bader (7) and Biquard and Vigier (18) developed mutants capable of undergoing full cycles of reproduction at 41°C, a temperature at which no transformation is apparent. However, at 37°C, transformation occurs that is comparable to that induced by wild virus. If the mutant transformed cells are shifted from 37 to 41°C, they revert to a normal morphology. Therefore one presumes that maintenance of the transformed state depends on the constant expression of some specific viral gene(s). This gene (or these genes) is unnecessary for viral reproduction. Experiments dealing with inhibition of macromolecular synthesis have suggested that the temperature-sensitive molecule is a protein. Only inhibitors of protein synthesis suppressed the appearance of new foci after the permissive temperature shift. At 41°C, inhibitors of macromolecular synthesis have no effect on reversion to normal morphology.

The product of the transforming viral gene(s) appears to alter the cell surface, since the number of foci of transformed cells is increased when cellular adhesion is reduced by polyanions (18). This surface alteration is also demonstrated by the increase in uptake of hexose that follows infection (7, 73). This could be due to the uncovering or rearrangement of preexisting sites involved in the regulation of cell division (18, 26, 55). It is known that protease stimulates cellular growth. Such an enzyme can promote rearrangement on the cell membrane and be responsible for the phenomenon of transformation. An enzyme of this type could be the gene product of the transforming RSV virus.

D. Theories of Viral Oncogenesis

As we have described, the genome of the oncogenic virus is integrated into the host cell genome. This fact does not give any information per se concerning the origin and natural occurrence of viral oncogenesis. Two main theories have been offered in explanation.

One is the "oncogene" theory, proposed by Huebner and Todaro (80). In this hypothesis, the oncogene part of the oncornavirus genome exists as a DNA provirus and is included in the normal gene pool of all vertebrates. It is transmitted vertically by the usual mechanism of inheritance. Normally repressed, it may be derepressed by different carcinogenic influences, such as radiations, chemicals, DNA viruses, hormones, or simply aging.

In the other hypothesis, as proposed by Temin (137), there exists a hypothetical gene system, the "protovirus," which is a portion of the host cell DNA and is susceptible to oncogenic alteration. This theory is based on the potential of somatic cells to undergo genetic evolution. In germ cells, only the usual modes of informational transfer, that is, DNA to DNA and DNA to RNA, exist. In somatic cells, informational transfer from DNA to RNA to DNA can also occur. RNA, which passes from cell to cell, can effect an alteration of the normal cell genes through the action of a reverse transcriptase that exists in normal cells. This alteration of the host cell genome may occur by insertion of the donor cell protovirus via reverse transcriptase. The insertion may occur in close proximity to the host protovirus, resulting in true gene amplification. Alternatively, the donor protovirus is inserted at a site distant from that of the host protovirus, in which case the new site may result in a modification of the information transcribed by the newly inserted protovirus. Finally, a chemically or physically induced mutation of such a protovirus may occur before its insertion, resulting in the insertion of a completely different piece of genetic information.

The two theories agree that RNA tumor viruses are responsible for natural neoplasms. According to Huebner and Todaro, DNA viruses can modify the normally functioning repressor of the oncogene and thus allow the neoplastic expression of the gene. Temin suggests that DNA virus oncogenesis is probably an experimental situation, an observation supported by the very low efficiency of transformation. It is most probable that a large number of viral genomes in a cell cause and maintain transformation. This mechanism would account for herpes-type viruses being responsible for oncogenesis, although their alterations of the cell membrane may argue for a more direct effect.

The main difference between the two theories is that, in the oncogene theory, the mechanism finally responsible for transformation is epigenetic (repressor system and switch in regulatory system). On the other hand, Temin's theory is really a genetic hypothesis and combines elements of somatic mutation and viral theories. The Temin theory also has the advantage of integrating the protovirus hypothesis into a more general view. The protovirus might be seen as the origin of viruses, the protovirus acquiring complementary gene(s) that permit(s) more autonomy, essentially in the form of the capsid. The protovirus might be seen also as a mechanism of the variability in the genomes of somatic cells and could play a role in cell differentiation and immunological memory.

E. Molecular Properties of GC Murine Leukemia Virus

The GC strain of murine leukemia virus was isolated from a cell-free filtrate of reticuloendothelial organs of noninbred mice. The virus induced lymphoid leukemia in all newborn and adult mice injected intraperitoneally after a latent period of 6 to 8 weeks. The disease especially affects the lymph nodes, spleen, and liver. The GC and S63 agents are both RNA viruses (23–25), as determined by Padget's method (109).

Electron microscope studies of infected mice reveal the presence of a large number of extracellular type C particles in the thymus or in the megakaryocytes (Fig. 1). The virus is antigenically related to the Friend, Moloney, and Rauscher viruses but not to the Gross virus. The GC and S63 viruses have had to be closely related to each other. The group-specific antigens common to all heterologous antiserum virus particles cross-react at low serum dilutions. On the basis of cross reactions between anti-GC serum and Rauscher virus and between anti-Rauscher serum and GC virus, it appears that these two agents also share antigenic determinants exclusive of the group-specific antigen. The GC and S63 viruses are closely related to the Moloney virus, the Friend virus, and particularly the Rauscher virus, but they are not identical with any of these agents by immunologic criteria.

The S63 and GC viruses were each established from infected spleen in tissue culture by the method of Wright and LasFargues (149). Bioassays of adult ICR resulted in lymphosarcomas which developed at the site of injection in about 30% of animals receiving 0.1 ml of the cell-free supernatant fluid, or 10^6 infected cells. The viruses were also grown in tissue cultures on mouse embryo cells.

Since the replication of murine leukemia virus does not involve any direct morphologic alteration of the cells, the assay used to demonstrate the replication of the virus depends on detection of the group-specific antigen and on a biological test in newborn mice.

The presence of the group-specific antigen, an internal component shared by all known murine leukemia viruses, was demonstrated by us using the CoMul test, which is the standard complement fixation described by Hartley et al. (78), utilizing anti-Moloney sarcoma virus rat serum. An anti-GC virus serum was also prepared in Fisher rats, using the GC cell-free filtrate as antigen to demonstrate a specific antibody against the GC virus.

The biological test of the replication of the virus was performed by inoculation of the viral pellet into newborn mice. The viral pellet induced leukemia in about 80% of the animals injected, a lower efficiency than the animal cell-free filtrate. The tests must be done early enough in the transfer of tissue culture to avoid the appearance of possible latent viruses, such as might occur in mouse embryo cells serially transferred for a long time.

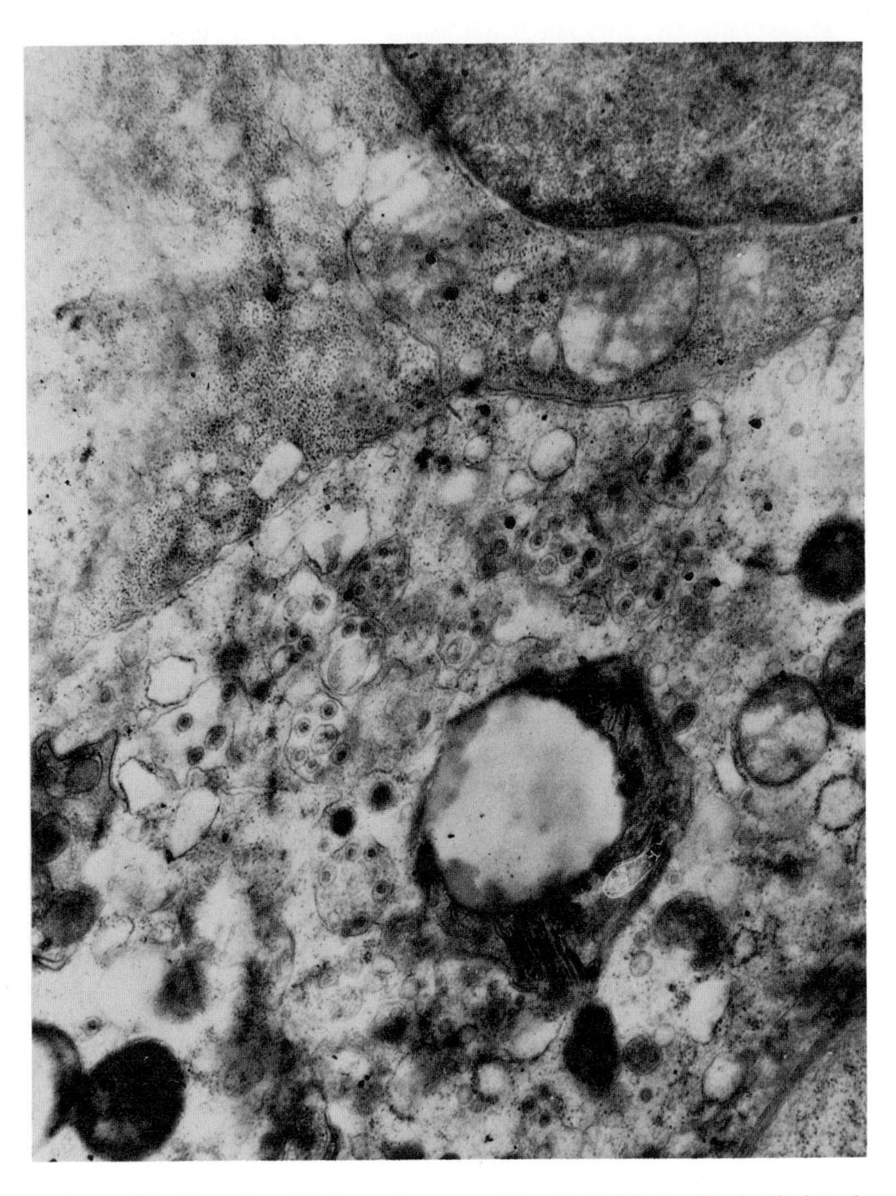

Fig. 1. Ultrathin section of the thymus from mouse infected with GC virus (fixed with glutaraldehyde; post fixed with OsO_4; stained with uranyl acetate followed by lead citrate; ×18,300). Note abundant C particles present in cytoplasmic vacuoles.

Three to five days after infection of mouse embryo tissue culture by the GC virus, considerably accelerated growth of the infected cells occurs. The production of foci of piled cells can be observed 4 to 5 days after infection (Fig. 2). The GC agent occasionally transforms hamster cells and induces a reticulum cell sarcoma in newborn hamsters, but is never expressed as full virion in hamster cells (63).

The experiments described below using "pulse labeling" (5 to 60 min) and "long-term labeling" (24 hr) experiments with uridine-5–H^3 ($U5H^3$) in mouse embryo cultures by autoradiography are subject to limitations in their interpretations. It is outside the scope of this chapter to delve into the details of these experiments and to describe such limitations. The reader is encouraged to review the papers listed in the bibliography by Albach, Gerber, and Brown (1). The review by Darnell (33) is especially pertinent to this subject, and additional information may be found in standard text books on autoradiography.

In experiments (1) on the labeling of RNA with radioactive precursors, an increase in the biosynthesis of all species of RNase-sensitive RNA ("rapidly labeled" and "long-term labeled") was found in mouse embryo tissue cultures infected with GC virus. This enhancement was observed with "rapidly labeled" RNA within 2 hr after virus infection. New synthesis of RNA viruses usually did not occur until at least 8 hr after infection, even in the best virus-replicating system. A stimulation of cellular nuclear RNA synthesis in infected culture is probably thus detected, since the infected cells multiply more rapidly than uninfected cultures. Because of certain problems inherent in the use of autoradiography, a degree of speculation based on experimental fact is unavoidable in the interpretation of results. The autoradiographic technique does not differentiate between DNA-dependent RNA synthesis and RNA-dependent RNA synthesis per se. One might argue that the increased amount of tritium detected in infected cultures may be a reflection of a change in the permeability of infected cells to the tritiated nucleoside and/or a change in the low-molecular-weight pool and therefore may not truly reflect an absolute increase in the amount of RNA in infected versus uninfected cultures. An independent method utilizing acridine-orange on the distribution of RNA and DNA in such cultures, however, did reveal much more extensive RNA in the infected than in uninfected cultures (Fig. 3).

It has been suggested that a very small amount of viral RNA nucleotides is made relative to cellular RNA nucleotides. Therefore, if cellular RNA synthesis occurs during viral RNA replication, viral RNA synthesis would not be detected by autoradiography. Under such circumstances, one must rely on biochemical extraction of all RNA from the cell. The sedimentation properties of such extracts are then used to determine whether or not a new viral species of RNA is present in the experimental, as compared to the control, cultures.

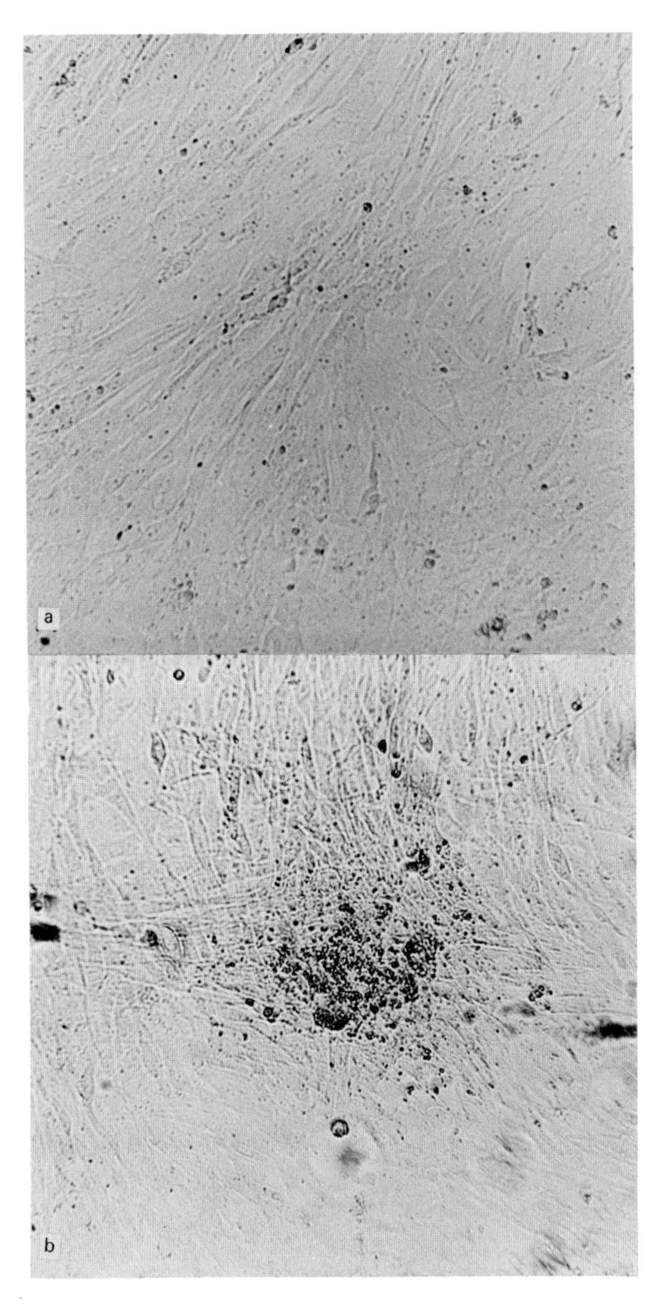

Fig. 2. Five day old mouse embryo tissue cultures; normal control showing monolayer (A) and GC virus infected culture showing foci of piled cells (B). (×14.2)

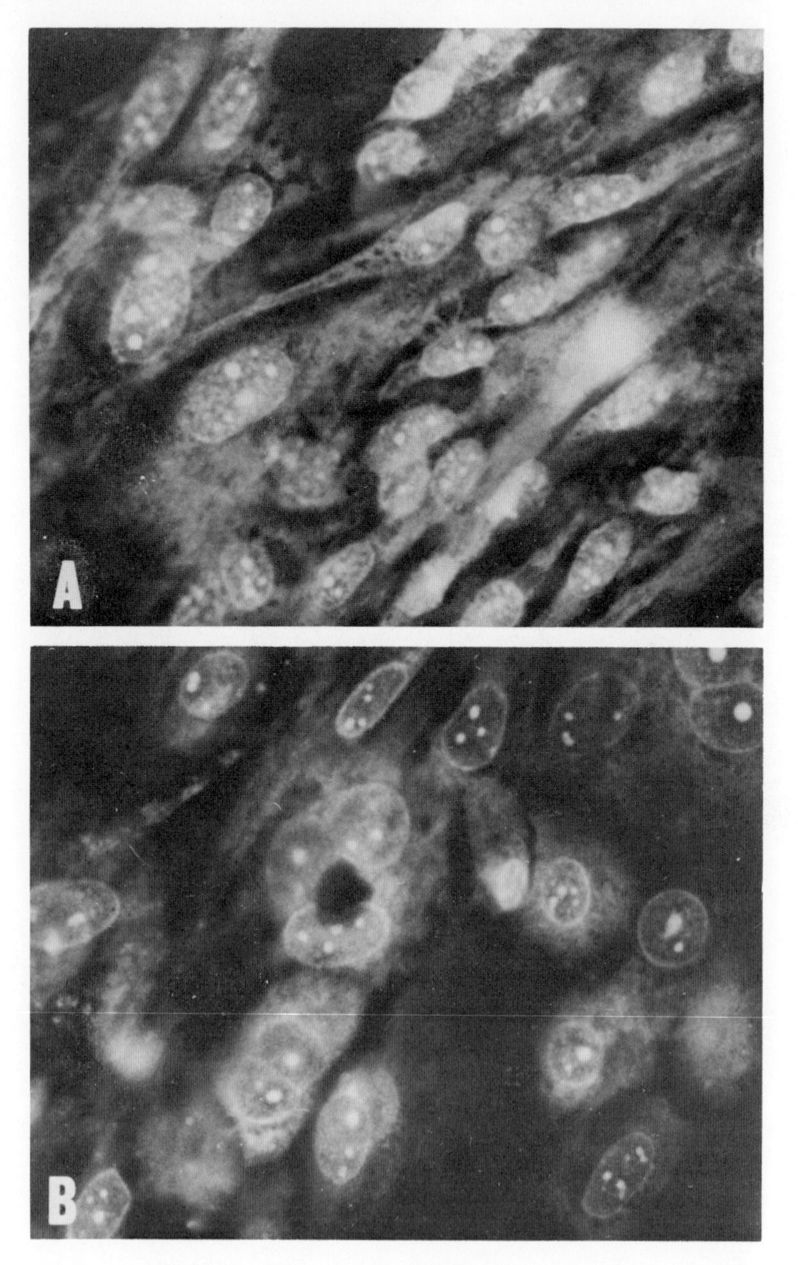

Fig. 3. Mouse embryo tissue cultures (ME) stained with acridine-orange (prepared by N. N. Sharma). A. ME infected with GC virus. B. ME control. Note the higher concentration of DNA and RNA in virus infected cells.

There are, however, exceptions. If viral RNA is synthesized in the cytoplasm with high efficiency in a system such as mouse embryo tissue culture, it may be detected by autoradiography. Since pulse labels (from 5 to 60 min) identify only nuclear RNA in mouse embryo tissue cultures, such a pulse label administered at the time of cytoplasmic viral RNA synthesis could be detected by autoradiography. Using the avian myeloblastic leukosis MC29 strain and the fluorescent antibody technique, Fritz, Langlois, and Beard (59) described the presence of a protein in the form of fluorescent cytoplasmic granules which appear as early as 2 hr after infection and which increase greatly during the first 24 hr. This protein appears to be specifically related to the cellular response to the injection of MC29. Our results showing the early increase in RNA synthesis appear to correlate with this finding, since the rise in RNA synthesis may reflect the synthesis of a new cellular protein, perhaps a regulatory protein involved in initiation of transformation.

4. CONCLUSIONS

The potentials for certain viruses to be intimately related to the oncogenic processes causing transformation have constituted a difficult transition point in the changing concepts of oncogenesis. Initially, the work of many investigators was either misunderstood or simply not accepted. As scientists began to uncover the molecular interactions going on within the cell, a composite picture emerged of the mechanisms by which a virus may incite a cell to be transformed into a neoplastic state. The application of the concept of a prophage-type virus as shown by Girardi et al. (66), and the discovery of murine leukemic viruses by Gross (72) and later of other viruses producing leukemia in higher animals by various workers, opened the way for the molecular biologist to explain the role of a virus as a direct agent capable of inducing transformation. Work on the polyoma virus indicated that such an agent may be involved in a lytic cycle or be maintained in a transforming state within the host cell. Transformed cells do not contain or produce infectious virus; rather a specific event resulting in a loss of cellular growth regulation takes place. The stimulation of a "DNA complex" of enzymes is detected early in the initiation of transformation. The role of the DNA complex is still subject to investigation and question, but strong evidence indicates that it plays a major role in the transformation process.

The transformation of a cell by polyoma virus is comparable to the phenomenon of "lysogeny" found among bacteriophage systems. Here we find the inheritance of transformation by progeny cells and the disappearance of the virus as an infectious particle as part of the transition to the tumor cell. This helps to explain why it is not possible to demonstrate the presence of the complete viral genome in polyoma- and adenovirus-transformed cells. It is now believed that the portion of the viral genome that is expressed and that persists in transformed cells is integrated in the cellular DNA. Many workers, including Ephrussi and his colleagues (34, 146), have supported the genetic concept to indicate that there is an integration of the viral genome within the DNA of the host cell. Studies on the role and action of RNA viruses in oncogenesis have furthered our general concept of this phenomena. Several workers have now demonstrated that RNA is the template for polymerase activity and that such an enzyme is localized in the virion. Thus much evidence can be marshaled to establish that the RNA-directed RNA polymerase is physiologically functioning in the transformation of RNA oncogenic virus systems. From the clinical point of view, work on milk obtained from Parsi Indians, as shown by Spiegelman et al., gives credence to the view that in a sound knowledge of the molecular mechanisms of the cell lies man's potential for diagnosing and possibly preventing and arresting the neoplastic diseases.

REFERENCES

1. Albach, R. A., M. J. Gerber, and E. R. Brown, *Can. J. Microbiol.* **16,** 1095 (1970).
2. Aurelian, L., I. Royston, and H. J. Davis, *J. Natl. Cancer Inst.* **45,** 455–464 (1970).
3. Bader, J. P., *Virology* **29,** 452 (1966).
4. Bader, J. P., and T. L. Teck, *J. Virol.* **4,** 454 (1969).
5. Bader, J. P., *Virology* **48,** 494–501 (1972).
6. Bader, J. P., *Virology* **48,** 485 (1972).
7. Bader, J. P., *J. Virol.* **10,** 276 (1972).
8. Balduzzi, P., and H. R. Morgan, *J. Virol.* **5,** 470 (1970).
9. Baltimore, D., *Nature* **226,** 1209 (1970).
10. Baltimore, D., and D. Smoler, *Proc. Natl. Acad. Sci. U.S.* **68,** 1507 (1971).
11. Baluda, M. A., and P. D. Markham, *Nature New Biol.* **231,** 90 (1971).
12. Basilico, C., M. Matsuya, and H. Green, *Virology* **41,** 295 (1970).
13. Basilico, C., and G. DiMayorca, *Proc. Natl. Acad. Sci. U.S.* **54,** 125 (1965).
14. Beard, J. W., *Proceedings of the Third National Cancer Conference*, Lippincott, Philadelphia, 1956.
15. Beaudreau, G. S., and J. C. R. Riman, *Acad. Sci. Paris,* **271,** 1728 (1970).
16. Benjamin, T. L., *Proc. Natl. Acad. Sci. U.S.* **54,** 121 (1965).
17. Benjamin, T. L., *Proc. Natl. Acad. Sci. U.S.* **67,** 394 (1970).
18. Biquard, J. M., and P. H. Vigier, *Virology* **47,** 444 (1972).
19. Bittner, J. J., *Science* **84,** 162 (1936).
20. Borrel, A., *Ann. Inst. Pasteur* **17, 81** (1903).
21. Borrel, A., *C.R. Sci. Biol.* **57,** 662 (1906).

22. Borrel, A., *Bull. Inst. Pasteur* **5,** 605 (1907).

23. Brown, E. R., P. Buinauskas, and S. O. Schwartz, *Can. J. Microbiol.* **13,** 957 (1967).

24. Brown, E. R., S. R. Rohlfing, and S. O. Schwartz, *Can. J. Microbiol.* **14,** 1347 (1968).

25. Brown, E. R., and J. G. Shaffer, *Med. Ecol. Clin. Res.* **2,** 8 (1969).

26. Burger, M. M., I. D. Noovan, J. R. Sheppard, J. O. Fox, and A. J. Levine, in *The Biology of Oncogenic Viruses* (2nd Le Petit Colloquim, 1970), North Holland, Amsterdam, p. 258.

27. Canaani, E., and P. Duesberg, *J. Virol.* **10,** 23 (1972).

28. Cassingena, R., P. Tournier, E. May, S. Estrade, and M. F. Bourali, *C.R. Acad. Sci.* **268,** 2834 (1969).

29. Cassingena, R., P. Tournier, S. Estrade, and M. F. Bourali, *C.R. Acad. Sci.* **269,** 261 (1969).

30. Casto, B. C., *J. Virol.* **2,** 376 (1968).

31. Churchill, A. E., and P. Biggs, *J. Gen. Virol.* **4,** 557 (1969).

32. Crick, F., *Nature* **227,** 561 (1970).

33. Darnell, J. E., Jr., *Bacteriol. Rev.* **32,** 262 (1968).

34. Defendi, V., B. Ephrussi, H. Koprowski, and N. C. Yoshida, *Proc. Natl. Acad. Sci.* **57,** 299 (1967).

35. De Harven, E., in *Experimental Leukemia*, M. Rich (ed.), Appleton-Century-Crofts, New York, 1968, p. 101.

36. De The, G., J. C. Ambrosioni, H. C. Ho, and H. C. Kwan, *Nature* **221,** 770 (1969).

37. DiMayorca, G., J. Ceillender, G. Marin, and R. Giordano, *Virology* **38,** 126 (1969).

38. Duesberg, P. H., *Proc. Natl. Acad. Sci. U.S.* **60,** 1511 (1968).

39. Duesberg, P. H., and P. K. Vogt, *Proc. Natl. Acad. Sci. U.S.* **67,** 1673 (1970).

40. Dulbecco, R., and W. Eckart, *Proc Natl. Acad. Sci. U.S.* **67,** 1775 (1971).

41. Dulbecco, R., *Perspect. Biol. Med.* **298** (Winter, 1966).

42. Dutcher, R. M., Proceedings of the 3rd International Symposium on Comparative Leukemia Research, Paris, 1967, *Bibl. Haematatol.* **31,** 116 (1968).

43. Eckart, W., *Ann. Rev. Biochem.* **41,** 503 (1972).

44. Eckart, W., *Virology* **38,** 120 (1969).

45. Eckart, W., R. Dulbecco, and M. M. Burger, *Proc. Natl. Acad. Sci.* **68,** 283 (1971).

46. Eddy, B. E., *Polyoma Virus*, Virology Monographs, Springer–Verlag, New York, 1969.

47. Ellerman, V., and O. Bang, *Centralbl. Bakteriol., Abt.* (I) **46,** 595 (1908).

48. Elliot, A. Y., E. E. Fialey, P. Cleveland, A. E. Castro, and M. Stein, *Science* **179,** 393 (1973).

49. Emanoil–Ravicovitch, R., M. F. Baudelaire, and M. Boiron, *C.R. Acad. Sci.* **269,** 1903 (1969).

50. Epstein, M. A., G. Henle, B. G. Achong, and Y. M. Ban, *J. Exp. Med.* **121,** 761 (1965).

51. Epstein, M. A., B. G. Achong, and Y. M. Ban, *Lancet* **1,** 702 (1964).

52. Evans, A. S. *J. Infect. Dis.* **126,** 330 (1971).

53. Ferrer, J. F., N. D. Stock, and Peck seen Lin. *J. Natl. Cancer Inst.* **47,** 613 (1971).

54. Fisher, H. W., H. Hatsumiya, and M. Azyma, *J. Bacteriol.* **91,** 1645 (1966).

55. Fox, T. O., J. R. Sheppard, and M. M. Burger, *Proc. Natl. Acad. Sci. U.S.* **67,** 244 (1971).
56. Frearson, P. M., S. Kit, and D. R. Dubbs, *Cancer Res.* **26,** 1653 (1966).
57. Fred, M., *Virology* **40,** 605 (1970).
58. Friend, C., *J. Exp. Med.* **105,** 307 (1957).
59. Fritz, R. B., A. J. Langlois, and J. W. Beard, *J. Virol.* **4,** 372 (1969).
60. Fuginaga, K., and M. Green, *Proc. Natl. Acad. Sci. U.S.* **55,** 1597 (1966).
61. Gallo, R. C., S. S. Young, and R. C. Ting, *Nature* **228,** 927 (1970).
62. Geering, G., L. J. Old, and E. A. Boyse, *J. Exp. Med.* **24,** 753 (1966).
63. Gerber, M. G., and E. R. Brown, *Cancer Res.* **32,** 2075 (1972).
64. Gerber, P., *Science* **145,** 833 (1964).
65. Giraldo, G., and E. Betv, *C. R. Acad. Sci.* **275,** 289 (1972).
66. Girardi, A. J., B. H. Sweet, B. Slotnick, and M. R. Hilleman, *Proc. Soc. Exp. Biol. Med.* **109,** 649 (1962).
67. Goldé, A., and R. Latarjet, *C.R. Acad. Sci.* **262,** 670 (1966).
68. Goodheart, C. *An Introduction to Virology*, Saunders, Philadelphia, 1969.
69. Goodheart, C., *J. Amer. Med. Assoc.* **211,** 91 (1970).
70. Goodman, N. C., and S. Spiegelmann, *Proc. Natl. Acad. Sci. U.S.* **68,** 2203 (1971).
71. Graffi, A., H. Bielka, F. Fey, F. Scharsach, and R. Weiss, *Wien. Med. Wehschr.* **105,** 61 (1955).
72. Gross, L., *Proc. Soc. Exp. Biol. Med.* **78,** 342 (1951), **94,** 767 (1957).
73. Habanaka, M., and G. Hanafusa, *Virology* **41,** 667 (1972).
74. Hanafusa, H., D. Baltimore, D. Smolder, K. F. Watson, A. Yaniv, and S. Spiegelman, *Science* **177,** 1188 (1972).
75. Hanafusa, H., T. Hanafusa, and T. Miyamoto, in *The Biology of Oncogenic Viruses*, L. Silvestri (ed.), North Holland, Amsterdam, 1971, p. 170.
76. Hanafusa, H., T. Hanafusa, and T. Miyamoto, *Proc. Natl. Acad. Sci. U.S.* **67,** 1797 (1970).
77. Harel, J., L. Harel, A. Goldé, and P. Vigier, *C. R. Acad. Sci.* **263,** 765 (1966).
78. Hartley, J. W., W. P. Rowe, W. J. Capps, and R. J. Huebner, *J. Virol.* **3,** 126 (1969).
79. Huebner, R. J., *Perspect. Virol.* **4,** 242 (1965).
80. Huebner, R. J., and G. C. Todaro, *Proc. Natl. Acad. Sci. U.S.* **64,** 1087 (1969).
81. Humphries, E. H., and H. M. Temin, *J. Virol.* **10,** 82 (1972).
82. Hunt, R. D., L. V. Melendez, N. W. King, C. E. Gilmore, M. D. Daniel, M. E. Williamson, and T. C. Jones, *J. Natl. Cancer Inst.* **44,** 447 (1970).
83. Jarrett, W. F. H., W. B. Martin, G. W. Crighton, R. G. Dalton, and M. F. Stewart, *Nature* **202,** 566 (1964).
84. Jerkofski, M. A., and F. Rapp, *Proc. Soc. Exp. Biol. Med.* **132,** 987 (1969).
85. Kakefuda, T., and J. P. Bader, *J. Virol.* **4,** 460 (1969).
86. Kara, J., *Biochem. Biophys. Res. Commun.* **32,** 817 (1968).
87. Kawai, S., and H. Hanafusa, *Virology* **49,** 37 (1972).
88. Kelly, T. J., and J. A. Rose, *Proc. Natl. Acad. Sci. U.S.* **68,** 1037 (1971).

89. Khoury, G., J. C. Byrne, and M. A. Martin, *Proc. Natl. Acad. Sci. U.S.* **69,** 1925 (1972).
90. Khoury, G., J. C. Byrne, K. K. Takemoto, and M. A. Martin, *J. Virol.* **11,** 54 (1973).
91. Kit, S., P. M. Frearson, and D. R. Dubbs, *Fed. Proc.* **24,** 596 (1965).
92. Kit, S., D. R. Dubbs, and P. M. Frearson, *Cancer Res.* **26,** 638 (1966).
93. Latarjet, R., and L. Chamaillard, *Bull. Cancer* **69,** 382 (1962).
94. Latarjet, R., and A. Goldé, *C.R. Acad. Sci.* **255,** 2866 (1962).
95. Lee, H. M., M. E. Kaighu, and J. D. Elbert, *Int. J. Cancer* **3,** 126 (1968).
96. Levinson, W., J. M. Bishop, N. Guinhill, and J. Jackson, *Nature* **27,** 1023 (1970).
97. Lieberman, M., and S. H. Kaplan, **130,** 387 (1959).
98. Lombard, L. S., J. B. Moloney, and C. Rickard, *Ann. N.Y. Acad. Sci.* **108,** 1086 (1963).
99. Lucke, B., *Amer. J. Cancer* **20,** 852 (1934).
100. Macieira-Coelho, H., I. J. Hiu, and E. Garcia-Giralt, *Nature* **222,** 1172 (1969).
101. MacPherson, I., *Science* **148,** 1731 (1965).
102. Martin, G. S., and P. Duesberg, *Virology* **47,** 494 (1972).
103. McAllister, R. M., M. Nicolson, M. B. Gardner, R. W. Rongey, S. Rasheed, P. S. Sarma, R. J. Huebner, M. Hatanaka, S. Oroszlan, R. V. Gilden, A. Kabigting, and L. Vernon, *Nature New Biol.* **235(53),** 3 (1972).
104. Melendez, L. V., M. D. Daniel, R. D. Hunt, C. E. O. Fraser, F. G. Garcia, N. W. King, and M. E. Williamson, *J. Natl. Cancer Inst.* **44,** 1175 (1970); **44,** 447 (1970).
105. Moloney, J. B., *Proc. Amer. Assoc. Cancer Res.* **3,** 44 (1959).
106. Montagnier, L., A. Goldé, and P. Vigier, *J. Gen. Virol.* **6,** 669 (1970).
107. Orth, G., Ph. Jeanteur, and O. Croissant, *Proc. Natl. Acad. Sci. U.S.* **68,** 1876 (1971).
108. Oxman, H. N., and P. H. Black, *Proc. Natl. Acad. Sci. U.S.* **55,** 1133 (1966).
109. Padgett, V., V. Kearns-Prestin, H. Voelz, and A. S. Levine, *J. Natl. Cancer Inst.* **36,** 465 (1966).
110. Priori, E. S., L. Dmochowsky, and B. Meyers, *Nature* **232,** 61 (1972).
111. Ptashne, M., *Proc. Natl. Acad. Sci. U.S.* **57,** 306 (1967).
112. Rapp. F., J. S. Butel, and J. L. Melnick, *Proc. Natl. Acad. Sci. U.S.* **54,** 717 (1965).
113. Rapp, F., *The Molecular Microbiology of Viruses* Cambridge University Press, 1968.
114. Robert, M. S., R. G. Smith, R. C. Gallo, P. S. Sarin, and J. W. Abrell, *Science* **176,** 798 (1972).
115. Rous, P., *J. Am. Med. Assoc.* **56,** 198 (1911).
116. Sarma, P. S., R. J. Huebner, H. C. Turner, R. V. Gilden, and T. Log, *Nature* **230,** 50 (1971).
117. Sarngadharan, M. G., P. S. Sarin, M. S. Reitz, and R. C. Gallo, *Nature New Biol.* **240,** 67 (1972).
118. Schlom, J., S. Spiegelman and D. Moore, *Nature* **231,** 97 (1971).
119. Schlom, J., S. Spiegelman, and D. H. Moore, *J. Natl. Cancer Inst.* **48,** 1197 (1972).
120. Schoolman, H. M., W. Spurrier, S. O. Schwartz, and P. B. Szanto, *Blood* **12,** 694 (1957).
121. Scolnick, E. M., E. Rands, S. A. Aaronson, and G. Todaro, *Proc. Natl. Acad. Sci. U.S.* **67,** 1789 (1970).
122. Scolnick, E. M., S. A. Aaronson, J. G. Todaro, and W. P. Parks, *Nature* **299,** 318 (1971).

123. Sheinin, R., (Abstract), *Fed. Proc.* **24,** 309 (1965).

124. Sheinin, R., *Virology* **28,** 47 (1966).

125. Shope, R. E., *J. Exp. Med.* **56,** 803 (1932); **58,** 607 (1933).

126. Sjögren, H. O., I. Hellstrom, and G. Klein, *Cancer Res.* **21,** 329 (1961).

127. Spiegelman, S., A. Burry, M. R. Das, J. Keydar, J. Schlom, M. Travnicek, and K. Watson, *Nature* **227,** 563 (1970).

128. Stewart, S. E., B. E. Eddy, A. M. Gochenour, N. G. Borgese, and G. E. Grubbs, *Virology* **3,** 380 (1957).

129. Stewart, S. E., G. Kasnic, C. Draycott, W. Feller, A. Golden, E. Mitchell, and T. Ben, *J. Natl. Cancer Inst.* **48,** 273 (1972).

130. Stocker, M., *Nature* **218,** 234 (1968).

131. Suarez, H. G., M. F. Bourali, R. Wicker, S. Estrade, and R. Cassingena, *Int. J. Cancer* **9,** 324 (1972).

132. Summers, J., and M. Vogt, in *Biology of Oncogenic Viruses*, L. Silurte (ed.), North Holland, Amsterdam, 1971, pp. 306–312.

133. Svoboda, J., in *Molecular Biology of Viruses* University Press, Cambridge, 1968, p. 249.

134. Svoboda, J., F. Hlozarek, and O. J. Machala, *Gen. Virol.* **3,** 461 (1968).

135. Temin, H. M., *Proc. Natl. Acad. Sci. U.S.* **52,** 323 (1966).

136. Temin, H. M., and S. Mizutani, *Nature* **226,** 1211 (1970).

137. Temin, H. M., *J. Natl. Cancer Inst.* **46,** III–VII (1971).

138. Todaro, G. J., and M. Green, *Proc. Natl. Acad. Sci. U.S.* **55,** 302 (1966).

139. Trentin, J. J., Y. Yake, and G. Taylor, *Proc. Amer. Assoc. Cancer Res.* **3,** 369 (1962).

140. Tsuei, D., K. Fujinaga, and M. Green, *Proc. Natl. Acad. Sci. U.S.* **69,** 427 (1972).

141. Vigier, P., *Int. J. Cancer* **9,** 150 (1972).

142. Vogt, M., and R. Dulbecco, *Virology* **16,** 41 (1962).

143. Vogt, P. K., *J. Natl. Cancer Inst.* **48,** 8 (1972).

144. Weil, R., and J. Vinograd, *Proc. Natl. Acad. Sci. U.S.* **50,** 730 (1963).

145. Weil, R., G. Petursson, J. Kara, and H. Diggelmna, *Molecular Biology of Viruses*, Academic Press, New York, 1967.

146. Weiss, M., B. Ephrussi, and I. Scaletta, *Proc. Natl. Acad. Sci. U.S.* **59,** 1132 (1968).

147. Westphal, N., *J. Mol. Biol.* **50,** 407 (1970).

148. Wever, G. H., S. Kit, and D. R. Dubbs, *J. Virol.* **5,** 578 (1970).

149. Wright, B. S., and J. C. LasFargues, *J. Natl. Cancer Inst.* **35,** 319 (1964).

CHAPTER

NINE

Molecular Rickettsiology

N. KORDOVÁ* and P. R. BURTON†

* Author of Sections 1, 3, 4, 5, and 6.
† Author of Section 2.

1. INTRODUCTION

According to present-day knowledge of molecular biology, the smallest free-living organisms approach sizes at which the structure of matter itself imposes limits on the complexity and function of the cells (1, 85). The size of an organism is inherent in the species (11, 44), and it has been suggested that retrograde evolution of microorganisms could proceed until only the molecules concerned with reproduction remained as the parasitic unit (21, 41, 75). The smallest free-living pleuropneumonia-like organism (*Mycoplasma*) has recently been considered in relation to the theoretical problem of the smallest possible self-replicating free-living cell (85). Interest in the possibility of synthesizing and reassembling living cells from isolated components greatly increases. Although the possibility exists of producing a cell in a test tube (52, 130), the ability of nature to produce organisms in ways other than by binary fission is generally disputed.

The class of Microtatobiotes, that is, the smallest living things, was considered in the past to encompass two orders: Rickettsiales and Virales (12). Viruses are now not considered to be microorganisms (78), and the nearest thing to a definition of virus would be "an infective agent that is not more convenient to classify as something else" (107). Rickettsiae have been stigmatized by evolution, that is, mutation, as small obligatory parasites and were thought in the past to be intermediate between bacteria and viruses (21, 106). At present, however, they are considered as true bacteria (14, 86, 96, 138, 140), and studies indicating a viruslike mode of replication in some members of the order Rickettsiales (108) have been met with scepticism (77, 78, 86). Yet it was not merely the size and obligate intracellular parasitism which tempted many microbiologists to think that rickettsiae are different from bacteria (21, 41, 75). Phenomena have been described in rickettsiae that are inconsistent with the known properties of bacteria; a few examples pertinent to the present discussion are spontaneous lysis and "regeneration" (140), the presence of "incomplete" or "inactive" rickettsiae in latent infections (110–112, 133), the disappearance of organisms during "adaptation" to new hosts (8), and the "reactivation" of hitherto invisible parasites after induced changes in the host (129). These descriptions were made in the days when rickettsioses were the domain of the observational scientist—the biologist.

In the last 15 years investigations on rickettsiae have reached the "fifth level" of microbiological studies (34). The biochemist liberated rickettsial cells from the host cell, studied their properties in the test tube, ruptured the particles, and examined cell-free systems. These studies revealed that rickettsiae contain chemical components characteristic of the cell wall of bacteria, synthesize their own enzymes, generate high-energy phosphate bonds with energy derived from oxidation of amino acids and carbohydrates, and possess both RNA and DNA. It is generally believed that rickettsiae multiply only by transverse binary fission. Why, despite their chemical complexity, rickettsiae are so dependent on a host cell remains a mystery, and the true nature of these pathogens is unknown. It was early recognized that transmission of rickettsiae to animal hosts is consistently biological, and that mechanical transmission has only an incidental or accidental role (104, 105). All attempts to grow rickettsiae outside living cells have failed, and the study of the parasite's multiplication cannot be dissociated from the study of parasite–host cell interactions. We may justifiably make the assumption, therefore, that molecular interactions underlie the parasite–host cell relationship.

Study of *Ricksettia prowazeki* (59, 69, 71, 72) and *Coxiella burneti* multiplication in the early stages of infection, at the cellular and subcellular levels, indicates a viruslike mode of replication. These studies have reached a point where an involement of molecular biologists in the study of intimate pathogen–host cell interactions is badly needed; in this chapter an attempt will be made to emphasize a few selected, unresolved problems of *C. burneti* and their host cell relationship. Recent comprehensive reviews of rickettsiae have covered the literature of the past years (14, 96, 139). The present chapter is concerned mainly with *C. burneti*—its composition, reproduction, fate in abortive infections, and interactions between variants of the parasite. An outline is presented of how *C. burneti* probably behaves in the host cells. Section 2 is concerned with the ultrastructure of *C. burneti* and the interaction with L cells, as observed recently by electron microscopy.

2. ULTRASTRUCTURE OF *C. burneti*

Most of the electron microscopic studies of *C. burneti* published thus far (4, 43, 83, 93, 121, 122, 135) are superficially descriptive and are incomplete in a number of respects, and no study has dealt with details of substructure at high resolution.

In our studies of the substructure of *C. burneti*, mouse L cells in culture provide the host cell system, and such problems as the entry of the rickettsia, a lysosomal response by host cytoplasm, and cytopathic changes are being studied in detail. Some of the results are presented here, in addition to detailed observations on the substructure of *C. burneti*.

A representative *C. burneti* cell is composed of (1) limiting surface layers, (2) a dense peripheral layer that is granular in appearance and contains the cell's ribosomes, and (3) a centrally located nucleoid mass of fibrous elements, the overall density of the mass being dependent on the packing of the filaments

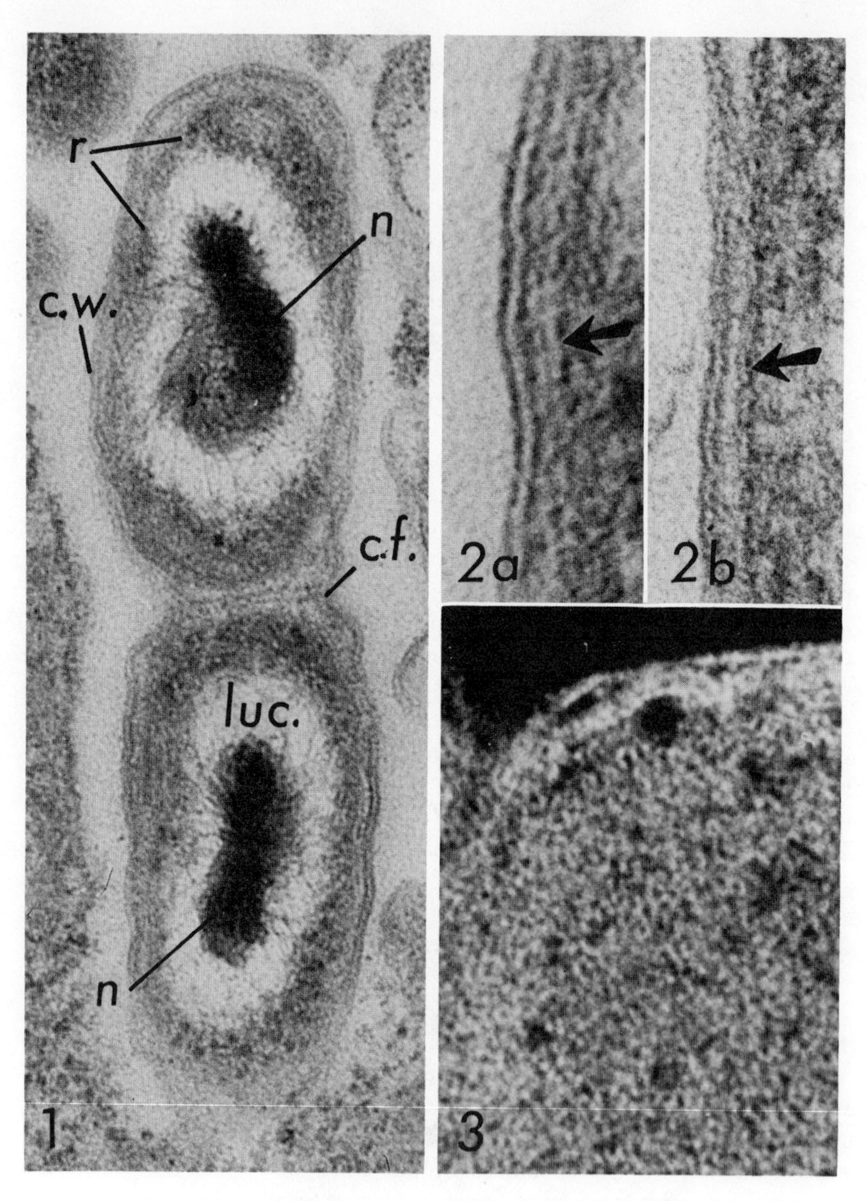

Fig. 1. *Coxiella burneti* undergoing binary fission. Abbreviations used: c.f., constriction furrow; c.w., cell wall; luc., lucent zone; n, nucleoid; r, ribosomal region. × 140,000.

Fig. 2*a*. Outer membrane of cell wall at high magnification (near focus). Note prominent trilaminar configuration; arrow indicates less prominent plasma membrane. × 450,000.

Fig. 2*b*. Surface of rickettsial cell, showing both outer membrane and plasma membrane. × 326,500.

Fig. 3. Surface of negatively stained *C. burneti*. Phosphotungstate; × 241,000.

(Fig. 1). Between the nucleoid mass and the peripheral ribosomal region there is usually a lucent zone of variable width.

During binary fission, the parent nucleoid mass appears to elongate in a deceptively direct manner into two daughter masses, with the formation of a constriction furrow. In some dividing cells, new cell walls can be seen along the midline internal to the peripheral furrow, but other cells may show deep furrows with no indication of precocious wall formation. Evidence obtained thus far indicates that, except for the division of one nucleoid mass into two masses, internal changes accompanying binary fission may be somewhat variable.

The arrangement of the surface layers of *C. burneti* is similar to that described for Gram-negative bacteria (37). There is a prominent outer membrane and an inner plasma membrane, both having a trilaminar "unit membrane" configuration (117, 118). Between the two membranes is a lucent region apparently devoid of formed material, comparable to the "dense intermediate layer" of Gram-negative bacteria; in accordance with conventional terminology (37), the lucent region and the outer membrane will be referred to collectively as the "cell wall" of *C. burneti*.

The outer membrane of the cell wall is generally more prominent than the plasma membrane, which is in close association with the peripheral ribosomal zone (Figs. 1, 2*a*, 2*b*, and 3), and the plasma membrane may be especially difficult to visualize after primary fixation in glutaraldehyde (Figs. 1 and 2*a*). Also, it has been noted that the plasma membrane may be more easily seen in damaged or swollen cells (22). The outermost dense layer of the outer trilaminar membrane is slightly the thickest of the three, with a mean thickness of about 35 Å. The plasma membrane has similar, if not identical, dimensions. Both outer and plasma membranes sometimes appear to be composed of repeating subunits, and the surfaces of many cells in negatively stained preparations also suggest the existence of subunits (Fig. 3). The diameters of such "subunits" in negatively stained material are in the 50 to 60 Å range, although the subunits are not precisely ordered as in certain Gram-negative (47) and Gram-positive (92) bacteria. In sectioned material, a series of alternating light and dark layers, stacked one upon the other, can sometimes be seen beneath all or only a portion of the plasma membrane. Although Nermut, Schramek, and Brezina (93) were not able to resolve a discrete plasma membrane in *C. burneti*, an outer "wall" and an inner plasma membrane, both trilaminar, have been described for *R. quintana* (50) and *R. prowazekii* (5).

Beneath the plasma membrane is a granular peripheral layer composed of ribosomes and smaller, dense (particulate?) elements (Figs. 1 and 5). The ribosomes are distributed in no discernible order, and they are generally separated from one another by a distance approximating their own dimensions. The small arrows at the lower left of Fig. 5 indicate a possible polysomal configuration, but isolation and gradient studies, combined with electron microscopy and negative staining, are needed to definitely establish such associations between ribosomes. Two subunits can often be seen to comprise the ribosome (curved arrows in Fig. 5); the larger probably corresponds to the classic 50S subunit, and the smaller member is probably the 30S subunit. The long axis of the ribosome has a mean length of about 170 Å. The small, dense elements found in the peripheral ribosomal region appear to vary in size, but most are from 50 to 60 Å in diameter. Although most of these appear to be particulate, some may be filaments seen in cross section. In Fig. 5, for example, two ribosomes appear to be associated with filamentous elements, as shown by the straight arrow. In both dimension and substructure, the bipartite ribosmes of *C. burneti* resemble those of *Escherichia coli* (19, 48). The previously described biochemical evidence of autonomous protein synthesis by *C. burneti* (81) predicts and confirms the presence of ribosomes in the organism.

Although cytochemical studies had indicated the presence of DNA in rickettsiae (50, 115), there was no information on its precise location within the cell until the utilization of electron microscopic autoradiography (23). The incorporation of ^{3}H–thymidine by the nucleoid region strongly suggests that the nucleoid filament is composed, wholly or partly, of DNA, with the former possibility the more likely one in view of its dimensions. In discussing the structure of the nucleoid region of *C. burneti*, the following assumptions are made: the nucleoid is the major repository of DNA in the cell, and the DNA in its unreplicated state forms a single continuous or circular filament, as is the case for so many microorganisms studied to date.

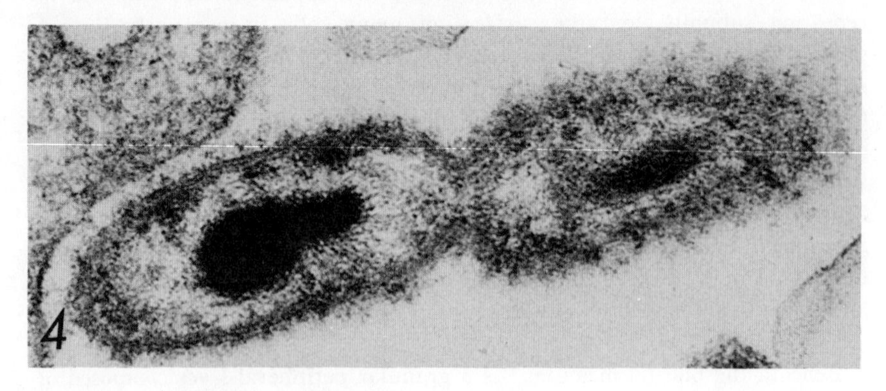

Fig. 4. An extraneous surface layer ("fuzz") of this *C. burneti* cell is associated with the surface of an L cell at left; note that the rickettsial particle is in a state of fission. × 97,000.

If the DNA of *C. burneti* is assumed to be in the form of a continuous filament, the major portion of the filament is usually tightly packed to form a prominent, electron-dense mass (Figs. 1, 4, and 7), although in some instances the filament may be sufficiently spread or "unraveled" to allow more detailed examination and measurement of portions of its length. Many measurements of nucleoid filaments from cells fixed in at least six different ways indicate a mean diameter of about 40 Å (ranging from about 20 to about 70 Å). In Fig. 5, for example, which represents a grazing section through a cell, the smallest portions of the filament measure about 20 Å in width. Considering that the mean diameter of the filament is about 40 Å, and that the diameter of a double-stranded DNA molecule is about 20 Å, it is suggested that the DNA of most cells is in the replicated state (i.e., with two DNA molecules). It is likely that the two strands are twined together to form an apparently single filamentous element, and the beaded appearance of many regions of the nucleoid filament probably reflects an underlying coiled configuration, with the "beads" representing gyres of the coil. In the region of the central nucleoid mass, which usually appears dense and compact, it is likely that the filament is supercoiled to provide for maximum packing in a restricted space. Stoker, Smith, and Fiset (135) indicated that the nucleoid mass of *C. burneti* might consist of irregularly twisted strands, although their micrographs provide little detailed information.

As yet, it has not been possible to directly correlate the state of packing of the nucleoid filament (or daughter filaments) with binary fission. Regardless of the compactness of the nucleoid, some portions of the filament always extend across the lucent zone into the peripheral ribosomal region (Figs. 1 and 5). Some of the small, dense profiles seen in the ribosomal region may be portions of the filament in cross section. Longitudinal profiles of the filament could easily be overlooked in the peripheral region because of its dense, particulate nature, although in Fig. 5 filamentous elements can be seen to extend into the ribosomal region.

The observations suggest that an intimate spatial relationship may exist between portions of the nucleoid filament and elements of the ribosomal region. In regard to a possible relation between the DNA filament and ribosomes, there is good evidence that, in cell-free extracts of *E. coli*, DNA can complex with ribosomes by means of RNA (7, 25). In exceedingly small cells, such as rickettsiae, where evolution has undoubtedly tended to select in favor of more efficient space utilization, there would be inherent adaptive advantage in synthetically functional DNA–ribosome complexes. Ormsbee (96), although presenting no micrographs to support his conclusions, claims to have noted in *C. burneti* a dense central body possessing radiating fibrils, connected to electron-dense spherical bodies in the cytoplasm. Ribonuclease was said to

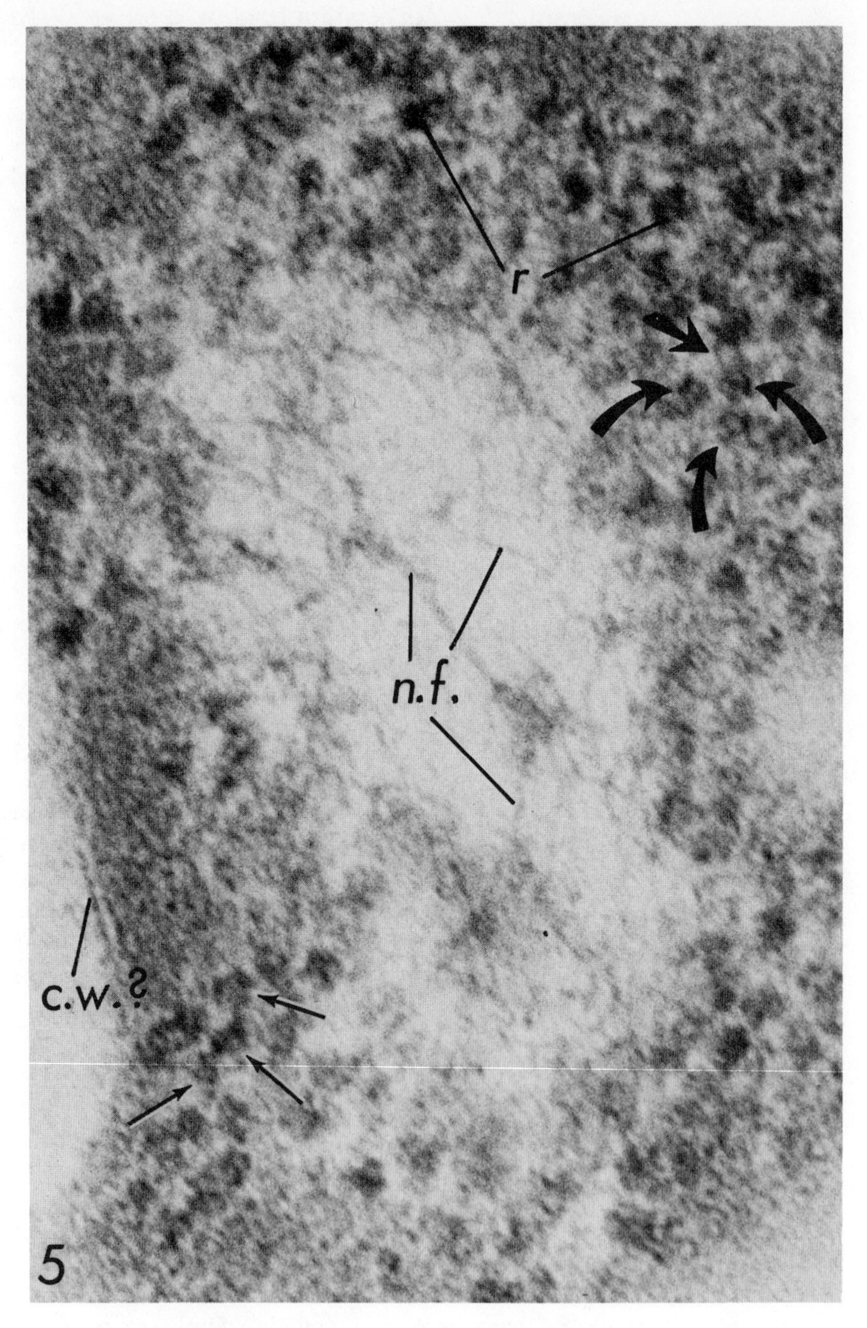

Fig. 5. Grazing section through *C. burneti* at high magnification (near focus, but slightly overfocused); note nucleoid filament (n.f.). See text for details. × 322,000.

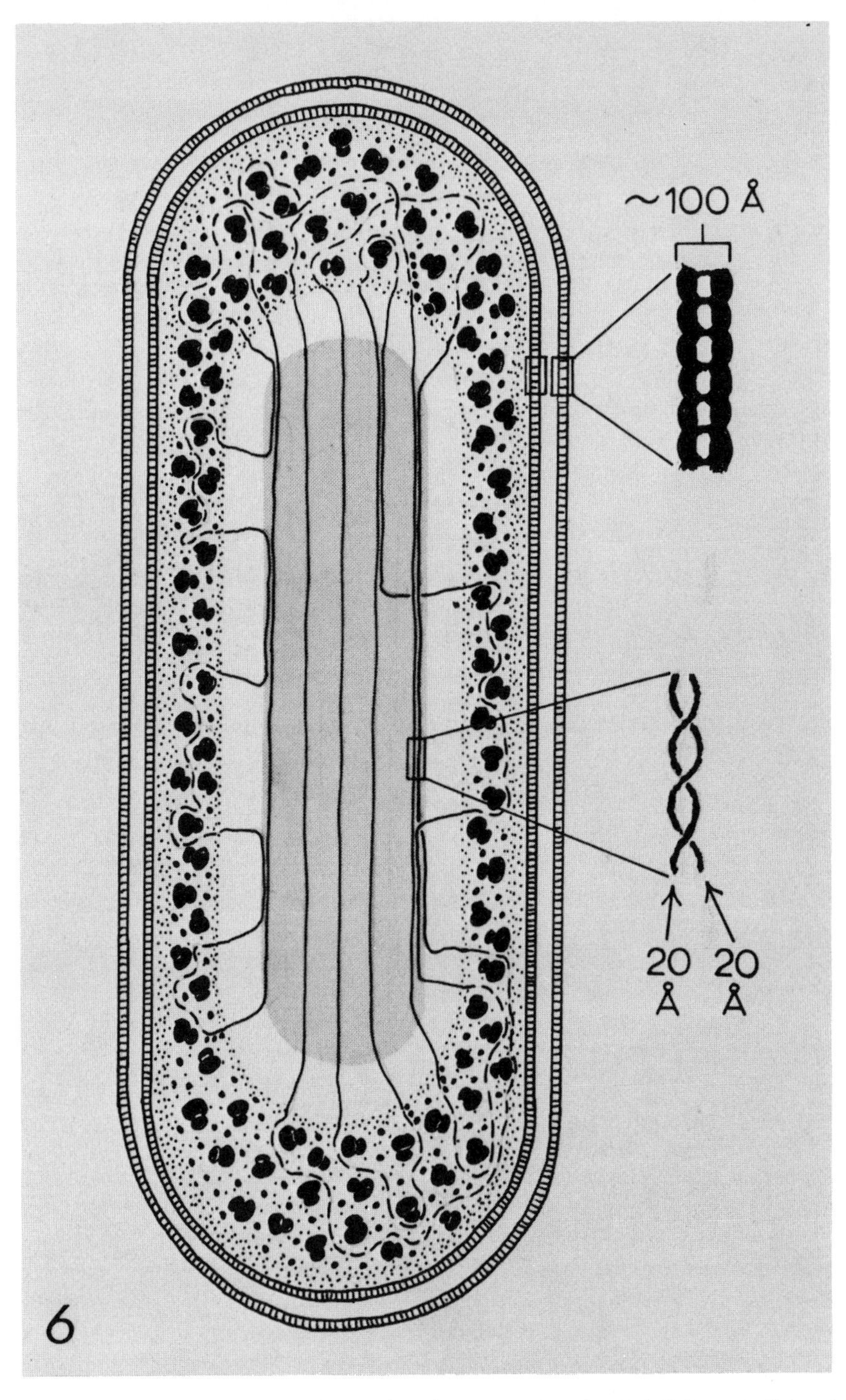

Fig. 6. Diagrammatic and partly hypothetical representation of typical *C. burneti* cell, based on observations discussed in text.

destroy the fibrils, suggesting that in *C. burneti* there exists an organized DNA center which may be physically connected with RNA-containing ribosomes in the cytoplasm (96).

On the basis of observations to date, it is possible to diagram, in a partly hypothetical manner, a typical *C. burneti* cell (Fig. 6). The gray region represents the central nucleoid region with its DNA filament generally oriented parallel to the long axis of the cell. Presumably, the basic 20–Å filament, or "chromosome," is a continuous, circular strand of DNA folded upon itself, and supercoiled, to form a dense nucleoid mass. The lower right inset shows two regions of the basic-20 Å filament twined together. Portions of nucleoid filament extend radially across the lucent zone, where they meander through the ribosomal region and then reenter the nucleoid zone. The outer and plasma membranes are shown in the inset at the upper right as a trilaminar configuration which may be composed of globular subunits. For completeness, one may wish to draw in a layer of extraneous "fuzz" associated with the cell surface (Fig. 4), for such a layer is sometimes observed (22).

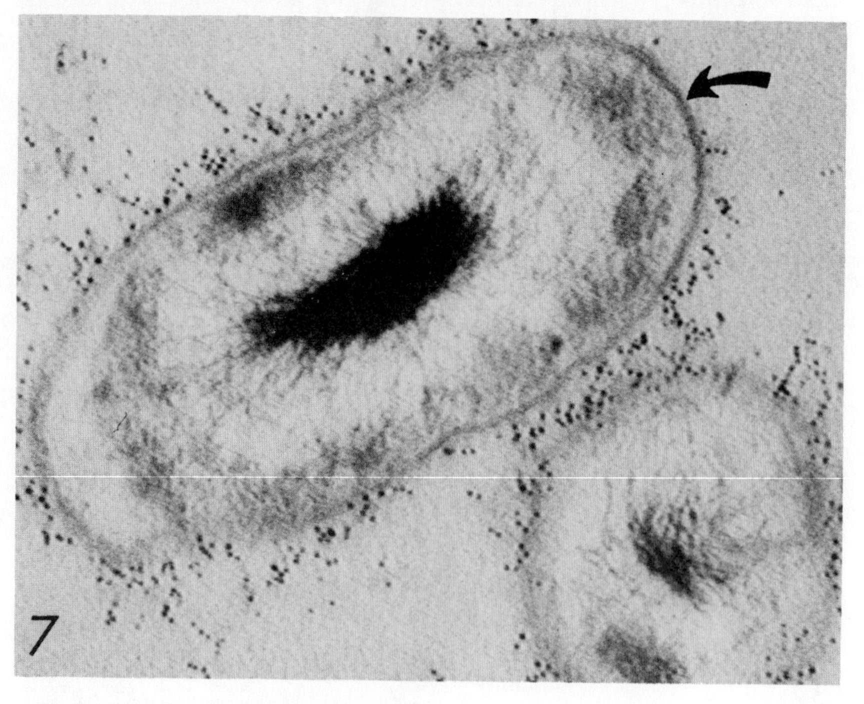

Fig. 7. Ferritin-conjugated antibody to phase II antigen bound to surface of rickettsial particles. Rickettsiae from L-cell culture 6 days after inoculation with phase II organisms. Note regions of surface of rickettsiae not binding ferritin (e.g., at arrow). × 128,000.

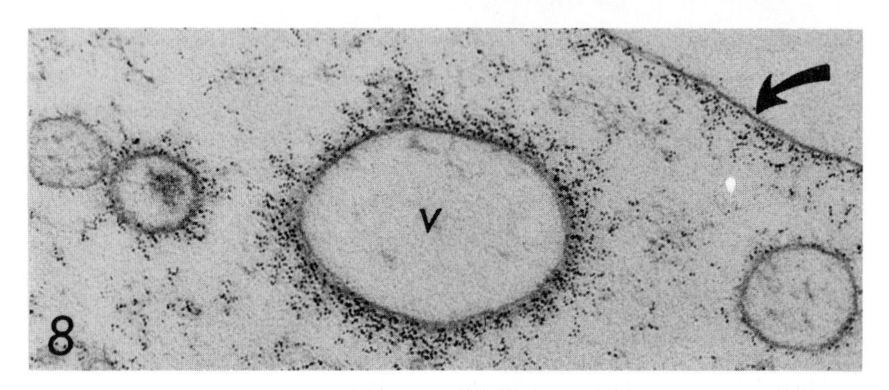

Fig. 8. Portion of infected L cell from experiment described for Fig. 7. Ferritin is associated with plasma membrane (arrow) and vacuoles (v) in cytoplasm. ×65,000.

Using a modified ferritin-labeled-antibody method (84, 114, 127), Burton, Kordová, and Paretsky (unpublished observation) have sought to localize *C. burneti*-specific antigens in infected mouse L cells *in vitro*. Since *C. burneti* organisms can exist in two antigenic phases (134), both antigenic phase I and phase II antibodies were conjugated to ferritin, using *m*-xylelene diisocyanate. Mouse L cells were inoculated with a phase II inoculum, then fixed, frozen to rupture the L-cell membranes, and incubated with ferritin-labeled antibody 24, 48, and 144 hr after inoculation. Although these studies are incomplete, some preliminary information can be presented in regard to the apparent localization of phase II antibodies 6 days (144 hr) after inoculation.

Ferritin (and presumably the antibody) was associated with the surfaces of some but not all *C. burneti*. In the rickettsiae coated with ferritin (Fig. 7), there were portions of the surface devoid of the marker (arrow, Fig. 7). Also, vacuoles in L cells were often densly coated with ferritin, and ferritin was sometimes seen along the inner surface of the host cell's plasma membrane (Fig. 8). In the only other ferritin-labeled antibody study utilizing *C. burneti*, the labeled antibody to phase I organisms was exposed to chick yolk-sac material infected with rickettsiae in the fifth egg passage (4). It is interesting to note that some "atypical" forms seen 4 to 5 days after infection did not react with the ferritin-labeled anitbody. Since our tentative findings indicate that the surfaces of some *C. burneti* bind the label whereas others do not, it is likely that the host system was infected with two antigenic kinds of organisms (13) or with a "mixed" population (63, 67). In L cells, the association between multiplying *C. burneti* and cytoplasmic vacuoles is well known (63, 116). From studies utilizing immunofluorescence, an apparently specific *C. burneti* antigen has been reported in association with the cytoplasmic vacuoles of infected L cells, although discrete rickettsial particles could not be visualized (63, 65). Our electron microscopic studies indicate an association between ferritin-labeled antibody and cytoplasmic vacuoles in infected L cells (Fig. 8), although no discrete *C. burneti* particles can be seen in the vicinity of the vacuoles. This can be interpreted as demonstrating the existence of protein that is specific to *C. burneti* (phase II) in the cytoplasm of the host cell, although additional studies are required to test this possibility.

Mouse L cells *in vitro* apparently become infected with *C. burneti* through phagocytosis, and L cells show a lysosomal response to the presence of rickettsiae in their cytoplasm (23). Although both antigenic phase I and phase II *C. burneti* produce cytoplasmic alterations in L cells, only the phase II organisms are found in early abundance in the cytoplasm (63). In comparing L cells infected with phase I *C. burneti* with cells infected with an equivalent number of phase II particles, after a given postinoculation period, cytoplasmic damage is much more extensive in the phase II infection. Cells inoculated with phase I *C. burneti* may show more endoplasmic reticulum than usual, often with dilated cisternae, and annulate lamellae are frequently seen; also, many cells display groups of smooth-surfaced vacuoles which may be related to their exposure to phase I rickettsiae.

Whereas cytoplasmic changes are subtle in cells inoculated with phase I *C. burneti*, the cytopathic effects of phase II organisms are striking. The cytoplasm shows accumulations of vacuoles of various sizes, many of which contain rickettsiae (Fig. 9). While vacuoles observed in phase I infections are usually of the "clear" type, those in phase II infections commonly contain dense or flocculent material, and many are probably secondary lysosomes (enzymes plus substrate). Large vacuoles containing rickettsiae often show masses of amorphous dense material (Fig. 9), as well as ingested mitochondria or even the nuclei of dissociated neighboring cells. Multinucleate infected cells are commonly seen, indicating fusion of cells (such nuclei can be distinguished from ingested nuclei, inasmuch as they are not contained within a vacuolar membrane but are intimately associated with the matrix cytoplasm). The nuclei of infected cells remain intact and appear relatively unchanged, even after disintegration of their cytoplasm (Fig. 10). After breakdown of the host cell's plasma membrane, endoplasmic reticulum becomes swollen and vesicular, and mitochondria are enlarged and swollen (Fig. 10). Disintegration of host cell cytoplasm provides, of course, for the release of *C. burneti* from vacuoles, for vacuolar membranes themselves tend to be disrupted upon exposure to the surrounding environment (Fig. 10). Neighboring cells that appear normal and uninfected (Fig. 10, upper left) may then ingest the released rickettsiae, and themselves become infected.

Without question, studies involving use of the electron microscope have expanded our knowledge of the structure and biology of *C. burneti*. The

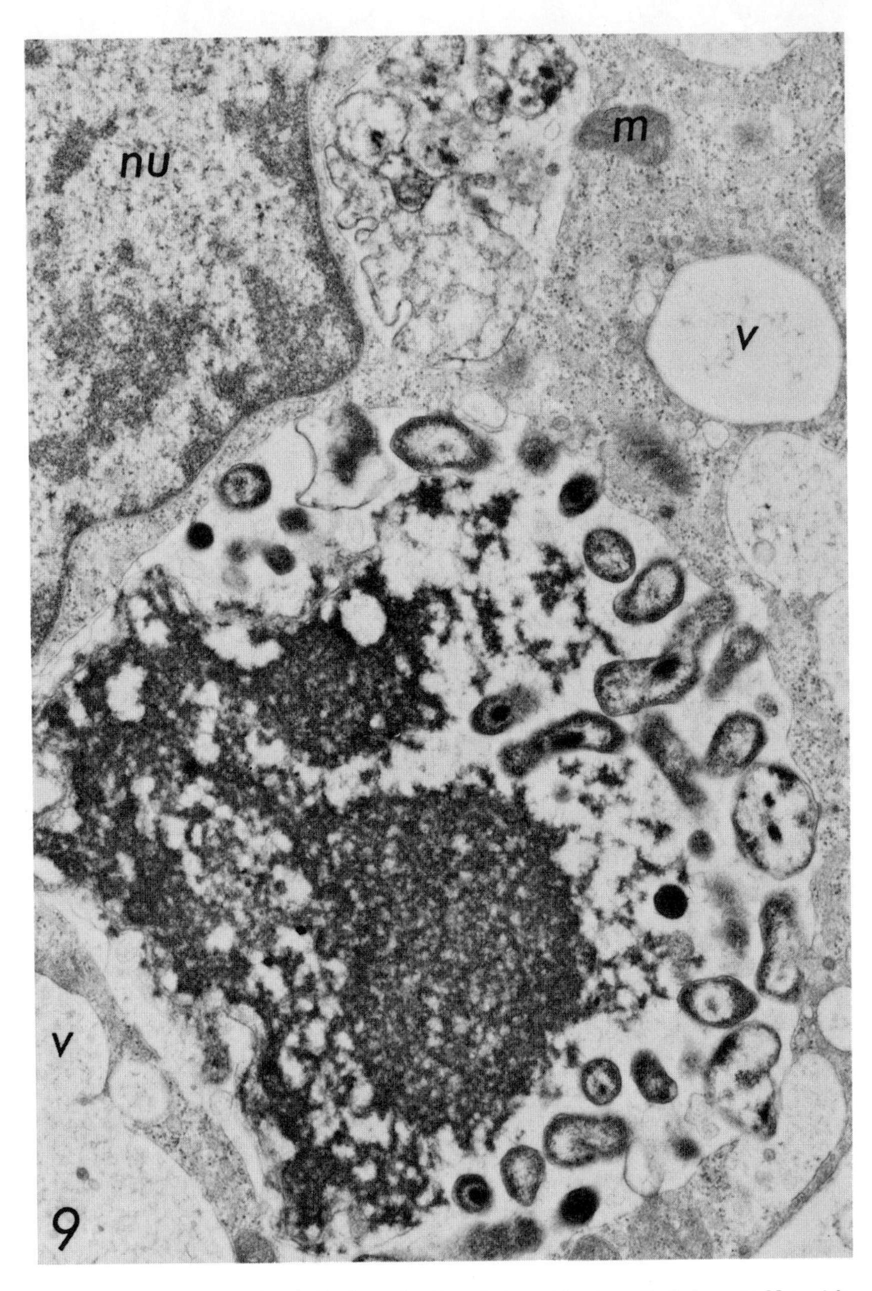

Fig. 9. Cytoplasm of L cell 10 days after inoculation with phase II *C. burneti*. Note rickettsiae in large vacuole and appearance of mitochondria (m) and host cell nucleus (nu). ×21,300.

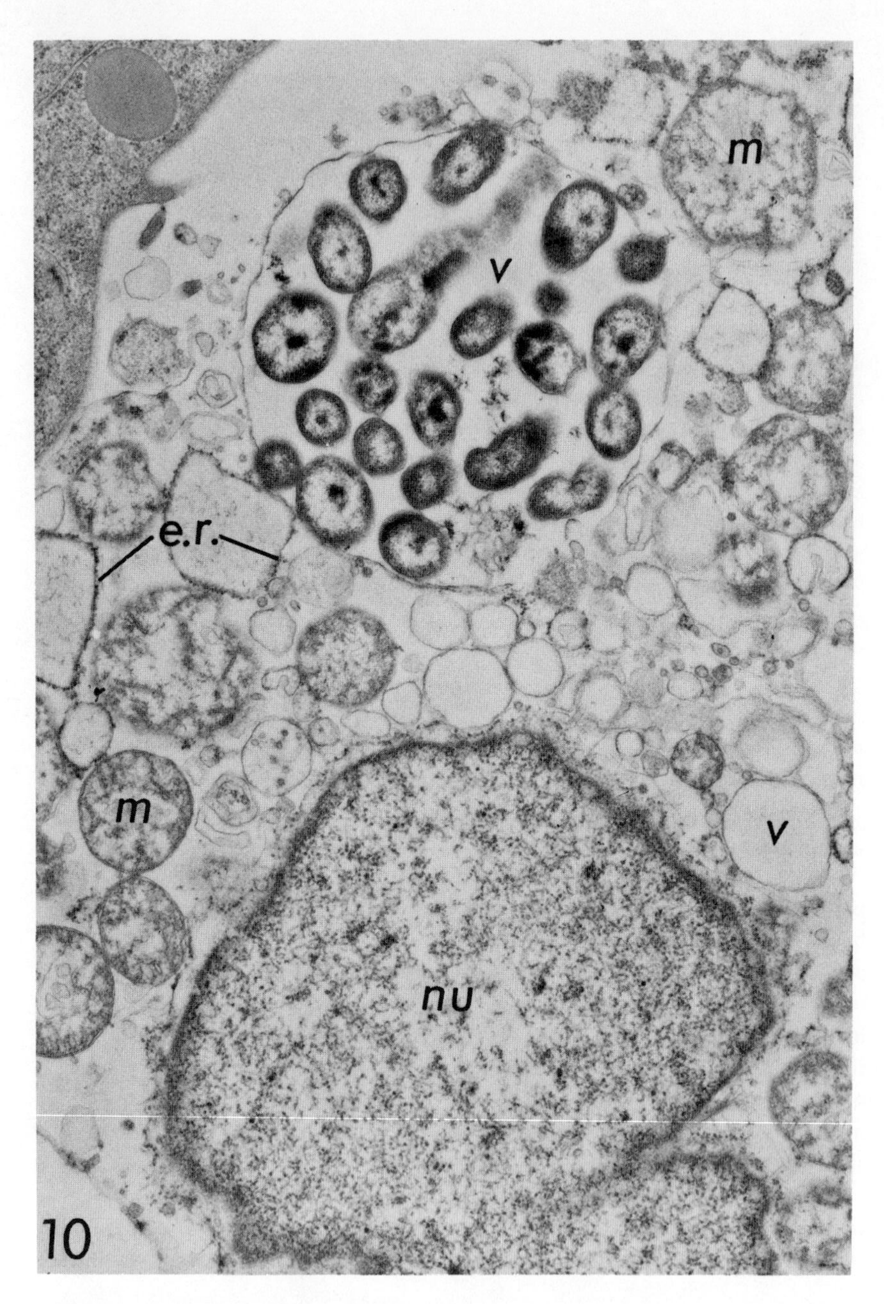

Fig. 10. Disrupted L cell from experiment described for Fig. 9. Note appearance of nucleus and mitochondria, as compared with cell in Fig. 9. Also note swollen endoplasmic reticulum (e.r.) and apparently uninfected cell at upper left. Membrane of vacuole containing *C. burneti* shows numerous points of discontinuity. ×21,300.

instrument has much more to contribute, however, particularly in conjunction with experimental and biochemical approaches. The structure and the biology of rickettsiae, in general, are poorly understood when compared with our knowledge of many bacteria and even viruses, and the scores of intriguing problems awaiting answers are well set forth in recent reviews of the biology of rickettsiae (14, 96, 140).

From the standpoint of future electron microscopic studies of *C. burneti*, certain problems are worthy of special mention. Clearly, more information is needed concerning the nature of the surface of the organism. Under what conditions can the extraneous filamentous layer be demonstrated? It is likely to be associated with the antigenic properties of *C. burneti* (6), and it may be the key to the phase variation in the organism. We, as well as others (93), have been unable to demonstrate consistent structural differences between phase I and phase II organisms, but the unpredictable appearance of a discrete surface "fuzz" could be the kind of structural variability with which phase variation may be associated. Some *C. burneti* appear to bind ferritin-labeled antibody, whereas others do not, and even in rickettsiae whose surfaces bind the marker there are conspicuous regions devoid of ferritin. If the extraneous surface layer is labile when conventional methods of preparation are employed, new methods must be sought to preserve and study it, particularly as related to immunological evidence and the organism's *in vitro* and *in vivo* environments.

Additional studies of the nature of the surface of *C. burneti* are much needed, particularly at high magnification, under different conditions of fixation, after growth in different kinds of host cells, and after various kinds of experimental treatment (e.g., extraction). The glycopeptide wall of the typical bacterial cell is characterized by its rigidity and by the presence of muramic acid as a basic structural component. In Gram-negative bacteria, to which *C. burneti* can most logically be compared if one wishes to argue the case for a bacterial-type cell wall in this rickettsia, the wall can be divided into an outer, trilaminar membrane, a dense intermediate layer (or layers), and the inner plasma membrane proper (37, 89). The outer membrane and the plasma membrane are conspicuous and remarkably similar in appearance and dimensions, both being of the "unit membrane" configuration. It is likely that the region between the two membranes, the intermediate layer, is the component that gives the integument of the Gram-negative bacterium many of its unique structural and chemical properties, for this is the lysozyme-sensitive layer and the site of the glycopeptide constituents that undoubtedly are responsible for the cell's rigidity (37, 89, 123).

Although *C. burneti* appears to contain muramic acid and is sensitive to lysozyme (3, 103), the precise location of the amino sugar and the site of action of lysozyme in this organism remain undetermined. Allison and Perkins (3) and Perkins and Allison (103), who described the presence of muramic acid in

C. burneti, noted that treatment of purified cell wall preparations with lysozyme caused the disappearance of cell walls recognizable as such in the electron microscope. Unfortunately, no electron micrographs were presented in either work for examination.

The simplest interpretation of the electron microscopic evidence dealing with the surface of *C. burneti* exposed to the environment is that the surface consists of a prominent trilaminar membrane which may show, under certain circumstances, an extraneous surface layer composed of short filaments (23). The cell wall of *C. burneti* is, in some ways, unlike that of the typical Gram-negative bacterium, a fact perhaps best illustrated by the work of Murray, Steed, and Elson (89): (*a*) the cell wall of *C. burneti* appears not to be as rigid, particularly *in situ* in host cytoplasm, (*b*) a dense intermediate layer between the outer membrane and the plasma membrane has not been demonstrated, and (*c*) penicillin does not suppress multiplication of cells in *C. burneti* as it does in Gram-positive and some Gram-negative bacteria (4, 30, 74, 95). If *C. burneti* possesses a muramic-acid-containing cell wall, an answer must be provided as to why penicillin does not inhibit the synthesis of new cell wall material. It seems likely that the structural and functional integrity of the surface of *C. burneti* is not dependent on the crosslinking of mucopeptides containing muramic acid, which is inhibited by penicillin (40); it is interesting to note that some *C. burneti* cells in infected HeLa cells have been reported to grow up to 10 times their normal sizes in the presence of penicillin (74).

Another aspect of the biology of *C. burneti* worthy of special consideration is its replication cycle, discussed fully in Section 4. To begin with, the amount of DNA accounts for about 10% of the cell dry weight in *C. burneti* (127), whereas it is but 1% in *E. coli* (136) and only 1 to 2% in most microorganisms (120). Thus, in respect to the amount of DNA, *C. burneti* appears to be comparable to the deoxyriboviruses (32) rather than the bacteria. Although it is generally agreed that *C. burneti* can multiply by binary fission, the question of whether the organism can replicate in a viruslike manner is intriguing.

If a developmental cycle, such as is known to occur in *Chlamydia* (87), were demonstrated, a new perspective would be provided in regard to the antigenic variation (phases I and II), the existence of a filamentous surface layer on some but not all cells, the apparent morphological differences between organisms found in different sedimentation bands obtained by density gradient centrifugation (139), and other aspects of the biology of *C. burneti*. The possibility that the DNA of *C. burneti* is infectious could be associated with the organism's resistance to environmental conditions and to many physical and chemical agents, as well as to its apparent ability to escape lysosomal degradation (23). Thus, although other components of the cell might be damaged or degraded, the "organism" would retain its infectious property as long as its DNA was intact.

3. MACROMOLECULAR COMPOSITION OF RICKETTSIAE

Analyses of the chemical composition of *C. burneti* have been made possible by the development of techniques for purifying large quantities of the organism; the results of these analyses have reflected the precision of the methods used at the time. Gross chemical studies (128) showed that purified coxiellae contained 13.2% nitrogen, 1.8% phosphorus, and most of the common amino acids, together with 9.7% (dry weight) DNA and 7.3% RNA. Lipid analyses have not been made.

A. Outer Layers

Most investigators studying ultrathin sections of rickettsiae have described the "cell wall" as trilamellar (45, 93). The rickettsiologist currently uses "cell wall" for the complex of surface layers of the rickettsiae particles as an essentially operational term. Some of the cell wall layers may be a mere dislocation in structure, and the nature of the association between capsular components and the underlying structure is not sufficiently known

The ultrastructure of the outer layers of *C. burneti* was discussed in Section 2 of this chapter. It is pertinent to the present topic, however, that particles of coxiellae have been observed in which the membranes were surrounded by a capsule (23, 122). There is clear indication of amorphous capsules around *R. prowazeki* (96).

Although no detailed analyses have been made of the macromolecular composition of the outer layers, some data can be deduced from chemical, biosynthetic, and antigenic studies in *C. burneti* and by correlation with newer findings described in some bacteria. The evidence so far available is in favor of mucopeptide being the principal, if not the only, mechanical support that maintains the shape of almost all free-living bacterial cells and protects the underlying membrane against mechanical damage. This evidence consists of the effect of lysozyme on a wide variety of organisms, the results of penicillin inhibition of mucopeptide synthesis, and the retention of the cell wall after various extraction procedures to remove substances other than the mucopeptides. It seems to be generally accepted that mucopeptides consist of chains of alternating *N*-acetylglucosamine and *N*-acetylmuramic acid, linked together by peptides. The apparent ease with which capsular material can sometimes be removed from bacteria is well recognized and needs to be carefully scrutinized before being cited as evidence contrary to the presence of a capsule. In many instances enzymes that degrade the capsular and wall substance have been shown in bacteria; it has been stated that the hydrolysis of only a single bond per very large molecule might be sufficient to remove the bulk of the capsular material from the surface (119).

"Cell wall" preparations of *C. burneti* were not changed morphologically by treatment with trypsin, but were lysed by lysozyme. Lysozyme was shown to liberate a material that gave a reaction for *N*-acetylamino sugars; muramic acid was isolated from coxiella particles (3). Hot sodium lauryl sulfate was effective in the purification and isolation of "cell walls" of *C. burneti*; hot formamide caused complete destruction of the coxiellae. These results suggest that *C. burneti* has some of the qualities of Gram-positive bacteria (96). Methods based on the prevention of the formation of cell wall glycopeptide and capsule in bacteria include treatment with penicillin (120). The influence of penicillin on the inhibition of capsule formation in the "smooth" coxiellae (33) has to be investigated. It is well recognized, however, that penicillin has no effect in preventing infectivity and multiplication of *C. burneti*. It is also well known that *C. burneti* is remarkably resistant to chemical and physical agents and in nature is more resistant than most nonsporogenic microorganisms (94); on the other hand, rickettsiae, separated from their host cells, behave more as cell organelles than do bacteria in regard to the regulatory behavior of the cell membrane (139). The specific functions of rickettsia membranes are unknown.

The "cell walls" of *R. mooseri* are resistant to trypsin; contain glucose, galactose, and glucuronic acid, the amino sugars glucosamine and muramic acid, and at least 15 amino acids; they lack techoic acid. Diaminopimelic acid was found in the "cell walls" of *R. mooseri* and in whole particles of *R. prowazeki, R. mooseri*, and *R. quintana* (90, 96).

B. Macromolecules and Enzymatic Properties

Another approach to the macromolecular analysis of rickettsiae involves the demonstration of host-independent enzyme systems. It is known that exorgenic reactions among rickettsiae show little variation and that these organisms possess enzymatic mechanisms for the breakdown of carbohydrates, the formation of high-energy phosphate bonds, and the synthesis of lipids and proteins (96).

In purified, intact coxiellae, pyruvate caused the greatest oxygen uptake. Glucose as substrate was not utilized by intact coxiellae (97). The adaptation of enzyme studies on disrupted coxiellae made it possible to show the broad host-independent biochemical properties of the agent; this study has recently been reviewed in detail (98, 137). Glucose dissimilation has been demonstrated in disrupted *C. burneti*, and the enzymes glutamate and malate dehydrogenases (99, 100), condensing enzyme, and isocitrate dehydrogenase (27) have been described. In disrupted, but not in whole cell preparations of *C. burneti*, the enzyme 6-phosphogluconic acid dehydrogenase oxidoreductase was detected (82). In connection with these finding the question was raised as to whether the glucose may be primarily used as a source for the sugar moieties of the "cell wall" in coxiellae (98). It has become apparent that sugar units are added to various macromolecules, including polysaccharides, by transfer from small molecules, such as UDP-glucose under the influence of specific glycosyl-transferring units. Investigations of the biosynthesis and genetic determination of pneumococcal capsular polysaccharides have revealed principles that seem to apply also to many other polysaccharides. The phenomenon of type transformation indicated that a specific polysaccharide or oligosaccharide is not an indispensable requirement for the synthesis of a new polysaccharide by whole bacteria cells (79).

Disrupted coxiellae have been shown to synthesize serine (91), citrulline, and ureidosuccinate (80) and to incorporate amino acids into a hot trichloracetic acid-insoluble (presumably protein) fraction (81). Polynucleotide synthesis from mixtures of the four ribonucleoside triphosphates has also been demonstrated (53). Studies indicate the presence of an autonomous DNA-dependent RNA polymerase (81).

Although there is little doubt that coxiellae possess ribosomes (23, 96), their role and functions have not been studied. Chloramphenicol has a remarkable selectivity in inhibiting protein synthesis. In general, the process is sensitive in bacteria and insensitive in yeasts, plants, and animals. It has been found that the antibiotic is bound to ribosomes from sensitive but not resistant organisms (83). The reproduction of coxiellae in eggs is only slightly affected by chloramphenicol and erythromycin; thyomycetin, a synthetic compound closely related to chloramphenicol, is completely without effect on coxiella infections at maximum tolerated doses (96).

Once removed from their host cells, rickettsiae are generally unstable, and inactivation of various functions of purified organisms has been described (140). Under certain conditions, activity could be restored by the addition of glutamate, DPN, or ATP. It has been suggested that relatively large molecules, like DPN, ATP, and the nucleic acids, can pass through the rickettsial membrane. In purified suspensions, rickettsiae, in contrast to free-living bacteria, probably have a unique, functionally defective cell membrane.

In relation to these speculations, it seems appropriate to refer here to some recently described findings concerning other intracellular pathogens. Data have been presented which indicate that all of the known vaccinia-associated enzyme activities are protected from heat inactivation by nucleotide triphosphates and that vaccinia incubated under conditions in which ATP phosphohydrolysis is known to occur does not inactivate virus infectivity (88). The adenosine triphosphatase activity of vaccinia *in vitro*, observed by different treatments and detected in thin sections by electron microscopy (38), has led to suggestions that with incomplete or otherwise defective or disrupted virus particles the enzyme could be catalytically active. On the basis of the mutual protection of adenosine triphosphatase, guanosine triphosphatase, cytosine triphosphatase, and uridine triphosphatase by ATP, GTP, and UTP in vaccinia, it has been hypothesized that there is a single enzymatic site capable of reacting with all of these substrates (88).

The participation of enzymatic processes has been demonstrated in the *in vitro* inactivation of rickettsiae, and Cohn et al. (26) have suggested that this inactivation represents a complex phenomenon. The survival of *C. burneti* after heat inactivation at 60°C for 30 min was studied *in vivo* in L cells. The survival of heat-inactivated rickettsiae was strongly influenced by the input multiplicities of infection. The "reactivation" of *C. burneti* in mixed infections of a heat-inactivated (at 80°C for 1 hr) mutant, and host-dependent parent coxiella in L cells, has been described recently (60). Although the mechanisms of this phenomenon in *C. burneti* are not known, the phenomenon may be similar to the Berry–Dedrick phenomenon in poxviruses (9).

There is little doubt that the macromolecular complexity of rickettsiae is similar to that of bacteria. Discrepancies exist in the data on the enzymatic capabilities of intact versus cell-free preparations of rickettsiae (96, 98, 139). The value of enzyme studies of extracts of both bacteria and viruses, however, has been firmly established (49, 77).

Rickettsiae are obligate intracellular parasites, yet the contribution of the host to rickettsial physiology remains unknown (98). The more a parasite is strictly dependent on its host, the more complex are its nutritional requirements and the more finely attuned is the growth of the parasite to the chemical environment of the host. A loss in the synthetic ability of the parasite occurs during the course of evolution because of reduction in the number of enzymes. At the same time, however, an increased concentration of the enzyme is maintained to utilize the preformed organic substances of the host. The enzymatic constitution of an organism is genotypically fixed; nevertheless, genotypic potentialities are expressed only as the environment of the organism provides a suitable substratum for growth and synthesis. The phenotypic or actual enzymatic constitution of an organism at any given moment, including both the kinds and the absolute and relative quantities of individual enzymes, varies with the nature of the environment (49).

"Undamaged" intact, and "damaged" disrupted, rickettsiae have been studied *in vitro*. Enzymic reactions underlie all biological phenomena, but lose much of their meaning when considered out of context with the environment in which they occur. The intimate association of a parasite like rickettsiae with the host cell imparts a new significance to the enzymes, which is lacking when they are considered purely in the context of an extracellular particle. It is hoped that future biochemical studies may help to answer the question of what activities of rickettsiae are essential for growth in the uncertain enzymic environment of a host cell.

C. Specific Polymers and Antigenic Variants of *C. burneti*

That coxiellae exist in two different antigenic forms was established in early studies (134), and attempts have been made by many workers to relate macromolecular structure with antigenic specificity. These studies have recently been reviewed in detail (13, 33). The "smooth" or phase I coxiella has a surface antigen, presumably a polysaccharide (14, 17, 33). When extracted with trichloracetic acid (TCA) and treated with phenol, this antigen is converted to a hapten, apparently because of the removal of a carrier protein (4). Data obtained from chromatographic analysis of TCA extracts of coxiallae in phase I showed this heterogeneity. A protective antigen was present in all fractions. Attempts to separate traces of the phase II antigenic components failed (18). References have been made to the lack of information about the chemical nature of the phase II antigenic component in coxiellae (13).

Although resemblance between the rough–smooth variations of pneumococci and the "phase variation" of coxiellae was stressed in early studies, it has been considered that *C. burneti* exists in nature only as a homogeneous population of the "smooth" coxiellae. The shift from the "smooth" phase or phase I to phase II is accompanied by a loss of the surface phase I antigen and was originally thought to represent a phenotypic phase variation during "adaptation" to eggs, rather than a true mutation (10, 13, 33, 96). In situations where no effective selective agent capable of eliminating the parent cells in favor of the mutant cells is known, the mutational origin of mutant cells is difficult to demonstrate (11).

Evidence indicative of mutational changes in coxiellae has been obtained recently (35, 36, 60, 63). It is well known from studies on bacteria that the spontaneous origin of mutants, independent of specific conditions, does not necessarily imply that the reactions of such mutants are identical in the absence and the presence of a specific environment. A loss of ability to perform biosynthetic steps such as capsule formation is well recognized in bacteria and has a genetic basis; it may be related to either a loss of information for the structural synthesis of the enzyme or to a change in the factors controlling the activity of the enzyme (11). There is good reason to believe that coxiellae in phase II originated by means of gene mutation and that the loss of the capsule of the parent type has a genetic origin. Complementation tests between the parent and the mutant have recently been performed and will be discussed later. It is known that complementation tests between closely related viruses make it possible to define functional units genetically in the absence of any information about the molecular nature of the function (32). The vast majority of determinants presently known are associated with chromosomal polynucleotides, and it is fairly safe to link most mutational agents with changes in chromosomal equivalents (11).

The easily achieved conversion of the egg-"adapted" coxiellae in phase II into the "smooth" phase I population by passage in laboratory animals is well known. However, the mechanism of this population change remains obscure. Nonsmooth–smooth population changes of some bacteria could be induced also *in vitro* provided that cultures were supplemented with enzymatic digests of DNA. These effects of breakdown products of DNA, which are independent of the source of the DNA, are now known to involve oligonucleotides. It appears that these effects are a general phenomenon in bacteria even though the mechanisms responsible for the selective effects of enzymatic DNA digests in Gram-positive and Gram-negative bacteria may be somewhat different. The significance of these observations lies in the proof they afford of interactions between genetically different sympatric bacterial populations because of the natural production of selectively inhibitory metabolic products (11)

The two antigenic variants of *C. burneti* would seem open to direct experimental approaches in different cell systems *in vitro* and to speculation about interfering and stimulating interactions of these variants in mixed infections, recently described in mammalian cells *in vitro* (63, 67). From natural sources the "smooth" phase of coxiellae is most frequently isolated (134), and this predominance of the "smooth" type in natural environment requires explanation.

D. Nucleic Acids in *C. burneti* and Some Unresolved Problems of Their "Extra"-Organismal Properties

Both types of nucleic acids have been found in relatively pure *C. burneti* preparations. Although the DNA content was constant, the amount of RNA varied with the particular purification procedure used. A similarity of coxiella and host-chick embryo nucleic acid has been reported (128). The low RNA/DNA ratio (1:3) in coxiellae (128), which has not been found in other rickettsiae, has led to the suggestion that losses of RNA occur in coxiellae and may be related to purification procedures and to the physicochemical lability of rickettsial RNA. Studies on *R. mooseri* indicated that the temperature during purification played an important role. The participation of enzymatic processes in the loss of nucleic acids from rickettsiae has been suggested. Since the omission of glutamic acid did not influence these results, the authors concluded that the "energy metabolism" did not seem to be of importance and that the rickettsial inactivation appeared to be a complex phenomenon (26).

The base compositions of coxiellal DNA studied by different authors ranged from 42.9 to 44.5% guanine plus cytosin (G+C) (125, 126, 128). The base composition, the nature of the melting curve, the buoyant density, and assays with formaldehyde indicated a double-stranded structure of DNA. The molecular weight was estimated at about 1.8×10^7 daltons (125).

The DNA of *C. burneti* has been found to be significantly different from that of *R. prowazeki* and indicates little genetic relatedness (96).

Two kinds of nucleic acids, DNA and RNA, exist in all organisms. The DNA is the repository of information identifying the organism and is largely or exclusively confined in bacteria to the chromatinic body. It has been suggested that the genetic structure of bacteria and DNA viruses may be represented by a single molecule of DNA, uncombined with histonelike proteins, and that the chromosomes of all organisms are basically a single DNA duplex (83). It appears that DNA acts as the repository of genetic information and that it has the innate characteristics necessary for such a repository, that is, of being copyable without the intervention of yet another, overriding source of information (46). It seems appropriate at this time to review the DNA properties of bacteria and DNA viruses in order to consider some of the phenomena that occur in coxiellae, regardless of our complete ignorance of the molecular mechanisms underlying the phenomena described.

Evidence of an ultrafilterable infectious agent obtained from lysates of cells after infection with coxiellae suggests that *C. burneti* was reduced to genetic material that differed from "organismal" coxiellae. Some differences between the "organismal" coxiellae and the "ultrafilterable" particles of this agent are summarized in Table 1. Fully infective and structurally characteristic coxiellae were produced when a sensitive host system such as a yolk sac, or ticks and their organs, were inoculated with the ultrafilterable material (55, 66, 69); all attempts to infect mammalian cells have failed (66). Recent experiments based on infectivity assays and immunofluorescence have indicated that the infectivity of the ultrafilterable material was lost after DNase treatment but was not affected by RNase and protease treatment (N. Kordová, D. Paretsky, and E. Kováčová, unpublished data). Hence these studies which are reminiscent of some experiences with the DNA viruses (32), indicate that rickettsial DNA alone could induce the development of *C. burneti* in a sensitive host system. It seems evident that such events would necessarily preclude a viruslike replication of the agent; we discuss in the next section the little that is known about multiplication in the coxiellae.

The coxiella DNA material passed membrane filters having an average pore size of 0.01 to 0.005 μ, which retain particles down to approximately 50,000 molecular weight, while the molecular weight of coxiella DNA is ap-

Table 1. Some Differences Between the Organismal Form and the "Ultrafilterable Particles" of C. *burneti* (from Egg-"Adapted" Strain)

Property	Organismal Form	"Ultrafilterable Particles"	
Filterability (collodion membranes)[a]	Retained by ultrafilters having mean porosity of 190 mμ	Passed ultrafilter having mean porosity of <25 mμ	Control by addition of baterioi-phage Tr4 of PFU 10^{11}/ml
Antigencity[b]	Reacted specifically in complement fixation reaction and fluorescent antibody reaction	*No reactivity*	
Immunogenicity[b]	Induced complement-fixing antibodies in animals	*No* specific antibodies detectable in immunized animals	
LD_{50} for chick embryo[a]	$4 \times 10^{9.5}$/ml	*No*	
ID_{50} in yolk-sac tissue culture and in ticks[c]	5×10^{9}/ml	$5 \times 10^{8.5}$/ml (after delay)	
Pathogenicity for adult guinea pigs[d]	In 94% animals Q-fever and specific complement-fixing antibodies	In 28% animals complement-fixing antibodies	
ID_{50} in mammalian cells in vitro[e]	$5 \times 10^{8.5}$/ml	*No production of C. burneti* particles (56)	
DNase treatment[f]	Infectivity maintained	*Infectivity lost*	Tested in yolk-sac tissue culture and ticks by fluorescent antibody reaction
RNase treatment[f]	Not tested	Infectivity maintained	
Pronase treatment[f]	Not tested	Infectivity maintained	

[a] Data from ref. 44.
[b] Data from ref. 46.
[c] Data from refs. 45 and 56.
[d] Data from ref. 47.
[e] Data from ref. 55.
[f] Kordová, Paretsky, Kováčová, unpublished data.

proximately 1.8 × 10⁷ daltons (125). The question arises, therefore, in which form is the DNA when it passes the dense filters?

Intermediate stages of DNA synthesis within cells have been reported, and intermediates much smaller than the final DNA have been described. Newly synthesized DNA consists chiefly of small units which are later incorporated into the larger DNA molecules. They are formed during both bacterial and phage growth; ligase-defective systems accumulated large quantities of the small pieces as precursors, and not split products, of DNA. A bacterial mutant that accumulated small pieces of single-stranded DNA under restrictive conditions has also been reported (46).

A shift from the "smooth" parent type to the mutant coxiellae population occurred in chick embryo (134); the former induced abortive infection in L cells (63, 67). The fate of *C. burneti* DNA was studied in L cells, and a shift of the ^{3}H-labeled coxiella DNA to the position of cellular DNA was described in L cells (23). It seems not accidental that ultrafiltrates from these host systems failed to induce development of coxiellae in mammalian cells *in vitro* (66) but, on the other hand, induced latent infections in the majority of experimental animals (58). A remarkable feature of rickettsiae is their "latency" (14, 110, 133). A "provirus" concept of the latency of coxiellae has been suggested (58), and consideration should be given here to the possibility that the ultrafilterable material mentioned above represented pieces of parent coxiella DNA accumulated under restrictive conditions.

4. C. *burneti* -HOST CELL INTERACTIONS

Studies of rickettsiae over the past 50 years have familiarized microbiologists with the growth of the pathogens; although a delay in the appearance of intracellular parasites after infection was observed in different host substrates (94, 106, 124, 131, 141), no attention was paid to this peculiar phenomenon.

A clear experimental demonstration of the occurrence of a viruslike stage in the early period of coxiellae and *R. prowazeki* (70–72) infection was lacking until methods of tissue culture were employed to study the rickettsiae–host cell relationship. Until recently, only egg-"adapted" strains of coxiellae in monolayer were assayed and the sequential events following initiation of infection were observed.

The following types of studies of various steps in *C. burneti* multiplication have been made: one-step growth curves (62, 65, 116); absorption and penetration studies (23); and studies of the single replicative cycle by phase contrast in living cells (68), as well as in fixed cells by light (68, 65, 116), by electron microscopy (121, 122) by immunofluorescence (65, 116), by histochemical methods (65), and by autoradiography (24).

The following aspects of the relationship between *C. burneti* and the host cell are considered below: (*a*) the reproduction of coxiellae in productive host cells; (*b*) the fate of coxiellae in abortive infection; and (*c*) the interactions of coxiella variants in a defective host cell system. The diversity and complexity of coxiellae–host cell interactions are emphasized here, in the hope of dispelling the notion that the host cell plays a nonspecific role in rickettsial growth and multiplication (86, 139). It seems appropriate to maintain for rickettsiae–host cell relationships the terminology currently used for virus–cell interactions.

A. *C. burneti* Development in Productive Host Cells

A productive virus–host cell system designates a relation in which the virus and cell interactions are relatively effective, leading to the production of mature infectious virus (113). The egg-"adapted" coxiellae of phase II grow relatively well in different cells *in vitro*, leading to the production of mature infectious coxiellae. Such a coxiellae–host cell relationship is designated here as a productive system.

Studies of the multiplication of coxiellae have been aimed at demonstrating the viruslike eclipse phase in early stages of infection when carried out by means of one-step growth experiments in different cell lines. A decrease in the infectivity of the agent occurred within 2 to 6 hr after infection. From about 12 hr after infection an increase in infectivity titers was established. The decrease in infectivity was marked in cells in which coxiellae multiplied well, whereas it was insignificant or nonexistent in cells showing poor multiplication (62, 65). Similar results have also been reported by other authors, but it was suggested that the decrease in infectivity was due to the failure of rickettsiae to penetrate into the host cell (116).

The combined techniques of observing coxiellae in living cells by phase contrast (68), and in fixed cells by light (56–58) and electron microscopy have proved very successful in providing information about the development of these organisms in productive host cells. These studies became even more informative when acridine-orange fluorochrome and immunofluorescence techniques were also performed in the one-step growth experiments. The systematic studies of coxiellae–host cell interactions at cellular and subcellular levels, although fragmentary, have led to the conclusion that replication of coxiellae during the *early* stages of cell infection has a viruslike pattern.

The attachment of coxiellae, when observed by phase contrast in living cells, was followed by an asynchronous engulfment of particles in small vesicles (phagosomes), which moved to and accumulated at the periphery of the cell nucleus (68). When stained with acridine-orange, these structures appeared as small RNA, RNase-sensitive, inclusion bodies.

Early changes in the nuclei of infected cells, as observed by acridine-orange staining (65), on the one hand, and by electron microscopy (122), on the other, indicated that events associated with coxiella replication, which takes place exclusively in the cytoplasm, involve in some way the nuclei of infected cells. The most pronounced feature, however, was the evidence of coxiellae-induced DNA synthesis in the cytoplasm of cells 4 to 6 hr after infection, that is, at a time when all methods failed to demonstrate coxiella particles inside infected cells (65). Electron-dense "matrices" observed in infected cells by electron microscopy (121, 122) were apparently one and the same macromolecular structure as the green, fluorescent, inclusionlike bodies, and probably represented the "factory" in which coxiellal DNA was synthesized, to be incorporated later in the mature particles of coxiellae. The latter appeared by light microscopy approximately 18 to 24 hr after infection in well-defined cytoplasmic vacuoles (65, 67).

The immunofluorescence technique provided evidence that a large number of fluorescent particles were engulfed by the cells. However, the particles shortly disappeared, and new coxiellae in well-defined cytoplasmic vacuoles were first demonstrated approximately 20 to 24 hr after infection. The study indicated that different antigens appeared during the early and late stages of infection (65, 66). Other authors have also pointed out that fluorescent coxiellae can be be detected in cells only after 21 hours after inoculation (116).

Almost nothing is known about the maturation and assembly process of coxiellae. The "matrix" observed by electron microscopy showed a progressively more complicated structure as the development process proceeded. "Intermediate" forms, with limiting membrane at one pole and a concentration of dense material in one half of the smaller particles, were noted. It appeared that the maturation of coxiellae occurred asynchronously and over a relatively long period (121, 122).

A schematic representation of the combined results of the studies of coxiella multiplication is summarized in Fig. 11. A sketch similar to the well-known diagram of the development of vaccinia virus (28) was arbitrarily chosen. An attempt has been made to emphasize the apparent general similarity of the sequence of events occurring during the development of vaccinia virus, on the one hand, and of coxiellae, on the other, when studied sequentially from the initiation of infection in productive host cells.

Although the synthesis and assembly of the virus represents, as far as we know, the last step in the development of the virus particle, the synthesis and assembly of the nonsmooth coxiella particle seems to be followed, under appropariate conditions, by subsequent division into daughter particles. Binary

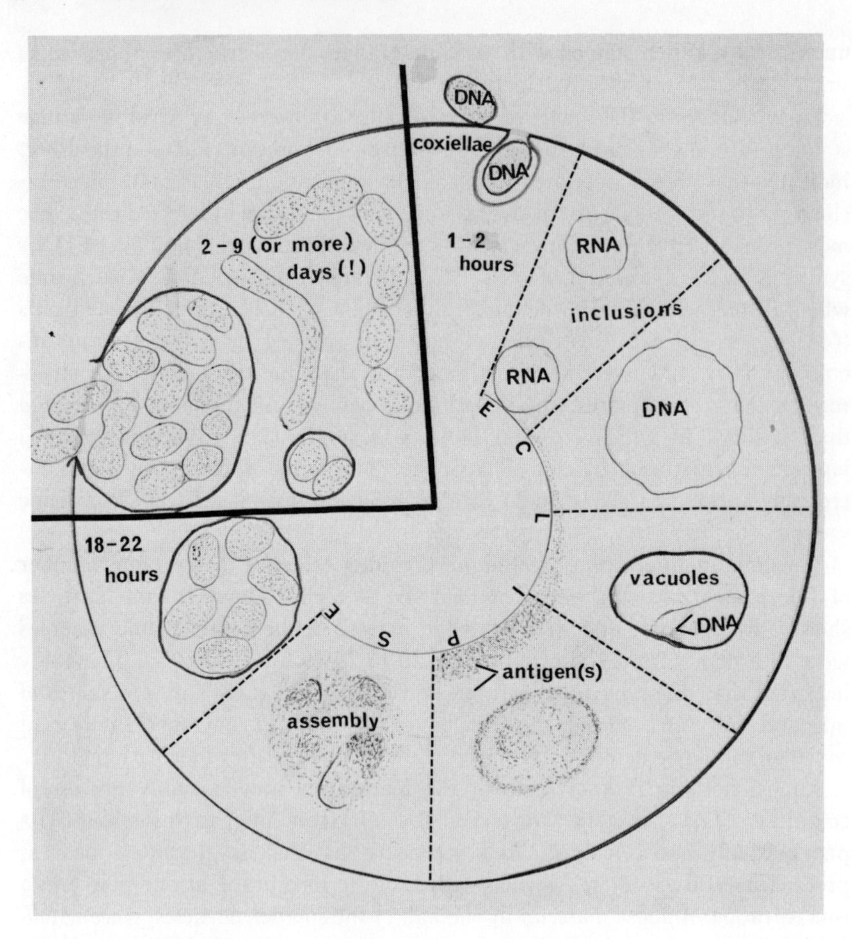

Fig. 11. Sequence of events associated with the development of the "nonsmooth" *C. burneti* variant, followed by its growth and segmentation.

fission and filament and chain formation have been repeatedly observed in different host systems infected with *C. burneti*. It should be pointed out, however, that binary fission in *C. burneti* (4, 24, 43, 54, 132), as well as in other rickettsiae (20, 54, 59, 124), has been described only in the late stages of infection, that is, from about 2 days and longer after infection.

Reference should be made here to the apparently different modes of cellular division of so-called "smooth" and "rough" types of Eubacteriales. Evidence has been presented that smooth-type bacteria are usually unicellular, having a tendency to separate completely at the time of cell division. In contrast, rough-type cells are usually multicellular; a wall and membranous septa separate the individual cells, which have a tendency to remain attached at the point of division. Experiments with *Pneumococcus* have indicated that the pattern of cellular separation after division and the production of capsular polysaccharide are independently heritable cellular properties and that both are subject to control throughout the transformation processes (11).

No studies have been performed on the multiplication of rickettsiae at the molecular level. Despite this deficiency further evidence of the viruslike pattern of coxiella replication was obtained when cultural conditions unfavorable for the growth of the parent strain of coxiellae were applied. These studies produced satisfactory evidence that the productive cycle is not the only possible outcome of coxiella infection and that the presence of *C. burneti* can have a variety of effects on the function of the host cell.

B. The Fate of *C. burneti* in Abortive Infection

The defective virus cell system is characterized by interruption of viral replication at a certain point. This makes it possible to examine the events that precede the interruption and to determine what is required to continue the process of viral replication (113).

The parent strain of coxiellae multiplied poorly in mammalian cells grown *in vitro* (16, 63), producing an early "toxic" effect, whereas the mutant coxiellae multiplied to high titer in the same system without producing obvious damage to infected cells; both variants propogated well in chick embryo and in guinea pigs. It has been established that the parent strain of coxiellae was absorbed and that it penetrated into the cell, where it induced the formation of specific antigens; it could not, however, complete its reproduction cycle as determined by the production of mature coxiellae. The amount of coxiella antigens increased during infection, but they appeared in an altered form as compared to the antigens occurring in productive coxiella infection (65–67). Intracellular events were further elucidated by utilization of electron microscopic autoradiography. A ^{3}H–thymidine-labeled parent strain of *C. burneti* was inoculated in the same defective system, and the subsequent fate of the pathogen in host cells was studied after initiation of infection. A substantial portion of ^{3}H-labeled coxiella DNA became degraded, and only a few labeled rickettsiae were noted in the cytoplasm. Part of the radioactive coxiella DNA, however, was found to shift to the position of cellular DNA (23). Thus the relationship of the parent strain of coxiellae in mammalian cells resembled that in cells abortively infected with some defective DNA viruses (32, 113).

We do not know where the block in the development of the virulent parent coxiellae in L cells is located. Immunofluorescent data indicate that the defect is apparently not involved in the intracellular steps which induce the production of "late" proteins. The presence of ^{3}H-labeled coxiella DNA in the dense nuclear chromatic material of abortively infected cells poses a problem; many possibilities exist as to the nature of the disturbances observed.

Virulence is the end result of the interactions between host cell and parasite and is governed by certain inherent properties of the agent and of the host. The term actually refers to the properties of the agent that determine its relation to the host. The virulence-determining properties of pathogens are obviously under the same sort of genetic control as the other properties of pathogens that depend on the formation and activity of specific enzymes (11).

The virulence of the "smooth" *C. burneti* is well known, but remarkably little information is available about the mechanisms involved (15, 109). Host cell damage is not a necessary consequence of multiplication of the pathogen. It seems that the early "toxic" effect produced by the virulent coxiella strain, which is not associated with multiplication of the organism, does not represent a unique phenomenon. Perhaps lysosomes that have been activated in L cells by different viruses (2) may contribute to the early "toxic" effect observed in the study of L cells (63), as well as to the damage of hepatic cells that occurs in coxiella infections in guinea pigs (101, 102). Bacterial endotoxins are known to activate lysosomes (2). No doubt a future study will yeild a detailed picture of the complex events connected with the "toxic" effect of coxiellae.

What makes the situation even more exiciting are the events occurring in cells after controlled mixed infection with the variant strains of coxiellae.

5. INTERACTIONS BETWEEN VARIANTS OF *C. burneti*

In view of the genetic basis of the restricted variability that occurs in "phase variation" in bacteria, which involves, as a rule, only two alternative and mutually exclusive sets of antigenic properties (11), as well as the fact that "phase variation" in *C. burneti* ordinarily involves the reproducible change from phase I to phase II and conversion back to phase I (10, 14, 94, 134), one might with confidence assume that the restricted antigenic variability associated with *C. burneti* has a similar genetic basis.

Bacterial and viral mutants arise at an "average" rate of 1 in 10^{9} particles; nonsmooth mutants, at an average of 1 in 10^{7} "smooth" cells (11). It could be expected that rickettsiae would behave in a somewhat similar way. Indeed, all attempts to devise an experimental system leading to a pure phase I antibody response have failed (10, 13, 33), and strains exclusively maintained in animals have been shown to consist of a heterogeneous population of *C. burneti* (35, 36, 63).

In nature, pathogens of different genotypes compete within populations. It seems evident that events associated with the replicative process of pathogens play a decisive role in the resulting interactions. Breakdown products of DNA and specific amino acids can affect population changes of bacteria both *in vivo* and *in vitro*. Lysed bacterial cultures or bacterial DNA is used as the transforming agent in bacteria; the transforming substance must nevertheless be provided during a stage of active growth of the recipient bacterium. The question as to whether transformation of bacteria ever occurs under natural conditions still lacks a decisive answer (11).

The Berry–Dedrick phenomenon, once called "transformation" in poxviruses, does not involve either transformation or genetic recombination between the viruses concerned, and represents, in fact, complementation. "Complementation" is used as a general term to describe situations in mixed infections with viruses that result in enhanced yield of one or both viruses in the mixture; many different mechanisms may operate in such interactions. Complementation between one-step mutants is useful as a preliminary index in genetic mapping. Several situations have been described in which a defective virus may be rescued by complementation. Intracellular interactions between viruses or their products result more often in interference. In addition to interferon, a wide variety of factors may cause interference; few of these have been investigated in detail. Another consequence of mixed infections is phenotypic mixing and transcapsidation, which is seen when two viruses with recognizably different coat antigens are used (32).

Studies have shown that interactions occur between the "smooth" parent and the "nonsmooth" mutant coxiellae in L cells and have a remarkable influence on the growth (63), as well as the antigenic makeup, of the organisms. Phenotypic mixing was indicated (67). "Reactivation" of a heat-inactivated host-cell-sensitive mutant strain of coxiellae in the presence of an active host-cell-dependent parent strain has been reported (60, 67). The rescued particles showed the antigenic makeup of the heat-inactivated mutant strain. The similarity of this phenomenon to the reactivation of poxviruses is striking (32).

Controlled mixed infections of the two active variants of coxiellae in the same host cell system (L cells) resulted either in interference or complementation, depending on the conditions (Fig. 12). In simultaneous infection with high doses of the mixture, complementation was achieved, and the rescued coxiellae acquired the antigenic makeup of the host-dependent parent type. With lower doses of the mixture, a marked interference and suppression of the growth of rickettsiae occurred. If, however, the host-dependent coxiellae were inoculated in high doses before the host-sensitive mutant, suppression of the growth of rickettsiae was observed (63, 67).

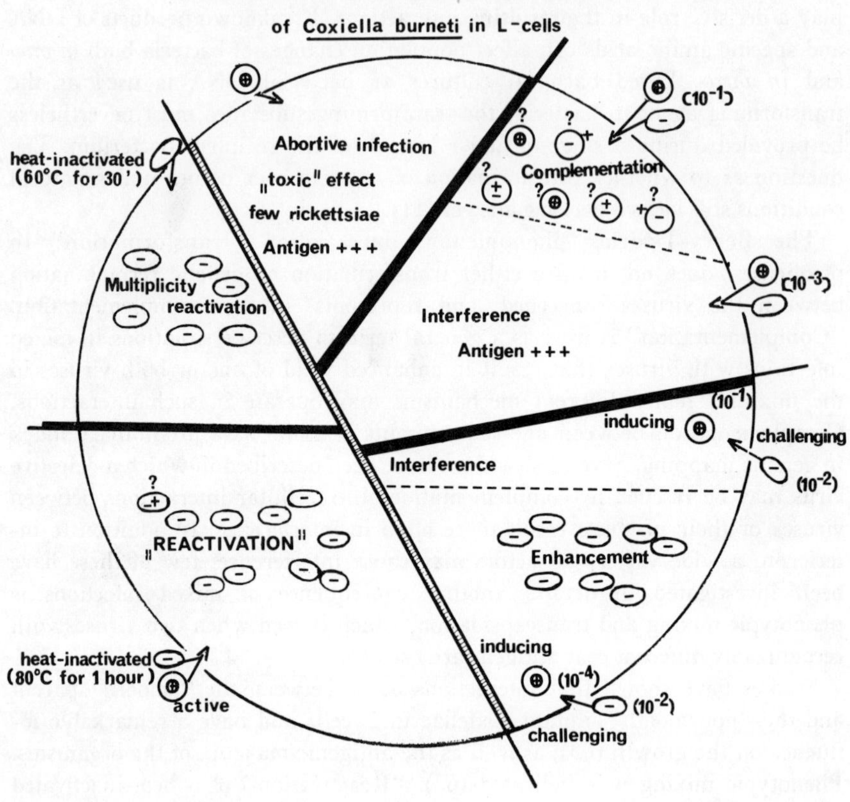

Fig. 12. Note: *Both strains, although antigenically monophasic, consisted of a mixed population of coxiellae particles (63, 67).

We are quite ignorant of the molecular events involved in the interaction of coxiellae. The fact is, however, that it has become possible to detect complementation and restoration under certain conditions of growth of the phase I phenotype of the host-dependent variant of *C. burneti*. Complementation, which is fundamentally a genetic technique, can be applied to the study of a variety of problems. Recent studies have produced evidence that the host cell is much more than a convenient medium for rickettsial growth and that the replication of this pathogen within a host cell is not an independent and isolated event.

"Transformation" with the psittacosis "virus" has been reported (39). Others have reported that a search for genetic interactions in the psittacosis group yielded no evidence of any kind of genetic interactions (42).

The execution of genetic studies demands certain precautions, and these were only recently recognized in the study of *C. burneti* (35, 36, 60, 63, 67). Many "unique" variational aspects in Gram-positive and Gram-negative bacteria have failed to survive under the impact of critical genetic studies. One can no longer arbitrarily neglect the "uniqueness" of phase variation in *C. burneti*, the concept of small, sporelike and large, vegetative forms in one and the same suspension of many rickettsiae, or the "unique" developmental cycle of the chlamydial "agent" with two distinctly small and large forms, thought to be homogenous.

6. CONCLUSIONS

For years the rickettsiae have been conceptualized as organisms occupying a position intermediate between bacteria and viruses. It can no longer be considered as coincidence that, in addition to the well-known segmentation of particles, a viruslike pattern of replication has been observed, in the early stages of infection, in psittacosis (108), *R. prowazeki* (70–72), *C. burneti*, and recently also *R. rickettsi* (73). Nor can it be accidental that "transformation" (39) and complementation occurred in the psittacosis "agent," and, on the other hand, "reactivation" (60) and complementation (63, 67) were described in *C. burneti*. The creation of a large group of minute, obligate intracellular pathogens that exhibit the well-known, but poorly understood, phenomena of latency and persistence, as well as many other puzzling properties, is not the work of a taxonomist. Rather, is was deduced by the taxonomist from an accomplishment of nature. The findings that rickettsiae and chlamydiae have biological properties that are in some ways similar to those of viruses should not be neglected. Even if they seem inconsistent with the present concept of viruses, there is no reason for thinking that the above-mentioned observations relating them to viruses are incorrect or exceptional (78, 86, 140). At present, from the molecular point of view, the cell is no longer considered the unit in biology, but is regarded as only one stage in a complicated pattern of transformation. There is no reason to think that a sequence of events leading from viruslike to bacterialike replication in a small, obligate, intracellular pathogen cannot occur. An oversimplified model of such events in the "nonsmooth" mutant of *C. burneti* is sketched in Fig. 13.

Rickettsiae possess both DNA and RNA, and the DNA of viruses is similar to the chromatinic bodies in bacteria. Yet the viral DNA obviously does not

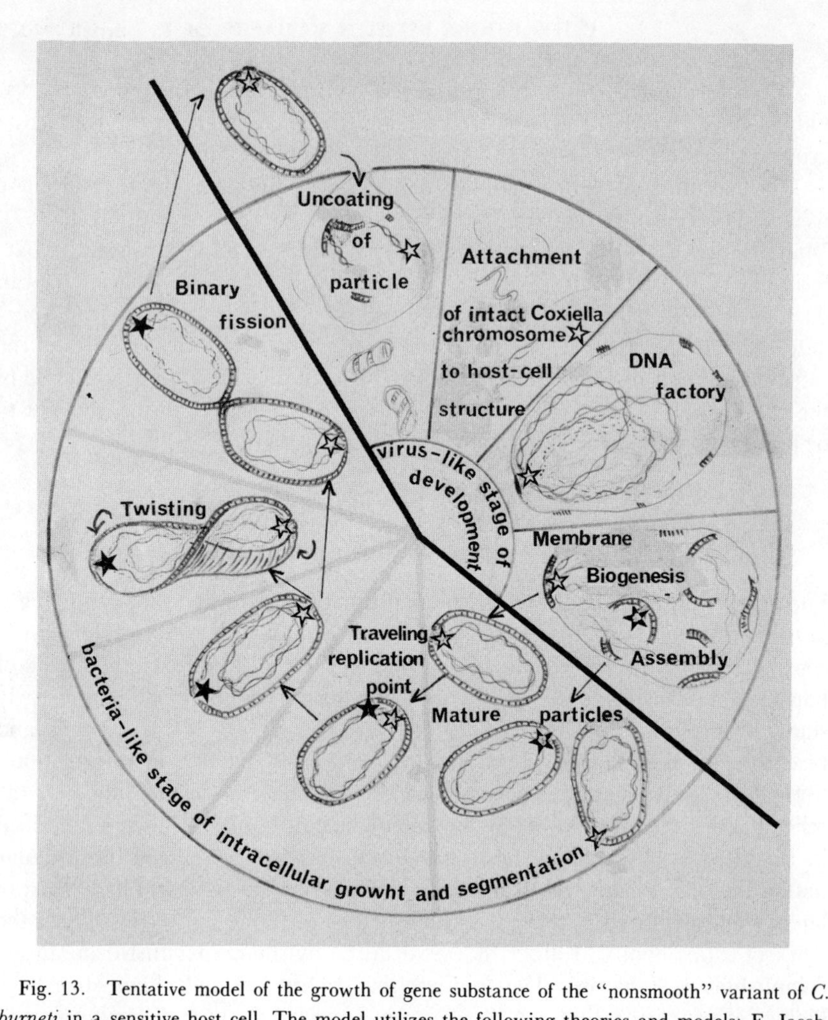

Fig. 13. Tentative model of the growth of gene substance of the "nonsmooth" variant of *C. burneti* in a sensitive host cell. The model utilizes the following theories and models: F. Jacob, S. Brenner, and F. Cuzin (51); K. G. Lark (76); data on the growth of gene substance of viruses and bacteria summarized by R. D. Hotchkiss (46); vaccinia as a model for membrane biogenesis by S. Dales and E. M. Mosbach (29); and observations of the twisted *C. burneti* by M. G. P. Stoker, K. U. Smith, and P. Fiset (135), and by P. Burton, D. Paretsky, and N. Kordová (unpublished data).

have the metabolic machinery characteristic of a bacterium. Evolution has provided bacteria with a chromosomal segregating apparatus, and recent evidence strongly indicates association of the cell membrane with the point of origin of DNA replication in bacteria. Not only DNA, but also DNA polymerase, seems to be particularly attached to the membrane. Evidence has also been presented that, when a host cell membrane is damaged, replication of phage DNA ceases, while some DNA synthesis continues, perhaps as a host-repair mechanism (44, 46). In as much as the presence of both DNA and RNA in rickettsiae constitutes a fundamental distinction from viruses, one ought to be alert to the possibility that even more complex combinations of nucleic acids occur in biological units (130).

ACKNOWLEDGMENTS

The senior author (N.K.) is indebted to Drs. D. Blaškovič, J. C. Wilt, L. Borecký and C. K. Hannan for criticism and comments on this chapter.

REFERENCES

1. Adler, H. I., W. D. Fisher, and A. A. Hardigree, *Trans. N.Y. Acad. Sci.* **31,** 1059 (1969).
2. Allison, A., *Perspect. Virol.* **5,** 29 (1967).
3. Allison, A. C., and H. R. Perkins, *Nature* **188,** 796 (1960).
4. Anacker, R. L., K. Fukushi, E. G. Pickens, and D. B. Lackman, *J. Bacteriol.* **88,** 1130 (1964).
5. Anacker, R. L., W. T. Haskins, D. B. Lackman, E. Ribi, and E. G. Pickens, *J. Bacteriol.* **85,** 1165 (1963).
6. Anacker, R. L., E. G. Pickens, and D. G. Lackman, *J. Bacteriol.* **94,** 260 (1967).
7. Bladen, H. A., R. Byrne, J. G. Levin, and M. W. Nirenberg, *J. Mol. Biol.* **11,** 78 (1965).
8. Begg, A. M., F. Fulton, and U. Van den Ende, *J. Pathol. Bacteriol.* **59,** 109 (1944).
9. Berry, G. P., and N. M. Dedrick, *J. Bacteriol.* **31,** 50 (1936).
10. Bobb, D., and C. M. Downs, *Can. J. Microbiol.* **8,** 689 (1962).
11. Braun, W., *Bacterial Genetics*, Saunders, Philadelphia, 1965.
12. Breed, R. S., E. G. D. Murray, and N. R. Smith, *Bergey's Manual of Determinative Bacteriology*, 7th ed., Williams & Wilkins, Baltimore, Md., 1957.
13. Brezina, R., Zentralbl. Bakteriol. Orig. **206,** 313 (1968).
14. Brezina, R., *Curr. Topics Microbiol.* **47,** (1969).
15. Brezina, R., J. Řeháček, and N. Kordová, *Acta Virol.* **7, 260 (1963).**
16. Brezina, R., J. Kazár, and V. F. Pospíšil, *Acta Virol.* **13,** 455 (1969).
17. Brezina, R., Š. Schramek, and J. Urvolgyi, *Acta Virol.* **9,** 180 (1965).
18. Brezina, R., and Š. Schramek, *Acta Virol.* **12,** 68 (1968).

19. Bruskov, V. I., and N. A. Kiselev, *J. Mol. Biol.* **37,** 367 (1968).
20. Burgdorfer, W., *Pathol. Microbiol. (Basel)* 24 (Supplem) 27, 1961.
21. Burnet, F. M., *Principles of Animal Virology*, 2nd ed., Academic Press, New York, 1960.
22. Burton, P. R., N. Kordová, and D. Paretsky, unpublished data.
23. Burton, P. R., D. Paretsky, and N. Kordová, *Can. J. Microbiol.* **17,** 143 (1971).
24. Burton, P. R., N. Kordová, and D. Paretsky, unpublished data.
25. Byrne, R., J. G. Levin, H. A. Bladen, and M. W. Nirenberg, *Proc. Natl. Acad. Sci. U.S.* **52,** 140 (1964).
26. Cohn, Z. A., F. E. Hahn, W. Ceglowski, and F. M. Bozeman, *Science* **127,** 282 (1958).
27. Consigli, R. A., and D. Paretsky, *J. Bacteriol.* **83,** 206 (1962).
28. Dales, S., and L. Siminovitch, *J. Biophys. Biochem. Cytol.* **10,** 475 (1961).
29. Dales, S., and E. M. Mosbach, *Virology* **35,** 564 (1968).
30. Downs, C. M., Personal communication, 1970.
31. Eakin, R. M., in Sensory Receptors, *Cold Spring Harbor Symp. Quant. Biol.* **30,** 363 (1965).
32. Fenner, F., *The Biology of Animal Viruses*, Academic Press, New York, 1968.
33. Fiset, P., and R. A. Ormsbee, *Zentralbl. Bakteriol. Orig.* **206,** 321 (1968).
34. Gale, E. F., *The Chemistry of Life*, Needham (ed.), United Press, Cambridge, 1970, p. 38.
35. Genig, V. A., *Vestn. Akad. Med. Nauk SSSR* **10,** 40 (1969).
36. Genig, V. A., *Bull. Soc. Pathol. Exp.* **62,** 476 (1969).
37. Glauert, A. M., and M. J. Thornley, *Ann. Rev. Microbiol.* **23,** 159 (1969).
38. Gold, P., and S. Dales, *Proc. Natl. Acad. Sci. U.S.* **60,** 845 (1968).
39. Gordon, F. B., R. W. Trimmer, and A. L. Quan, *Bacteriol. Proc.* **58,** 64 (1958).
40. Gottlieb, D., and P. D. Shaw, *Antibiotics*, Vol. I, Springer-Verlag, New York, 1967.
41. Green, R. G., *Science* **82,** 443 (1935).
42. Greenland, R. M., and J. W. Moulder, *J. Infect. Dis.* **108,** 293 (1961).
43. Handley, J., D. Paretsky, and J. Stueckemann, *J. Bacteriol.* **94,** 263 (1967).
44. Hayes, W., *The Genetics of Bacteria and Their Viruses*, Wiley, New York, 1964.
45. Higashi, N., *Zentralbl. Bakteriol. Orig.* **206,** 277 (1968).
46. Hotchkiss, R. D., *Cold Spring Harbor Symp. Quant. Biol.* **33,** 857 (1968).
47. Houwink, A. L., Biochim. Biophys. Acta **10,** 360 (1953).
48. Huxley, H. E., and G. Zubay, *J. Mol. Biol.* **2,** 10 (1960).
49. Ingraham, L. L., and A. B. Pardee, *Metabolic Pathways, 3rd ed., Vol. I, D. M. Greenberg (ed.), Academic Press, New York, 1967.*
50. Ito, S., and J. W. Vinson, *J. Bacteriol.* **89,** 481 (1965).
51. Jacob, F., S. Brenner, and F. Cuzin, *Cold Spring Harbor Symp. Quant. Biol.* **28,** 324 (1963).
52. Jeon, K. W., I. J. Lorch, and Danielli, *Science* **167,** 1626 (1970).
53. Jones, F., Jr., and D. Paretsky, *J. Bacteriol.* **93,** 1063 (1967).
54. Kokorin, I. N., *Acta Virol.* **12,** 31 (1968).
55. Kordová, N., *Acta Virol.* **3,** 25 (1959).
56. Kordová, N., *Folia Microbiol.* **4,** 237 (1959).

57. Kordová, N., *Acta Virol.* **4,** 56 (1960).
58. Kordová, N., *Acta Virol.* **4,** 173 (1960).
59. Kordová, N., *Arch. Ges. Virusforsch.* **15,** 697 (1965).
60. Kordová, N., *Acta Virol.* **14,** 36 (1970).
61. Kordová, N., *Zentralbl. Bakteriol. Parisotol.* **226,** 107 (1971).
62. Kordová, N., and R. Brezina, *Acta Virol.* **7,** 84 (1963).
63. Kordová, N., P. R. Burton, C. M. Downs, D. Paretsky, and E. Kováčová, *Can. J. Microbiol.* **16,** 125 (1970).
64. Kordová, N., and E. Kováčová, *Acta Virol.* **11,** 252 (1967).
65. Kordová, N., and E. Kováčová, *Acta Virol.* **12,** 25 (1968).
66. Kordová, N., and E. Kováčová, *Acta Virol.* **12,** 460 (1968).
67. Kordová, N., E. Kováčová, and J. C. Wilt, *Can. J. Microbiol.* **16,** 561 (1970).
68. Kordová, N., and P. Kvíčala, *Folia Microbiol.* **7,** 89 (1962).
69. Kordová, N., and J. Řeháček, *Acta Virol.* **3,** 201 (1959).
70. Kordová, N., and J. Řeháček, *Acta Virol.* **3,** 465 (1964).
71. Kordová, N., M. Rosenberg, E. Mrena, *Arch. Ges. Virusforsch.* **15,** 707 (1965).
72. Kováčová, E., and N. Kordová, *Arch. Ges. Virusforsch.* **19,** 57 (1966).
73. Kováčová, E., *personal communication.*
74. Kravchenko, A. T., O. S. Gudima, and V. N. Milyutin, *Vopr. Virusol.* **6,** 300 (1961).
75. Laidlaw, P. P., *Virus Diseases and Viruses,* University Press, Cambridge, 1939.
76. Lark, K. G., *Bacteriol. Rev.* **30,** 3 (1966).
77. Luria, S. E., and J. E. Darnell, Jr., *General Virology,* 2nd ed., Wiley, New York, 1967.
78. Lwoff, A., *Gen. Microbiol.* **17,** 239 (1957).
79. Mäkelä, P. M., and B. A. D. Stocker, *Ann. Rev. Genet.* **3,** 291 (1969).
80. Mallavia, L. P., and D. Paretsky, *J. Bacteriol.* **86,** 232 (1963).
81. Mallavia, L. P., and D. Paretsky, *J. Bacteriol.* **93,** 1479 (1967).
82. McDonald, T. L., and L. Mallavia, *J. Bacteriol.* **102,** 1 (1970).
83. McQuillen, K., *XVth Symposium of the Society for General Microbiology,* University Press, London, 1965.
84. Morgan, C., R. A. Rifkind, K. C. Hsu, M. Holden, B. C. Seegal, and H. M. Rose, *Virology* **14,** 292 (1961).
85. Morowitz, H. J., in *Ciba Foundation Symposium,* G. E. W. Wolstenholme and U. O'Connor (eds.), Churchill, London 1966.
86. Moulder, J. W., *The Biochemistry of Intracellular Parasitism,* University of Chicago Press, Chicago, 1962, pp. 43–83.
87. Moulder, J. W., *Bioscience* **19,** 875 (1969).
88. Munyon, W., J. Mann, and J. T. Grace, *J. Virol.* **5,** 32 (1970).
89. Murray, R. G. E., P. Steed, and H. E. Elson, *Can. J. Microbiol.* **11,** 547 (1965).
90. Myers, W. F., R. A. Ormsbee, J. V. Osterman, and C. L. Wisseman, Jr., *Proc. Soc. Exp. Biol. Med.* **125,** 459 (1967).
91. Myers, W. F., and D. Paretsky, *J. Bacteriol.* **82,** 761 (1961).
92. Nermut, M. V., and R. G. W. Murray, *J. Bacteriol.* **93,** 1949 (1967).
93. Nermut, M. V., Š. Schramek, and R. Brezina, *Acta Virol.* **12,** 446 (1968).

94. Ormsbee, R. A., *J. Bacteriol.* **63,** 73 (1952).

95. Ormsbee, R. A., in *Viral and Rickettsial Infections of Man*, Lippincott, Philadelphia, 1965.

96. Ormsbee, R. A., *Ann. Rev. Microbiol.* **23,** 275 (1969).

97. Ormsbee, R. A., and M. G. Peacock, *J. Bacteriol.* **88,** 1205 (1964).

98. Paretsky, D., *Zentralbl. Bakteriol. Orig.* **206,** 283 (1968).

99. Paretsky, D., R. A. Consigli, and C. M. Downs, *J. Bacteriol.* **83,** 538 (1962).

100. Paretsky, D., C. M. Downs, R. A. Consigli, and B. K. Jonce, *J. Infect. Dis.* **103,** 6 (1958).

101. Paretsky, D., C. M. Downs, and C. M. Salmon, *J. Bacteriol.* **88,** 137 (1964).

102. Paretsky, D., and J. Stuckemann, *J. Bacteriol.* **102,** 337 (1970).

103. Perkins, H. R., and A. C. Allison, *J. Gen. Microbiol.* **30,** 469 (1963).

104. Philip, C. B., *Ann. Rev. Entomol.* **6,** 391 (1961).

105. Philip, C. B., *J. Egypt. Public Health Assoc.* **38,** 61 (1963).

106. Pinkerton, H., *Bacteriol. Rev.* **6,** 37 (1942).

107. Pirie, N. W., Principles of Biomolecular Organization, in *Ciba Foundation Symposium, G. E. W. Wolstenholme and M. O'Connor (eds.), Churchill, London, 1966, p. 136.*

108. Pollard, M., and Y. Tanami, *Ann. N.Y. Acad. Sci.* **98,** 50 (1962).

109. Pospíšil, V. F., *Acta Virol.* **10,** 542 (1966).

110. Price, W. H., *Dynamics of Virus and Rickettsial Infections*, Blakiston, New York, 1954.

111. Price, W. H., *J. Bacteriol.* **69,** 106 (1955).

112. Price, W. H., H. Emerson, H. Nagel, R. Blumberg, and S. Talmalge, *Amer. J. Hyg.* **67,** 154 (1958).

113. Rapp, F., *Ann. Rev. Microbiol.* **23,** 293 (1969).

114. Rifkind, R. A., K. C. Hsu, and C. Morgan, *J. Histochem. Cytochem.* **12,** 131 (1964).

115. Ris, H., and J. P. Fox, *J. Exp. Med.* **89,** 681 (1949).

116. Roberts, A. N., and C. M. Downs, *J. Bacteriol.* **77,** 194 (1959).

117. Robertson, J. D., *Biochem. Soc. Symp.* **16,** 3 (1959).

118. Robertson, J. D., in *Principles of Biomolecular Organization*, Little, Brown, Boston, 1966.

119. Rogers, H. J., *Symp. Soc. Gen. Microbiol.* **15,** 186 (1965).

120. Rose, A. H., *Chemical Microbiology*, Plenum, New York, 1968.

121. Rosenberg, M., and N. Kordová, *Acta Virol.,* **4,** 52 (1960).

122. Rosenberg, M., and N. Kordová, *Acta Virol.,* **6,** 176 (1962).

123. Salton, M. R. J., *Microbial Cell Walls*, Wiley, New York, 1960.

124. Schaechter, M., F. M. Bozeman, and J. E. Smadel, *Virology* **3,** 160 (1957).

125. Schramek, Š., *Acta Virol.* **12,** 18 (1968).

126. Shankel, D. M., N. J. Tungerius, and C. M. Downs, *Bacteriol. Proc.* **19,** (1965).

127. Singer, S. J., and A. F. Schick, *J. Biophys. Biochem. Cytol.* **9,** 519 (1961).

128. Smith, J. D., and M. G. P. Stoker, *Brit. J. Exp. Pathol.* **32,** 433 (1951).

129. Spencer, R. R., and R. R. Parker, *Hyg. Lab. Bull.* **154,** 1 (1930).

130. Stanley, W. M., in *Molecular Basis of Virology*, ACS Monograph 164, H. Fraenkel-Conrat, (ed.), Reinhold, New York, 1968.

131. Steinhaus, E. A., *Insect Microbiology*, Comstock, Ithaca, 1946.

132; Stelzner, A., and W. Linss, *Nature* **218,** 1069 (1968).

133. Stoker, M. G. P., *Brit. Med. J.* **1,** 963 (1957).

134. Stoker, M. G. P., and P. Fiset, *Can. J. Microbiol.* **2,**310 (1956).

135. Stoker, M. G. P., K. M. Smith, and P. Fiset, *J. Gen. Microbiol.* **15,** 632 (1956).

136. Watson, J. D., *Molecular Biology of the Gene*, Benjamin, New York, 1965.

137. Weiss, E., Zentralbl. Bakteriol. Orig. **206,** 292 (1968).

138. Weiss, A., B. L. Elisberg, F. M. Bozeman, R. A. Ormsbee, C. B. Philip, C. L. Wisseman, Jr., P. Fiset, D. Paretsky, C. M. Downs, and J. W. Vinson, *Science* **159,** 553 (1968).

139. Wiebe, M. E., P. R. Burton, and D. M. Shankel, *J. Bacteriol.* **110,** 368 (1972).

140. Wisseman, C. L. Jr., *Zentralbl. Bakt. Orig.* **206,** 299 (1968).

141. Zdrodowsky, P. F., and E. H. Golinevich, *The Rickettsial Diseases*, Pergamon Press, New York, 1960.

CHAPTER

TEN

Molecular Bacteriology

M. R. J. SALTON

"Molecular bacteriology" is such a broad topic that it could encompass everything related to the existence and multiplication of the bacterial cell. In its strictest sense it would thus cover the chemistry of the cell from the simplest molecules to the most complex macromolecules, as well as the energy and biosynthetic reactions and the complex factors involved in the continuity of the bacterial genome and its expression in terms of control and regulation of cellular processes.

The prokaryotic bacterial cell has a number of distinctive features that separate it from its more "complex" eukaryotic counterpart (73). The major differences between eukaryotic and prokaryotic cells are now quite well understood and specified in terms of their anatomy and biochemistry. The energy-yielding reactions, the nature of the genetic material and the machinery of its replication, and the mechanisms of protein biosynthesis are "universal." The essential differences between eukaryotic and prokaryotic cells relate to a limited number of major features, including the type of ribosomes (70S type in bacteria), the anatomical organization of the nucleus, the respiratory equipment (mitochondria), the endoplasmic reticulum, the inability of bacteria to synthesize sterols, and the possession of unique heteropolymers in their outer cell walls and envelope layers (65).

Obviously this chapter on "molecular bacteriology" cannot cover the entire scope of bacterial chemistry, biochemistry, and regulation of growth and division. Many of the unique features of the bacterial cell relate to the chemical structure, biosynthesis and organization of the cell surface components (walls and membranes), and the chromosome. Therefore this chapter will concentrate on selected aspects of these unique features, in the hope that it will serve to emphasize these fascinating areas of comparative cellular biology and biochemistry.

1. CHEMICAL COMPOSITION OF BACTERIAL CELL

The overall chemical composition of the bacterial cell is very similar to that of all other types of cells of animal, plant, and microbial origins. Bacteria thus possess the major classes of chemical constituents, namely, the proteins, nucleic acids (ribonucleic acid, RNA, and deoxyribonucleic acid, DNA), polysaccharides, and lipids. These are the constituents common to all types of cells and are components of the macromolecular structures required for the cell's genetic continuity (DNA), biochemical activities, and permeability characteristics (lipids and proteins). In addition to these four major classes of chemical constituents, complexes of proteins, polysaccharides, and lipids are found in bacteria as in other types of cells. Such complexes include glycoproteins, lipoproteins, and lipopolysaccharides. The lipopolysaccharides constitute a rather unique type of cellular component, characteristic of the Gram-negative bacteria. Many of the unusual polymeric components or macromolecular complexes of the bacterial cell are present in the cell wall structures. These unique components found only in bacteria are the peptidoglycans (mucopeptides, murein) and teichoic acids, the former structures being present also in certain blue-green algae.

As would be expected, the building blocks of the major classes of substances found in bacteria are the same as those present in all cells. Thus bacterial proteins contain the same variety of amino acids universal to all proteins. It is only the "unusual" amino acids, such as diaminopimelic acid, ornithine, aminobutyric acid, and D-isomers of alanine, glutamic acid, and several other amino acids, which occur in peptide form in the cell wall peptidoglycan structures. There are, however, some quantitative differences in the amino acid compositions of proteins; in particular, the exocellular proteins of Gram-positive bacteria are generally lower in cysteine content than most proteins (55). Other marked quantitative differences are seen in the base ratios of bacterial DNA. Thus bacterial DNAs span a wide range of guanine plus cytosine content (mole per cent G + C), from the Gram-negative *Spirillum linum* with 28 to 30% G + C to micrococci and *Nocardia* species in the 70 to 80% range. Various bacterial groups are represented throughout the range between these two extremes. This quantitative chemical difference in the base compositions of bacterial deoxyribonucleic acids contrasts with the situation in regard to the DNA content of vertebrates, which is essentially the same in all species (approximately 40% G + C).

The monomeric components of bacterial polysaccharides are generally similar to those of higher animal and plant cells. These include the common aldohexoses such as glucose, galactose, and mannose. Bacterial polysaccharides also contain deoxyhexoses such as rhamnose and fucose, found too in animal and plant polysaccharides and glycoproteins. Uronic acids and the two amino sugars, glucosamine and galactosamine, are present in complex polysaccharides of bacterial, animal, and plant origins. Some the unusual sugar building blocks of bacterial polysaccharides are found in the lipopolysaccharide complexes. These include the dideoxy sugars (colitose, abequose, tyvelose, etc.), discovered by Westphal and his colleagues (79; see the review in ref. 40). In addition to the occurrence of dideoxy sugars in certain lipopolysaccharides, these structures also contain aldoheptoses and the keto sugar, 2-keto-3-deoxyoctonic acid (KDO). Although the amino sugars *N*-acetylglucosamine and *N*-acetylgalactosamine occur widely in Nature, bacterial polysaccharides and lipopolysaccharides contain a great variety of additional amino sugars that have not so far been detected in carbohydrates isolated from animal and plant sources. These include fucosamine, viosamine, diamino sugars, and a number of other compounds found in bacterial polysaccharides and certain antibiotics (see the reviews in refs. 63 and 69). One of the first unique bacterial amino sugars to be found was of course, muramic acid, discovered by Strange in spore peptidoglycan (74). The *N*-acetylamino sugar, together with *N*-acetylglucosamine, occurs in the glycan backbone of the wall peptidoglycan, where muramic acid, by virtue of its carboxyl group, serves as the linkage site for the peptide moieties.

In common with higher cells, the dominant class of lipids in bacteria consists of the phospholipids. The variety of phospholipids present in bacteria is rather similar to that found in other cells, except that lecithin (phosphatidylcholine) occurs in but a few species and is not as widely distributed as in plant and animal species. Phosphatidylglycerol, cardiolipin, phosphatidylethanolamine, phosphatidylserine, and phosphatidylinositol are commonly present in various bacterial species. Although the chemical structures of the bacterial phospholipids are basically similar to those from other sources, the main point of departure from the phospholipids derived from plants and animals is usually in the nature of the fatty acid constituents. Thus many bacterial phospholipids have branched-chain fatty acids of 15 and 17 carbon atoms as the dominant acids in contrast to the prevalent types of saturated and unsaturated straight-chain fatty acids found in other phospholipids.

In addition to the more "classical" variety, several unusual phospholipids have been found in bacteria. These include the amino acid acyl derivatives of phospholipids and the phytanyl ethers of phosphatidylglycerol, the latter having been found in the halophilic bacteria (for reviews see refs. 35, 41, and 51). Until quite recently it was believed that sphingolipids were absent from bacteria. However, a ceramide phospholipid has been found in the obligate anaerobe *Bacteroides melaninogenicus* by White, Tucker, and Sweeley (82). The occurrence of this type of lipid in bacteria is extremely rare. One other unusual feature of this bacterial sphingolipid was the presence of what appeared to be the branched forms of the sphinganines.

Other lipids and "lipid-soluble" compounds occurring in bacteria are the glycosyl diglycerides, other glycolipids, sulfolipids in several rare instances, carotenoids, vitamin K_2 isoprenologs (menaquinones), ubiquinones, and squalene in a halophile. A glucosamine derivative of phosphatidylglycerol has also been found in *Bacillus* species and appears to make the membrane a more rigid structure (52). Plasmalogens have been found in some of the clostridia (23).

Unusual lipids present in bacteria include the "storage polymer," poly β-hydroxybutyrate, and the "lipid A" of the Gram-negative lipopolysaccharide structures. In many of the enteric bacteria, β-hydroxymyristic acid is linked to glucosamine to constitute the lipid A component. Hydroxy acids of shorter chain lengths have been reported in pseudomonads.

The only other unusual bacterial constitutents that should be mentioned are the teichoic acids, the polyol phosphate polymers of ribitol or glycerol (1). These phosphodiester-linked polymers have in addition sugars (e.g., glucose, *N*-acetylglucosamine, *N*-acetylgalactosamine) and D-alanine in ester linkage, joined to the polyol phosphate backbone. These compounds appear to be quite unique for Gram-positive bacteria and have not as yet been found in any other organisms.

Some of the unique chemical constituents found in bacterial cells are summarized in Table 1.

Table 1. Unique Features of Certain Chemical Constituents Found in Bacteria

Constituent	
Amino acids	
Diaminopimelic acid, aminobutyric acid, ornithine, homoserine (in cell wall peptidoglycans)	
D-Alanine (in peptidoglycans, teichoic acids)	
D-Glutamic acid (in peptidoglycan, capsular polypeptide)	
Sugars and amino sugars	
Dideoxy sugars	(in lipopolysaccharides)
Aldoheptoses	(in lipopolysaccharides)
KDO	(in lipopolysaccharides)
Muramic acid (in peptidoglycan)	
Fucosamine	(in polysaccharides or lipopolysaccharides)
Viosamine	(in polysaccharides or lipopolysaccharides)
Diamino sugars	(in polysaccharides or lipopolysaccharides)
Polyols	
Ribitol	(in wall teichoic acids)
Glycerol	(in wall teichoic acids)
Phospholipids	
Phytanylglycerol diethers (in halophiles)	
Glucosaminylphosphatidylglycerol, ceramide phospholipid	

Apart from the chemical constituents of the major macromolecular structures and components, the bacterial cell also contains essential mineral constituents such as K^+, Na^+, Ca^{2+}, Mg^{2+}, and smaller amounts of metal ions, including Fe^{3+}, Cu^{2+}, Mn^{2+}, and Zn^{2+}, required for specific catalytic processes in the cells. It is evident that PO_4 and SO_4 will also be present as inorganic substances needed for a variety of biosynthetic processes and cellular structures. Some of the mineral constituents of the bacterial cell may be in the bound form when they are required for the stability of surface structures such as the lipopolysaccharide layers or membranes. Mineral constituents, together with other low-molecular-weight compounds (polyamines, amino acids, vitamins, nucleotides, metabolites), may occur in the "cell sap" and constitute the "metabolic pool" of the cell. Such soluble components of the metabolic pool can constitute about 5 to 10% of the dry weight of the bacterial cell.

It is evident that the overall chemical composition of the bacterial cell will vary tremendously, depending on such factors as the presence or absence of capsular substances, storage inclusions, the amount and thickness of the cell wall, the phase of growth of the cells, and the nutritional status of the medium.

Many other factors will affect the contents of individual compounds. Thus the rate of turnover of specific substances will influence their contents at a given time in the life of the cell. For these reasons no attempt will be made to summarize the gross chemical composition of the bacterial cell. In this chapter, the nature and distribution of specific substances in recognizable structures of the cell will be emphasized by illustration with selected examples. The treatment of this topic will be far from exhaustive, however, since the body of data available on the chemistry and biochemistry of bacterial walls, capsules, membranes, flagella, lipopolysaccharides, and ribosomes is now enormous.

2. STRUCTURAL ORGANIZATION OF BACTERIAL CELL AND ITS MOLECULAR ARCHITECTURE

In discussing the anatomy of the bacterial cell it is important to understand the distinctions between the two broad groups separated by the Gram-stain reaction into the Gram-positive and Gram-negative bacteria. The essential structural differences between these two groups are presented diagrammatically in Figs. 1 and 2, where the bacterial cells are depicted as rod-shaped organisms. The literature abounds in illustrations of thin sections of both

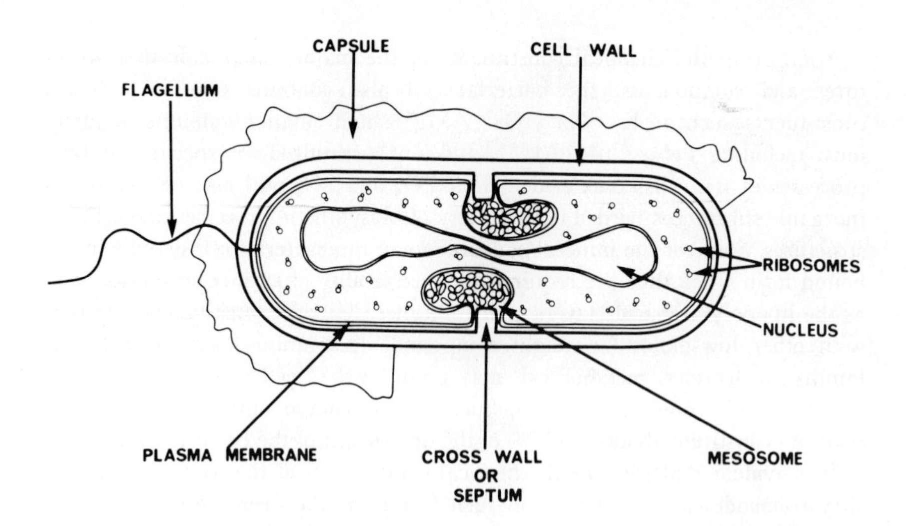

Fig. 1. A diagrammatic representation of the anatomy of a Gram-positive bacterium, indicating surface appendages, surface layers (capsules, walls and plasma membranes), and internal structures. The discontinuity of the capsule indicates that not all species possess this surface component.

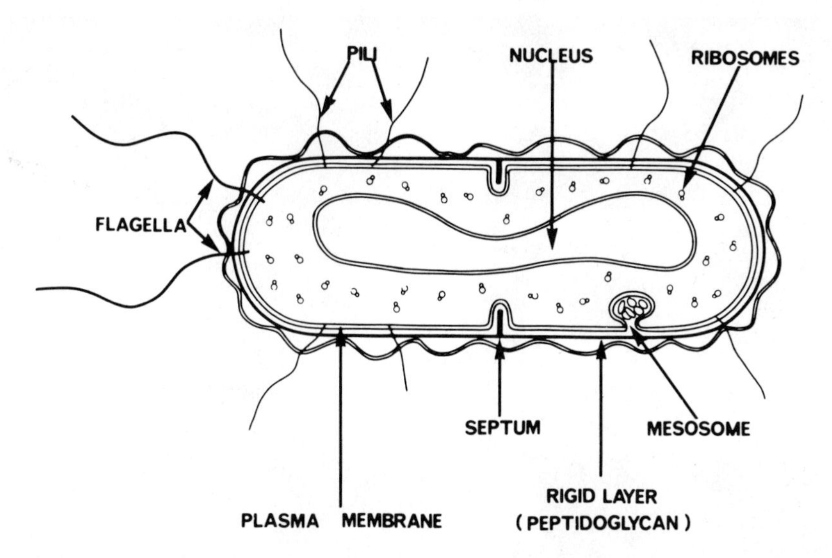

Fig. 2. A diagrammatic representation of the anatomy of a Gram-negative bacterium. Note the two types of surface appendages, flagella and pili. The outer envelope component (lipopolysaccharide, protein, lipid complex) is illustrated as a convoluted double-track structure touching the peptidoglycan layer. Capsules have not been shown, although they occur in some species.

Gram-positive and Gram-negative bacteria, and the diagrammatic representations presented in Figs. 1 and 2 serve to emphasize that the major differences between the two groups reside in the details of their surface structures.

Flagella, the organs of locomotion of the bacterial cell, are found in various species of both groups. However, the pili appear to be restricted to certain Gram-negative organisms. Both types of structures can be classed as "surface appendages." Bacterial flagella from both Gram-positive and Gram-negative organisms have been isolated in homogeneous states, and analysis has shown that they are composed almost exclusively of protein. Small amounts of carbohydrate detected in earlier preparations from *Proteus vulgaris* were probably derived from contaminating fragments of the lipopolysaccharide layer. The proteins belong to the class termed "flagellins," which have a subunit molecular weight of about 20,000. In both groups of bacteria, the flagella pass through the cell walls and terminate in a complex, specialized connecting structure that is retained in position in the region of the plasma or cytoplasmic membrane. Thus the "end" of the flagellum is a highly differentiated structure and probably possesses proteins differing from the flagellin which makes up the external appendage. Although the functions of the flagella are known, the precise functions of some of the pili have yet to be

Fig. 3. Isolated cell walls of *Streptococcus pyogenes* group A, negatively stained with phosphotungstic acid. The flattened walls are about 1 μ in diameter. Unpublished electron micrograph of Nachbar, Freer, and Salton.

defined. Brinton (6) has investigated the pili of *Escherichia coli* and has shown the existence of a number of distinct types. These can be distinguished serologically as well as on the basis of size. A function for one pilus as the male-specific structure in *E. coli* has been suggested, and there is strong evidence that it is involved in the conduction of DNA during mating. Certain pili may have a role in the adhesion of Gram-negative bacteria to other cellular structures. Brinton (6) has also established the protein nature of pili.

Capsules may be formed by both Gram-positive and Gram-negative bacteria, and such structures can be removed enzymatically from either group without impairing viability. Thus there does not appear to be any essential difference between the two groups in the nature of the association of the capsules with the underlying cell wall surface. Other than protection against phagocytosis, a clear function for the capsule has not been established. The "colloidal" nature of high-molecular-weight surface capsules could afford the cell some protection against desiccation and perhaps also protect it by sequestering toxic metal ions from its immediate environment.

Comparison of the organizations of the cell wall structures in Figs. 1 and 2 shows the major difference between the two groups. In Gram-positive bacteria, the cell wall structure is generally thick and amorphous with little evidence of fine-structured elements. This type of structure is illustrated in the electron micrograph of the isolated cell walls of streptococci seen in the negatively stained preparation presented in Fig. 3. However, it should be pointed out that in a few instances certain Gram-positive species also have highly structured layers external to the rigid peptidoglycan (see ref. 47).

The profiles of the surfaces of Gram-negative bacteria are basically very similar, an observation first clearly emphasized by Murray, Steed, and Elson (48). In general, the outermost part of the envelope surface presents a typical "unit membrane" profile, as seen in thin sections, and presumably represents the organized "outer membrane" (16, 22) containing the lipopolysaccharide (LPS)–protein–lipid complex. It is possible that the polysaccharide chains of the LPS extend out beyond the surface as defined by the outer track of the envelope layer. The outer membrane and globular structures are anchored to the rigid peptidoglycan layer. It is quite likely that the "globular" components of the *E. coli* envelope are the lipoprotein structures isolated and characterized by Braun and Rehn (4). This lipoprotein has been described as the "murein lipoprotein" and has been shown to be covalently linked to the peptidoglycan (4). Although the peptidoglycan layer is considerably thinner than it is in the majority of Gram-positive species, there is little doubt that it functions as the rigid, shape-conferring structure of Gram-negative bacteria. One of the fascinating consequences of the existence of this more complex envelope structure in Gram-negative organisms is that it confers an additional "barrier function" on these cells, as shown by the work of Leive (38) and Heppel (27).

Plasma membranes and mesosomes are present as the "surface" structures underlying the walls of both Gram-positive and Gram-negative bacteria, although the mesosome intrusions in the latter group are fewer and generally smaller and less conspicuous. The membrane systems of bacteria are multifunctional structures and are discussed in further detail in Section 5.

The principal functions and chemical compositions of the surface structures of both groups of bacteria are summarized in Table 2.

The internal organization of Gram-positive bacteria is indistinguishable from that of Gram-negative organisms. Thin sections of representative species of both groups reveal the nucleus or chromatin as a fibrous network in the central portion of the cell. Moreover, the nucleus lacks a surrounding membrane, a feature that constitutes one of the outstanding characteristics of

Table 2. Functions of Surface Structures of Bacterial Cells

Structure	Function	Chemical Composition
Flagella	Locomotion	Protein
Pili	Conjugation tube	Protein
	Cell adhesion	
Capsules and extracellular material	Protective(?)	Polysaccharides, Polypeptide
	Phage receptors	
	Cell adhesion	Polysaccharide
Cell wall (Gram-positive)	Mechanical protection	Peptidoglycan, teichoic acids
	Phage receptors	Polysaccharides
"Wall," outer envelope	Mechanical protection	Peptidoglycan
(Gram-negative)	Permeability	Lipopolysaccharide,
	Phage receptors	Lipid, and protein
Plasma membranes + mesosomes	Permeability	Lipid, Protein
	Biosynthesis	
	Electron transport	
	Chromosome anchoring and partition	

From ref. 66.

the prokaryotic cell. In order to ensure the anchoring and separation of nuclei during growth and division, the bacterial membrane assumes an important role in these processes (60). Such a function, noted in Table 2, is thus one of the many functions performed by the membrane system. The precise mode of attachment of the DNA to the plasma–mesosome membrane structures has yet to be elucidated. At least it can be assumed that such an attachment site would be of a highly specific nature, and attempts have been made to identify the structures involved.

The overall appearances of thin sections of bacteria and related blue-green algae are thus very similar. Ribosomes are packed into the remaining spaces in the cells of most species. There are of course obvious species differences in the ribosomal populations in bacterial cells. In some the ribosomes are sparsely scattered throughout the "cytoplasm"; others are densely packed with ribosome particles. The bacterial ribosomes are unlike their mammalian and plant cell counterparts in that they are not organized on a membranous endoplasmic reticulum. There have been suggestions, however, that the ribosomes may be attached to a fine matrix in the bacterial cell.

Apart from complex "internal" structures such as prespores and mature endospores, bacteria have a variety of granules (β-hydroxybutyrate polymer, glycogen, starchlike polysaccharides, and polymetaphosphate), core and microtubular structures of unknown functions, and crystalline bodies and crystal-like arrays of particles. Such internal inclusions are found in both groups differentiated by the Gram and acid-fast stains.

3. MOLECULAR STRUCTURE AND FUNCTIONS OF WALLS, ENVELOPES, AND MEMBRANES

Much is now known about the chemical structures of the major components of the cell walls of Gram-positive bacteria. The peptidoglycan and teichoic acids have been intensively studied and reviewed in considerable detail by Ghuysen (21), Rogers and Perkins (58), Rogers (56), Baddiley (1), and Salton (62). The peptidoglycan structure has been largely elucidated by the use of specific enzymes, many of which have been isolated as exocellular enzymes from *Streptomyces* spp. (21). Enzymes specific for the degradation of the glycan backbone, breaking the glycosidic bonds between *N*-acetylmuramic acid and *N*-acetylglucosamine (β—1→4 linkages), have established the chemical structure of this part of the peptidoglycan. Amidases specific for the bond between muramic acid and L-alanine were used in the chemical characterization of the mode of linkage of the peptide moiety to the glycan (21). A variety of endopeptidases have been found to be highly specific for the interpeptide bridges in the various wall peptidoglycans (21).

Table 3. Peptide Bridges in Peptidoglycans of Walls of Gram-Positive Bacteria[a]

Bacterium	Peptide Bridges
M. lysodeikticus	Polymerized peptide subunits
S. lutea	Polymerized peptide subunits
M. roseus	(L-Ala-L-Ala-L-Ala-L-Thr)-L-Lys
M. roseus	(L-Ala-L-Ala-L-Ala)-L-Lys
M. radiodurans	(Gly-Gly)-L-Orn
Staph. aureus	(Gly-Gly-Gly-Gly-Gly)-L-Lys
Strep. pyogenes	(L-Ala-L-Ala)-L-Lys
Strep. faecalis	(L-Ala-L-Ala-L-Ala)-L-Lys
Strep. faecalis	(D-iso-AsN)-L-Lys
Lact. casei	(D-iso-AsN)-L-Lys
L. mesenteroides	(L-Ala-L-Ser)-L-Lys

[a] Data from ref. 21.

Table 4. G + C Contents and Types of Interpeptide Linkage in Peptidoglycans of Various Micrococci

Organism	G + C Content (mole %)[a]	Interpeptide Linkage in Peptidoglycan[b]
M. lysodikticus	73.3	*M. lysodeikticus*—type (21)
S. flava	72.4	*M. lysodeikticus*—type (21)
M. tetragenus	73.0	Not determined
S. lutea	73.0	*M. lysodeikticus*—type (21)
M. roseus	73.5	L-Lys-L-Ala_3—type (21)
M. roseus R27	———	L-Lys-Thr-L-Ala_3—type (21)
M. conglomeratus	69.4	L-Lys-L-Ala_3-type (34)
M. varians	72.4	*meso*-DAP-D-Glu_2—type (34)
Micrococcus spp.	———	*meso*-DAP-D-Glu_2—type (34)
C. coelicolor	65.3	L-Lys-(Ser, Ala, Thr)[c]
M. caseolyticus	44.4	L-Lys-(Gly_4, Ser_1)
M. rhodochrous	70.4	Not determined
S. ventriculi	30.6[d]	Not determined
S. maxima	28.6[d]	Not determined
S. ureae	42.9	L-Lys-Gly-D-Glu (34)
B. subtilis	43.0	Direct (11)

[a] Determinations performed by Dr. Manley Mandel.
[b] References in parentheses.
[c] Determinations performed in Dr. O. Kandler's laboratory.
[d] Data from ref. 7.
[e] Data from ref. 83.

Apart from a few rare exceptions, the basic structure of the peptidoglycan involves a tetrapeptide of L-alanine, D-isoglutamic acid, L-lysine or α-ϵ-diaminopimelic acid and D-alanine, linked to the muramic acid, with a variable interpeptide cross-linking structure between the diamino acid (lysine or DAP) and the D-alanine of an adjacent tetrapeptide. Ghuysen (21) has distinguished the principal varieties of these cross-linked structures, and a few examples are given in Table 3. It is of interest to note that the mode of interpeptide linking is often characteristic of certain taxonomic groups. An example of the relationship of this chemical parameter to G + C content is given in Table 4 and has been taken from a recent comparison of some of the micrococci by Whiteside, De Siervo, and Salton (83). These results show that there is considerable individuality in the exact type of cross bridge in micrococci possessing G + C contents in the 70 to 74 mole % range. Similar variations within other groups of Gram-positive bacteria have also been observed (e.g., peptidoglycans of lactic acid bacteria; see ref. 21). Thus the peptidoglycan–interpeptide bridging can be used as a supplementary taxonomic character rather than an absolute chemical parameter. These results will undoubtedly be extended by the comprehensive survey of peptidoglycans carried out by Kandler and his colleagues (see, e.g., ref. 34).

Although the chemical structure of *E. coli* peptidoglycan has been fully established and shown to be similar in fundamental character to that of the peptidoglycans of Gram-positive bacterial walls, much less is known about other Gram-negative species. The peptidoglycan of *E. coli* has a direct bridging between the ϵ—NH_2 group of DAP and D-alanine of the adjacent peptide; and, as Ghuysen (21) has pointed out, available evidence suggests that a similar interpeptide linkage is present in peptidoglycans of other Gram-negative species.

As already mentioned, the peptidoglycans provide both Gram-positive and Gram-negative bacteria with a rigid structure that confers shape and affords considerable mechanical protection of bacterial cells. Although other wall polymers may be covalently linked to the peptidoglycans of Gram-positive bacteria (e.g., teichoic acids and polysaccharides, the chemical structures of which are reviewed in detail elsewhere: refs. 1, 58, 62), the evidence suggests that

Table 5. Composition of Isolated Cell Walls (Envelopes) of Gram-Negative Bacteria[a]

Organism	N (%)	Reducing Sugar (%)	Total Lipid (%)	Peptidoglycan (Mucopeptide) (%)	Reference
E. coli	10.1	16	20	*ca.* 10	62
E. coli			15.6	7.2	23a
N. animalis			16.8	6.4	23a
N. caviae			30.0	6.4	23a
N. pharyngis-flavus			11.4	8.3	23a
V. parvula			27.0	24.5	23a
P. aeruginosa	8.4	8.0	15		14
R. rubrum	8.3		23		62
S. bethesda	8.0	18.5	12		14
S. pullorum	6.4	46	19		62
S. typhimurium (S)	10.4	15.6	14.2, 18.4		27a

[a] Data as summarized in ref. 19.

Table 6. Amino Acid Compositions of Cell Walls (Envelopes) and Wall Protein Fractions from Gram-Negative Bacteria[a]

Amino Acid	*E. coli* Envelopes[b] (g/100 g)	*E. coli* Envelopes[c] (g/100 g)	*E. coli* Phenol-Soluble Protein[c] (g/100 g)	*Ps. aeruginosa* Wall Protein[d] (%)	*Ps. aeruginosa* Walls[e] (g/100 g)	*S. bethesda* Walls[e] (g/100 g)
Asparatic acid	7.1	7.8	10.5	10.2	9.3	14.5
Threonine	3.8	2.5	2.9	4.7	—	—
Serine	3.7	2.0	2.5	4.7	5.4	2.6
Glutamic acid	6.9	6.6	6.5	13.1	5.8	4.9
Proline	1.5	1.7	1.5	4.5	—	—
Glycine	3.1	3.1	3.9	10.0	7.1	5.5
Alanine	5.6	4.3	4.0	11.4	5.1	4.3
Valine	3.4	3.7	3.7	7.6	3.9	2.0
Methionine	0.7	1.4	1.4	+	—	—
Isolencine	3.7	2.7	2.6	5.2	}2.8	}1.6
Leucine	5.3	4.9	4.6	9.9		
Tyrosine	3.3	3.5	5.2	1.3	4.2	2.5
Phenylalanine	3.0	3.1	3.1	3.2	7.3	4.1
Lysine	4.0	3.5	4.3	4.9	1.7	0.5
Histidine	0.9	1.1	0.7	1.8	—	—
Arginine	3.8	4.0	3.7	6.7	1.3	0.4

[a] Data as summarized in ref. 19.
[b] From ref. 62.
[c] From ref 50.
[d] From ref. 30.
[e] From ref. 14; - not determined.

these are less intimately involved in the mechanical rigidity of the wall structures. The associations of other structures with the walls of Gram-negative bacteria are less clearly understood, although it is now known that the murien lipoprotein of *E. coli* is covalently joined to the peptidoglycan (4).

It has been known for a considerable period of time that the walls or envelopes of Gram-negative bacteria are chemically more complex (61). There is, however, a good body of information on the overall chemical composition, amino acid composition, and lipid and fatty acid components, and indeed even some quantitative estimates of the peptidoglycan contents of these complex structures. A selection of analytical data for the walls or envelopes of a number of Gram-negative species is presented in Table 5, and the amino acid compositions for several species are summarized in Table 6. The walls or envelopes of Gram-negative bacteria are generally richer in lipid material than those of Gram-positive organisms. Thus, as shown in Table 5, the lipid contents range from about 10 to 30% of the weight of the isolated structures, compared to low values of 1 to 2% for many Gram-positive walls. It should be noted that the

Table 7. Fatty Acid Compositions (as Per Cent of Total) of Wall or Envelope Lipid of Gram-Negative Bacteria[a]

Fatty Acid Chain Length	*E. coli*[b]	*S. gallinarum*[b]	*Ps. aeruginosa*[c]	*B. abortus*[c]
Saturated straight chain				
10:0			Trace	0.4
12:0		2.9	0.4	0.3
13:0		5.9		
14:0	6.8	9.1	1.5	0.2
15:0	0.8	3.8	3.7	
16:0	42.0	49.0	55.1	6.9
18:0	1.7	2.0	0.3	10.8
19:0				18.9
22:0				4.0
Unsaturated straight chain				
14:1	2.1	3.8		
16:1	2.3	4.7	12.8	0.7
18:1	25.7	2.0	11.7	10.1
18:2				19.7

[a] Data as summarized in ref. 19.
[b] Data from ref. 9 for total envelope lipid;
[c] Data from ref. 3 for readily extractable cell wall lipid. In both reports there were unknown fatty acid methyl esters.

structures isolated from members of the mycobacteria and corynebacteria may also contain as much as 30% lipid. Although there is abundant information on the phospholipid compositions of whole cells of Gram-negative bacteria, very few studies have been made on the lipids of the isolated envelope structures. The distribution of phospholipids between wall and membranes is poorly documented because of the difficulties of separating these structures as homogeneous entities. Moreover, the distinction between wall and membrane material in Gram-negative bacteria is difficult to assess from analysis of fractions isolated by mechanical or osmotic rupture of the cells. However, there have been a number of studies of the fatty acid compositions of isolated envelopes of Gram-negative bacteria, and analyses for several species are given in Table 7. Further aspects of the lipid constituents of Gram-negative organisms are discussed in Section 4.

From the investigations of the chemistry of the cell wall or envelope structures isolated from Gram-negative bacteria, the chemical structures of only three type of components have been characterized, namely, the lipopolysaccharides, the peptidoglycan in a few species, and the lipoprotein in *E. coli*. The nature and variety of other protein or lipoprotein components remain to be established. It is also evident that the envelopes of Gram-negative bacteria contain the characteristic compounds of the lipopolysaccharides. The extent to which lipopolysaccharides are detached from the cell surfacce and are lost during isolation of the walls or envelopes is largely unknown; this topic has been discussed in more detail by Freer and Salton (19).

4. CHEMICAL COMPOSITION OF BACTERIAL MEMBRANES

Membrane structures can be readily isolated from a variety of Gram-positive bacteria because of the availability of wall-degrading (muralytic) enzymes which can be used to selectively digest away the cell-wall structure. When the wall can be degraded completely by such an enzyme, the plasma membrane can be isolated free of any insoluble wall residues. Characteristic compounds of wall polymers can be used as guides to the extent of contamination of membrane preparations with wall material. Thus analyses of membrane fractions for muramic acid, D-alanine, D-glutamic acid, and DAP (where it is present in the wall peptidoglycan), and ribitol for teichoic acids or for certain wall polysaccharide sugars (e.g., rhamnose, arabinose) will provide indices of residual wall contents.

Investigations of the chemical compositions of isolated membranes from a number of Gram-positive bacteria have been performed in several laboratories, and a selection of the data is summarized in Table 8 and compared to the

Table 8. The Chemical Compositions of Membranes, Envelopes, and Particle Fractions Isolated from Bacteria[a]

Organism	Protein (%)	Total Lipid (%)	Hexose (or Carbo-hydrate) (%)	RNA (%)	Description of Fraction
Gram-positive					
B. licheniformis	75	28	—	0.8	Membranes
B. subtilis	62.6	18.4	—	21.3	Membranes
B. megaterium KM	75	23	0.2–1	—	Membranes
	65	20	8	5.1	(Ghosts)
	58	27	2.3	1.2	(Fraction 2)
B. megaterium M	66.7	18.5	4.8	1.3	Membranes
B. megaterium	70–85	6–9	1.5	10–15	Membranes
B. stearothermophilus	65–74	13–21	—	9–13	Membranes
M. lysodeikticus	52.5	28	16–19	—	Membranes
	68	23	—	2.3	Membranes
S. lutea	57	23	—	5.4	Membranes
	39.8	28.9	—	1.2	Membranes
Staph. aureus	41	22.5	1.7	2.4	Small particle fraction
Staph. aureus	69	30	—	3.0	Membranes
Strep. faecalis	45.5	32.1	—	2.7	Membranes
Streptococci group A	68	25	2.1	2.0	Protoplast membranes
	67	19	2.2	0.6	Membrane particles
Strep. pyogenes	68	15.3	1.7	—	Membranes
Strep. pyogenes L form	59	36	7.9	—	L-form membranes
Gram-negative					
M. laidlawii	47.3	36.0	6.0	2.8	Membranes
M. bovigenitalium	59.2	37.3	6.8	2.0	Membranes
A. agilis	75	19.5	—	9.4	Small particle fraction
Ps. aeruginosa	60.2	35.1	1.4	0.5	Membranes[b]
Ps. aeruginosa	52.0	45	2.3	—	Membrane particles
M. denitrificans	60.2	32.0	0.6	—	Cytoplasmic membrane[b]

[a] Data summarized in ref. 67.

[b] Preparations probably of envelope type.

Table 9. Features of the Lipid Constituents of Membranes or Envelopes of Gram-Positive and Gram-Negative Bacteria

1. *Phospholipids*: for example, phosphatidylglycerol, cardiolipin (diphosphatidylglycerol), phosphatidylethanolamine, phosphatidylinositol, phosphatidylserine, glucosaminyl-phosphatidylglycerol, phytanylglycerol diethers of phosphatidylglycerol (in halophiles), phosphatidylcholine in a few species, sphingolipid (ceramide phospholipid) in *Bacteroides melaninogenicus*.
2. *Glycosyl diglycerides*: for example, mono-, di-, and triglycosyl diglycerides; diglycosyl diglycerides dominant class (e.g., mainly in *Mycoplasma* and Gram-positive bacteria).
3. *Carotenoids, menaquinones, ubiquinones, polyisoprenoid derivatives*: accessory "lipid-soluble" compounds usually localized in membranes, envelopes, or particles derived from membranes. Polyisoprenoid alcohol phosphate as "wall" intermediates; squalene in halophile.
4. *Sterols*: absent from bacterial membranes or envelopes except in *Mycoplasma* spp., where they can be taken up from exogenous source. Trace amounts in some blue-green algae.
5. *Hydrocarbons*: distributed in whole cells of the Micrococcaceae.

results obtained with a more limited group of Gram-negative organisms. For full details of the types of preparations and methods used, the reader is referred to reviews by Salton (64, 67) for all original references. Apart from the mycoplasmas, where the problem of wall structures associated with membranes does not arise, the preparations from the Gram-negative organisms listed in Table 8 are probably of the envelope type possessing wall components other than peptidoglycans or the small-particle type derived from fragmented membranes. Preparations of the latter kind have been isolated from *Pseudomonas aeruginosa* and found to be devoid of wall lipopolysaccharide sugars (24).

Thus the chemical analysis of isolated bacterial membranes has shown that they are essentially lipid–protein structures. In this respect the membranes of bacteria are basically similar to structures isolated from mammalian or plant cells. Perhaps the only major difference from other cellular membranes is that the bacterial membranes have a somewhat higher proportion of proteins than that observed in many animal cell membranes (see ref. 36).

Phospholipids constitute the major class of lipid constituent in the bacterial membrane, in common with other biomembranes. Indeed, it was shown by Vorbeck and Marinetti (78) that in the bacterial cell the phospholipids are localized exclusively in the membrane structures. Moreover, it has been established that all of the other lipid constituents found in bacteria are also localized in the membranes (or envelopes in the case of Gram-negative species). Thus glycolipids, carotenoids, menaquinones, ubiquinones, and so on are all found in the membranes and are not present in the cytoplasmic fractions devoid of contaminating membrane fragments. Some features of the variety of lipid constituents found in bacterial membranes and envelope or small membrane particle fractions of Gram-negative organisms are presented in Table 9. It will be noted that sterols are generally absent in bacteria, except in the special case of the mycoplasmas grown on media containing these compounds. The mycoplasmas "scavenge" exogenously supplied sterol but apparently lack the ability to synthesize sterol compounds. Squalene, a precursor of lanosterol on the biosynthetic pathway leading to cholesterol, has been found in the halophiles. Hydrocarbons have been reported recently in a number of micrococci, and although whole cells were used in the analysis (75) it is likely that these compounds are also exclusively present in the membranes.

The fatty acids present in bacterial lipids have been investigated extensively and have been reviewed in detail by Kates (35) and O'Leary (51). Some of the more prominent features of the fatty acid compositions of membrane or envelope lipids are listed in Table 10.

As pointed out above, the phospholipids constitute the major lipid class in baccteria or isolated bacterial membranes. They may account for approximately 80% of the total lipid of the cell or membrane. Although a great number of phospholipids have been found in bacteria (see Table 9), there is a considerable degree of individuality in the varieties found in a particular species. This is clearly illustrated in a recent survey of the phospholipid compositions of a number of *Micrococcus* species and other Gram-positive organisms carried out by Whiteside, De Siervo, and Salton (83). The data presented in Table 11 show that the patterns of phospholipid composition are

Table 10. Features of the Fatty Acid Compositions of Bacterial Membranes

Gram-positive	
Micrococci Bacilli	High proportions of C_{15} and/or C_{17} branched-chain acids.
Streptococci Lactobacilli	Predominantly C_{16}, C_{16}-, and C_{18}- monoolefinic (C_{18} and *cis*-vaccenic in some).
Gram-negative	
Predominantly C_{16} and C_{18} straight-chain fatty acids, saturated and unsaturated; cyclopropane acids in some; β-hydroxymyristate in lipopolysaccharides, C_{12} hydroxy acids in *Pseudomonas* lipopolysaccharide. Fatty acids in only trace amounts in halophile envelopes. Sphinganines (predominantly 17 and 19 carbon atoms) found in ceramide phospholipids of a *Bacteroides*.	
Polyunsaturated acids absent in most bacteria.	

Table 11. Composition of Lipids from Stationary-Phase Cells of Various Micrococci and Sarcinae[a]

Organism		Relative Mole Per Cent Compositions of Phospholipids[b]									Non-P- Lipids		
	R_f range:	CL .73–.87	X_1 .63–.72	PG .43–.63	X_3[c] .20–.34	PI .22	AAPG .19	X_4[c] .13–.26	X_5 .08	PE[d,e] .70	GL_1 .78	GL_2 .66	X_2 .33
M. lysodeikticus		38.8	2.4	46.1		12.7					±	+	+
M. tetragenus		31.0	1.3	58.8		8.9						+	
S. flava		36.5		55.6		7.9						+	
S. lutea		1.1		89.6		9.3					+	+	
M. roseus		25.5		74.5								+	
M. roseus R27		34.8		64.6								+	
M. varians		9.6	2.3	88.1									+
Micrococcus sp.		31.0	5.0	62.7									+
M. conglomeratus		27.4	1.2	69.5			1.3					+	
C. coelicolor		59.8		24.3		15.6					++	++	
M. caseolyticus		10.0	1.7	52.5			35.8						
M. rhodochrous		2.0		61.7	24.8			11.5					
S. ventriculi		14.5	7.9	31.4	16.5			29.7				±	
S. maxima		3.0	6.5	32.2	40.7			17.6				±	
S. ureae		27.9		72.1						+			
B. subtilis		3.6	1.8	77.8			16.8			+			

[a] Data from ref. 83.

[b] To calculate relative mole per cent compositions we assumed 2 moles P/mole for cardiolipin, and 1 mole P/mole for other phospholipids. Abbreviations used are as follows: CL, cardiolipin (diphosphatidylglycerol); PG, phosphatidyl glycerol; PI, phosphatidylinositol; AAPG, aminoacylphosphatidylglycerol; GL, glycolipid; PE, phosphatidylethanolamine; X, unidentified compounds.

[c] These unidentified phospholipids were positive for vicinal hydroxyl groups. Components within the given R_f range may not necessarily be identical.

[d] Detected by ninhydrin and autoradiography. Since the small amount of phosphatidylethanolamine cochromatographed in this system with phosphatidylglycerol, it was not quantitated separately.

[e] Symbols used: +, present; ++, present in large amounts; ±, appear to be present in small amounts

Table 12. Principal Fatty Acids Detected in Lipids Extracted from Stationary-Phase Cells of Various Micrococci and Sarcinae[a]

Organism	Per Cent Composition of Fatty Acids[b]											
	Saturated Straight Chain					Branched Chain						
	14:0	15:0	16:0	17:0	18:0	13:0	14:0	15:0	16:0	17:0	18:0	19:0
M. lysodeikticus	1.3		0.5		Tr	2.4	2.0	91.3	0.7	1.7	Tr	
S. flava	0.8		1.4			3.2	6.8	89.0				
M. tetragenus						3.3	3.6	93.0				
M. roseus	3.1		7.0				0.9	80.8	8.2	Tr		
M. roseus R27	3.2		6.2				3.0	79.5	6.2	1.8		
S. lutea							1.9	88.0	5.1	2.8		
M. conglomeratus			Tr		1.2		1.6	83.4	5.6	8.2		1.2
M. varians	Tr		1.6				1.1	76.6	9.5	11.2		
Micrococcus sp.	Tr		2.7	4.0			1.7	80.6	4.6	6.4		
C. coelicolor[c]	1.8		7.0		Tr		0.6	57.7	5.9	26.6	Tr	
M. caseolyticus[d]	4.0	2.8	15.5	2.9	13.1		10.7	3.2	13.3			
M. rhodochrous[e]		12.5	19.5	16.0						15.6		13.5
S. ventriculi	5.3	Tr	65.2	Tr	5.6					Tr		20.1
S. maxima	3.4		69.0		10.0						Tr	17.5
S. ureae	Tr		10.4		Tr		2.9	61.4	8.7	16.4	Tr	
B. subtilis	Tr		4.2	18.2			Tr	74.3	3.2			

[a] Data from ref. 83.
[b] The per cent composition of fatty acids calculated by triangulation of peak areas of the methyl esters separated by gas chromatography.
[c] No hydroxy fatty acid methyl esters were detected by thin-layer chromatography.
[d] In addition, 27.7% 18:1 fatty acid and small amounts of two unidentified fatty acid methyl esters were also detected.
[e] The 18:1 acid accounted for 22.8% of the fatty acids in this organism. Nocardic acids were also detected.

very similar in closely related micrococci. Thus cardiolipin, phosphatidylglycerol, and phosphatidylinositol are the three major phospholipids present in *M. lysodeikticus* and close relatives. These organisms lack phosphatidylethanolamine, which has been found in substantial amount in certain *Bacillus* species (35).

The fatty acid compositions of the bacterial membrane lipids can also be highly characteristic for certain species. Thus, as shown in Table 12, one type of fatty acid (C_{15} branched-chain) may account for as much as 90% of the lipid fatty acids. These data also show that the C_{15} branched-chain fatty acid is the principal acid in a number of the Gram-positive aerobes. However, it should be kept in mind that both fatty acid composition and phospholipid composition can be varied by changes in media and growth conditions (35).

Although protein accounts for the major portion of the bacterial membrane, the proteins as a class have been less intensively studied than the lipid constituents. Like other cell membranes, the bacterial membrane shows a complexity of protein composition, as judged by polyacrylamide gel electrophoresis of dissociated membrane fractions (see the review in ref. 67). That a great variety of proteins are present in bacterial membranes is not surprising in view of the multiplicity of functions that have to be performed by the membrane system. Further aspects of this topic are dealt with in Section 5. It is difficult to say anything meaningful about bacterial membrane proteins until the properties of well-characterized functional proteins have been established after isolation in the pure state. The amino acid compositions of defatted membranes have been determined in several laboratories (e.g., 26, 70), and apart from low or trace amounts of cystine there does not appear to be anything unusual about the composition of the membrane protein fraction. Nor has analysis of the amino acid composition of the *M. lysodeikticus* membrane protein (26) revealed an unduly high proportion of hydrophobic amino acids. It is evident that neither mere amino acid analysis nor analysis of membrane proteins by polyacrylamide gel electrophoresis will tell us much about the special structural features of membrane proteins. Elucidation must await the isolation and characterization of individual membrane enzymes and studies of their functions in relation to protein–lipid and protein–substrate interactions.

5. FUNCTIONS OF BACTERIAL MEMBRANES

Because of the greater ease of isolation of the membranes of Gram-positive organisms, it has been a much simpler task to determine the localization of certain enzymes in their membranes. The absence of separate membranous organelles such as mitochondria, golgi, and endoplasmic reticulum has clearly

Table 13. Plasma Membrane–Mesome Functions in Gram-Positive Bacteria[a]

Active transport.
Electron transport and oxidative phosphorylation.
Secretion of exocellular protein toxins and enzymes.
Cell wall biosynthesis—translocation of "lipid intermediates".
Protein synthesis and membrane-associated ribosomes.
Phospholipid biosynthesis.
Chromosome (DNA) anchoring. replication, and amitotic division.

[a] Data summarized in ref. 67.

indicated that these functions must be "packaged" in the multifunctional plasma membrane–mesosome system of the bacterial cell. Some of the functions that must be performed by these structures are listed in Table 13 for Gram-positive organisms and Table 14 for Gram-negative bacteria.

Although the membranes in Gram-positive bacteria can be distinguished anatomically as the plasma membrane and mesosome membrane regions, the distribution of specific functions between the two has not been conclusively established. From time to time, a number of the functions listed in Table 13 have been specifically assigned to the mesosome structures (64, 68). it is still largely a matter of speculation, however, as to what specific functions the mesosome possesses, although a role in cell wall growth or assembly was one of the first possibilities suggested and may yet prove to be one of its major assignments in the cell. Some differences in the levels of activity of certain enzymes in plasma and mesosome membranes have been recorded (56), but the

Table 14. Membrane Functions in Gram-Negative Bacteria[a]

Active and inducible transport systems.
Electron transport and oxidative phosphorylation.
Transport of proteins into periplasm or for surface location.
Cell wall (peptidoglycan) biosynthesis and translocation of "lipid intermediates" for wall and outer membrane (lipopolysaccharide) synthesis.
Protein synthesis and membrane-associated ribosomes.
Phospholipid biosynthesis.
Chromosome (DNA) anchoring, replication, and amitotic division.
Photosynthetic membranes, nitrogenases in N-fixing bacteria, and special functions in intracellular membranes.

[a] Data summarized in ref. 67.

Table 15. Some Enzymatic Activities Associated with Membranes Isolated from Gram-Positive Bacteria[a]

Activity	Organisms
Cytochromes NADH oxidase Succinic dehydrogenase NADH dehydrogenase Malic dehydrogenase Lactic dehydrogenase	e.g., *B. licheniformis*, *B. subtilis*, *B. megaterium*, *M. lysodeikticus*, *L. monocytogenes*, *Staph. aureus*, *S. lutea*
ATPase, Mg^{2+} or Ca^{2+} activated	e.g., *Strep. faecalis*, *M. lysodeikticus*, *Staph. aureus*, *Bacillus* spp., Lactobacilli
Enzymes involving phospholipid synthesis, diglyceride synthesis, utilization of polyisoprenoid intermediates Ribosomes	*M. lysodeikticus*, *Staph. aureus* *B. megaterium*, *Strep. faecalis*

[a] Data summarized in ref. 67.

full significance of these data has not yet been assessed. Further investigations are clearly needed to pinpoint the unique functions of the mesosome.

There have been many studies of the enzymatic activities associated with the membranes of Gram-positive bacteria, and Table 15 lists some of the enzymes detected in isolated membrane preparations. It is very difficult to know precisely to what extent these membrane preparations represent plasma membranes or fractions of plasma membranes with coseparated mesosomes.

The difficulties of separating the plasma membrane from the outer membrane component of the complex Gram-negative cell envelope have already been mentioned. Because of the technical difficulties in isolating homogeneous perparations of plasma and outer membranes, there is little direct experimental evidence concerning the localization and distribution of enzymes between the membrane and external envelope components. One of the interesting functional consequences of the existence of the outer membrane (organized lipopolysaccharide–protein–lipid structure) in Gram-negative bacteria is that it confers on these cells an additional "barrier" function. This led to the discovery of "surface-located" enzymes and transport proteins in Gram-negative organisms. The concept of a periplasmic region in the peripheral structures of these organisms has emerged from the induced sensitivity to actinomycin D (38) and the release of certain enzymes by osmotic shock (27), which follow from the disruption of the outer membrane structure. Thus the latter may act as a protective barrier by excluding molecules such as actinomycin D and other antibacterial agents and by retaining the surface enzymes in the periplasmic region of the cell. Whether this barrier is, strictly speaking, a permeability barrier in the active transport sense cannot be decided with certainty at the present time. The bulk of the evidence, however, is against the outer membrane being a site of active transport, since the "gene products" controlling these processes are either present in the periplasmic region as "binding proteins" or firmly anchored to material of membrane origin.

Some of the periplasmic enzymes and binding proteins believed to be involved in transport processes are listed in Tables 16 and 17, respectively. It is of considerable interest and significance that these proteins and enzymes are selectively released from the Gram-negative cell and that the process does not involve gross leakage of intercellular constituents (nucleic acids, metabolites) or cytoplasmic enzymes. That the release of these enzymes from the periplasm involves damage to the external "outer membrane" appears certain from the observed concomitant relesase of lipopolysaccharide complexes (39).

The evidence from the study of Nisonson, Tannenbaum, and Neu (49) suggests that at least the 5′-nucleotidase may be anchored to the external surface of *E. coli*. This was indicated by cytochemical localization with intact bacteria.

Table 16. Periplasmic or Surface Enzymes of Gram-Negative Bacteria[a]

Enzymes released by osmotic shock or detected on surface[b]
Alkaline phosphatase
Aspariginase II
Acid hexose phosphatase
Adenosine diphosphoglucose pyrophosphatase
Cyclic phosphodiesterase
Adenylation of streptomycin[c]
Phosphorylation of streptomycin[d]
Nonspecific acid phosphatase
Penicillinase
Ribonuclease I
5′-Nucleotidase[b,e]

[a] See ref. 67 for original references.
[b] Data summarized in ref. 27 and recent additions.
[c] From ref. 2.
[d] From ref. 53.
[e] From ref. 49.

Table 17. Periplasmic- or Surface-Located Transport Proteins in Gram-Negative Bacteria[a,b]

1. Periplasmic
Leucine-binding protein
Galactose-binding protein
Sulfate-binding protein
Arginine-binding protein
2. Membrane "anchored"
Phosphotransferase enzyme II
Lactose transport system
Proline transport system

[a] See ref. 67 for original references.
[b] Original references given in text except for arginine-binding protein: ref. 85.

However, it was clear that the SO_4-binding protein was not detectable on the cell surface. Thus it is possible that certain enzymes in *E. coli* may be truly surface (externally) located, and others may be retained in the periplasmic space bounded by the plasma membrane as the internal face and the outer membrane as the outer face.

The molecular architecture of the bacterial membrane remains to be established. The profiles as seen in thin sections are very similar to those of membranes from other sources; but, as Korn (36) has pointed out, the heavy-metal staining used in electron microscopy cannot yield specific information on the molecular architecture of the membranes. Localization of enzymes on outer or inner faces of the membranes would be of great value in determining the membrane architecture in terms of specific proteins. Cytochemical staining, as well as specific protein localization with ferritin-labeled antibody, will undoubtedly be used in elucidating the precise arrangement of enzymes and functions in bacterial membranes. The molecular orientations of lipids in membranes have been investigated by a variety of physicochemical techniques and studied indirectly by the accessibility to externally added phospholipases attacking the polar and nonpolar regions of the phospholipids. Many molecular models have been proposed for cell membranes, but much remains to be done in the precise mapping of proteins in and on the membranes of bacteria and other cells.

6. GROWTH AND DIVISION IN RELATION TO BACTERIAL STRUCTURE

The foregoing sections of this chapter on the all-embracing topic of molecular bacteriology have focused primarily on the chemical nature and molecular architecture of the bacterial cell. It should be abundantly clear from this brief discussion of selected aspects of this enormous and rapidly expanding field of bacterial physiology that some of the unique features of the bacterial cell, distinguishing the prokaryotic from the eukaryotic type, reside in the chemistry and organization of the cell surface structures. Indeed, it follows from this that bacteria have "evolved" unique biosynthetic pathways for the synthesis of compounds such as the cell wall peptidoglycan, lipopolysaccharides, and certain surface capsular substances. The specific enzymes involved in the assembly and transport of intermediate building blocks of these complex structures have been investigated in some detail during the past decade, and in more recent years the regulation of these processes during cell growth and division has become more amenable to study with the use of a great variety of bacterial mutants.

The organization of the nucleus, its anchoring, and its site of replication in the prokaryotic cell differ from the corresponding features in the eukaryotes, and many of the details of these differences are emerging from work with "growth and division" mutants. The biochemical lesions in many such mutants are still far from being fully understood, but it is evident that wall–membrane–nuclear relationships are involved, the perturbation of which leads to abnormal growth and division processes. Some progress has been made in two studies of mutants of *E. coli* that are defective in DNA replication (31, 72). In both investigations evidence of alterations in membrane proteins has been presented, but the functions of the proteins affected have not yet been elucidated. These are the types of studies that are completing our knowledge and description of the bacterial cell as a dynamic, coordinated, growing structure. Thus the regulated biosynthesis and replication of the major structural elements and organelles of the cell is a prerequisite for division into daughter cells, thereby ensuring the continuity and expression of the bacterial genome.

The description of the major types of molecular and macromolecular components of the bacterial cell, which has been the substance of the preceding sections, should not lead to the impression of metabolic inertness. Indeed, it is apparent that the bacterial cell is in a dynamic state and that even the more structurally rigid parts (walls, membranes, etc.) are subject to turnover and constant change. It is obvious that the extension of preexisting macromolecules of the cell must involve modification of these structures to permit insertion of new "units" during growth and division (29). The description of the sequence of events during growth and division of the bacterial cell is so voluminous that a review of this major segment of microbial physiology is considered to be outside the scope of this chapter emphasizing the molecular architecture of the bacterial cell. In this concluding section, therefore, only selected examples indicating the dynamic aspects of the development of the major structures, cell walls, membranes, and the nucleus, will be discussed.

The first attempts to elucidate the manner of replication of the bacterial cell wall were made by Cole and Hahn (12) and by May (43), using fluorescent-labeled antibody to follow the fate of cell surface antigens during growth and division. The fluorescent-antibody labeling of group A streptococcal wall was spectacular, and from this Cole and Hahn were able to deduce that new wall was not diffusely intercalated but was formed equatorially at the site of the next cross-wall septum. Equatorial wall growth with peripheral and centripetal development was thus suggested as the mode of wall formation in streptococci (10, 12). In contrast to this zonal type of growth, where regions of new wall may be conserved, a dispersive extension of wall was found in *Salmonella typhimurium* and *E. coli* (May, quoted in ref. 62), and similar results were obtained with other Gram-negative bacteria by the antibody-labeling method (11). One of the principal problems arising from the use of this technique is that the labeling pattern refers to the extension of the surface antigens of the wall and not specifically to the rigid peptidoglycan structure (11, 29).

However, despite some ot the inadequacies of the fluorescent-antibody labeling method, essentially similar results have been obtained by radioactive labeling. Thus van Tubergen and Setlow, (77) found dispersion of ^{3}H–diaminopimelic acid in the wall of *E. coli*, and Briles and Tomasz (5a) observed the segregation pattern of ^{3}H–choline-labeled teichoic acid in pneumococci, which indicated a zonal mode of growth of this wall polymer.

The fate of newly formed cross wall during elongation and thickening has been studied extensively in *Streptococcus faecalis* by Shockman and Higgins and their coworkers (29). By electron microscopy of *Strep. faecalis*, the growth of which could be manipulated by amino acid starvation, Higgins and Shockman (28, 29) were able to distinguish between old and new peripheral wall by the presence of the equatorial "wall band" or "wall notch" and to follow the events of wall formation during the division cycle. This excellent model system has provided an insight into the dynamics of wall development during exponential growth, and from their observations they propose that cross wall is made through a balanced control of the following processes: (*a*) linear extension, which occurs at the leading edges of the cross wall; (*b*) wall thickening, which brings the newly extended wall to its normal thickness; and (*c*) wall separation, which results from the peeling apart of the cross-wall base into two layers of peripheral wall (29). Considerable "postexponential phase" thickening of the walls of Gram-positive bacteria can be induced by amino acid starvation or by inhibitors of protein synthesis such as chloramphenicol or tetracycline (29).

The association of autolysins with the bacterial cell wall has been known for some time (62), but the physiological significance of these wall-degrading enzymes has not been clearly established. The autolysin of *Strep. faecalis* has been investigated in some detail by Shockman and his colleagues and has been characterized as an *N*-acetylmuramidase (71). Active, wall-bound and latent, trypsin-activatable forms of the autolysin have been detected, and the site of autolysis has been established by ultrastructural studies. The initial sites of autolytic action in *Strep. faecalis* were identified as the leading edges and tip of the centripetally growing cross wall (29). The autolysin appears to remain active in the region of wall biosynthesis during the entire cell division cycle. Wall thickening does not appear to require the autolytic enzyme, since thickening occurs in autolysis-resistant cells. An interaction between autolysin activity, cross bridging of the peptidoglycan, and perhaps cell separation has been suggested for one of the *Strep. faecalis* mutants (29). There seems to be little doubt that autolytic enzymes play a role in wall growth, division, and cell separation. The isolation and characterization of phage-resistant mutants of *Staph. aureus* (8), morphologically disturbed rod mutants of *Bacillus* species (13, 57), and autolysin mutants of *E. coli* (5) and *B. subtilis* (18) should lead to a understanding of the precise biochemical function of the autolysins.

In addition to the fluorescent-antibody and radioisotope labeling methods, anatomical abnormalities have also been utilized in following wall extension. May (unpublished observations quoted in ref. 62) was one of the first investigators to use such a method. He followed the fate of the knobs on a "knobbed" cell mutant of *E. coli* by time-lapse phase-contrast photomicrography and confirmed a dispersive mode of growth by uniform extension of the cell surface. More recent studies of Donachie and Begg (17) have involved the use of penicillin to form wall bulges at the penicillin-sensitive site of wall synthesis in *E. coli*. The movement of the wall bulge from the pole to the mid-point of the rod and its segregation were followed, and a model for wall growth and extension was proposed.

The extent to which cell wall components turn over is still a matter for conjecture since there have been conflicting reports on this problem. Turnover of both peptidoglycan and teichoic acid in *B. subtilis* wall was reported by Mauck and Glaser (42); the rate of turnover of the teichoic acid was about twice that observed for the peptidoglycan. On the other hand, Pitel and Gilvarg (54) found no evidence for peptidoglycan turnover in *B. megaterium*, and Higgins and Shockman (29) indicated that little, if any, peptidoglycan turnover occurs in *Strep. faecalis*. Turnover in the strictest sense has not been reported for the outer lipopolysaccharide–lipoprotein complex of the Gram-negative cell envelope. However, overproduction and "excretion" (loss from the surface or defective anchoring) of LPS complexes has been reported (59, 86).

Our knowledge of the dynamics of plasma membrane growth and turnover in bacteria is much more rudimentary than that relating wall or envelope structure. Kahane and Razin (33) have presented evidence that the membrane protein of *Mycoplasma laidlawii* turns over with a half-life of approximately 3 hr. Membrane lipid also exhibited turnover, but in younger cultures this occurred only after a lag period. Soluble, cytoplasmic proteins of *M. laidlawii* did not show turnover under these conditions. White and his colleagues have studied the phospholipid metabolism (turnover and synthesis) in *Haemophilus influenzae* and *Staphylococcus aureus* during formation of membrane-bound electron transport systems (20, 80, 81). Turnover and marked shifts in the phospholipid compositions of the membranes are evident from all of these studies and point to coordinate changes in membrane lipids during the formation of specific functional membrane proteins such as those of the electron transport system. In the shift from anaerobic to aerobic growth, extensive changes in *Staph. aureus* membrane phospholipid composition were observed, and in addition marked rearrangement of the fatty acids was found (81). All of these studies point to the dynamic state of the membrane structure.

The extent of the "tightness" of the control of membrane synthesis has been examined more recently with a variety of techniques, including the use of specific lipid mutants and inhibitors of protein synthesis. It is obvious from earlier studies that there is some degree of coordinated synthesis of proteins and lipids during the induction of membrane-bound electron transport, permease, and photosynthetic systems (see the review in ref. 67). However, the use of chloramphenicol, which blocks membrane protein synthesis, leads to the formation of a lipid-enriched membrane of lower density because of continued synthesis of the membrane lipid (33). The more recent studies of Mindich (44, 45) with glycerol auxotrophs of *B. subtilis* enabled him to "uncouple" membrane lipid and protein synthesis. Mindich concluded that the control of membrane composition is, therefore, not very stringent.

The study of inducible transport systems is also yielding new information on the dynamic state of the membrane, although some of the reports to date are conflicting. A requirement for the synthesis of lipid containing unsaturated fatty acids during the induction of lactose transport in *E. coli* indicates a high degree of interdependence between lipid and protein syntheses for this particular function. No such strict dependence on phospholipid synthesis and citrate transport function could be observed in *B. subtilis* (84), although it is possible that the efficiency of the system is influenced by lipid synthesis.

Attempts have been made to determine the site of synthesis of lipids in the membrane in order to gain some insight into the mode of growth of the membrane. Morrison and Morowitz (46) concluded from studies of ^{3}H–palmitic-acid-labeled membranes of *B. megaterium* that membrane growth was primarily "end-localized" and conserved in cell growth and division. This supported a similar conclusion deduced from tellurite labeling of the membrane (see ref. 60). The opposite conclusion, however, has been reached by Green and Schaechter (25). Their data indicate multiple-site deposition and subsequent dispersion of newly made membrane material, the conserved regions being smaller than 1% of the membrane area.

It is evident from the foregoing that the membrane is a dynamic structure, but the modes of growth and extension and of segregation of membrane regions have yet to be specified by further studies. There is little doubt that the selection of mutants, which have already yielded a great deal of information about the control of membrane composition and the integration of certain transport function, will rapidly broaden our understanding of the biogenesis of this important cellular structure.

The bacterial "chromosome" or "nucleus" differs from its eukaryotic equivalent in that it is not organized within a discrete nuclear membrane and it undergoes an amitotic division into two nuclear bodies, which gradually separate into the two daughter cells during division (60). Alhough the bacterial nucleus is organized as a single circular molecule of DNA and is therefore infinitely more simple than the complex organization of chromosomes in nuclei of the eukaryotic cell, the mechanisms of its anchoring, replication, and division are still incompletely understood (37, 60). That a specific association of the replicating DNA with the bacterial membrane exists seems to be unquestionable (15, 37, 76). However, the nature of the membrane components involved in the "replicon" have not been identified, although the DNA-replication mutants have given some clues as to the involvement of membrane proteins (31, 72). The replicon hypothesis of Jacob, Brenner, and Cuzin (32) proposed that new membrane synthesis brings about separation of the two chromosomes. The events during DNA replication, membrane growth, wall septation, and nuclear segregation are still poorly understood, and as pointed out by Lark (37) attempts to relate these various processes to one another have led to contradictory results. There appears to be little evidence for a permanent association of the bacterial nuclei with septa which segregate these bodies (37). It is suggested that septation can occur at sites that need not be adjacent to the replicating chromosomes.

The circularity of the bacterial chromosome confers certain advantages and disadvantages, as pointed out by Lark (37). One advantage of such a circular system would be ease of control of cell division, which could be triggered by the termination of a round of DNA replication. The mechanical problems arising from the circularity of the replication process involve the need for rotation of the DNA complex to facilitate its separation. The formation of breaks and subsequent rejoining have been suggested as a possible solution to replication of a closed, continuous DNA circle (37). The existence of multiple replication forks presents further difficulties in attempts to understand the duplication of the "closed" DNA nuclear body.

After the replication of the bacterial DNA, the daughter chromatids are segregated into each daughter cell. Termination of DNA replication triggers cell division, and there is good evidence that cell division is controlled by replication. Studies with mutants show that the cell division process alone does not regulate replication or segregation, unless it involves the regulation of protein synthesis. Thus during cell division increasing the synthesis of protein initiates cycles of replication, which in turn initiate new divisions (37).

The dynamic aspects of the "molecular bacteriology" of replication, growth, and division of the major cellular structures are still under active investigation, and many facets of these problems are poorly understood. Nonetheless we are beginning to understand the sequence of events during growth and division, and the isolation of mutants has greatly facilitated studies in these areas. The regulation of intermediary metabolism has been extensively investigated, and some insight into the control and regulation of the biosynthesis of the major structural elements of the bacterial cell has been gained.

REFERENCES

1. Baddiley, J., *Proc. Roy. Soc.* **B170,** 331 (1968).
2. Benveniste, R. E., B. W. Ozanne, and J. Davies, *J. Bacteriol. Proc.*, p. 48 (1969).
3. Boro, R. A., and R. G. Eagon, *Can. J. Microbiol.* **14,** 503 (1968).
4. Braun, V., and K. Rehn, *Eur. J. Biochem.* **10,** 426 (1969).
5. Braun, V., and U. Schwarz, *J. Gen. Microbiol.* **57,** iii (1969).

5a. Briles, E. B., and A. Tomasz, *J. Cell Biol.* **47,** 786 (1970).

6. Brinton, C. C., *Trans. N. Y. Acad. Sci.* **27,** 1003 (1965).
7. Canale-Parola, E., M. Mandel, and D. G., Kupfer, *Arch. Microbiol.* **58,** 30 (1967).
8. Chatterjee, A. N., D. Mirelman, H. J. Singer, and J. T. Park, *J. Bacteriol.* **100,** 846 (1969).

9. Cho, K. Y., and M. R. J. Salton, *Biochim. Biophys. Acta* **116,** 73 (1966).
10. Chung, K. L., R. Z. Hawirko, and P. K. Isaac, *Can. J. Microbiol.* **10,** 473 (1964).
11. Cole, R. M., *Bacteriol. Rev.* **29,** 326 (1965).
12. Cole, R. M., and J. J. Hahn, *Science* **135,** 722 (1962).
13. Cole, R. M., T. J. Popkin, R. J. Boylan, and N. H. Mendelson, *J. Bacteriol.* **103,** 793 (1970).
14. Collins, F. M., *J. Gen. Microbiol.* **34,** 379 (1964).
15. Daniels, M. J., *Biochem. J.* **122,** 197 (1971).
16. De Petris, S., *J. Ultrastruct. Res.* **19,** 45 (1967).
17. Donachie, W. D., and R. J. Begg, *Nature* **227,** 1220 (1970).
18. Fan, D. P., and M. M. Beckman, *J. Bacteriol.* **105,** 629 (1971).
19. Freer, J. H., and M. R. J. Salton, *Toxins*, Vol. IV, Chap. 2, Acadmic Press, New York, 1971.
20. Frerman, F. E., and D. White, *J. Bacteriol.* **94,** 1868 (1967).
21. Ghuysen, J. M., *Bacteriol. Rev.* **32,** 425 (1968).
22. Glauert, A. M., and M. J. Thornley, *Ann. Rev. Microbiol.* **23,** 159 (1969).
23. Goldfine, H., *Ann. Rev. Biochem.* **37,** 303 (1968).
23a. Graham, R. K., and J. W. May, *J. Gen. Microbiol.* **41,** 243 (1965).
24. Gray, G. W., and P. F. Thurman, *Biochim. Biophys. Acta* **135,** 947 (1967).
25. Green, E. W., and M. Schaechter, *Bacteriol. Proc.*, p. 158 (1971).
26. Grula, E. A., T. F. Butler, R. D. King, and G. L. Smith, *Can. J. Microbiol.* **13,** 1499 (1967).
27. Heppel, L. A., *J. Gen. Physiol.* **54,** 95s (1969).
27a. Herzberg, M., and J. H. Green, *J. Gen. Microbiol.* **35,** 421 (1964).
28. Higgins, M. L., and G. D. Shockman, *J. Bacteriol.* **103,** 244 (1970).
29. Higgins, M. L., and G. D. Shockman, *CRC Crit. Rev. Microbiol.* **1,** 29 (1971).
30. Homma, J. Y., and N. Suzuki, *Ann. N. Y. Acad. Sci.* **133,** Art. 2, 508 (1966).
31. Inouye, M., and J. P. Guthrie, *Proc. Natl. Acad. Sci. U.S.* **64,** 957 (1969).
32. Jacob, F., S. Brenner, and F. Cuzin, *Cold Spring Harbor Symp. Quant. Biol.* **28,** 329 (1964).
33. Kahane, I., and S. Razin, *Biochim. Biophys. Acta* **183,** 79 (1969).
34. Kandler, O., *Int. J. Syst. Bacteriol.* **20,** 491 (1970).
35. Kates, M., *Advan. Lipid Res.* **2,** 17 (1964).
36. Korn, E. D., *J. Gen. Physiol.* **52,** 257s (1968).
37. Lark, K. G., *Ann. Rev. Biochem.* **38,** 569 (1969).
38. Leive, L., *Biochem. Biophys. Res. Commun.* **21,** 290 (1965).
39. Leive, L., V. K. Shovlin, and S. E. Merhenhagen, *J. Biol. Chem.* **243,** 6384 (1968).
40. Lüderitz, O., J. Jann, and R. Wheat, *Compr. Biochem.* **26A,** 105 (1968).
41. MacFarlane, M. G., *Advan. Lipid Res.* **2,** 91 (1964).
42. Mauck, S., and L. Glaser, *Biochem. Biophys. Res. Commun.* **39,** 699 (1970).
43. May, J. W., *Exp. Cell Res.* **27,** 170 (1962).
44. Mindich, L., *J. Mol. Biol.* **49,** 415 (1970).

45. Mindich, L., *J. Mol. Biol.* **49,** 433 (1970).
46. Morrison, D. C., and H. J. Morowitz, *J. Mol. Biol.* **49,** 441 (1970).
47. Murray, R. G. E., *Microbial Protoplasts, Spheroplasts and L–Forms,* Williams and Wilkins, Baltimore, Md., 1968, p. 1.
48. Murray, R. G. E., P. Steed, and H. E. Elson, *Can. J. Microbiol.* **11,** 547 (1965).
49. Nisonson, I., M. Tannenbaum, and H. C. Neu, *J. Bacteriol.* **100,** 1083 (1969).
50. Okuda, S., and G. Weinbaum, *Biochemistry* **7,** 2819 (1968).
51. O'Leary, W. M., *The Chemistry and Metabolism of Microbial Lipids*, World, Cleveland, 1967, p. 93.
52. Op den Kamp, J. A. F., W. van Iterson, and L. L. M. van Deenen, *Biochim. Biophys. Acta* **135,** 862 (1967).
53. Ozanne, B., R. Benveniste, D. Tipper, and J. Davies, *J. Bacteriol.* **100,** 1144 (1969).
54. Pitel, D. W., and C. Gilwarg, *J. Biol. Chem.* **245,** 6711 (1970).
55. Pollock, M. R., *The Bacteria,* Vol. IV, Academic Press, New York, 1962, p. 121.
56. Rogers, H. J., *Bacteriol. Rev.* **34,** 194 (1970).
57. Rogers, H. J., M. McConnell, and I. D. J. Burdett, *J. Gen. Microbiol.* **61,** 155 (1970).
58. Rogers, H. J., and H. R. Perkins, *Cell Walls and Membranes*, Chap. 21, E. F. N. Spon, London, 1968.
59. Rothfield, L., and M. Pearlman-Kothencz, *J. Mol. Biol.* **44,** 477 (1969).
60. Ryter, A., *Bacteriol. Rev.* **32,** 39 (1968).
61. Salton, M. R. J., *Biochim. Biophys. Acta* **10,** 512 (1953).
62. Salton, M. R. J., *The Bacterial Cell Wall,* Elsevier, Amsterdam, 1964.
63. Salton, M. R. J., *Ann. Rev. Biochem.* **34,** 143 (1965).
64. Salton, M. R. J., *Ann. Rev. Microbiol.* **21,** 417 (1967).
65. Salton, M. R. J., *Comprehensive Biochemistry*, Vol. 23, Elsevier, Amsterdam, 1968, p. 127.
66. Salton, M. R. J., *The Future of the Brain Sciences*, Plenum, New York, 1969, p. 1.
67. Salton, M. R. J., *CRC Crit. Rev. Microbiol.* **1,** 161 (1971).
68. Salton, M. R. J., *Biomembranes,* Vol. 1, Plenum, New York, 1972, p. 1.
69. Sharon, N., *Amino Sugars*, Vol. IIA, Academic Press, New York, 1965, p. 1.
70. Shockman, G. D., J. J. Kolb, B. Bakay, M. J. Conover, and G. Toennies, *J. Bacteriol.* **85,** 168 (1963).
71. Shockman, G. D., J. S. Thompson, and M. J. Conover, *Biochemistry* **6,** 1054 (1967).
72. Siccardi, A. G., B. M. Shapiro, Y. Hirota, and F. Jacob, *J. Mol. Biol.* **56,** 475 (1971).
73. Stanier, R. Y., *Organization and Control in Prokaryotic and Eukaryotic Cells*, Cambridge University Press, Cambridge, England, 1970, p. 1.
74. Strange, R. E., and J. F. Powell, *Biochem. J.* **58,** 80 (1954).
75. Tornabene, T. G., S. J. Morrison, and W. E. Kloos, *Lipids* **5,** 929 (1970).
76. Tremblay, G. Y., M. J. Daniels, and M. Schaechter, *J. Mol. Biol.* **40,** 65 (1969).
77. Van Tubergen, R. P., and R. B. Setlow, *Biophys. J.* **1,** 589 (1961).
78. Vorbeck, M. L., and G. V. Marinetti, *Biochemistry* **4,** 296 (1965).
79. Westphal, O., O. Lüderitz, and F. Bister, *Z. Naturforsch.* **7b,** 148 (1952).

80. White, D. C., *J. Bacteriol.* **89,** 299 (1965).
81. White, D. C., and F. E. Frerman, *J. Bacteriol.* **95,** 2198 (1968).
82. White, D. C., A. N. Tucker, and C. C. Sweeley, *Biochim. Biophys. Acta* **187,** 527 (1969).
83. Whiteside, T. L., A. J. De Siervo, and M. R. J. Salton, *J. Bacteriol.* **105,** 957 (1971).
84. Willecke, K., and L. Mindich, *J. Bacteriol.* **106,** 514 (1971).
85. Wilson, O. H, and J. T. Holden, *J. Biol. Chem.* **244,** 2743 (1969).
86. Work, E., K. W. Knox, and M. Vesk, *Ann. N.Y. Acad. Sci.* **133,** 438 (1966).

CHAPTER

ELEVEN

Molecular Mycology

R. STORCK

1. INTRODUCTION

Fungi are eukaryotic organisms that undergo, in their natural hibitat, remarkable morphological changes. These changes occur either during asexual and sexual life cycles or as a result of modifications of environmental conditions. Many of these phenomena can be studied quantitatively under laboratory conditions that are no more difficult to control than those required for experimentation on prokaryotic cells. In addition, some fungi are facultative anaerobes, and some are easily amenable to genetic analysis. Also, in the scale of genome size, or amount of DNA per haploid nucleus, which ranges from 10^{-15} g in prokaryotes to 10^{-12} g in mammals, fungi occupy a position closer to the prokaryotes than do algae and protozoa (36, 37, 148, 176). In spite of these interesting features, it is only recently that organisms outside of what has been called "the *Neurospora, Saccharomyces, Aspergillus* axis" (79) have been studied with the tools of the molecular biologist.

The results of some investigations on molecular mycology are presented in this chapter. In the first part an attempt has been made to establish a synopsis of the properties of fungal nucleic acids and to evaluate these properties for the study of taxonomic and phylogenetic relationships. Studies on other macromolecules are not discussed, and the reader is referred to other chapters in this book and to refs. 136, 317, and 318 and refs. 15 and 266, which deal with proteins and polysaccharides, respectively. In the second part an account is given of the progress of research aimed at determining the role that nucleic acids play in the control of fungal morphogenesis and differentiation. Other biochemical and physiological aspects of fungal differentiation have been recently and extensively reviewed by others (263). Because acellular and cellular slime molds are regarded as organisms with uncertain affinities, they have not been discussed in detail here. Their case has been abundantly documented in recent monographs and reviews (30, 106, 240).

2. NUCLEAR GENOME

A. Base Composition

The introduction 12 years ago of simple and reliable methods for the determination of the guanine (G) plus cytosine (C) content (in mole %) of DNA made it possible to survey a large number of organisms. For the bacteria, which were the first to be studied intensively, it was soon discovered that there is a great diversity in GC content, ranging from 25 to 75%. In contrast, the diversity was found to be much smaller within groups of higher forms of life such as plants and animals. It was further established in the case of bacteria that organisms described under the same specific epithet display, as a rule, a minimum range in the base composition of their DNA. As a result, %GC has become a valuable tool for the taxonomy of these organisms and is now routinely included in the list of attributes of taxonomic groups, even in introductory microbiology textbooks (38, 269). Later surveys have revealed that a similar wide range in %GC characterizes the eukaryotic groups of the microbial world as well as the prokaryotic ones (see Table 1).

Among the eukaryotic microorganisms, the fungi have been the most extensively studied. The results of these studies have been the object of two reviews. One (279), which deals primarily with filamentous fungi, is summarized in Table 2, and the other (185), which includes data pertaining exclusively to yeasts and yeastlike fungi, is presented in Table 3. Since the publication of the survey of yeasts and yeastlike fungi, additional reports on the following genera have become available: *Pichia* (202), *Debaryomyces* (203), *Hansenula* (204, 327), *Cryptococcus* and *Rhodotorula* (205), *Torulopsis* (206), *Saccharomyces* (207), *Candida* (208), and *Trichosporon* (68). These reports on several hundred additional species generate averages and ranges similar to those listed in Table 3. Whereas the values used for the construction of Table 2 were obtained mostly from buoyant density determinations, many of those in Table 3 were derived from T_m determinations. The overall GC content ranges from 27.5 to 70%. The diversity of DNA composition is thus as large in fungi as it is in bacteria. On the average, the Zygomycetes have the lowest GC content and the Basidiomycetes the highest, while the Ascomycetes fall into two groups, the first comprising the Hemiascomycetidae (primarily sporogenous yeasts) and showing a range of 29 to 50%, similar to that of the Zygomycetes, and the second comprising the Euascomycetidae (Table 2) and showing a range of approximately 50 to 60%. The average GC content for most Deuteromycetes is close to that for the Ascomycetes. The %GC of the perfect forms is the same as that of their imperfect counterparts for Hemiascomycetidae, Euascomycetidae, and Sporobolomycetaceae. Since there is an enormous difference

Table 1. Ranges of GC Content

Taxonomic Group	%GC	Reference
Bacteria	25–75	123, 255
Blue-green algae	35–71	75
Algae	37–68	181
Protozoa	22–68	181
Fungi	27–70	279

between the total number of existing fungal species and the number that can possibly be analyzed, the sampling is limited and biased. This limitation should be kept in mind while attempting to evaluate the significance of these surveys.

Intrataxonomic heterogeneity in DNA composition can be expressed in terms of the standard deviation of the frequency distribution of GC in taxons. If this is done, values equal to or less than 2% GC are found for species and less than 10% for genera. Higher values are characteristic for larger taxonomic groups, such as families, orders, and classes. Whereas the frequency distributions of %GC are almost symmetrical and have standard deviations equal to 2.42 and 1.96% for *Penicillium* and *Aspergillus*, respectively, for genera such as *Chaetomium* and *Sporobolomyces* the distributions are skewed and bimodal and have respective standard deviations equal to 3.45 and 4.24% GC. Such a bimodality is also found for the genera *Candida, Cryptococcus*, and *Rhodotorula*. Several species of *Candida* have %GC values falling into the range

Table 2. Statistical Indices of the Frequency Distribution of GC Content Values for Fungal Classes[a]

Class	N_s	N_i	$\bar{X}$	R	s^2	s	SE	PE
Oömycetes	24	27	53.0	40.5–62	28.15	5.31	1.08	3.58
Zygomycetes	66	155	42.6	27.5–59	58.02	7.62	0.94	5.14
Ascomycetes	69	90	53.4	48.5–60	6.10	2.47	0.30	1.67
Deuteromycetes	163	220	52.1	35.5–64.5	10.67	3.27	0.26	2.21
Basidiomycetes[b]	42	62	55.0	44.0–59.5	12.88	3.59	0.55	2.42

[a] Abbreviations: N_s, number of species analyzed; N_i, number of isolates analyzed; $\bar{X}$, arithmetic average; R, range; s^2, variance; s, standard deviation; SE, standard error; PE, probable error.
[b] From ref. 281.

Table 3. Statistical Indices of the Frequency Distribution of GC Content Values for Yeasts[a, b]

Taxon[c]	N_s	N_i	X	R	s^2	s	SE	PE
Dipodascus, Endomyces, Endomycopsis, Schizosaccharomyces, Schwanniomyces	13	13	38.6	32.5–43.0	11.42	3.38	0.936	2.28
Pichia	20	27	39.7	29.0–45.0	19.69	4.44	0.854	2.99
Saccharomyces	18	35	40.2	32.0–46.3	11.15	3.34	0.564	2.25
Kluyveromyces	6	13	40.0	34.0–43.0	6.83	2.61	0.724	1.76
Debaryomyces	4	6	36.1	32.0–40.0	7.62	2.76	1.13	1.86
Metschnikowia	5	11	45.8	42.2–48.3	5.15	2.27	0.684	1.53
Miscellaneous sporogenous genera	7	7	40.8	33.0–48.5	23.49	4.85	1.83	3.27
Candida	51	87	42.6	31.0–63.0	55.13	7.43	0.797	5.01
Torulopsis	10	11	44.8	36.5–61.0	62.69	7.92	2.39	5.34
Cryptococcus, Rhodotorula, Sporobolomyces	24	57	57.1	49.0–70.0	32.39	5.69	0.754	3.84
Miscellaneous asporogenous genera	6	7	51.3	33.0–62.0	97.99	9.90	3.74	6.68
All sporogenous genera	73	112	40.2	29.0–48.5	17.06	4.13	0.390	2.79
All asporogenous genera	91	162	48.2	31.0–70.0	95.65	9.78	0.768	6.60

[a] For abbreviations, see Table 2.
[b] This table includes data from refs 199–201, 276.
[c] The organisms are grouped according to ref. 185.

characteristic for the ascosporogenous yeasts. This could indicate that they are the imperfect forms of such yeasts. The same is true when the %GC values found among members of the genera *Rhodotorula* and *Sporobolomyces* are compared. The case of *Rhodotorula* is very interesting in this respect, since it was shown to have a perfect stage (12, 213).

Yeast and yeastlike fungi pose a special problem for the taxonomist, since most of the morphological and life-cycle features used for filamentous fungi are not applicable in their case. As a result, zymologists have relied heavily on biochemical criteria. These, with morphological and cultural traits, have been used with various degrees of success. Although yeasts have long been considered to be true fungi, in many cases there have been difficulties in assigning them to existing taxonomic groups. The studies on DNA base composition have helped to improve this situation. As a result, the analysis of DNA base composition appears to have been more profitable for the taxonomy of yeast and yeastlike fungi than for that of the filamentous forms.

With respect to phylogeny, these studies have unveiled two main features that are worth mentioning. One pertains to the monophyletic origin of fungi, and the other to the origin of yeasts. According to the monophyletic scheme (99), fungi, with the exception of the Oömycetes, arose from ancestral flagellates. The Zygomycetes were, following this scheme, the direct precursors of the Ascomycetes, which in turn preceded the Basidiomycetes. As the results shown in Tables 2 and 3 suggest, it would appear that evolution was marked by a progressive increase in DNA GC content. The fact that the average %GC value for the Oömycetes is 53% could be accepted on the basis that these organisms are not on the main evolutionary line. Of significance is the fact that the average %GC for Zygomycetes is closer to the value for the Hemiascomycetidae than to that for the Euascomycetidae. This could be taken as an indication that the yeasts are primitive rather than reduced forms. It should be pointed out that the origin of yeasts has been and remains a source of controversy. Some view these organisms as derived from the myceloid ones by reduction, and thus as degenerate forms, whereas others regard them as intermediary forms between the Zygomycetes and the Ascomycetes. Although the available %GC values do not provide a solution for this controversy, it is of relevance to note that the range of GC content is the same among the organisms regarded as "yeasts" as among the filamentous fungi. Yeast forms suspected of being related to the Heterobasidiomycetes have as high GC contents as the Homobasidiomycetes, whereas, as already indicated, yeasts with ascomycetous affinity have significantly lower %GC. This maximum diversity in the composition of yeast DNA might be taken as an indication that yeasts are degenerate forms derived from members of some classes of Eumycotina. One could also invoke the fact that most yeasts exist as unicellular organisms and that this could favor mutant selection and thus increase the probability for greater variation in DNA composition.

In order to explain the apparent increase in the average %GC from Zygomycetes to Basidiomycetes, the suggestion has been made that, since such an increase would favor thermal stability of the DNA molecule, it would have a selective advantage (279). Admittedly this is a weak argument, since this protection does not appear to be needed under normal physiological conditions. In bacteria a correlation appears to exist between the GC content and the extent of exposure to sunlight. For example, actinomycetes with airborne conidia, myxobacteria with fruiting bodies, aquatic bacteria living close to the surface of the water, and carotenoid-containing organisms have almost without exception a high %GC in their DNA, whereas bacteria such as non-photosynthetic obligate anaerobes and obligate parasites all have low GC contents. This has been interpreted (259) as an indication that increased GC content would protect against specific damage of thymine from ultraviolet radiation in sunlight. Such a theory does not apply, unfortunately, to fungi.

Increasing evidence suggests that the monophyletic scheme may have to be modified. First of all, the analysis of 8 species (admittedly a very small sample) of Chytridiomycetes, a class placed between the ancestors and the Zygomycetes in that scheme, shows an average %GC value of 51 (277). More important are the results of surveys of chemical and biochemical criteria other than DNA base composition.

These criteria are based on the characterization of biosynthetic pathways, enzymes and enzyme complexes, and the chemical composition of the cell wall. The distribution among 26 fungal species of the two biosynthetic pathways of lysine has been investigated (317, 318). One pathway, which is via diaminopimelic acid (DAP), was found only in species representative of the classes Hyphochitridiomycetes and Oömycetes. The other, which is routed through aminoadipic acid (AAA), is present in all other fungal classes. In another study it was found that the sedimentation patterns of the enzymes involved in the biosynthesis of tryptophan fell into four groups (136). The assignment of these patterns to fungal groups is shown in Table 4. (It should be mentioned that the slime mold *Physarum polycephalum* has type I.) As can be seen, the results of this study link the Chytridiomycetes to the Euascomycetidae and the Homobasidiomycetidae. The Zygomycetes and Heterobasidiomycetidae share in common another pattern, while the Hemiascomycetidae and Oömycetes are each isolated into another group. This has led to the suggestion to abandon the Zygomycetes and the Hemiascomycetidae as intermediates between the Chytridiomycetes and the Euascomycetidae. The only features that are left from the monophyletic scheme are the connections between the Euascomycetidae and the Homobasidiomycetidae and the independent origin

Table 4. Sedimentation Patterns of Tryptophan Biosynthetic Enzymes[a]

Fungal Group	Sedimentation Pattern Type
Hyphochytridiomycetes	—
Chytridiomycetes	I
Oömycetes	IV
Zygomycetes	III
Hemiascomycetidae	II
Euascomycetidae	I
Heterobasidiomycetidae	III
Homobasidiomycetidae	I

[a] From ref. 136.

Table 5. Cell Wall Composition[a]

Fungal Group	Polysaccharide Pair[b]
Hyphochytridiomycetes	Cellulose–chitin (III)
Chytridiomycetes	Chitin–β-glucan (V)
Oömycetes	Cellulose–β-glucan (II)
Zygomycetes	Chitin–chitosan (IV)
Hemiascomycetidae	Mannan–β-glucan (VI)
Euascomycetidae	Chitin–β-glucan (V)
Heterobasidiomycetidae	Chitin–mannan (VII)
Homobasidiomycetidae	Chitin–β-glucan (V)

[a] From ref. 15.
[b] The Roman numerals are those used by this author to designate cell wall category (15).

of the Oömycetes. It might be mentioned in passing that, in agreement with the results of the DNA base composition studies discussed earlier, the imperfect yeast forms such as *Cryptococcus laurentii* and *Rhodotorula glutinis* have the same enzyme pattern as *Sporobolomyces salmonicolor, Tremella mesenterica,* and *Ustilago maydis,* which are Heterobasidiomycetidae.

The nature of the polysaccharides present in cell walls has been extensively investigated, and the data obtained have been recently compiled and discussed (15). The phylogenetic approach rests on recognition of the predominance of two types of polysaccharides in each category of cell wall. The distribution of these pairs of macromolecules among the major classes of fungi is described in Table 5. As can be seen, it also suggests breaking away from the monophyletic scheme by side-lining the Zygomycetes and the Hemiascomycetidae from the main line going to the Euascomycetidae and the Homobasidiomycetidae. A new scheme, combining these results and those on the enzymes of the tryptophan pathways and on the pathways of lysine biosynthesis, has been presented (15). This scheme suggests that ancestral Chytridiomycetes gave rise to ancestral Ascomycetes and the latter, possibly, to ancestral Basidiomycetes. From each of these ancestral classes would have originated, respectively, the Zygomycetes and the Chytridiomycetes, the Hemi- and the Euascomycetidae, and the Hetero- and the Homobasidiomycetidae.

More recently, an attempt has been made to build an evolutionary scheme combining the %GC of DNA with the three other criteria discussed above (278). This scheme maintains one monophyletic series: Chytridiomycetes to Euascomycetidae to Homobasidiomycetidae. It suggests that the Heterobasidiomycetidae evolved from the Homobasidiomycetidae, and the Hemiascomycetidae from the Euascomycetidae and therefore favors the degeneracy theory of the origin of yeasts. It presents the Zygomycetes as an isolated group which might have evolved from an "unknown" that would also have been the precursor of the Chytridiomycetes.

The properties of glutamic dehydrogenases and other regulatory enzymes of different species of fungi have been compared (159). These properties, together with the criteria discussed above (except %GC of DNA) and additional results on cell wall chemical composition, have been combined to produce yet another evolutionary pathway. This study places emphasis on the lower fungi and suggests divergent pathways from the Chytridiomycetes for the Blastocladiales and a few Mucorales, on one hand, and the Chytridiales and most Mucorales, on the other.

From what has been said so far, it should be apparent that the understanding of the taxonomic and phylogenetic relationships in the fungi has improved with the introduction of molecular criteria. There are thus good reasons to believe that, as more organisms are surveyed and as other macromolecules are studied, it will become possible to express these relationships in a quantitative manner. Although the determination of DNA GC content has a limited "resolving power," it has nevertheless helped in the understanding of these relationships. There are other reasons why the practice of %GC determination of fungal DNA should not be abandoned. Indeed, DNA base composition is very stable. It is not affected by changes in environmental conditions. Claims for significant variations due to induced mutation have been shown to be false (299). The procedures used give reproducible results and are not affected by such factors as the production of extracellular deoxyribonuclease (DNase), as is the case for some species of yeasts (46) and hyphomycetes (325). It follows that an unexpected %GC value for a given organism immediately creates suspicion about its identity. As will be seen later, this reasoning applies to nuclear DNA only, since mitochondrial DNA base composition can be drastically affected as a result of mutation.

B. Sequence Homologies

Knowledge of nucleotide sequences is possible at present for molecules of relatively low molecular weight, such as transfer RNA and 5S RNA (60). For larger molecules, information about sequence can be obtained from hybridization studies. Sequential homologies can be measured by comparing DNA molecules among themselves or by comparing them with RNA molecules. Hybridization of DNA is possible with molecules of high molecular weight, such as ribosomal RNA, as well as with smaller RNA molecules. In all instances the degree of hybridizability can be taken as a quantitative expression of genetic relatedness. The phylogenetic and taxonomic value of such studies is, of course, superior to that of nucleotide composition studies, provided that a large number of organisms, adequately sampled, are investigated and that there is an understanding of the molecular events involved.

For fungi only a few studies have as yet been made. With the DNA–agar method (134) several fungal species were investigated (72). It was found that the extent of DNA–DNA binding was constant between species of *Neurospora* and was the same as between DNA samples originating from a single species *(Neurospora crassa)*. The DNA of *Coprinus lagopus* hybridized better with itself than with the DNA of either *N. crassa* or *Neurospora intermedia*. Similarly, *N. crassa* DNA hybridized better with itself than with the DNA of *Fusarium oxysporium* or *Aspergillus nidulans*, *Rhizopus nigricans*, or *C. lagopus*. More recently, DNA–DNA homologies within the genera *Saccharomyces* (24, 25) and *Candida* (163) have been measured by the filter membrane procedure (100), which represents an improvement over the agar method. These studies have revealed that DNAs with identical %GC can be heterologous in terms of nucleotide sequences and that there is a higher diversity among these yeast species than among the filamentous forms. Similarities in DNA nucleotide sequence have also been measured in the genus *Neurospora* (70). The per cent homology, when normalized to 100 for *N. crassa*, was 84 for *N. intermedia*, 88 for *Neurospora sitophila*, and 85 for *Neurospora tetrasperma*. There were only 19% homology between the DNA sequences of *N. crassa* and those of *C. lagopus*. There is no doubt that such types of studies, which are now being pursued in several laboratories, will considerably improve our understanding of the taxonomic and phylogenetic relationships among fungi.

C. Genome Size

The genome size represents the sum total of the genetic information of an organism. It can be expressed as the amount of DNA per haploid nucleus. As already mentioned in the Introduction, this amount increases from 10^{-15} to 10^{-12} g from bacteria to mammals. In bacteria it corresponds to that of a unique chromosome which itself is a single, closed, circular DNA molecule having a molecular weight of 2 to 3×10^9 daltons and corresponding to about 3×10^6 nucleotide pairs. With the exception of repeated cistrons coding for ribosomal RNA, which amount to less than 1% of the total DNA (27), the DNA of bacteria appears to be composed exclusively of unique or nonrepetitive sequences of nucleotides. In contradistinction, the greater amount of DNA in higher forms of life is characterized by the existence of repeated sequences that may amount to 50% or more of the total genome (36, 37, 148).

Before reviewing the available information on fungal genome size, it is worth while to inquire about the degree of complexity of the nuclear apparatus of fungi. The minuteness of the nucleus in most fungi has been a major rcason for the slow progress in our understanding of nuclear division in the somatic structures of these organisms. It is now established that the fungal nucleus undergoes a true mitotic division; proof can be found in the references listed in Table 6. For a long time the counting of fungal chromosomes has been rendered difficult because of the facts that their staining is difficult and that when they are stainable, that is, at diakinesis and metaphase, they tend to be clumped. As a result, the haploid number of chromosomes for a given organism has often had to be revised as improved staining methods became available. Except for the few organisms, such as *N. crassa*, where the chromosome number was arrived at both by cytology and by determination of linkage groups by genetic analysis, in many instances the published chromosome numbers may represent underestimates of the true values.

Some of the results of chromosome counts that appear reliable are listed in Table 6. This tabulation was made in order to find out whether the chromosomal formulas of fungi might bear some relationship to established taxonomic and phylogenetic positions. As a rapid inspection of Table 6 indicates, this turns out not to be the case, since the species representative of the Chytridiomycetes and the Zygomycetes have haploid numbers of chromosomes equal to or greater than those of the Ascomycetes and the Basidiomycetes. These numbers vary among taxons from classes to genera. The latter type of variation appears, however, to be rare. Needless to say, many more data will be required in to order to disprove or confirm this observation. It might be pointed out that sexual behavior may be dependent on chromosome number (or vice versa), since in the case of *Puccinia* all the homothallic species have 4 chromosomes whereas the heterothallic ones have either 3 or 6. Polyploidy has been responsible for very high numbers; this is especially true of slime molds (106). The correct haploid chromosome number is 10 in *P. polycephalum* (111) and 7 in the acrasiale *Dictyostelium discoideum* (284). The major feature of Table 6 is that, as a rule, the numbers of chromosomes tend to be smaller in fungi than in algae (101).

At present no information is available concerning the size of DNA molecules as they exist in fungal chromosomes (126). Since in most cases the disruption of cell walls requires brutal mechanical treatment, the chances for obtaining intact nuclei and intact DNA molecules are small. There have been reports of the isolation of intact nuclei from a few organisms. Among these are *D. discoideum* (285), *P. polycephalum* (189), *Saccharomyces cerevisiae* (23), and *Schizosaccharomyces pombe* (66). In the last two cases, advantage was taken of the possibililty to make protoplasts (315). In three instances, listed in Table 7, the amount of DNA could be expressed not only per nucleus but also per

Table 6. Chromosome Numbers

Organism[a]	Haploid Number of Chromosomes	Reference
Chytridiomycetes		
Allomyces spp. (1)	14	333
(2)	16	333
(3)	28	333
Zygomycetes		
Phycomyces blakesleeanus	14	88
Rhizopus nigricans	16	88
Ascomycetes		
Arthroderma flavescens	4	324
A. simii	4	324
A. uncinatum	4	324
Ascobolus immersus	16	83
A. stercorarius	16	83
Aspergillus nidulans	8	83
Bombardia lunata	7	83
Ceratocystis fagacearum	4	1
Emericellopsis spp.	2–4	18
Fusarium oxysporum	4	133
Gelasinospora tetrasperma	7	83
Glomerella cingulata	4	83
Lipomyces lipofer	6	235
Nannizzia fulva	4	324
N. grubyia	4	324
N. gypsea	4	324
N. incurvata	4	324
Neurospora crassa	7	218
N. sitophila	7	83
N. tetrasperma	7	83
Penicillium expansum	4–5	154
Podospora anserina	7	83
P. arizonensis	2	180
Poronia oedipus	7	236
P. punctata	7	236
Saccharomyces cerevisiae	18	83, 286
Schizosaccharomyces pombe	6	89
Sordaria fimicola	7	218
S. macrospora	7	83
Trichophyton mentagrophytes	4	324

Table 6. (*Continued*)

Organism[a]	Haploid Number of Chromosomes	Reference
Basidiomycetes		
Agaricus campestris	12	218
Amanita fulva	8	218
Coleosporium vernoniae	8	218
Coprinus atrameutarices	16	172
C. comatus	14	172
C. lagopus	12	172
C. micaceus	12	172
C. radiatus	8–9	83
Cyathus stercoreus	12	218
Puccinia arenaria	4	177
P. asteris	4	177
P. carthomi	6	177
P. coronata calamagrosistis	3	177
P. coronata secalis	3	177
P. graminis	6	177, 329
P. helianthi	6	177
P. malvacearum	4	177
P. sorghi	6	177
P. xanthii	4	177
Schizophyllum commune	3	218

[a] Organisms are assigned to classes according to ref. 3, and in each class are listed in alphabetical order for convenience.

chromosome. More organisms need to be studied in order to make significant comparisons of such values possible.

The amount of DNA per spore has been determined in several fungal species; the available data are presented in Table 8. It should be pointed out that in most instances the number of nuclei per spore was unknown. Also, the procedures used for the extraction of DNA and its chemical determination have not always been the same. This should be kept in mind before scrutinizing the content of this table. In the case of several organisms, the amount of DNA was determined in both haploid and diploid cells, and it can be seen that there is proportionality. The same is true for *S. cerevisiae*, where the proportionality applies also for polyploid forms. Advantage has been taken of this proportionality to certify the existence of a perfect stage in the case of *Candida*

Table 7. DNA per Nucleus and Chromosome[a]

Organism	DNA/ nucleus (pg)	Nucleotide Pairs	
		Per Nucleus	Per Chromosome
Dictyostelium discoideum	0.06	5.4×10^7	7.7×10^6
Physarum polycephalum	0.3	2.7×10^8	2.7×10^7
Saccharomyces cerevisiae	0.05	4.4×10^7	2.5×10^6

[a] For references see text.

albicans (301), *Cryptococcus albidus* (300), and *Rhodosporidium toruloides* (301). *Aspergillus sojae* appears to be an exception because its haploid spores contain twice the number of nuclei present in its diploid ones. The DNA contents of the sporangiospores of *Mucor* do not vary greatly from species to species, with the exception of *Mucor bacilliformis*. It is not known whether this is a reflection of a difference in ploidy or in number of nuclei. Noteworthy is the fact that the volume of *M. bacilliformis* spores is much smaller than that of the spores of the other *Mucor* species. With the limited number of values available, it is impossible at the present time to decide whether the evolution of fungi has been marked by a measurable increase in the amount of DNA per nucleus.

Encouraging signs indicate that an answer to this question will be provided by the results of DNA reassociation experiments. In these, single-stranded DNA molecules of a given length are placed under conditions that favor their reassociation. It has been demonstrated that this reassociation has a second-order kinetics and that the rate of reassociation is inversely related to the length of the polynucleotides and directly related to the genome complexity (36, 326). This complexity is the sum of all unique sequences present in the DNA. Redundant DNA, composed of repeated sequences, can be distinguished from the unique sequences by the fact that its reassociation rate does not fit that expected for a second-order reaction. Clearly, this method offers a possibility to obtain first-hand information about the genome of any organism. So far, however, its application to the genomes of fungi has been limited. In Table 9 are presented the few values available in the literature and preliminary results obtained in the author's laboratory. The first significant feature of this table is that the genome sizes of related organisms tend, as a rule, to vary little. The second is that the genome size of the mucors and that of yeasts appear

Table 8. DNA Content per Cell

Organism[a]	DNA Content[b]	Reference
Zygomycetes		
Mucor bacilliformis	4.50	[c]
M. fragilis	9.90	[c]
M. hiemalis	12.00	[c]
M. mucedo	9.30	[c]
M. racemosus	8.90	[c]
M. rouxii	10.50	[c]
Ascomycetes		
Aspergillus sojae	37.10	
	37.30 (2*n*)	
	8.8/nucleus	138
	16.9/nucleus (2*n*)	
Candida albicans	0.82	301
	1.90 (2*n*)	
Cryptococcus albidus	0.11	300
	0.23 (2*n*)	
Neurospora crassa	1.70	187
	4.60	124
Ophiostoma multiannulatum	4.80/nucleus	124
Protomyces inundatus	0.154	298
Saccharomyces cerevisiae	2.45	217
	4.50 (2*n*)	332
	4.95 (2*n*)	217
	6.70 (3*n*)	217
	10.20 (4*n*)	217
Basidiomycetes		
Rhodosporidium toruloides	10.00	12
	20.00 (2*n*)	
Schizophyllum commune	13.20	2
Sporobolomyces salmonicolor	0.60	
	1.30 (2*n*)	301

[a] Organisms are assigned to classes according to ref. 3, and in each class are listed in alphabetical order for convenience.
[b] The DNA content is expressed in grams $\times 10^{-14}$ per spore or per cell in the case of the yeasts, unless otherwise indicated. When the ploidy is not indicated, the cells analyzed were either haploid or assumed to be so.
[c] R. Storck, unpublished data.

Table 9. Genome Size of Fungi, Calculated from Rates of Reassociation

Organism	Genome Size (daltons)	Reference
Zygomycetes		
Mucor bacilliformis	7.4×10^9	[a]
M. fragilis	6.4×10^9	[a]
M. hiemalis	6.6×10^9	[a]
M. mucedo	7.2×10^9	[a]
M. racemosus	7.8×10^9	[a]
M. rouxii	5.2×10^9	[a]
Ascomycetes		
Candida catenulata	1.2×10^{10}	49
C. macedoniensis	8.3×10^9	49
Debaryomyces hansenii	6.0×10^9	49
Hansenula holstii	6.5×10^9	49
Kluyveromyces lactis	8.5×10^9	49
Neurospora crassa	2.2×10^{10}	71
Saccharomyces carlsbergensis	9.4×10^9	49
S. cerevisiae	9.2×10^9	25
S. exiguus	1.1×10^{10}	49
Talaromyces vermiculatus	3.0×10^{10}	254
Torulopsis holmii	1.4×10^{10}	49
T. candida	1.2×10^{10}	49
Basidiomycetes		
Coprinus lagopus	3×10^{10}	219
Cellular slime molds		
Dictyostelium discoideum	3.6×10^{10}	285

[a] R. Storck, unpublished observations.

to be somewhat smaller than those of the Euascomycetidae representatives and of *C. lagopas*. It was observed that *N. crassa* and *Agaricus bisporus* (Basidiomycete) had genomes about 2 to 4 times larger than the genome of *S. cerevisiae* (24). The third important feature of Table 9 is that the genome sizes of fungi range from 6×10^9 to 3×10^{10} daltons. Since *Escherichia coli* and *Bacillus subtilis* have genome sizes of 2.9×10^9 and 2.0×10^9 daltons, respectively (155), it can be concluded that the informational content or genome complexity of fungi is about 3 to 10 times greater than that of bacteria.

Is there in fungal DNA a significant proportion of redundant sequences? In order to answer this question, we might first go back to Table 8. The amount of DNA per haploid nucleus varies from 2.45×10^{-14} g (haploid strain of *S. cerevisiae*) to 8.8×10^{-14} g (*A. sojae*). These values correspond to about 1.5×10^{10} and 5×10^{10} daltons, respectively. Assuming that chemical determination of DNA offers the same sensitivity as reassociation measurement, it would appear that a certain proportion of DNA might have a rate of reassociation that does not obey the kinetics of a second-order reaction. However, the reassociation profiles for *Mucor, S. cerevisiae, N. crassa, Thalaromyces vermiculatus, C. lagopus*, and *Agaricus campestris* (24) gave no indications of the existence of repetitive DNA. Furthermore, the latest estimates of DNA content for haploid *S. cerevisiae* cells of 8.6×10^{9} daltons (24) and 1.27×10^{10} daltons (250) are closer to those listed in Table 9 than to those given in Table 8. These comparisons are invalidated, nevertheless, by a recent report (49) claiming the existence in yeasts of 5.3 to 16.2% of repetitive DNA. We will have to wait for more analyses of fungal DNAs before it can be stated with confidence that redundant DNA is present in all eukaryotes regardless of the DNA content of their haploid nucleus. Similarly, analyses of more prokaryotes are equally needed in order to establish without doubt that none of these contains redundant information in addition to that for ribosomal RNAs. The distinction between prokaryotes and eukaryotes is real and useful and in many aspects intellectually appealing. Care should nevertheless be exercised before it is assumed that none of the criteria now used is distributed in an all-or-none fashion between these two groups of organisms. For example, there are contradictory reports regarding the presence of histones in fungi. Although the existence of these proteins has been claimed in the cases of *P. polycephalum* (188), *S. cerevisiae* (23, 294), and *Saccharomyces carlsbergensis* (190), their absence has been reported in *Microsporum gypsum, N. tetrasperma, Phycomyces blakesleeanus* (158), *N. crassa* (73, 158), and *Allomyces arbuscula* (282).

3. MITOCHONDRIAL GENOME

The existence of DNA in mitochondria is now well established (32, 209, 225, 238, 243).As a result, three important questions have been raised. What is the mitochondrial genome size? What processes underlie the biogenesis of mitochondria? Are mitochondria endosymbiotes derived from distant prokaryotic ancestors? Attempts to answer these questions have resulted in the last few years in an impressive number of publications. The body of this literature is so large, even that pertaining to fungal systems, that no attempt has been made here to cover it in its totality, and the reader is referred to some recent reviews (5, 52, 210, 230, 320) which, in addition to the references cited above, cover most of the publications.

The following discussion, as well as that in Section 6, is intended to show why some fungi constitute good experimental systems for the study of these problems.

A. Base Composition

In CsCl profiles of total DNA, bands satellite to the main band of nuclear DNA are often detected. In some instances such bands correspond to mitochondrial DNA, and in others they do not. In the latter case, the buoyant density is often less than 1.680 g/cm^3. Such bands are insensitive to DNase, and their buoyant density shows considerable variation from one preparation of the same isolate to another. In the specific case of *Mucor subtillisimus* it was shown that the satellite is a polysaccharide contaminated with DNA (195). Such a type of satellite has since been described in DNA preparations from *D. discoideum* (285), *A. nidulans* (77), bacteria (339), blue-green algae (75), algae, protozoa, plants, and animals (182). This satellite, which might be regarded as an artifact of DNA extraction, can be eliminated with various degrees of success by treatment with amylase, chitinase, and other appropriate polysaccharide-hydrolyzing enzymes. Satellite bands that appear to be neither those containing DNA linked with polysaccharides nor those representing other macromolecules have also been detected. In a recent compilation (58), it was shown that in many plants and animals this type is, in fact, repetitious DNA. Its abundance (30% or more of the totality of the DNA) has raised questions about its role. Several theories have been proposed, and the reader is referred to one of the most recent ones, according to which the chromosomes that carry such a DNA type have "an improved ability to stand up to the rigours of meiosis" (322). With regard to *S. cerevisiae*, there is a recent claim for the existence of a satellite DNA that appears to have a nuclear origin, since it is found in lysates devoid of mitochondria (194, 274). Whereas nuclear DNA bands at 1.699 g/cm^3 and mitochondrial DNA at 1.684 g/cm^3, this satellite DNA, designated as gamma, has a buoyant density in CsCl corresponding to 1.703 g/cm^3. In order to ascertain the nuclear origin of this gamma satellite, it will be necessary to study DNA preparations from isolated nuclei. In *P. polycephalum* a satellite DNA with buoyant density 0.014 g/cm^3 higher than that of nuclear DNA (1.700 g/cm^3) has also been described (34). In *Blas-*

Table 10. Comparison of Mitochondrial and Nuclear DNA

Organism[a]	Buoyant Density (g/cc)[b]		GC Content[c]		Reference
	Nu-DNA	Mit-DNA	Nu-DNA	Mit-DNA	
Zygomycetes					
Cumminghamella echinulata	1.693	1.693	34	34	314
Mucor fragilis	1.698	1.697	39	38	314
M. rouxii	1.696	1.697	37	38	314
Ascomycetes					
Aspergillus nidulans	1.711	1.689	51	30	77
Candida lipolytica	1.709	1.687	50	28	162
C. pelliculosa	1.696	1.688	37	29	162
C. pseudotropicalis	1.701	1.684	42	25	162
C. utilis	1.704	1.685	46	26	162
C. zeylanoides	1.715	1.692	56	33	162
Ceratocystis ulmi	1.715	1.699	56	40	314
Chaetomium globosum	1.717	1.693	58	34	314
Gelasinospora autosteria	1.713	1.700	54	41	314
G. calospora	1.714	1.700	55	41	314
G. cerealis	1.714	1.701	55	42	314
Kluyveromyces lactis	1.699	1.691	40	32	175
Neurospora crassa	1.713	1.701	54	42	314
N. sitophila	1.714	1.702	55	43	314
Saccharomyces carlsbergensis	1.699	1.683	40	24	194
	1.699	1.684	40	25	57
S. cerevisiae	1.697	1.682	38	22	248
	1.700	1.685	41	26	292
	1.699	1.684	40	24.5	162
Sordaria macrospora	1.713	1.701	54	42	314
Basidiomycetes					
Deadalea confragosa	1.716	1.690	57	31	314
Schizophyllum commune	1.720	1.687	61	28	314
Slime molds					
Dictyostelium discoideum	1.683	1.688	23	28	285
Physarum polycephalum	1.700	1.685	41	26	84

[a] Organisms were classified according to ref. 3. In each class they are listed alphabetically for convenience.
[b] Nu-DNA, nuclear DNA; Mit-DNA, mitochondrial DNA.
[c] The GC contents were calculated according to ref. 245.

tocladiella emersonii whole-spore DNA contains four components distinguishable by their buoyant densities. It appears that one of these bands corresponds to DNA that originates from the gamma particle, an organelle found in the motile cell of this organism (197).

When intact mitochondria have been treated with DNase in order to eliminate contaminating nuclear DNA (174, 231), they contain, as a rule, a population of DNA molecules that form a single band in CsCl equilibrium centrifugation. The buoyant density of such a DNA has been determined for several fungi, as shown in Table 10. With the exception of the three species of *Mucorales* and *D. discoideum*, the buoyant density of mitochondrial DNA is always less than that of its nuclear counterpart. Comparisons of the GC contents calculated from the buoyant density values do not suggest a correlation between the two types of DNA. Worth noting is the fact that the GC content of mitochondrial DNA, like that of nuclear DNA, varies little for related species. The overall range is from 22 to 43%GC for mitochondrial DNA, as compared to 23 to 61%GC for nuclear DNA.

The conversion of buoyant density values into %GC is possible for DNAs that contain no appreciable amount of sugars other than deoxyribose or bases other than adenine, cytosine, guanine, and thymine. When such is the case, %GC values calculated from buoyant density and T_m measurements are identical. This condition was satisfied in the case of *N. crassa* (173) and *P. polycephalum* (84). For *D. discoideum* nuclear DNA has a value of 23% when calculated by both methods, whereas for mitochondrial DNA density and T_m measurements yield 28% and 26%, respectively (285). Even greater discrepancies have been observed for mitochondrial DNA from *S. cerevisiae* and *S. carlsbergensis* (20, 32, 292) and from *Candida* species (162). In the latter genus, differences as high as 22% were recorded. This unexpected behavior is due not to an abnormal chemical composition but apparently to the presence of repetitive sequences, especially alternating and nonalternating A–T pairs (20, 21).

B. Sequence Homologies

Such studies ought to provide us, as they do for nuclear DNA, with a measurement of the genetic relatedness of mitochondria from various sources and, in addition, of the degree of homology existing between mitochondrial DNA and its nuclear counterpart. The few available studies, made primarily on yeasts (92–94) and on *Neurospora* (337), indicate that nuclear and mitochondrial DNA have either few or no common sequences.

C. Genome Size

In most organisms the cell content in mitochondrial DNA represents a small proportion of the total DNA, for example, 2% in *A. campestris* (316), 7% in *P. polycephalum* (34, 84), and 15% in *S. cerevisiae* (32, 94). A notable exception is *D. discoideum*, in which the proportion may be as high as 40% (285)! This unexpected result makes this organism, already remarkable in other respects, the system of choice for study of the interactions between the nuclear and the mitochondrial genomes. Since the physicochemical properties of the two types of DNA in this organism are different, direct studies of total DNA would lead to erroneous inference about the properties of nuclear DNA. In the case of *D. discoideum* and perhaps in that of other organisms as well, further studies will require a separation of nuclear and mitochondrial DNAs by isolation of intact nuclei and mitochondria. There is at the present time no explanation for the differences in the proportion of mitochondrial DNA existing between various organisms. It is possible that in some cases they reflect variations in procedure.

Does the amount of mitochondrial DNA in a given organism vary with changes in environmental conditions? An example is found in organisms such as yeast, where the mitochondrial respiratory activity can be repressed by anaerobiosis. At first it was thought that under such conditions mitochondrial DNA could be eliminated altogether (57, 194) or be greatly reduced (293), but later improved methodology revealed that this type of DNA never amounted to much less than about 15% of the total DNA (32, 59, 94). Under conditions such as transfer of yeast cells into a medium favoring respiration and containing cycloheximide (108, 224), or inhibition of protein synthesis by amino acid starvation or temperature sensitivity (108), it is possible to observe a rapid increase in the rate of synthesis of mitochondrial DNA so that its level is raised above the normal one. As will be discussed in Section 6 on the autonomy of the mitochondrial genome, the amount of DNA can be decreased to the point of complete elimination as a result of treatment with mutagens such as ethidium bromide.

Perhaps of greater significance for assessing the size of mitochondrial genomes are direct measurements of the length of the DNA molecules. These measurements were made either *in situ* [with the DNA spreading technique (145)] or on isolated molecules. The information pertaining to fungi is summarized in Table 11. Additional information about nonfungal systems can be found in a recent review (5). As a rule, a great heterogeneity of sizes was found in fungal mitochondrial DNA preparations (6, 7, 19, 20, 32, 109). It now appears that this heterogeneity is primarily the result of the breakage of

Table 11. Characteristics of Mitochondrial DNA Molecules

Organism	Size (μm)	Weight (daltons)	Geometry	Reference
Neurospora crassa	25	6.6×10^7	Linear	153, 174, 337
Physarum polycephalum	37	4×10^7	Linear	
	26	—	Open circles	265
	28	—	Linear	144
	11	$2\text{–}3 \times 10^7$	Linear	84
Saccharomyces carlsbergensis	25.6	—	Circular	125
S. cerevisiae	Various	$>4 \times 10^7$	Some molecules are circular	109
	—	6×10^6		194
	—	6×10^7	—	292
	4.5–5.5	Variable	Circular and linear	7
	0.1–5.5	1×10^7	Linear	258
	25	5×10^7	Circular	125
	25–28	—	Linear	242

molecules that takes place during extraction and purification, since a recent report states that for *N. crassa* 90% of the mitochondrial DNA molecules were linear and had a length of 25 μm only when pipetting and shaking were avoided during purification (242). As a result, it apppears that the smaller values listed in Table 11 should be regarded with suspicion. That small molecules of nuclear origin such as the gamma type described above contaminate mitochondrial preparation is possible. This might also be the case for molecules of unknown origin that are homogeneous in size, are circular, and have lengths of 2.2 μm and multiples thereof (7, 26, 110, 125, 258). Of relevance is the fact that the geometry of the molecules is probably a reflection of the procedure used. For example, in yeast *in situ* mitochondrial DNA appeared as closed circles with a perimeter of 25 μm, and in DNA preparations as linear molecules with lengths ranging from 15 to 27 μm (125). If 25 μm is to be taken as the best estimate of the size of mitochondrial DNA molecules in fungi (see Table 11), it would follow that these molecules are 5 times longer than those found in the mitochondria of animal cells (5, 209). Their average molecular weight would, as a result, be equal to about 5×10^7 daltons. Such molecules would contain about 7.5×10^4 nucleotide pairs and have the capacity to code for 2.5×10^4 amino acids. Clearly, this coding capacity is too small for all the different types of proteins found in mitochondria (5, 320). It follows that some of these proteins must have the information for their amino acid sequences stored in nuclear DNA.

Why should the DNA molecules of fungi be larger than those of higher forms? There is at present no explanation in sight. It might be added that mitochondrial DNA molecules of fungi, like those of other systems, are double-stranded.

More work needs to be done to find out whether circularity is a general property of these molecules. There are more and more examples of circularity for DNA of various origins, and the reader is referred to a recent review (117) for additional information. Before closing this section, it should be pointed out that the low-molecular-weight DNA molecules found in crystals of yeast lactate dehydrogenase and thought to be unique have been shown to be degradation products and thus to be artifacts (94).

4. RIBOSOMES

Sedimentation analyses were first performed on ribonucleoprotein particles extracted from *S. cerevisiae* (48). Reports that followed on the sedimentation of ribosomes from other fungi often failed to give values for the sedimentation coefficient(s) corrected for variation in temperature, concentration, and viscosity of the suspending medium. Since these factors affect sedimentation significantly, comparisons between analyses originating from various laboratories were not possible. Fortunately, this situation has since been corrected. A survey of 26 species of fungi, 25 of bacteria, and 2 of blue-green algae demonstrated that their ribosomes were distributed between two distinct classes characterized, respectively, by mean values of 68.4 and 81.3$S^{\circ}_{20,w}$ (289). Thus, without doubt, fungal ribosomes are of the 80S type, like those of other eukaryotic cells (221, 268). This has been confirmed by more recent determinations of sedimentation values, for example, on *Rhizopus arrhizus* and *Pythium debaryanus* (165), *Nematospora coryli, C. laurentii, Nadsonia fulvescens* and *Endomycopsis capsularis* (64), *Physarum rigidum* (118), *D. discoideum* (47, 139, 140), and *Dictyostelium purpureum* (47). The differences between the sedimentation coefficients are extremely small, and these values appear therefore to have no taxonomic value when determined by analytical ultracentrifugation. The possibility is not ruled out, however, that preparative isokinetic ultracentrifugation (214), which is claimed to have a resolving power of about 5% of the observed sedimentation value, might permit the grouping of organisms into distinct categories (215).

The first mitochondrial ribosomes to be characterized by sedimentation analysis were those extracted from *N. crassa* (152, 233). A sodium deoxycholate lysate of mitochondria yielded three peaks in a sucrose density gradient centrifugation profile. The ribonucleoproteinic nature of these peaks

Table 12. Sedimentation of Mitochondrial Ribosomes

Organism	Method[a]	Sedimentation Coefficient(s)[b]	Reference
Aspergillus nidulans	SDG	67	77
Candida utilis	SDG	77–80	312
Mucor rouxii	AUC	79	313
Neurospora crassa	AUC and SDG	81	233
	SDGI	73	152
	SDGI	50, 37	149, 151
Saccharomyces cerevisiae	SDG	80	246
	SDG	75	271
	AUC	80	191

[a] Abbreviations: AUC, analytical ultracentrifugation; SDG, sucrose density gradient centrifugation; SDGI, sucrose density gradient isokinetic centrifugation.
[b] The sedimentation coefficient(s) is expressed in svedberg units (S).

was ascertained by specific radioactive labeling. The cosedimentation of the heavy particle with cytoplasmic 80S monomers indicated that these two types of ribosomes could not be separated by sedimentation analysis. Mitochondrial ribonucleoprotein particles could, however, be distinguished from their cytoplasmic cosedimenting homologs by the fact that they dissociated more readily as a result of a decrease of the Mg^{2+} concentration. As shown in Table 12, in several instances, including *N. crassa*, sedimentation values for monomers lower than 80S were reported from a number of laboratories. In view of the fact that this is also the case for a few systems other than fungi, it is generally assumed that mitochondrial ribosomes are of the 70S or prokaryotic type. Worth mentioning is a recent claim that 55S ribosomes are miniature monosomes which are peculiar for mammalian mitochondria (216). Since the ribosomes found in chloroplasts are also of the prokaryotic type (28), many authors have concluded that this situation can be taken as an indication of a possible parasitic origin of both mitochondria and chloroplasts (52, 283, 328). As far as fungi are concerned, it is clear that the sedimentation coefficient values have a range including prokaryotic and eukaryotic ribosomes. No doubt some of the discrepancies found for the same species could be resolved by using rigorously standardized analytical ultracentrifugation. It should be pointed out that in some cases, for example, that of *S. cerevisiae*, the two types of ribosomes can be distinguished from each other by differences in nucleotide composition, electrophoretic mobility of the proteins, and thermal denaturation profiles, even though their sedimentation coefficients are the same (191). More importantly, in all organisms tested so far, the two types of ribonucleoprotein particles can be distinguished, as will be seen later, by their different sensitivities to antibiotics inhibiting protein synthesis.

5. RIBONUCLEIC ACIDS

Ribosomal monomers of the 70S and 80S types contain two molecules of RNA. Sedimentation analyses have demonstrated that, on the average and regardless of the nature and concentration of the ions in the solvent, the sedimentation coefficients of the RNA molecules in the 80S ribosomes are greater than those in the 70S ribosomes (268, 288). In view of this, it has been suggested that the difference between the sedimentation velocities of eukaryotic and prokaryotic ribosomes might be due to their RNA constituents (268, 288). The sedimentation coefficients of these molecules are shown in Table 13. As one can

Table 13. Sedimentation of Cytoplasmic Ribsomal RNA

Organism	Method[a]	Sedimentation Coefficient(s)[b]		Reference
Aspergillus nidulans	AUC	26.5	17.0	77
Blastocladiella emersonii	E	25.0	18.0	171
Candida pulcherrima	AUC	22.4	15.9	288
C. utilis	SDG	24.0	16.0	311
Cunninghamella echinulata	AUC	23.7	16.7	288
Mucor rouxii	AUC	23.2	16.7	288
Neurospora crassa	AUC	24.3	16.3	288
	AUC	27.9	18.0	311
	SDG	25.0	17.4	152
	SDG	25.6	16.0	69
Saccharomyces cerevisiae	E	26.0	19.0	271
	AUC	24.6	16.2	237
	AUC	25.1	15.2	16
	AUC	23.3	17.0	161
	SDG	24.7	17.4	161
Schizophyllum commune	AUC	23.3	16.4	288

[a] Abbreviations: AUC, analytical ultracentrifugation; E, polyacrylamide gel electrophoresis; SDG, sucrose density gradient centrifugation.
[b] The sedimentation coefficient(s) is expressed in svedberg units (S) for molecules suspended in a medium devoid of divalent cations with the apparent exception of *Candida utilis*.

Table 14. Sedimentation of Mitochondrial Ribsomal RNA

Organism	Method[a]	Sedimentation Coefficient(s)[b]		Reference
Aspergillus nidulans	AUC	23.5	15.5	77
Candida utilis	SDG	21.0	16.0	311
Neurospora crassa	AUC	24.5	18.9	233
	SDG	23.0	16.0	69
	SDG	20.5	16.4	152
Saccharomyces cerevisiae	AUC	21.9	15.1	161
	AUC	22.4	17.8, 12.7	237
	SDG	22.7	15.6	161
	SDG	25.0	15.5	272
	E	25.0	17.0	271

[a] Abbreviations: AUC, analytical ultracentrifugation; E, polyacrylamide gel electrophoresis; SDG, sucrose density gradient centrifugation.
[b] The sedimentation coefficient(s) is expressed in svedberg units (S) for molecules suspended in a medium devoid of divalent cations with the apparent exception of *Candida utilis.*

see, there is no indication of systematic differences between the fungal species listed. In contrast, measurements of the electrophoretic mobilities of ribosomal RNAs in polyacrylamide gels indicate that molecular weights vary in agreement with a monophyletic evolutionary scheme. Indeed, for the small ribonucleic acid molecules the weights range from 7.1 to 7.5 $\times 10^5$ daltons. The smallest values were found to be characteristic for the primitive fungi, and the highest ones for the higher fungi (168). It would seem, therefore, that polyacrylamide gel electrophoresis may turn out to be a profitable tool for the student of fungal phylogeny and taxonomy.

Ribosomal RNAs have been extracted from mitochondria; their sedimentation coefficients are listed in Table 14. If one compares these values with those listed in Table 13, it appears that the ribosomal RNAs of these organelles are smaller molecules than their respective cytoplasmic counterparts. Direct evidence for this has been obtained in the case of *A. nidulans* by measurements with the electron microscope of the length of these molecules. The average values are 0.47 and 0.9 μm for the mitochondrial ribosomal RNAs and 0.52 and 1.10 μm for the cytoplasmic ones (309). These values are in good agreement with those obtained by sedimentation and electrophoresis techniques for the RNAs of *A. nidulans* and other fungi (62, 76, 191). Mitochondrial and cytoplasmic ribosomal RNAs can further be distinguished by their thermal denaturation profiles (76, 309).

Ribosomes contain, in addition to the pair of large molecules discussed above, a smaller molecular type known as 5S RNA. These molecules have been identified in thc ribosomes of a variety of organisms, prokaryotes and eukaryotes (55, 56, 98, 183, 239, 296). Although the biological function of this type of RNA is still unknown, the complete nucleotide sequences for at least two species have already been determined (60, 179).

A. Nucleotide Composition

Transfer RNA, which was called soluble before the discovery of its function, is a heterogenous population of molecules having an average sedimentation coefficient of 4S. It can also be distinguished from ribosomal RNA by its base composition (39) (see Table 15). Analyses of purified transfer RNA species distinguishable by their amino acid acceptor specificities indicate that their base compositions are not uniform (60, 251). The nucleotide sequences of several of these transfer RNAs from yeast have already been determined (60). It is hoped that such types of determination will be extended to other organisms, since these data could provide us with a quantitative measurement of evolution (178). Small-molecular-weight RNAs are also found in mitochondria; and, as is the case for ribosomal RNA, these molecules differ from their cytoplasmic homologs. This is best illustrated by the results of investigations on *N. crassa* (13, 14, 81), which have shown that there exist 15 species of transfer RNA molecules in mitochondria that differ from their cytoplasmic counterparts. It will be very interesting to compare the sequences of these two types of transfer RNAs in order to determine their degrees of evolutionary divergence, since

Table 15. Comparison between the Guanine Plus Cytosine Contents (GC Content) of Transfer and Ribsomal RNA

	GC Content (moles %)		
Organism	Ribosomal RNA	Transfer RNA	Reference
Aspergillus niger	53.0	58.5	196
Blastocladiella emersonii	50.3	60.0	54, 169
Mucor rouxii	48.1	54.6	290
Neurospora crassa	50.2	61.6	119
Saccharomyces carlsbergensis	47.0	56.4	63
S. cerevisiae	45.2	54.7	251

the divergence of eukaryotes from prokaryotes has been calculated as "2.6 times more remote in evolution than the divergence of the nucleated organisms into separate kingdoms" (178).

Since the nucleotide compositions of the RNAs of several fungal species have been analyzed, it is interesting to inquire about the diversity found and to compare it to that of DNA. From Table 15, it can be seen that the %GC for transfer RNA varies among six organisms from 54.6 to 61.6 and averages 57.6. The corresponding maximum variations in the individual bases are 16.2, 7.6, 6.9, and 9.2% for cytosine, adenine, uridine, and guanine, respectively. Variations in %GC and in per cent of each base in ribosomal RNA are listed in Table 16. Some of the data in Tables 15 and 16 have been combined in Table 17. For the six organisms, the range in %GC is 37 to 66 for DNA, whereas it is only 47.0 to 53.0 for ribosomal RNA and 54.6 to 61.6 for transfer RNA. It can thus be concluded that in fungi, as well as in other oganisms, diversity in composition is much more restricted in RNA than in DNA. It follows that such analyses of RNA have little taxonomic and phylogenetic value. It is interesting, however, to note that the %GC values of both types of RNA are the lowest for a Zygomycete and two Hemiascomycetidae. This is indicative of a possible correlation between the composition of total RNA, which is the sum total of all RNA molecules present in a cell extract, and that of DNA. The results of a survey of many fungal species are presented in detail in Table 18 and in a condensed form in Table 19. As one can see, the lowest GC contents are found in the Zygomycetes and the highest ones in the Basidiomycetes, while those for the Ascomycetes are intermediate between the values for these two classes. As was the case for DNA, the data also suggest the existence of a dichotomy among the Ascomycetes between the Hemiascomycetidae and the Euascomycetidae. It has been calculated that for fungal organisms having less than 50%GC in their DNA there is a positive linear correlation coefficient with a probability value lying between 0.005 and 0.001. A negative linear coefficient with a probability lying between 0.02 to 0.01 has been found for organisms with more than 50% GC in DNA (276). This apparent contradiction between the two types of correlations remains unresolved. A weak positive linear correlation has also been described for bacteria (17).

The availability of investigations of mitochondrial RNA makes possible comparisons with cytoplasmic RNA. As shown in Table 20, the %GC of mitochondrial RNA is always less than that of its cytoplasmic homolog. Also, the compositional diversity of the RNA of these organelles is greater than that of cytoplasmic RNA. This situation is in contrast with that found for DNA (Table 10), where the compositional diversity was less for mitochondrial than for nuclear DNA.

Table 16. Nucleotide Composition of Ribosomal RNA

Organism	Moles/100 Moles of Identified Nucleotides[a]					Reference
	C	A	U	G	G + C	
Aspergillus nidulans	23.0	24.0	23.0	28.0	51.0	77
A. niger	22.9	24.3	22.7	30.1	53.0	196
Blastocladiella emersonii	18.5	26.2	23.5	31.8	50.3	169
Mucor rouxii	20.4	26.1	25.8	27.7	48.1	290
Neurospora crassa	22.8	24.9	24.9	27.4	50.2	119
Pythium debaryanum	17.4	27.7	29.3	25.6	43.0	165
Rhizopus arrhizus	16.4	28.1	27.4	28.3	44.7	165
Saccharomyces carlsbergensis	19.3	26.3	26.6	27.7	47.0	63
S. cerevisiae						
F[b]	19.6	25.6	26.8	28.0	47.6	16
S	20.1	26.1	28.1	25.7	45.8	16
	19.7	25.2	27.9	27.4	47.1	183

[a] Abbreviations: C, cytidylic acid; A, adenylic acid; U, uridylic acid; G, guanylic acid.
[b] F, fast-moving ribosomal RNA molecules; S, slow-moving ones.

Table 17. Comparison of GC Contents of DNA, Ribosomal RNA, and Transfer RNA

Organism	GC Content (Mole %)		
	Transfer RNA	Ribosomal RNA	DNA
Aspergillus niger	58.5	53.0	52.0[a]
Blastocladiella emersonii	60.0	50.3	66.0[b]
Mucor rouxii	54.6	48.1	37.0[c]
Neurospora crassa	61.6	50.2	54.0[d]
Saccharomyces carlsbergensis	56.4	47.0	40.0
S. cerevisiae	54.7	47.0[e]	39.6[f]

[a] From ref. 196.
[b] From ref. 54.
[c] From Table 10.
[d] From Table 10.
[e] Average of the three values listed in Table 16.
[f] Average of the three values listed in Table 10.

Table 18. Nucleotide Composition of Total[a] RNA From Fungi

Organism[b]	Moles/100 Moles of Identified Nucleotides[c]					Reference
	C	A	U	G	G + C	
Zygomycetes						
Absidia glauca	19.2	26.1	26.3	28.4	47.6	275
Cunninghamella echinulata	20.1	26.7	25.3	27.7	47.8	313
Lichtheimia spp.	19.5	28.8	26.4	25.3	44.8	297
Mucor racemosus (1608)[d]	19.2	27.9	27.5	25.4	44.6	275
M. rouxianus (4855)	20.4	27.3	26.4	25.9	46.3	275
(8097)	19.5	28.3	27.6	24.6	44.1	275
M. rouxii (1894)	19.6	27.1	26.6	26.7	46.3	275
M. subtilissimus (1743)	19.6	27.7	26.0	26.7	46.3	275
(1909)	19.8	26.9	27.3	26.0	45.8	275
Phycomyces blakesleeanus	20.8	27.8	23.6	27.8	48.6	275
(−)	21.1	28.2	21.0	29.7	50.8	297
(+)	21.2	27.4	21.5	29.9	51.1	297
Zygorhynchus moelleri	20.0	27.4	26.9	25.7	45.7	275
Deuteromycetes						
Aspergillus niger	23.9	24.2	21.9	30.0	53.9	196
	25.0	25.0	19.9	30.1	55.1	297
Botrytis cinerea	22.7	28.0	24.0	27.3	50.0	307
Candida pulcherrima	23.1	26.0	22.8	28.1	51.2	275
Cryptococcus albidus	19.9	22.1	25.5	32.4	52.3	275
C. laurentii	21.6	25.6	23.3	29.5	51.1	275
Penicillium chrysogenum	21.5	23.5	24.5	30.4	51.9	275
P. notatum	22.8	23.5	23.8	29.9	52.7	275
P. stolonifer	23.6	24.9	24.7	27.0	50.6	146
Rhodotorula mucilaginosa	24.8	30.9	20.8	23.4	48.2	275
Torulopsis stellata	22.7	27.5	21.8	27.6	50.4	275
Trichothecium roseum	23.7	25.8	20.8	29.7	51.4	307
Ascomycetes						
Ceratocystis ulmi	22.6	22.1	23.0	32.4	55.0	313
Chaetomium globosom	23.9	23.7	21.1	29.2	53.1	313
Debaryomyces kloeckeri	20.2	27.2	25.4	27.2	47.4	275
Endomyces reesii	19.6	26.8	25.4	28.2	47.8	275
Helvella esculenta	22.3	25.7	23.6	28.4	50.7	307
Lipomyces starkeyi	23.2	23.0	22.6	31.3	54.4	275
Neurospora crassa	23.7	24.7	24.2	27.4	51.0	119
	25.9	23.7	24.2	26.2	52.1	187
Pichia membranaefaciens	24.0	24.8	22.6	28.6	52.6	275
Saccharomyces cerevisiae	20.1	26.6	26.5	26.7	46.8	275
S. fragilis	21.9	27.1	25.4	25.6	47.4	275
Schizosaccharomyces octosporus	23.1	24.0	24.4	28.4	51.5	275
Sclerotinia libertinia	21.9	28.0	21.8	28.3	50.2	297

Table 18. (Continued)

Organism[b]	Moles/100 Moles of Identified Nucleotides[c] C	A	U	G	G + C	Reference
Basidiomycetes						
Agaricus bisporus	21.6	24.2	22.4	28.8	50.4	297
Amanita muscaria	21.3	27.7	21.6	29.4	50.7	307
A. strobiliformis	22.2	24.1	24.9	28.8	51.0	308
Lycoperdon spp.	22.7	27.0	22.3	28.0	50.7	307
Polyporus versicolor	25.9	20.0	19.4	34.6	60.5	308
Schizophyllum commune	23.4	25.4	22.5	28.6	52.0	275
Sporobolomyces roseus	23.0	25.0	22.9	29.1	52.1	275
S. salmonicolor	22.8	26.8	21.6	28.8	51.7	275

[a] See text for definition.
[b] Organisms were classified according to ref. 3. In each class they are listed alphabetically for convenience.
[c] Abbreviations: C, cytidylic acid; A, adenylic acid; U, uridylic acid; G, guanylic acid.
[d] The numbers in parentheses after the *Mucor* species refer to strains from the Northern Regional Research Laboratory, Peoria, Ill.

Table 19. Range of GC Content of Total RNA[a] among Classes and Subclasses

Taxonomic Group	N_s[b]	Range[c]
Zygomycetes	9	44.1–51.1
Ascomycetes	12	46.8–55.0
Deuteromycetes	11	48.2–55.1
Basidiomycetes	8	50.4–60.5
Hemiascomycetidae	7	46.8–54.4
Euascomycetidae	5	50.2–55.0
All classes	40	44.1–60.5

[a] See text for definition.
[b] N_s = number of species analyzed.
[c] Calculated from the values listed in Table 18.

Table 20. Comparison between the Nucleotide Compositions of Mitochondrial and Cytoplasmic RNA

Organism	Moles/100 Moles of identified Nucleotides[a]					
	C	A	U	G	G + C	Reference
Aspergillus nidulans[b]						
Cyto.	23.0	24.0	23.0	28.0	51.0	77
Mito.	13.0	30.0	38.0	19.0	32.0	
Ceratocystis ulmi[c]						
Cyto.	22.6	22.1	23.0	32.4	55.0	313
Mito.	22.2	26.2	23.0	28.6	50.8	
Chaetomium globosum[c]						
Cyto.	23.9	23.7	21.1	29.2	53.1	313
Mito.	22.6	26.2	23.5	27.6	50.2	
Cunninghamella echinulata[c]						
Cyto.	20.1	26.7	25.3	27.7	47.8	313
Mito.	18.4	29.2	26.0	24.9	43.3	
Mucor rouxii[c]						
Cyto.	22.2	26.9	26.3	24.5	46.7	313
Mito.	22.0	32.9	28.2	16.9	38.9	
Neurospora crassa[b]						
Cyto.	21.1	24.1	24.3	28.1	49.2	152
Mito.	14.8	27.2	29.8	22.9	37.7	
Cyto. } F	21.9	24.8	23.9	29.4	51.3	233
Mito.	15.0	33.9	31.9	19.1	34.1	
Cyto. } S	21.6	25.3	25.4	27.7	49.3	
Mito.	16.0	31.8	31.7	20.4	36.4	
Saccharomyces cerevisiae[b]						
Cyto. } F	19.2	26.4	26.0	28.4	47.6	85
Mito.	11.0	40.3	34.6	14.0	25.0	
Cyto. } S	19.1	26.6	28.1	26.1	45.2	
Mito.	11.0	38.4	34.5	16.1	27.1	
Schizophyllum commune[c]						
Cyto.	22.8	26.0	21.7	29.1	51.9	313
Mito.	20.3	26.5	23.8	28.6	48.6	

[a] For abbreviations see Tables 18 and 19.
[b] Ribosomal RNA was analyzed.
[c] Total RNA was analyzed.

B. Sequence Homologies

Studies on DNA–RNA hybridization offer the possibility not only of measuring homologies, as is the case in DNA–DNA hybridization, but, in addition, of estimating in a genome the number of cistrons for a given class of RNA (267). Such studies are also a fine tool for the analysis of taxonomic and phylogenetic relationships. For example, the degree of hybridizability of ^{32}P-labeled RNA from *Mucor rouxii* with DNA from other fungal species was measured by the membrane technique. The results are presented in Table 21. The second column gives the %GC of DNA of the various organisms for reference. In the third column are listed the percentages of RNA input that hybridized with DNA in individual experiments. In these, 10 to 15 different RNA concentrations were used. An average value for all these experiments is given in the fourth column, together with an accuracy index. The ratios (expressed in percentages) of the saturation for heterologous DNAs to that for the homologous DNA [that of *M. rouxii* (1894)] are listed in the last column. These ratios should thus be regarded as an expression of the homologies between the organisms listed. It is not surprising (and therefore encouraging) to find that the ribosomal RNA of *M. rouxii* (1894) does not hybridize at all with either *E. coli* or chick DNA. Among fungi, intergeneric hybridization takes place as shown for the three *Mucor* isolates. Unfortunately the amount of hybrid formed between the two isolates of *M. rouxii* is the same as that formed between *M. rouxii* (1894) and *Mucor genevensis*, two species that are easily distinguished phenotypically; in addition, the hybridization ratios are highest for *Rhizopus oligosporus* and *Penicillium chrysogenum* (287).

These preliminary results suggest, however, that with technical refinement, reciprocal hybridization (i.e., with more than one type of ribosomal RNA), and knowledge of the genome size, this type of study might become profitable. For example, in the case of the genus *Candida*, a survey of the base composition of the DNA revealed the existence of considerable heterogeneity (273), but subsequent homology studies have helped to identify the ascosporogenic counterparts of three of the species. It was further possible to differentiate species that could not be distinguished on the basis of biological properties and DNA base composition (163). A similar investigation was performed on the genus *Saccharomyces* (24, 25). In this case relatedness was measured by hybridization between DNA and high-molecular-weight ribosomal RNA (25S). The results demonstrated that the existing classification, based on morphological and physiological characters of the species studied, was in most instances in agreement with the molecular homologies. They further indicated that the sequences in this class of ribosomal RNA had been conserved.

Table 21. Hybridization of *Mucor rouxii* (1894)[a] Ribsomal RNA with Various DNAs[b]

Organism	GC (mole %)	RNA(%)/DNA	$m^c \pm \bar{z}^d$	Homology (%)
Mucor rouxii (1894)[a]	39	0.5300	0.4745 ± 0.0554	100.0
		0.4464		
		0.5259		
		0.5241		
		0.2912		
		0.4714		
		0.4185		
		0.5816		
		0.4824		
		0.4738		
M. rouxii (8097)[a]	43	0.3045	0.3003 ± 0.0462	63.4
		0.3767		
		0.2161		
		0.3366		
		0.2700		
M. genevensis	40	0.2748	0.3138 ± 0.0390	66.1
		0.3529		
Rhizopus oligosporus	39	0.3662	0.4040 ± 0.0416	85.1
		0.4873		
		0.3638		
Cunninghamella echinulata	26	0.2183		46.0
Penicillium chrysogenum	51	0.4254		89.6
Schizophyllum commune	59	0.2705		57.0
Neurospora crassa	54	0.1684	0.2473 ± 0.0526	52.1
		0.2491		
		0.3245		
Escherichia coli	50	0.0000		0.0
Chick embryo cell	42	0.0187		3.9

[a] The number in parentheses refers to the strain from the Northern Utilization Research and Development Division of the U. S. Department of Agriculture, Peoria, Ill.

[b] The data presented in this table are from ref. 287.

[c] m = mean.

[d] $\bar{z}$ = average deviation from the mean.

Whereas the percentages of homology between DNAs ranged from 0 to 40, they ranged from 83 to 100 for ribosomal RNA.

As mentioned above, DNA–RNA hybridization can be performed in order to determine the number of cistrons for each type of RNA molecule. For *N. crassa* it was found that 25S and 19S mitochondrial RNAs were complementary to 6.1 and 2.8% of the mitochondrial DNA, respectively. The 28S and 18S cytoplasmic RNAs were complementary to 0.67 and 0.33% of the nuclear DNA, and no hybridization of mitochondrial rRNA with nuclear DNA could be detected. From these results it could be inferred that nuclear DNA contains equal numbers of genes for the two types of cytoplasmic ribosomal RNA. The same applies to mitochondrial DNA and ribosomal RNAs. Measurements of the kinetics of renaturation yielded a value of 66×10^6 daltons for the molecular weight of mitochondrial DNA. On the assumption that the combined weight of the two mitochondrial ribosomal RNA molecules is 1.5×10^6 daltons, it was calculated that there must be at least four copies of each mitochondrial gene in order to account for the 9% hybridization observed (337). Redundancy of genes has also been observed in *S. cerevisiae,* not only for ribosomal RNA but also for transfer RNA. The 26S and 18S cytoplasmic RNAs were complementary to 1.6 and 0.8% of nuclear DNA. Between 0.064 and 0.08% of nuclear DNA was complementary to transfer RNA. No hybridization could be detected between mitochondrial DNA and cytoplasmic ribosomal or transfer RNA. It was calculated on the basis of these results and the assumption of a genome size of 1.25×10^{10} daltons that there are 140 cistrons for ribosomal RNA and between 320 and 400 cistrons for transfer RNA (251). In another work on *S. cerevisiae*, it was estimated that nuclear DNA contains a fraction of $2.5 \pm 0.5\%$ which is complementary to cytoplasmic ribosomal RNAs. Mitochondrial DNA appeared not to contain sequences homologous to those of these RNA molecules (92, 93). Similar experiments have been performed with *S. carlsbergensis*. In this instance 1.3 and 0.7% of the nuclear DNA was found to be homologous to 26S and 17S cytoplasmic ribosomal RNAs, respectively (61).

These experiments reinforce confidence in the importance of molecular hybridization. They indicate that, in the three species of fungi analyzed, about the same percentages of nuclear DNA correspond to the genes for ribosomal RNAs. They also stress the necessity to know the genome sizes of both nucleus and mitochondria in order to arrive at an absolute value for the number of cistrons involved. Worth mentioning is the fact that two-dimensional fingerprinting after digestion with T1 ribonuclease shows the nucleotide sequence of mitochondrial ribosomal RNAs to differ from that of their cytoplasmic homologs in the case of *A. nidulans* (310).

6. AUTONOMY OF THE MITOCHONDRIAL GENOME

The incapacity of the mitochondrial genome to code for all the different proteins of the organelles raises many questions. Which of these proteins are coded for in mitochondrial DNA? Are all such proteins synthesized within the mitochondria? Does the biosynthesis of the nucleic acids and ribosomes take place in the organelles? As will be shown in the following discussion, the answers to some of these and to related questions are still incomplete.

A. DNA Synthesis

Three different DNA polymerases have been identified in yeast homogenates. One of these was shown to be associated with mitochondria (335, 336). This mitochondrial polymerase could be distinguished from its nuclear homologs by DEAE-cellulose chromatography and the specificity of the Mg^{2+} concentration required for maximum activity. Also, mitochondrial DNA was a better template and primer for the mitochondrial enzyme than for the two other polymerases. A polymerase with similar properties was also isolated from a respiratory-deficient mutant (336). This finding points toward the probability that mitochondrial DNA polymerase is synthesized in the cytoplasm. Further support for independent mitochondrial DNA synthesis comes from the discovery that isolated mitochondria from *P. polycephalum* synthesized DNA when incubated with the four nucleotide phosphate precursors and that the rate of synthesis was independent of the period of the mitotic cycle at which the mitochondria were isolated (35). *In vivo* experiments have shown that in this organism (34) and in *S. cerevisiae* (330) mitochondrial DNA is synthesized continuously during the mitotic cycle. Discontinuous patterns of synthesis have been reported, however, in the cases of *S. lactis* (262) and *N. crassa* (115). As yet, no complete explanation has been found for this discrepancy (330).

There is now a large collection of antibiotics that inhibit the function and the synthesis of nucleic acids (102). These antibiotics, plus those that inhibit protein synthesis and some mutagens, have been used with great profit in mitochondrial research. In yeast, for example, two types of antibiotic-resistant mutants have been isolated. In the first type, as is the case for cycloheximide resistance, the genes involved are on the nuclear chromosomes (192). In the second type, which consists of mutants resistant to antibiotics such as chloramphenicol (43) and oligomycin (321), the hereditary transmission is non-Mendelian. Studies of crosses between some of these mutants have suggested the existence not only of mitochondrial recombination but also of a polarity in the transmission of these traits to the progeny. This polarity has been taken as evidence for the presence of sex in mitochondria (29, 51, 261). Crosses of yeast strains having mitochondrial DNA with different buoyant densities produced in the progeny mitochondrial DNA with intermediate buoyant density, indicating that recombination of mitochondrial DNA takes place in such crosses (45). Worth mentioning are experiments with yeasts (57) and *Neurospora* (174) indicating that mitochondrial DNA is replicated semiconservatively.

In some genera of yeasts such as *Saccharomyes*, cytoplasmic mutations occur which result in the loss of respiratory activity. Such a loss can be traced back to the absence of cytochrome oxidase activity. The mutants are called "petites" because they form small colonies (238). The frequency of this type of mutation varies greatly among species (42) and can be increased up to 100% by compounds such as acriflavin (80) and ethidium bromide (260). These mutants are called p^- and can be suppressive. Suppressiveness is established when the zygote resulting from a cross between a petite strain and a wild-type strain has a petite phenotype. The frequency of transmission of this phenotype varies from 0 to 100%. The petites that do not transmit this phenotype are called neutral. It was first thought that a spontaneous petite mutation resulted in the loss of mitochondrial DNA (57, 194). It has since been shown, however, that what may happen is an alteration of the buoyant density of this mitochondrial DNA (19, 184, 193). In one instance, the buoyant density was equal to 1.678 g/cm^3, which would correspond to poly(dAT–dAT) (19). Chemical analysis revealed, however, that the GC content was equal to 12.6%. Respiratory-deficient mutants obtained from treatment of wild-type cells with either acriflavin or ethidium bromide also contained mitochondrial DNA with lower buoyant density. Furthermore, treatment of these mutant cells with these agents resulted in a further lowering of the buoyant density (45, 84, 220) corresponding to poly(dAT–dAT). Treatment with ethidium bromide may have another consequence, namely, the disappearance of mitochondrial DNA (103, 104, 108). It was shown that this disappearance could be progressive and accompanied by decreasing size of the mitochondrial DNA molecules (104). Suppressiveness and propensity to lose DNA appear to be unrelated because, when treated in the same manner with ethidium bromide, neutral petite mutants lose mitochondrial DNA, whereas suppressive petite mutants do not (186, 198). Degree of suppressiveness could not, however, be correlated with mitochondrial DNA buoyant density (186).

The effect of ethidium bromide on other fungi has also been studied. Mitochondrial DNA was lost after treatment of *Kluyveromyces lactis* cells not only with ethidium bromide but also with acriflavin and nalidixate (175). In *P. polycephalum*, however, DNA breakdown was not observed, despite the fact that ethidium bromide specifically inhibited the biosynthesis of mitochondrial DNA (131). This mutagen appeared to have no permanent effect on cells of *S. pombe* since respiratory competence was regained after its removal from the growth medium (249). Modification and elimination of mitochondrial DNA of *S. cerevisiae* have also been accomplished by treatment with either erythromycin or chloramphenicol (331). These antibiotics do not react directly with DNA as acriflavin and ethidium bromide do, but are known to be specific inhibitors of bacterial and mitochondrial protein synthesis.

Cells of *S. cerevisiae* grown in anaerobiosis contain mitochondria-like organelles that have been designated as protomitochondria (59). Although they differ in some structural and enzymatic aspects from the mitochondria themselves, they are converted to the latter after exposure to air (223) and contain DNA with a buoyant density undistinguishable from that of their aerobic counterparts (59). The available evidence suggests that mitochondria or mitochondrialike structures are also present in petite yeasts, at least those in which mitochondrial DNA is present. It is not certain yet, however, that the yeasts which have lost such DNA still contain mitochondria. If the presence of these organelles can be demonstrated by the appropriate electron microscopy procedure (59), it would clearly follow that the nucleus and the cytoplasm play a major role in mitochondriogenesis (198). A similar conclusion would have to be reached in the case of organisms having mitochondrial DNA that has been mutated into a poly(dAT–dAT) polymer. Involvement of the nuclear genome in mitochondrial biogenesis is also suggested by the existence of petite mutants that have a nuclear origin. These mutants are not suppressive (257), and the buoyant density of their DNA is not altered (193). In addition, it is of relevance to note that there is often accumulation of cytoplasmic petite mutations in these nuclear petites (256).

B. RNA Synthesis

Acriflavin and ethidium bromide also interfere with *in vivo* transcription of mitochondrial DNA in yeast (97). Indeed, RNA synthesized during respiratory adaptation normally contains two classes of molecules. Some hybridize with nuclear DNA, and others with mitochondrial DNA (95). After treatment with the mutagens , the synthesized RNA hybridized only with nuclear DNA. That altered mitochondrial DNA from a ρ^- petite mutant is transcribed has been inferred from the fact that the RNA of such an organism hybridizes better with its homologous mitochondrial DNA than with the DNA of a ρ^+. The reciprocal is also true, namely, the percent hybridization is greater between ρ^+ DNA and ρ^+ RNA than between ρ^- DNA and ρ^+ DNA (96). By a similar method, it was shown in the same study that transfer RNA hybridizes with mitochondrial DNA and that again mitochondrial DNA of ρ^+ can be distinguished from that of ρ^-. Thus mitochondrial DNA contains cistrons not only for ribosomal RNAs but also for transfer RNA.

The existence of a DNA-dependent RNA polymerase that can be distinguished from the nuclear homolog by several criteria has been demonstrated in the case of *S. cerevisiae* (295, 334). This enzyme, which is sensitive to actinomycin D and insensitive to rifampicin and α-amanitin, was present in petite mutants—even those in which mitochondrial DNA could not be detected. It should be added, however, that the activity of the enzyme was very low in the latter type of mutant. A rifampicin-insensitive RNA polymerase has also been extracted from *N. crassa* mitochondria (122, 153). Since bacterial RNA synthesis is sensitive to this antibiotic, it appears that mitochondria cannot, in this case, be represented as descendants of prokaryotic ancestors.

C. Protein Synthesis

A large number of experiments performed with a variety of organisms, including fungi, have demonstrated that *in vitro*, as well as *in vivo*, amino acids are incorporated into mitochondrial proteins (5, 90, 320). For *S. cerevisiae* it has recently been possible to show that extracts of mitochondria free of cytoplasmic components had all the ingredients required to synthesize protein under the direction of an appropriate message. The messenger RNA used was either polyuridylic acid or R17 virus. Comparison with the cytoplasm indicated that the mitochondrial system required a high Mg^{2+} concentration and that, unlike the cytoplasmic system, it responded to the addition of R17 RNA. Cycloheximide inhibited specifically amino acid incorporation in the cytoplasmic system, while for the mitochondrial system, the specific inhibitors were chloramphenicol, erythromycin, and streptomycin (252). In the case of *N. crassa* an *in vitro* polyuridylic-directed system has also been developed (151). Ribosomes and supernatant enzymes were prepared not only from the cytoplasm and the mitochondria from *N. crassa*, but also from *E. coli*, and rat liver. These two components of the biosynthetic system were mixed in homologous and heterologous combinations. There was a good incorporation of ^{14}C–phenylalanine in the *E. coli* ribosome and *N. crassa* mitochondrial enzyme mixtures, but a low one in the combinations of *N. crassa* and rat liver with either component of *N. crassa* and *E. coli*. These results were interpreted as an indication that a functional relationship exists between mitochondrial and bacterial ribosomes and chain elongation factors. The existence of two peptide chain elongation factors in *N. crassa* that are specific for 70S ribosomes, including those of *E. coli*, was later claimed (105). Additional evidence for the fact that bacteria and mitochondria are similar with respect to the protein-synthesizing system is provided by the finding that methionyl–tRNA isolated from *N. crassa* mitochondria could be formylated with an extract of *E. coli* but not with one from the cytoplasm of the fungus. In conjunction, a formylase reacting specifically with mitochondrial formylmethionyl–tRNA was identified in mitochondrial extracts (82).

As indicated above, it would appear that mitochondrial DNA polymerase of yeast is not synthesized within the mitochondria. The same situation appears to prevail for the proteins of mitochondrial ribosomes, since incorporation of labeled amino acids into these proteins was prevented by cycloheximide but not by chloramphenicol (150, 211). Extramitochondrial synthesis appears also to be the case for the mitochondrial outer membrane proteins, as opposed to the proteins of the inner membrane (212). Since the outcome of experiments with antibiotics often depends on whether they are performed *in vivo* or *in vitro* (114, 212, 253), additional work is needed in order to establish firmly which ones of the mitochondrial proteins are synthesized in the organelles.

7. GERMINATION AND SPORULATION

A. Germination

Spores are differentiated cells that are formed during asexual and sexual life cycles. They are found in or on morphological structures that are distinct from the vegetative mycelium. In the majority of terrestrial fungi these structures are aerial and are not formed in liquid cultures. This constitutes a serious handicap for the quantitative study of sporulation. The situation is different in regard to germination. It follows, therefore, that in these fungi germination has been studied with greater intensity than sporulation. In addition, spore germination is intriguing because it is a transition from the state of low metabolic activity to a state of active growth. The level of metabolic activity varies greatly from one type of spore to another. Thus, whereas some readily germinate, even in the absence of nutrients, others, which might be called truly dormant, require a heat shock or a special chemical treatment in order to germinate.

Germination is accompanied by a general and rapid increase in metabolic activities (4). Among these are the syntheses of DNA, RNA, and proteins (2, 50, 137, 244, 303). What triggers these syntheses, and in what order do they occur? Two experimental approaches have been followed in order to answer these questions. The first consists of comparing the properties of ungerminated spores with those of germlings and growing hyphae, and the other in following the fate of these macromolecules during the germination process itself. The results of investigations performed up to 1969 have been adequately reviewed (303). The present analysis is therefore limited primarily to more recent studies.

About 10 years ago it was demonstrated, primarily with bacteria, that the number of ribosomes contained within a cell is proportional to the rate of protein synthesis. It was therefore reasonable to hypothesize that, in spores where this synthesis either is absent or occurs at a very low level, either ribosomes might be absent or the ribosomal population might be qualitatively or/and quantitatively different from that found in growing cells. A detailed comparison of the physicochemical properties of conidia, ascospores, and growing hyphae of *N. crassa* showed, however, that neither of these two hypotheses was valid. Indeed, the sedimentation properties and the relative proportions of ribosomal RNAs, as well as of ribosomes and their subunits, were the same in these three morphological states (120), and the base compositions of ribosomal and transfer RNAs were also found to be constant (119, 121). In agreement with these findings was the demonstration of the presence of 80S ribosomes in the nuclear caps of the zoospores of *Blastocladiella emersonii* (169) and in the conidia of *Aspergillus oryzae* (130). A question was therefore raised: Are these sporal ribosomes active? This was affirmatively answered by the demonstration that ribosomes from ungerminated conidia of *Botryodiplodia theobromae* were functional in the polyuridylic acid-directed synthesis of polyphenylalanine (302, 304). Such ribosomes were less active, however, than those extracted from germinated spores. In contrast, the ribosomes from the zoospores of *B. emersonii* were inactive when assayed in extracts of growing cells (247).

The demonstration by analytical ultracentrifugation that polyribosomes were present in extracts of growing hyphae and germinating spores, but not in those of ungerminated spores, suggested that the latter did not store messenger RNA, at least in the form of polysomes (121). The same conclusion was reached in the cases of *B. emersonii* (247) and *A. oryzae* (130) and more recently in regard to *D. purpureum* (87). In the last organism the appearance of polyribosomes during germination was prevented by the addition of cycloheximide or puromycin. Polyribosomes have, however, been detected in ungerminated uredospores of *Uromyces phaseoli* (270), macroconida of *Fusarium solani* (226), and conidiospores of *B. theobromae* (33). It is still too early to decide whether this apparent contradiction reflects a physiological difference between these two groups of fungi or simply stems from variations in extraction procedure. The latter supposition finds support from work done with bacterial endospores (86) and with spores of *F. solani*. For this fungus, the only method that gave positive results was sand grinding (50). It should be added, however, that polysomes could be detected in this organism only after 15 minutes of incubation.

Claims for conservation of messenger RNA in ungerminated spores have also been made on the basis of experiments on RNA and protein syntheses. For example, a population of RNA molecules with sedimentation coefficients ranging from 4S to 19S was found to stimulate amino acid incorporation into an acid-insoluble precipitate when added to an *E. coli* cell-free protein-synthesizing system (228, 229). Similarly, extracts from *F. solani* ungerminated spores stimulated leucine incorporation. However, these were 10 times less efficient than extracts from germinated spores. The levels of activity of these two types of extracts were nevertheless the same in a polyuridylic acid-directed system. Since the latter system does not require initiation factors, it was suggested that ungerminated spores are probably deficient in such factors (227). Several workers found that, whereas cycloheximide inhibited germination, actinomycin D did not. This was taken as an indication that RNA synthesis is not necessary for the early steps in the germination process (67, 128, 170) and that the messenger RNA molecules needed for these steps are therefore stored in spores. A similar conclusion based on DNA–RNA hybridization was reached in the case of *N. crassa* (22). A recent comparative study suggests that the synthesis of RNA seems to be necessary for some fungi but not for others (129).

A third component of the protein-synthesizing machinery, namely, the transfer RNAs, was found to have, when extracted from ungerminated spores, acceptor activity for all 20 amino acids when tested with an enzymatic system prepared from germinates spores (306, 338). In the case of *B. theobromae*, the activity of 13 aminoacyl–tRNA synthetases was either nonexistent or very low in ungerminated spores. This activity increased rapidly, however, as germination proceeded (305).

The possibility that certain structural events occurring during zoospore germination do not require protein synthesis has been raised (264).

It is obvious from the above discussion that more work is needed in order to understand the role that nucleic acids play in the control of germination. Generalizations will be difficult to make until it is recognized that in fungi, unlike in bacteria, not all spores are dormant and also that valid comparisons between diverse spore systems require a better definition of the criteria used to distinguish ungerminated spores from those that have already entered the germination process.

B. Sporulation

As mentioned above, quantitative studies on sporulation are limited in number. As examples, two recent investigations might be mentioned. In one on the water mold *Achlya*, actinomycin D inhibited the differentiation of sporangia and the incorporation of labeled precursors into RNA. Sucrose density gradient centrifugations revealed the presence of a labeled RNA fraction that could be distinguished from ribosomal and transfer RNAs (107).

Similarly, a new species of RNA was detected and isolated during sporulation in *S. cerevisiae* (142, 143). This RNA was isolated by sucrose density gradient centrifugation and polyacrylamide gel electrophoresis. It had a sedimentation coefficient of about 20S and could be further distinguished from the other known RNA species by its nucleotide composition. This RNA species, specific for sporulating cells, hybridized with nuclear DNA but not with mitochondrial DNA (143). It is hoped that these investigations will stimulate other workers to study the sporulating systems of these and other fungi.

8. FUNGAL VIRUSES

There are in the mycological literature several well-documented cases of fungal diseases that are not the result of parasitism by bacteria or by other fungi. It has therefore been suggested that the causative agent might be a virus. This appeared to be the case for "vegetative death" in *Aspergillus* (141), "senescence" in *Podospora* (234), and the diseases of cultivated mushrooms (135). The first two phenomena occur in cultures that are maintained by hyphal transfer. It has not been possible, however, to infect healthy mycelium with extracts of diseased hyphae, and no virus particles could be detected by electron microscopy. It has recently been suggested that this type of disease might be due to an accumulation of translation errors which, when large enough, would include defective key enzymes such as those involved in the biosynthesis of nucleic acids (164). In *A. bisporus*, the suspicion that a virus infection was associated with diseases such as the "die-back" of the stipe has, however, been proved to be correct, since two types of isodiametric viruslike particles with respective diameters of 25 and 29 nm have been identified in the filaments of diseased mushrooms (127). These particles have been isolated by sucrose density gradient centrifugation. It was found that they did not react with an antiserum against "Lucerne mosaic viruses," although they are morphologically similar to these viruses. These viruslike particles are present only in diseased organisms (157). Incubation at 33°C for 2 weeks cures the organism. Viruslike particles could not be detected in the hyphae of the cured fungi. Although the disease can be transmitted to healthy filaments by dipping them into a maceration of diseased ones, it has not been possible to infect the mushroom with isolated virus particles (157). The same is true for the viruslike particles recently isolated from the apothecia of a species of *Peziza (Ascomycete)* (65). These particles are morphologically different from those found in *Agaricus*. They are rod-shaped and strikingly similar to tobacco mosaic viruses, but they cannot infect tobacco plants.

Viruslike particles have also been found in several species of *Aspergillus* and *Penicillium*. These "viruses" were discovered almost by accident, as the following historical sketch will show. Since the early 1950s it has been known that extracts of *Penicillium stoloniferum* and *Penicillium funiculosum* can induce antiviral activity in animals. The principle responsible for this activity is called "statolon" in the case of the first organism, and "helenin" in the case of the other. It was soon discovered that helenin is a mixture of ribonucleoprotein particles with sedimentation characteristics similar to those of yeast ribosomes (166). Later these particles were found to elicit the production in cell cultures and in mice of a viral inhibitor with properties similar to those of interferon (241). Further studies revealed that the RNA molecules present in helenin are double-stranded, and that this double-strandedness is essential for the induction of interferon (156). Since double-stranded RNA is identifiable with some viruses and with viral replicative forms, it was suggested that helenin is a fungal virus (156). In *P. stoloniferum* statolon was identified *in situ* by electron microscopy as viruslike particles, polyhedral and regular in shape with a diameter of 30 to 35 nm (78). As in the case of helenin, these particles contain double-stranded RNA. It was in this connection that the term "mycophage" was coined (147).

The presence of virus particles has since been detected in several strains of *Penicillium chrysogenum* (8, 9), all derived from the famous (NRRL 1951) penicillin-producing strain and from *Penicillium cyaneo-fulvum* (10). In both cases, polyhedral particles about 30 nm in diameter were identified *in situ* and isolated: antiserum reactions demonstrated that these viruses are species specific. Further studies on *P. stoloniferum* revealed that the cell wall of the infected organisms contains 18% galactosamine, as opposed to 1% for the uninfected ones. The galactosamine content was normal in cultures obtained from spores that had been heated for 30 min at 74°C. Such a treatment also resulted in the disappearance of the "mycophages" (40). This observation led to the suggestion that galactosamine is a component of a receptor site located in the cell wall. Virus particles containing double-stranded RNA have also been seen in *A. niger* and *A. foetidus* (11). Three different types of particles have been found in *P. stoloniferum* (41), suggesting possible multiple infection. The RNA molecules in the mycophage of *P. chrysogenum* have an average contour length of 0.5 nm, corresponding to a molecular weight of the order of 10^6 daltons (160). None of these viruses so far tested has failed to elicit interferon production (53, 222). Claims for the demonstration of viral transmission, either by direct artificial inoculation or through mycelium fusion (31, 167), are still not very convincing.

Suggestions have been made at various times that there are "zymophages." The experimental data presented, however, have not been sufficient to substantiate these suggestions. This is the case also for more recent studies on yeast (319).

9. CONCLUDING REMARKS

Study of the properties of some macromolecules will help in our understanding of taxonomic and phylogenetic relationships among fungi. Although it is difficult to evaluate at the present time which macromolecule will be most profitable for this purpose, there is no doubt that the elaboration of trustworthy phylogenetic schemes will require the survey of a large number of well-selected organisms. If such surveys are extended so as to include other groups of protists, they might, in addition, provide us with some clues about the origin of not only fungi but also of the other eukaryotic microorganisms. Additional work is also needed to answer some important questions raised in this chapter: Is the genome of fungi really smaller than that of the other protists and, in respect to size, closest to the bacterial genome? Is the presence of repeated sequences in nuclear DNA limited to some yeasts? Is the absence of nuclear histones widespread in the fungi? Is the mitochondrial genome of fungi significantly larger than that of higher forms? It is obvious that affirmative answers to these questions would throw new light on some of the mysteries of evolutionary pathways.

Yeasts have been, as we have seen, intensively studied with regard to the biogenesis of mitochondria. As a result, it is often believed that they are the only fungal type endowed with facultative anaerobiosis.

It is therefore relevant to point out here that this property is also found among true filamentous fungi. For example, it has been shown that respiratory adaptation can be studied with organisms belonging to the order of the Mucorales (112, 291) and that in one instance respiratory-deficient mutants similar to some "petite" yeast were isolated (28). Also facultative or even obligate anaerobiosis is found among aquatic fungi (116). Finally, in the case of *Neurospora* it has recently been shown that, in addition to the well-known respiratory-deficient mutants of the *poky* type (74, 232), there are strains which can grow anaerobically (132). Since in these organisms morphological changes accompany alteration of mitochondrial structure and function, we are provided with new approaches for the study of morphogenesis and differentiation.

In spite of the apparent complexity of germination and sporulation, encouraging signs indicate that they will attract the attention of molecular biologists to an increasing degree. Also, the discovery that, after all, fungi also have viruses will hopefully stimulate virologists to apply their methods to this group of organisms.

REFERENCES

1. Aist, J. R., *J. Cell Biol.* **40,** 120 (1969).
2. Aitken, W., and D. J. Niederpruem, *J. Bacteriol.* **104,** 981 (1970).
3. Alexopoulos, C. J., *Introductory Mycology*, Wiley, New York, 1962.
4. Allen, P. J., *Ann. Rev. Phytopathol.* **3,** 313 (1965).
5. Ashwell, M., and T. S. Work., *Ann. Rev. Biochem.* **39,** 251 (1970).
6. Avers, C. J., *Proc. Natl. Acad. Sci. U.S.* **58,** 620 (1967).
7. Avers, C. J., F. E. Billheimer, H. P. Hoffmann, and R. M. Pauli, *Proc. Natl. Acad. Sci. U.S.* **61,** 90 (1968).
8. Banks, G. T., K. W. Buck, E. B. Chain, F. Himmelweit, J. E. Marks, J. M. Tyler, M. Hollings, F. T. Last, and O. M. Stone, *Nature* **218,** 542 (1968).
9. Banks, G. T., K. W. Buck, E. B. Chain, J. E. Darbyshire, and F. Himmelweit, *Nature* **222,** 89 (1969).
10. Banks, G. T., K. W. Buck, E. B. Chain, J. E. Darbyshire, and F. Himmelweit, *Nature* **223,** 155 (1969).
11. Banks, G. T., K. W. Buck, E. B. Chain, J. E. Darbyshire, F. Himmelweit, and G. Ratti, *Nature* **227,** 505, (1970).
12. Banno, I., *J. Gen. Appl. Microbiol.* **13,** 167 (1967).
13. Barnett, W. E., and D. H. Brown, *Proc. Natl. Acad. Sci. U.S.* **57,** 452 (1967).
14. Barnett, W. E., D. H. Brown, and J. L. Epler, *Proc. Natl. Acad. Sci. U.S.* **57,** 1775 (1967).
15. Bartnicki-Garcia, S., in *Phytochemical Phylogeny*, B. Harborne, (ed.), Academic Press, New York, 1970, p. 81.
16. Beck, G., J. Duval, G. Aubel-Sadron, and J. Ebel, *Bull. Soc. Chim. Biol.* **48,** 1205 (1966).
17. Belozersky, A. N., and A. S. Spirin, *Nature* **182,** 111 (1958).
18. Benson, B. W., and J. H. Grosklags, *Mycologia* **61,** 718 (1969).
19. Bernardi, G., F. Carnevali, A. Nicolaieff, G. Piperno, and G. Tecce, *J. Mol. Biol.* **37,** 493 (1968).
20. Bernardi, G., M. Faures, G. Piperno, and P. P. Slonimski, *J. Mol. Biol.* **48,** 23 (1970).
21. Bernardi, G., and S. N. Timasheff, *J. Mol. Biol.* **48,** 43 (1970).
22. Bhagwat, A. S., and P. R. Mahadevan, *Mol. Gen. Genetics* **109,** 142 (1970).
23. Bhargava, M. M., and H. O. Halvorson, *J. Cell Biol.* **49,** 423 (1971).
24. Bicknell, J. N., Ph.D. Thesis, University of Washington, University Microfilms, Inc., Ann Arbor, Mich., 1969.
25. Bicknell, J. N., and H. C. Douglas, *J. Bacteriol.* **101,** 505 (1970).
26. Billheimer, F. E., and C. J. Avers, *Proc. Natl. Acad. Sci. U.S.* **64,** 739 (1969).
27. Birnstiel, M. L., M. Chipchase, and J. Speirs, in *Progress in Nucleic Acid Research and Molecular Biology*, Vol. 11, J. N. Davidson, and W. E. Cohn (eds.), Academic Press, New York, 1971, p. 351.
28. Boardman, N. K., R. I. B. Francki, and S. G. Wildman, *J. Mol. Biol.* **17,** 470 (1966).
29. Bolotin, M., D. Coen, J. Deutsch, B. Dujon, P. Netter, E. Petrochilo, and P. P. Slonimski, Reunion Soc. Fran. de Genetique, Toulouse, April 24–25, 1970.

30. Bonner, J. T., *The Cellular Slime Molds*, Princeton University Press, Princeton, N.J., 1967.
31. Borre´, E., L. E. Morgantini, V. Ortali, and A. Tonolo, *Nature* **229,** 568, 1971.
32. Borst, P., and A. M. Kroon, *Int. Rev. Cytol.* **26,** 108 (1969).
33. Brambl, R. M., and J. L. Van Etten, *Arch. Biochem. Biophys.* **137,** 442 (1970).
34. Braun, R., and T. E. Evans, *Biochim. Biophys. Acta* **182,** 511, (1969).
35. Brewer, E. N., A. De Vries, and H. Rusch, *Biochim. Biophys. Acta* **145,** 686 (1967).
36. Britten, R. J., and D. E. Kohne, *Science* **161,** 529 (1968).
37. Britten, R. J., and D. E. Kohne, *Sci. Amer.* **222,** 24 (1970).
38. Brock, T. D., *Biology of Microorganisms*, Prentice-Hall, Englewood Cliffs, N.J., 1970.
39. Brown, G. L., in *Progress in Nucleic Acid Research*, Vol. 2, J. N. Davidson, and W. E. Cohn (eds.), Academic Press, New York, 1963, p. 260.
40. Buck, K. W., E. B. Chain, and J. E. Darbyshire, *Nature* **223,** 1273 (1969).
41. Buck, K. W., and G. F. Kempson-Jones, *Nature* **225,** 945 (1970).
42. Bulder, C. J. E. A., *Antonie van Leeuwenhoek J. Microbiol. Serol.* **30,** 1 (1964).
43. Bunn, C. L., C. H. Mitchell, H. B. Lukins, and A. W. Linnane, *Proc. Natl. Acad. Sci. U.S.* **67,** 1233 (1970).
44. Burgoyne, L. A., and R. H. Symons, *Biochim. Biophys. Acta* **129,** 502 (1966).
45. Carnevali, F., G. Morpurgo, and G. Tecce, *Science* **163,** 1331 (1969).
46. Cazin, J., Jr., T. R. Kozel, D. M. Lupan, and W. R. Burt, *J. Bacteriol.* **100,** 760 (1969).
47. Ceccarini, C., and R. Maggio, *Biochim. Biophys. Acta* **166,** 134 (1968).
48. Chao, F. C., and H. K. Schachman, *Arch. Biochem. Biophys.* **61,** 220 (1956).
49. Christiansen, C., A. Leth Bak, A. Stenderup, and G. Christiansen, *Nature New Biol.* **231,** 176 (1971).
50. Cochrane, J. C., T. A. Rado, and V. W. Cochrane, *J. Gen. Microbiol.* **65,** 45 (1971).
51. Coen, D., J. Deutsch, P. Netter, E. Petrochilo, and P. P. Slonimski, *Symposium 24, Society for Experimental Biology*, Cambridge University Press, 1969, p. 449.
52. Cohen, S. S., *Amer. Sci.* **58,** 281 (1970).
53. Colby, C., and M. J. Chamberlin, *Proc. Natl. Acad. Sci. U.S.* **63,** 160 (1969).
54. Comb, D. G., R. Brown, and S. Katz, *J. Mol. Biol.* **8,** 781 (1964).
55. Comb, D. G., and S. Katz, *J. Mol. Biol.* **8,** 790 (1964).
56. Comb, D. G., and T. Zehavi-Willner, *J. Mol. Biol.* **23,** 441 (1967).
57. Corneo, G., C. Moore, D. R. Sanadi, L. I. Grossman, and J. Marmur, *Science* **151,** 687 (1966).
58. Coudray, Y., F. Quetier, and E. Guille, *Biochim. Biophys. Acta* **217,** 259 (1970).
59. Criddle, R. S., and G. Schatz, *Biochemistry,* **8,** 322 (1969).
60. Dayhoff, M. O., in *Atlas of Protein Sequence and Structure*, National Biomedical Research Foundation, Silver Spring, Md., 1969.
61. De Kloet, S. R., *Arch. Biochem. Biophys.* **136,** 402 (1970).
62. De Kloet, S. R., B. A. G. Andrean, and V. S. Mayo, *Arch. Biochem. Biophys.* **143,** 175 (1971).
63. De Kloet, S. R., and P. J. Strijkert, *Biochem. Biophys. Res. Commun.* **23,** 49 (1966).

64. De Ley, J., *J. Gen. Microbiol.* **37,** 153 (1964).
65. Dieleman-van Zaayen, A., O. Igesz, and J. T. Finch, *Virology* **42,** 534 (1970).
66. Duffus, J. H., *Biochim. Biophys. Acta* **195,** 230, (1969).
67. Dunkle, L. D., R. Maheshwari, and P. J. Allen, *Science* **163,** 481 (1969).
68. Dupont, P. F., and L. R. Hedrick, *J. Gen. Microbiol.* **66,** 349 (1971).
69. Dure, L. S., J. L. Epler, and W. E. Barnett, *Proc. Natl. Acad. Sci. U.S.* **58,** 1883 (1967).
70. Dutta, S. K., and P. K. Chakrabartty, *Neurosp. News* **18,** (1971).
71. Dutta, S. K., Personal communication.
72. Dutta, S. K., N. Richman, V. W. Woodward, and M. Mandel, *Genetics* **57,** 719 (1967).
73. Dwivedi, R. S., S. K. Dutta, and D. P. Bloch, *J. Cell Biol.* **43,** 51 (1969).
74. Eakin, R. T., and H. K. Mitchell, *J. Bacteriol.* **104,** 74 (1970).
75. Edelman, M., D. Swinton, J. A. Schiff, H. T. Epstein, and B. Zeldin, *Bacteriol. Rev.* **31,** 315 (1967).
76. Edelman, M., I. M. Verma, R. Herzog, E. Galun, and U. Z. Littauer, *Eur. J. Biochem.* **19,** 372 (1971).
77. Edelman, M., I. M. Verma, and U. Z. Littauer, *J. Mol. Biol.* **49,** 67 (1970).
78. Ellis, L. F., and W. J. Kleinschmidt, *Nature* **215,** 649 (1967).
79. Emerson, R., and M. S. Fuller, *Quart. Rev. Biol.* **41,** 303 (1966).
80. Ephrussi, B., H. Hottinguer, and A. M. Chimenes, *Ann. Inst. Pasteur* **76,** 351 (1949).
81. Epler, J. L., *Biochemistry* **8,** 2285 (1969).
82. Epler, J. L., L. R. Shugart, and W. E. Barnett, *Biochemistry* **9,** 3575 (1970).
83. Esser, K., and R. Kuenen, *Genetics of Fungi*, Springer-Verlag, New York, 1967.
84. Evans, T. E., and D. Suskind, *Biochim. Biophys. Acta* **228,** 350 (1971).
85. Fauman, M., M. Rabinowitz, and G. S. Getz, *Biochim. Biophys. Acta* **182,** 355 (1969).
86. Feinsod, F. M., and H. A. Douthit, *Science* **168,** 991 (1970).
87. Feit, I. N., L. K. Chu, and R. M. Iverson, *Exp. Cell Res.* **65,** 439 (1971).
88. Flanagan, P. W., *Can. J. Bot.* **47,** 2055 (1969).
89. Flores da Cunha, M., *Genet. Res. Camb.* **16,** 127 (1970).
90. Fournier, M. J., and M. V. Simpson, in *Biochemical Aspect of the Biogenesis of Mitochondria*, E. C. Slater, J. M. Tager, S. Papa, and E. Quagliariello (eds.), Adriatica Editrice, Bari, 1968, p. 227.
91. Fukuhara, H., *Biochem. Biophys. Res. Commun.* **18,** 297 (1965).
92. Fukuhara, H., *Proc. Natl. Acad. Sci. U.S.* **58,** 1065 (1967).
93. Fukuhara, H., in *Biochemical Aspects of the Biogenesis of Mitochondria*, E. C. Slater, J. M. Tager, S. Papa, and E. Quagliariello (eds.), Adriatica Editrice, Bari, 1968, p. 303.
94. Fukuhara, H., *Eur. J. Biochem.* **11,** 135 (1969).
95. Fukuhara, H., *Mol. Gen. Genet.* **107,** 58 (1970).
96. Fukuhara, H., M. Faures, and C. Genin, *Mol. Gen. Genet.* **104,** 264 (1969).
97. Fukuhara, H., and C. Kujawa, *Biochem. Biophys. Res. Commun.* **41,** 1002 (1970).
98. Galibert, F., C. J. Larsen, J. C. Lelong, and M. Boiron, Nature **207,** 1039 (1965).
99. Gäumann, E., *Die Pilze,* Birkhauser, Basel, 1964.
100. Gillespie, D., and S. Spiegelman, *J. Mol. Biol.* **12,** 829 (1965).

101. Godward, M. B. E. (ed.), *The Chromosomes of the Algae*, Edward Arnold, London, 1966.
102. Goldberg, I. H., and P. A. Friedman, *Ann. Rev. Biochem.* **40,** 775 (1971).
103. Goldring, E. S., L. I. Grossman, D. Krupnick, D. R. Cryer, and J. Marmur, *J. Mol. Biol.* **52,** 323 (1970).
104. Goldring, E. S., L. I. Grossman, and J. Marmur, *J. Bacteriol.* **107,** 377 (1971).
105. Grandi, M., and H. Küntzel, *FEBS Lett.* **10,** 25, (1970).
106. Gray, W. D., and C. J. Alexopoulos, *Biology of the Myxomycetes*, Ronald Press, New York, 1968.
107. Griffin, D. H., and C. Breuker, *J. Bacteriol.* **98,** 689 (1969).
108. Grossman, L. I., E. S. Goldring, and J. Marmur, *J. Mol. Biol.* **46,** 367 (1969).
109. Gúerineau, M., C. Grandchamp, Y. Yotsuyanagi, and P. P. Slonimski, *C. R. Acad. Sci. Paris* **266,** 1884 (1968).
110. Gúerineau, M., C. Grandchamp, C. Paoletti, and P. Slonimski, *Biochem. Biophys. Res. Commun.* **42,** 550 (1971).
111. Guttes, E., Personal communication.
112. Haidle, C. W., and R. Storck, *J. Bacteriol.* **92,** 1236 (1966).
113. Hartwell, L. H., *Ann. Rev. Genet.* **4,** 373 (1970).
114. Hawley, E. S., and J. W. Greenawalt, *J. Biol. Chem.* **245,** 3574 (1970).
115. Hawley, E. S., and R. P. Wagner, *J. Cell. Biol.* **35,** 489 (1967).
116. Held, A. A., R. Emerson, M. S. Fuller, and F. H. Gleason, *Science* **165,** 706 (1969).
117. Helinski, D. R., and D. B. Clewell, *Ann. Rev. Biochem.* **40,** 899 (1971).
118. Henney, H. R., Jr., and D. Jungkind, *J. Bacteriol* **98,** 249 (1969).
119. Henney, H. R., Jr., and R. Storck, *J. Bacteriol.* **85,** 822 (1963).
120. Henney, H. R., Jr., and R. Storck, *Science* **142,** 1675 (1963).
121. Henney, H. R., Jr., and R. Storck, *Proc. Natl. Acad. Sci.* **51,** 1051 (1964).
122. Herzfeld, F., *Hoppe-Seyler's Z. Physiol. Chem.* **351,** 658 (1970).
123. Hill, L. R., *J. Gen. Microbiol.* **44,** 419 (1966).
124. Hofsten, A. V., *Physiol. Plant.* **16,** 709 (1963).
125. Hollenberg, C. P., P. Borst, and E. F. J. Van Bruggen, *Biochim. Biophys. Acta* **209,** 1 (1970).
126. Holliday, R., in *Organization and Control in Prokaryotic and Eurkaryotic Cells*, H. P. Charles, and B. C. J. G. Knight (eds.), University Press, Cambridge, England, 1970, p. 359.
127. Hollings, M., *Nature* **196,** 962 (1962).
128. Hollomon, D. W., *J. Gen. Microbiol.* **55,** 267 (1969).
129. Hollomon, D. W., *J. Gen. Microbiol.* **62,** 75 (1970).
130. Horikoshi, K., Y. Ohtaka, and Y. Ikeda, *Agr. Biol. Chem.* **29,** 724 (1965).
131. Horwitz, H. B., and C. E. Holt, *J. Cell Biol.* **49,** 546 (1971).
132. Howell, N., C. A. Zuiches, and K. D. Munkres, *J. Cell Biol.* **50,** 721 (1971).
133. Howson, W. T., R. C. McGinnis, and W. L. Gordon, *Can. J. Genet. Cytol.* **5,** 60 (1963).
134. Hoyer, B. H., B. J. McCarthy, and E. T. Bolton, *Science* **144,** 959 (1964).
135. Huhnke, W., *Sci. J.* **6,** 62 (1970).
136. Hütter, R., and J. A. DeMoss, *J. Bacteriol.* **94,** 1896 (1967).

137. Inoue, H., and T. Ishikawa, *Japan J. Genet.* **45,** 357 (1970).

138. Ishitani, G., K. Uchida, and Y. Ikeda, *Exp. Cell. Res.* **10,** 737 (1956).

139. Iwabuchi, M., and H. Ochiai, *Biochim. Biophys. Acta* **190,** 211 (1969).

140. Iwabuchi, M., K. Ito, and H. Ochiai, *J. Biochem.* **68,** 549 (1970).

141. Jinks, J. L., *J. Gen. Microbiol.* **21,** 397 (1959).

142. Kadowaki, K., and H. O. Halvorson, *J. Bacteriol.* **105,** 826 (1971).

143. Kadowaki, K., and H. O. Halvorson, *J. Bacteriol.* **105,** 831 (1971).

144. Kessler, D., *J. Cell Biol.* **43,** 68A (1969).

145. Kleinschmidt, A. K., in *Molecular Genetics*, Vol. 2, J. H. Taylor, (ed.), Academic Press, New York, 1967, p. 47.

146. Kleinschmidt, W. J., and J. A. Manthey, *Arch. Biochem. Biophys.* **73,** 52 (1958).

147. Kleinschmidt, W. J., L. F. Ellis, R. M. Van Frank, and E. B. Murphy, *Nature* **220,** 167 (1968).

148. Kohne, D. E., *Quart. Rev. Biophys.* **3,** 327 (1970).

149. Küntzel, H., *J. Mol. Biol.* **40,** 315 (1969).

150. Küntzel, H., *Nature* **222** 142 (1969).

151. Küntzel, H., *FEBS Lett.* **4,** 140 (1969).

152. Küntzel, H., and H. Noll, *Nature* **215,** 1340 (1967).

153. Küntzel, H., and K. P. Schafer, *Nature* **231,** *265 (1971).*

154. Laane, M. M., *Can. J. Genet. Cytol.* **9,** 342 (1967).

155. Laird, C. D., *Chromosoma (Berl.)* **32,** 378 (1971).

156. Lampson, G. P., A. A. Tytell, A. K. Field, M. M. Nemes, and M. R. Hilleman, *Proc. Natl. Acad. Sci. U.S.* **58,** 782 (1967).

157. Last, F. T., M. Hollings, and O. M. Stone, *Ann. Appl. Biol.* **59,** 451 (1967).

158. Leighton, T. J., B. C. Dill, J. J. Stock, and C. Phillips, *Proc. Natl. Acad. Sci. U.S.* **68,** 677 (1971).

159. LéJohn, H. B., *Nature* **231,** 164 (1971).

160. Lemke, P. A., and T. M. Ness, *J. Virol.* **6,** 813 (1970).

161. Leon, S. A., and H. R. Mahler, *Arch. Biochem. Biophys.* **126,** 305 (1968).

162. Leth Bak, A., C. Christiansen, and A. Stenderup, *Nature* **224,** 270 (1969).

163. Leth Bak, A., and A. Stenderup, *J. Gen. Microbiol.* **59,** 21 (1969).

164. Lewis, C. M., and R. Holliday, *Nature* **228,** 877 (1970).

165. Lewis, J., K. Raghu, C. L. Keswani, and D. J. Weber, *Biochim. Biophys. Acta* **199,** 194 (1970).

166. Lewis, V. J., E. L. Rickes, D. E. Williams, L. McClelland, and N. G. Brink, *J. Amer. Chem. Soc.* **82,** 5178 (1960).

167. Lhoas, P., *Nature* **230,** 248 (1971).

168. Lovett, J. S., Personal communication.

169. Lovett, J. S., *J. Bacteriol.* **85,** 1235 (1963).

170. Lovett, J. S., *J. Bacteriol.* **96,** 962 (1968).

171. Lovett, J. S., and C. J. Leaver, *Biochim. Biophys. Acta* **195,** 319 (1969).

172. Lu, B. C., and N. B. Raju, *Chromosoma (Berl.)* **29,** 305 (1970).

173. Luck, D. J. L., *Amer. Nat.* **99,** 241 (1965).

174. Luck, D. J. L., and E. Reich, *Proc. Natl. Acad. Sci. U.S.* **52,** 931 (1964).

175. Luha, A. A., L. E. Sarcoe, and P. A. Whittaker, *Biochem. Biophys. Res. Commun.* **44,** 396 (1971).

176. McCarthy, B. J., in *Progress in Nucleic Acid Research and Molecular Biology*, Vol. 4, J. N. Davidson, and W. E. Cohn (eds.), Academic Press, New York, 1965, p. 129.

177. McGinnis, R. C., **J. Hered. 47,** 254 (1956).

178. McLaughlin, P. J., and M. O. Dayhoff, *Science* **168,** 1469 (1970).

179. Madison, J. T., in *Ann. Rev. Biochem.* **37**, 131 (1968).

180. Mainwaring, H. R., *Arch. Mikrobiol.* **75,** 296 (1971).

181. Mandel, M., in *Chemical Zoology*, Vol. I, M. Florkin, B. T. Scheer, and G. W. Kidder, (eds.), Academic Press, New York, 1967, p. 541.

182. Mandel, M., C. L. Schildkraut, and J. Marmur, in *Methods in Enzymology*, Vol. XII, S. P. Colowick, and N. O. Kaplan (eds.), Academic Press, New York, 1968, p. 184.

183. Marcot-Queiroz, J., J. Julein, R. Rosset, and R. Monier, *Bull. Soc. Chim. Biol.* **XLVII,** 183 (1965).

184. Mehrotra, B. D., and H. R. Mahler, *Arch. Biochem. Biophys.* **128,** 685 (1968).

185. Meyer, S. A., and H. J. Phaff, in Recent Trends in Yeast Research, D. G. Ahearn, (ed.), *Spectrum, Monograph Series in Arts and Sciences,* Vol. 1, Georgia State University, Atlanta, Ga., 1970, p. 1.

186. Michaelis, G., S. Douglass, M. Tsai, and R. Criddle, *Biochem. Genet.* **5,** 487 (1971).

187. Minagawa, T., B. Wagner, and B. Strauss, *Arch. Biochem. Biophys.* **80,** 442 (1959).

188. Mohberg, J., and H. P. Rusch, *Arch. Biochem. Biophys.* **134,** 577 (1969).

189. Mohberg, J., and H. P. Rusch, *Exp. Cell Res.* **66,** 305, (1971).

190. Molenaar, I., W. W. Sillevis Smitt, T. H. Rizyn, and A. J. M. Tonino, *Exp. Cell. Res.* **60,** 148 (1970).

191. Morimoto, H., and H. O. Halvorson, *Proc. Natl. Acad. Sci. U.S.* **68,** 324 (1971).

192. Mortimer, R. K., and D. C. Hawthorne, *Genetics* **53,** 165 (1966).

193. Mounolou, J., H. Jakob, and P. Slonimski, *Biochem. Biophys. Res. Commun.* **24,** 218 (1966).

194. Moustacchi, E., and D. H. Williamson, *Biochem. Biophys. Res. Commun.* **23,** 56 (1966).

195. Moyer, R. C., Ph.D. Thesis, University of Texas, Austin, Tex., 1965.

196. Moyer, R. C., and R. Storck, *Arch. Biochem. Biophys.* **104.** 193 (1964).

197. Myers, R. B., and E. C. Cantino, *Arch. Mikrobiol.* **78,** 252, 1971.

198. Nagley, P., and A. W. Linnane, *Biochem. Biophys. Res. Commun.* **39,** 989 (1970).

199. Nakase, T., and K. Komagata, *J. Gen. Appl. Microbiol.* **14,** 345 (1968).

200. Nakase, T., and K. Komagata, *Yeast News Lett.* **17,** 4 (1968).

201. Nakase, T., and K. Komagata, *J. Gen. Appl. Microbiol.* **15,** 85 (1969).

202. Nakase, T., and K. Komagata, *J. Gen. Appl. Microbiol.* **16,** 511 (1970).

203. Nakase, T., and K. Komagata, *J. Gen. Appl. Microbiol.* **17,** 43, (1971).

204. Nakase, T., and K. Komagata, *J. Gen. Appl. Microbiol.* **17,** 77 (1971).

205. Nakase, T., and K. Komagata, *J. Gen. Appl. Microbiol.* **17,** 121 (1971).

206. Nakase, T., and K. Komagata, *J. Gen. Appl. Microbiol.* **17,** 161 (1971).

207. Nakase, T., and K. Komagata, *J. Gen. Appl. Microbiol.* **17,** 227 (1971).

208. Nakase, T., and K. Komagata, *J. Gen. Appl. Microbiol.* **17,** 259 (1971).

209. Nass, M. M. K., *Science* **165,** 25 (1969).

210. Nass, S., *Int. Rev. Cytol.* **25,** 55 (1969).

211. Neupert, W., W. Sebald, A. J. Schwab, P. Massinger, and Th. Bücher, *Eur. J. Biochem.* **10,** 589 (1969).

212. Neupert, W., and G. D. Ludwig, *Eur. J. Biochem.* **19,** 523 (1971).

213. Newell, S. Y., and J. W. Fell, *Mycologia* **62,** 272 (1970).

214. Noll, H., *Nature* **215,** 360 (1967).

215. Noll, H., *Anal. Biochem.* **27,** 130, (1969).

216. O'Brien, T. W., *J. Biol. Chem.* **246,** 3409 (1971).

217. Ogur, M., S. Minckler, G. Lindegren, and C. C. Lindegren, *Arch. Biochem. Biophys. 40,* 175 (1952).

218. Olive, L. S., in *The Fungi*, Vol. I, G. C. Ainsworth, and A. S. Sussman (eds.), Academic Press, New York, 1965, p. 143.

219. Penn, S., and S. K. Dutta, *Genetics* **64,** s50 (1970).

220. Perlman, P. S., and H. R. Mahler, *Nature* **231,** 12 (1971).

221. Petermann, M. L., *The Physical and Chemical Properties of Ribosomes*, Elsevier, New York, 1964.

222. Planterose, D. N., P. J. Birch, D. J. F. Pilch, and T. J. Sharpe, *Nature* **227,** 504 (1970).

223. Plattner, H., M. M. Salpeter, J. Saltzgaber, and G. Schatz, *Proc. Natl. Acad. Sci. U.S.* **66,** 1252 (1970).

224. Rabinowitz, M., G. S. Getz, J. Casey, and H. Swift, *J. Mol. Biol.* **41,** 381 (1969).

225. Rabinowitz, M., and H. Swift, *Physiol. Rev.* **50,** 376 (1970).

226. Rado, T. A., and V. W. Cochrane, *Fed. Proc.,* **28,** 890 (1969).

227. Rado, T. A., and V. W. Cochrane, *J. Bacteriol.* **106,** 301 (1971).

228. Ramakrishnan, L., and R. C. Staples, *Phytopathol.* **58,** 886 (1968).

229. Ramakrishnan, L., and R. C. Staples, *Contrib. Boyce Thompson Inst.* **24,** 197 (1970).

230. Raven, P. H., *Science* **169,** 641 (1970).

231. Reich, E., and D. J. L. Luck, *Proc. Natl. Acad. Sci. U.S.* **55,** 1600 (1966).

232. Rifkin, M. R., and D. J. L. Luck, *Proc. Natl. Acad. Sci. U.S.* **68,** 287 (1971).

233. Rifkin, M. R., D. D. Wood, and D. J. L. Luck, *Proc. Natl. Acad. Sci. U.S.* **58,** 1025 (1967).

234. Rizet, G., D. Marcou, and J. Schecroun, *Bull. Soc. Fran. Physiol. Végét.* **4,** 136 (1958).

235. Robinow, C. F., *J. Biophys. Biochem. Cytol.* **9,** 879 (1961).

236. Rogers, J. D., *Can. J. Bot.* **48,** 1665 (1970).

237. Rogers, P. J., B. N. Preston, E. B. Titchener, and A. W. Linnane, *Biochem. Biophys. Res. Commun.* **27,** 405 (1967).

238. Roodyn, D. B., and D. Wilkie, *The Biogenesis of Mitochondria.* Methuen, London, England, 1968.

239. Rosset, R., R. Monier, and J. Julien, *Bull. Soc. Chim. Biol.* **46,** 87 (1964).

240. Rusch, H. P., in *Advances in Cell Biology*, Vol. 1, D. M. Prescott, L. Goldstein, and E. McConkey (eds.), Appleton-Century-Crofts, Meredith Corp., New York, 1970, p. 297.

241. Rytel, M. W., R. E. Shope, and E. D. Kilbourne, *J. Exp. Med.* **123,** 577 (1966).

242. Schafer, K. P., G. Bugge, M. Grandi, and H. Küntzel, *Eur. J. Biochem.* **21,** 478 (1971).

243. Schatz, G., E. Haslbrunner, and H. Tuppy, *Biochem. Biophys. Res. Commun.* **15,** 127 (1964).

244. Scheld, H. W., and J. J. Perry, *J. Gen. Microbiol.* **60,** 9 (1970).

245. Schildkraut, C. L., J. Marmur, and P. Doty, *J. Mol. Biol.* **4,** 430 (1962).

246. Schmitt, H., *FEBS Lett.* **4,** 234 (1969).

247. Schmoyer, I. R., and J. S. Lovett, *J. Bacteriol.* **100,** 854 (1969).

248. Schneider, W. C., and E. L. Kuff, *Proc. Natl. Acad. Sci. U.S.* **54,** 1650 (1965).

249. Schwab, R., M. Sebald, and F. Kaudewitz, *Mol. Gen. Genet.* **110,** 361 (1971).

250. Schweizer, E., and H. O. Halvorson, *Exp. Cell. Res.* **56,** 239 (1969).

251. Schweizer, E., C. MacKechnic, and H. O. Halvorson, *J. Mol. Biol.* **40,** 261 (1969).

252. Scragg, A. H., H. Morimoto, V. Villa, J. Nekhorocheff, and H. O. Halvorson, *Science* **171,** 908 (1971).

253. Sebald, W., A. J. Schwab, and Th. Bücher, *FEBS Lett.* **4,** 243 (1969).

254. Seidler, R., Personal communication.

255. Shapiro, H. S., in *Handbook of Biochemistry*, H. A. Sober, (ed.), Chemical Rubber Co., Cleveland, Ohio, 1968, p H-31.

256. Sherman, F., *Genetics* **48,** 375 (1963).

257. Sherman, F., and B. Ephrussi, *Genetics* **47,** 695 (1962).

258. Sinclair, J. H., B. J. Stevens, P. Sanghavi, and M. Rabinowitz, *Science* **156,** 1234 (1967).

259. Singer, C. E., and B. N. Ames, *Science* **170,** 822 (1970).

260. Slonimski, P., Perrodin, G, and J. Croft, *Biochem. Biophys. Res. Commun.* **30,** 232 (1968).

261. Slonimski, P., E. Petrochilo, P. Netter, B. Dujon, J. Deutsch, D. Coen, and M. Bolotin, Abstract: Genetical Society Meeting, Oxford, March 24–26, 1970.

262. Smith, D., P. Tauro, E. Schweizer, and H. O. Halvorson, *Proc. Natl. Acad. Sci. U.S.* **60,** 936 (1968).

263. Smith, J. E., and J. C. Galbraith, in *Advances in Microbial Physiology*, Vol. 5, A. H. Rose, and J. F. Wilkinson (eds.), Academic Press, New York, 1971, p. 45.

264. Söll, D. R., and D. R. Sonneborn, *Proc. Natl. Acad. Sci. U.S.* **68,** 459 (1971).

265. Sonenshein, G. E., and C. E. Holt, *Biochem. Biophys. Res. Commun.* **33,** 361 (1968).

266. Spencer, J. F. T., and P. A. J. Gorin, *Antonie van Leeuwenhoek, J. Microbiol. Serol.* **35,** 33 (1969).

267. Spiegelman, S., in *Ideas in Modern Biology,* J. A. Moore, (ed.), Natural History Press, Garden City, N.Y. 1965, p. 19.

268. Spirin, A. S., and L. P. Gavrilova, *The Ribosome*, Springer-Verlag, New York, 1969.

269. Stanier, R. Y., M. Doudoroff, and E. A. Adelberg, in *The Microbial World*, 3rd ed. Prentice-Hall, Englewood Cliffs, N.J., 1970.

270. Staples, R. C., D. Bedigian, and P. H. Williams, *Phytopathology* **58,** 151 (1968).

271. Stegeman, W. J., C. S. Cooper, and C. J. Avers, *Biochem. Biophys. Res. Commun.* **39,** 69 (1970).

272. Steinschneider, A., *Biochem. Biophys. Acta* **186,** 405 (1969).

273. Stenderup, A., and A. Leth Bak, *J. Gen. Microbiol.* **52,** 231 (1968).

274. Stevens, B. J., and E. Moustacchi, *Exp. Cell Res.* **64,** 259 (1971).
275. Storck, R., *J. Bacteriol.* **90,** 1260 (1965).
276. Storck, R., *J. Bacteriol.* **91,** 227 (1966).
277. Storck, R., Unpublished observations.
278. Storck, R., in *Evolution of Genetic Systems,* Brookhaven Symposia **23,** 371 (1972).
279. Storck, R., and C. J. Alexopoulos, *Bacteriol. Rev.* **34,** 126 (1970).
280. Storck, R., and R. C. Morrill, *Biochem. Genet.* **5,** 467 (1971).
281. Storck, R., M. K. Nobles. and C. J. Alexopoulos, *Mycologia* **LXIII**, 38 (1971).
282. Stumm, C., and J. L. van Went, *Experentia (Basel)* **24,** 1112 (1968).
283. Stutz, E., and H. Noll, *Proc. Natl. Acad. Sci. U.S.* **57,** 774 (1967).
284. Sussman, R. R., *Exp. Cell Res.* **24,** 154 (1961).
285. Sussman, R., and E. P. Rayner, *Arch. Biochem. Biophys.* **144,** 127 (1971).
286. Tamaki, H., *J. Gen. Microbiol.* **41,** 93 (1965).
287. Taylor, M. M., Ph.D. Thesis, University of Texas, Austin, Tex., 1966.
288. Taylor, M. M., J. E. Glasgow, and R. Storck, *Proc. Natl. Acad. Sci. U.S.* **57,** 164 (1967).
289. Taylor, M. M., and R. Storck, *Proc. Natl. Acad. Sci. U.S.* **52,** 958 (1964).
290. Terenzi, H. F., Ph.D. Thesis, Rice University, Houston, Tex., 1969.
291. Terenzi, H. F., and R. Storck, *J. Bacteriol.* **97,** 1248 (1969).
292. Tewari, K. K., J. Jayaraman, and H. R. Mahler, *Biochem. Biophys. Res. Commun.* **21,** 141 (1965).
293. Tewari, K., W. Vötsch, H. R. Mahler, and B. Mackler, *J. Mol. Biol.* **20,** 453 (1966).
294. Tonino, G. J. M., and T. H. Rozyn, *Biochim. Biophys. Acta* **124,** 427 (1966).
295. Tsai, M., G. Michaelis, and R. S. Criddle, *Proc. Natl. Acad. Sci. U.S.* **68,** 473 (1971).
296. Udem, S. A., K. Kaufman, and J. R. Warner, *J. Bacteriol.* **105,** 101 (1971).
297. Uryson, S. O., and A. N. Belozersky, *Dokl. Akad. Nauk. SSSR* **133,** 708 (1961).
298. Valadon, L. R. G., J. G. Manners, and A. Myers, *Nature* **190,** 836 (1961).
299. Van der Plaat, J. B., P. Apontoweil, and W. Berends, *Mutation Res.* **7,** 13 (1969).
300. Van der Walt, J. P., and J. de Leeuw, *Mycopathol. Mycol. Appl.* **42,** 17 (1970).
301. Van der Walt, J. P., and M. J. Pitout, *Antonie van Leeuwenhoek, J. Microbiol. Serol.* **35,** 227 (1969).
302. Van Etten, J. L., *Arch. Biochem. Biophys.* **125,** 13 (1968).
303. Van Etten, J. L., *Phytopathology* **59,** 1060 (1969).
304. Van Etten, J. L., *J. Bacteriol.* **106,** 704 (1971).
305. Van Etten, J. L., and R. M. Brambl, *J. Bacteriol.* **96,** 1042 (1968).
306. Van Etten, J. L., R. K. Koski, and M. M. El-Olemy, *J. Bacteriol.* **100,** 1182 (1969).
307. Vanyushin, B. F., A. N. Belozersky, and S. L. Bogdanova, *Dokl. Akad. Nauk. SSSR* **134,** 1222 (1960).
308. Venner, H., *Z. Physiol. Chem.* **333,** 5 (1963).
309. Verma, I. M., M. Edelman, M. Herzberg, and U. Z. Littauer, *J. Mol. Biol.* **52,** 137 (1970).
310. Verma, I. M., M. Edelman, and U. Z. Littauer, *Eur. J. Biochem.* **19,** 124 (1971).
311. Vignais, P. V., and J. Huet, *Proc. Biochem. Soc.* **116,** 26 (1970).

312. Vignais, P. V., J. Huet, and J. Andre′, *FEBS Lett.* **3,** 177 (1969).

313. Villa, V. D., Ph.D. Thesis, Rice University, Houston, Tex., 1970.

314. Villa, V. D., and R. Storck, *J. Bacteriol.* **96,** 184 (1968).

315. Villanueva, J. R., in *The Fungi,* Vol. II, G. C. Ainsworth and A. S. Sussman (eds.), Academic Press, New York, 1966, p. 3.

316. Vogel, F. S., *Lab. Invest.* **14,** 1849 (1965).

317. Vogel, H. J., in *Evolving Genes and Proteins*, V. Bryson, and H. J. Vogel, (eds.), Academic Press, New York, 1965, p. 25.

318. Vogel, H. J., J. S. Thompson, and G. D. Shockman, in *Organization and Control in Prokaryotic and Eukaryotic Cells*, H.P. Charles and B. C. J. G. Knight (eds.), University Press, Cambridge, England, 1970, p. 107.

319. Volkoff, O., and T. Walters, *Can. J. Genet. Cytol.* **12,** 621 (1970).

320. Wagner, R. P., *Science* **163,** 1026 (1969).

321. Wakabayashi, K., and N. Gunge, *FEBS Lett.* **6,** 302 (1970).

322. Walker, P. M. B., *Nature* **229,** 306 (1971).

323. Weislogel, P. O., and R. A. Butow, *J. Biol. Chem.* **246,** 5113 (1971).

324. Weitzman, I., P. W. Allderdice, and M. Silva-Hutner, *Mycologia* **62,** 89 (1970).

325. Werner, H., *Zentralbl. Bakteriol.* **210,** 397 (1969).

326. Wetmur, J. G., and N. Davidson, *J. Mol. Biol.* **31,** 349 (1968).

327. Wickerham, L. J., in Recent Trends in Yeast Research, D. G. Ahearn, (ed.), *Spectrum*, Monograph Series in Arts and Sciences, Vol. 1, Georgia State University, Atlanta, Ga., 1970, p. 31

328. Wilkie, D., in *Organization and Control in Prokaryotic and Eukaryotic Cells*, H. P. Charles, and B. C. J. G. Knight (eds.), University Press, Cambridge, England, 1970, p. 381.

329. Williams, P. G., and M. J. Hartley, *Nature New Biol.* **229,** 181, 1971.

330. Williamson, D. H., and E. Moustacchi, *Biochem. Biophys. Res. Commun.* **42,** 195 (1971).

331. Williamson, D. H., N. G. Maroudas, and D. Wilkie, *Mol. Gen. Genet.* **111,** 209 (1971).

332. Williamson, D. H., and A. W. Scopes, *Exp. Cell. Res.* **24,** 151 (1961).

333. Wilson, C. M., *Bull. Torrey Botan. Club* **79,** 139 (1952).

334. Wintersberger, E., *Biochem. Biophys. Res. Commun.* **40,** 1179 (1970).

335. Wintersberger, U., and E. Wintersberger, *Eur. J. Biochem.* **13,** 11 (1970).

336. Wintersberger, U., and E. Wintersberger, *Eur. J. Biochem.* **13,** 20 (1970).

337. Wood, D. D., and D. J. L. Luck, *J. Mol. Biol.* **41,** 211 (1969).

338. Yaniv, Z. and R. C. Staples, *Contrib. Boyce Thompson Inst.* **24,** 103 (1968).

339. Young, F., and A. P. Jackson, *Biochem. Biophys. Res. Commun.* **23,** 490 (1966).

INDEX